Rapid Assessment Program
Programa de Evaluación Rápida

A Biological Assessment of the Aquatic Ecosystems of the Caura River Basin, Bolívar State, Venezuela

Una Evaluación Rápida de los Ecosistemas Acuáticos de la Cuenca del Río Caura, Estado Bolívar, Venezuela

RAP
Bulletin of Biological Assessment

Boletín RAP de Evaluación Biológica

28

Editors/Editores
Barry Chernoff, Antonio Machado-Allison, Karen Riseng, and Jensen R. Montambault

Center for Applied Biodiversity Science (CABS)

Conservation International

The Field Museum

Instituto de Zoología Tropical, Universidad Central de Venezuela

Organization Kuyujani

The *RAP Bulletin of Biological Assessment* is published by:
Conservation International
Center for Applied Biodiversity Science
Department of Conservation Biology
1919 M St. NW, Suite 600
Washington, DC 20036
USA

202-912-1000 telephone
202-912-0773 fax
www.conservation.org
www.biodiversityscience.org

Conservation International is a private, non-profit organization exempt from federal income tax under section 501c(3) of the Internal Revenue Code.

Editors: Barry Chernoff, Antonio Machado-Allison, Karen Riseng, and Jensen R. Montambault
Design/production: Kim Meek
Map: Mark Denil
Photos: Barry Chernoff, Antonio Machado-Allison, and Jensen R. Montambault
Translations: [Spanish] Antonio Machado-Allison and Ana Liz Flores, [Kuyujani] Alberto Rodriguez

***RAP Bulletin of Biological Assessment* Series Editors:**
Terrestrial and AquaRAP: Leeanne E. Alonso and Jennifer McCullough
Marine RAP: Sheila A. McKenna

ISBN: 1-881173-69-0

Library of Congress Catalog Card Number: 2002116698

RAP Bulletin of Biological Assessment was formerly *RAP Working Papers*. Numbers 1–13 of this series were published under the previous title.

Suggested citation:
Chernoff, B., A. Machado-Allison, K. Riseng, and J. R. Montambault (eds.). 2003. A Biological Assessment of the Aquatic Ecosystems of the Caura River Basin, Bolívar State, Venezuela. RAP Bulletin of Biological Assessment 28. Conservation International, Washington, DC.

Chernoff, B., A. Machado-Allison, K. Riseng, and J. R. Montambault (eds.). 2003. Una evaluación rápida de los ecosistemas acuáticos de la Cuenca del Río Caura, Estado Bolívar, Venezuela. Boletín RAP de Evaluación Biologíca 28. Conservation International, Washington, DC.

The Rufford Foundation and the Gordon and Betty Moore Foundation generously supported publication of this report.

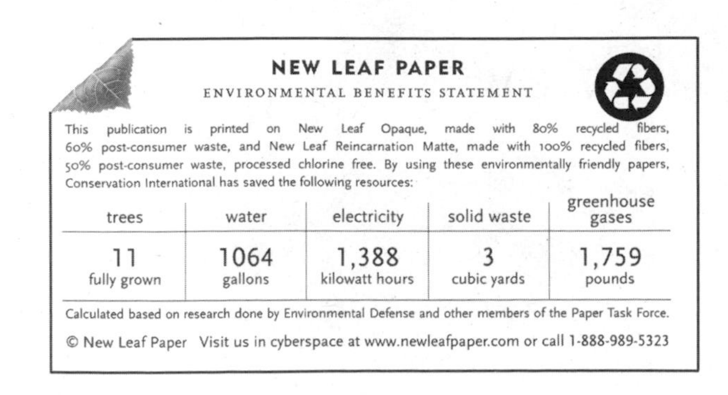

Table of Contents/ Tabla de Contenidos

ENGLISH VERSION

KUYUJANI

SPANISH

APPENDICES

Preface

The Caura River is more than 700 km long, originating in the highlands of the Guayana Shield at 2,000 meters above mean sea level. The area of the basin is approximately 45,336 km^2, which is 20% of the total surface of Bolívar State or 5% of Venezuela. The watershed of the Caura River contains thousands of species of plants and animals, many endemic to the Guayana Shield or the Orinoco River Basin. At the same time, the Caura River is the lifeblood of many human communities who depend upon the river for food, transportation, recreation, their livelihoods and their cultures. The basin forms a part of the homelands of the Ye'kuana people who are extremely careful in their management of and appreciation for their natural resources. The high environmental quality of the Caura River Basin, especially from the Raudal Cinco Mil to the headlands, is due in large measure to their excellent stewardship. In short, the Caura River and its basin serve as critical habitat for all species of the region, including humans.

The Caura River Basin is one of the largest tracts of pristine Guayana Shield forest remaining in the world. The aquatic environments of the Caura River, especially above Raudal Cinco Mil, are also in excellent environmental condition. The Caura River watershed provides an excellent opportunity to develop an integrated plan for the basin that promotes sustainable use along with species and ecosystem conservation.

Nonetheless, the region is facing immediate threats that could easily change the pristine nature of the environment. Deforestation due to logging in the southern part of the basin, increasing population growth, colonization and habitat degradation in the northern part of the basin, a plan to divert significant amounts of water from the Caura for a hydroelectric facility in the Caroní, and illegal hunting and over-exploitation of fisheries below Salto Pará by non-indigenous fishermen are all exerting serious pressures upon this unique watershed. In order to investigate the potential effects of these threats upon the biodiversity and the aquatic habitats of the Caura River, a team of 17 Kuyujani experts, 13 scientists, two logistical coordinators, two technicians, one student and one web reporter undertook an AquaRAP in the Caura River from 25 November through 12 December, 2000. The team was comprised of individuals from Venezuela, Brazil and the United States. The scientists were specialists in aquatic and terrestrial botany, decapod crustaceans, macroinvertebrates, ichthyology and limnology, including water chemistry and plankton. The expedition was focused primarily on the biological and conservation value of the region and how integrated solutions can work to preserve the maximum amount of biodiversity in the face of current and future threats.

The organization of this report begins with an executive summary, including a brief overview of the physical and terrestrial characteristics of the region, summaries of the technical reports and recommendations for a conservation strategy. We next present the biological results from botany, limnology, benthic invertebrates, decapod crustaceans and fishes. The final chapters highlight the commonalties among different floral or faunal elements in relation to geography and ecology. The disciplinary reports are written as scientific papers, each self-contained with its own literature citations for ease of use. After the scientific papers, we include a glossary of terms and appendices containing much of the raw data.

We would like to explain our use of the terms *diversity* and *species richness*. In vernacular usage *diversity*, when used in the context of biological organisms, refers to the numbers of and variety of types of species, organisms, taxa, etc. In ecological literature, *diversity* takes on a slightly different, more specific meaning, referring to the number of entities in combination with their relative abundances. In this volume *diversity* is used in two ways: i) in the general, vernacular, sense we occasionally refer to the "*diversity* of organisms," meaning the number and variety of organisms; but ii) in the ecological sense, we refer to "low or high *diversity*," meaning *"diversity"* as calculated from a specific formula. Note that in the vernacular usage, *diversity* is never modified by high or low or used in a comparative way. From the ecological literature, the term *species richness* means the number of species. We use *species richness* in several chapters (e.g., Chapter 4) when referring to the number of species present in a habitat or river basin.

This report is intended for decision makers, environmental managers, governmental and non-governmental agencies, students and scientists. The original information and analyses presented herein have two aims: (i) to present a compelling case and cogent strategy for conservation efforts within the region; and (ii) to provide scientific data and analyses that will stimulate future scientific research of this critical region. We have attempted in this volume not only to present an inventory of the organisms that we encountered during our expedition, but also to use that information to evaluate conservation strategies under different scenarios of environmental threat. We welcome comments and criticism as we continue to evolve AquaRAP protocols and the methods used for evaluating conservation strategies from biological data.

Barry Chernoff
Antonio Machado-Allison
Karen Riseng
Jensen Montambault

Participants and Authors

Mariapia Bevilacqua (logistic coordination, botany)
Asociación Venezolana para la Conservación de Áreas Naturales (ACOANA)
Av. Humboldt con calle Coromoto, Edif. Breto oficina 5, planta baja, Bello Monte Norte
Caracas, 1063-A
VENEZUELA
Email: mariapia@cantv.net

Barry Chernoff (ichthyology, editor)
Dept. Zoology
Field Museum
1400 S. Lakeshore Dr.
Chicago, IL 60605
USA
Email: chernoff@fmnh.org

Wilmer Díaz (botany)
Jardín Botánico del Orinoco
Calle Bolívar, Ciudad Bolívar
Edo. Bolívar
VENEZUELA
Email: jbov@telcel.net.ve

José V. García D. (benthic invertebrates)
Instituto de Zoología Tropical, Facultad de Ciencias
Universidad Central de Venezuela
Apto. Correos 47058
Caracas, 1041-A
VENEZUELA
Email: jvgarcia@strix.ciens.ucv.ve

Antonio Machado-Allison (ichthyology, editor)
Instituto de Zoología Tropical
Universidad Central de Venezuela
Apto. Correos 47058
Caracas, 1041-A
VENEZUELA
Email: amachado@strix.ciens.ucv.ve

Celio Magalhaes (decapod crustaceans)
Instituto Nacional de Pesquisas Amazonicas
INPA/CPBA, Cx. Postal 478
69011-970 Manaus
BRAZIL
Email: celiomag@inpa.gov.br

Nigel Maxted (author)
School of Biosciences
The Birmingham University
Edgbaston B15 2TT
Birmingham
UNITED KINGDOM

Alberto Marcano (ichthyology)
Instituto de Zoología Tropical
Universidad Central de Venezuela
Apto. Correos 47058
Caracas, 1041-A
VENEZUELA

Jensen R. Montambault (web coordination, editor)
Conservation International
Rapid Assessment Program
1919 M Street NW, Suite 600
Washington, DC 20036
USA

Guido A. Pereira S. (benthic invertebrates, decapod crustaceans)
Instituto de Zoología Tropical, Facultad de Ciencias
Universidad Central de Venezuela
Apto. Correos 47058
Caracas, 1041-A
VENEZUELA
Email: gpereira@strix.ciens.ucv.ve

Geoffrey Petts (author)
School of Geography and Environmental Sciences
The Birmingham University
Edgbaston B15 2TT
Birmingham
UNITED KINGDOM

Francisco Provenzano-Rizzi (ichthyology)
Instituto de Zoología Tropical
Universidad Central de Venezuela
Apto. Correos 47058
Caracas, 1041-A
VENEZUELA
Email: fprovenz@strix.ciens.ucv.ve

Lourdes Rico-Arce (author)
Herbarium
Royal Botanic Gardens
Kew, Richmond, Surrey, TW9 3AB
UNITED KINGDOM

Karen J. Riseng (report coordination, limnology, editor)
Department of Ecology and Evolutionary Biology
University of Michigan
830 N. University
Ann Arbor, MI 48109
USA
Email: kjriseng@umich.edu

Ángel Rojas (ichthyology)
Instituto de Zoología Tropical
Universidad Central de Venezuela
Apto. Correos 47058
Caracas, 1041-A
VENEZUELA

Judith Rosales (botany)
Universidad Nacional Experimental de Guayana (UNEG)
Urbanización Chilemex, Calle Chile, Sede Uneg-Investigacion y Postgrado
Puerto Ordáz, Edo. Bolívar
VENEZUELA
Email: jrosales@uneg.edu.ve

Luzmila Sanchez (limnology)
Fundación La Salle
Estación de Investigaciones Hidrobiológicas de Guayana
UD-104 El Roble, Apto. 51
San Félix, Edo. Bolívar
VENEZUELA
Email: luzsanchez@cantv.net

Brian Sidlauskas (ichthyology)
Dept. Zoology
Field Museum
1400 S. Lakeshore Dr.
Chicago, IL 60605
USA
Email: bls@midway.uchicago.edu

John S. Sparks (limnology)
Museum of Zoology
Division of Fishes
University of Michigan
Ann Arbor, MI 48109-1079
USA
Email: jsparks@umich.edu

Philip W. Willink (ichthyology)
Dept. Zoology
Field Museum
1400 S. Lakeshore Dr.
Chicago, IL 60605
USA
Email: pwillink@fmnh.org

Kuyujani team:
Upper Caura
Pilots: Luís Flores, Wilfredo Flores and Carmelo Castro
Boat drivers: Rogelio Pérez, José Sarmiento, Justino Castro Bonifacio, Lucas González and Juan Núñez (also Kuyujani Coordinator for the expeditions)

Lower Caura
Pilots: Miguel Estaba, Wilfredo Flores, José Sosa Silva and Simón Caura
Boat drivers: Juan Núñez, Nelson Espinoza, Ernesto Sarmiento, Delfín Rivas, Eugenio García and Alberto Sarmiento (also cook's assistant)

Organizational Profiles

CONSERVATION INTERNATIONAL

Conservation International (CI) is an international, nonprofit organization based in Washington, DC. CI believes that the Earth's natural heritage must be maintained if future generations are to thrive spiritually, culturally and economically. Our mission is to conserve the Earth's living heritage, our global biodiversity, and to demonstrate that human societies are able to live harmoniously with nature.

Conservation International
1919 M Street NW, Suite 600
Washington, DC 20036
USA
Tel. 800-406-2306
Fax. 202-912-0772
Web. www.conservation.org
www.biodiversityscience.org

CONSERVATION INTERNATIONAL-VENEZUELA

CI-Venezuela was established in 2000 to conserve biodiversity in the Tropical Andes Hotspot and to demonstrate that human societies are able to live in harmony with nature. CI's experience indicates that successful conservation occurs within the frame of sustainable development that includes local communities in alternative and creative activities, builds the local capacity for the conservation and appropriate use of natural resources, advances environmental education, and seeks to avoid the destructive use of the Earth, the contamination of water, and the loss of biodiversity. We work through strategic alliances with social and institutional partners to develop conservation activities, based on scientific and technical criteria, that respect sociocultural diversity, develop local creativity, evaluate habitat damage, identify threats, and create alternative income sources.

Avenida Las Acacias
Torre La Previsora, Piso 15
Ofina Noreste, Noroeste
Urbanización Los Caobas
Caracas, 1050
VENEZUELA

THE FIELD MUSEUM

The Field Museum (FMNH) is an educational institution concerned with the diversity and relationships in nature and among cultures. Combining the fields of Anthropology, Botany, Geology, Paleontology and Zoology, the Museum uses an interdisciplinary approach to increasing knowledge about the past, present, and future of the physical earth, its plants, animals, people and their cultures. In doing so, it seeks to uncover the extent and character of the biological and cultural diversity, similarities and interdependencies so that we may better understand, respect and celebrate nature and other people. Its collections, with more than 20 million specimens from around the world, public learning programs, and research are inseparably linked to serve a diverse public of varied ages, backgrounds and knowledge. The Museum publishes a peer-reviewed scientific journal, *Fieldiana.*

The Field Museum
1400 South Lake Shore Dr.
Chicago, IL 60657
USA
Tel. 312-922-9410
Fax. 312-665-7932
Web. www.fieldmuseum.org

INSTITUTO DE ZOOLOGÍA TROPICAL, UNIVERSIDAD CENTRAL DE VENEZUELA

The Institute of Tropical Zoology (IZT) is a research institute of the Faculty of Sciences, Universidad Central de Venezuela (UCV). Within the broad disciplines of Zoology and Ecology, the IZT emphasizes education and research in systematic zoology, parasitology, theoretical and applied ecology, environmental studies and conservation. The IZT is responsible for UCV's Museum of Biology, which houses some of the most valuable zoological collections in the world. Among the collections of note are the freshwater fish collection, one of the largest in Latin America, and the mammal collection, the most comprehensive in Venezuela. The IZT also runs the Aquarium "Agustín Codazzi," which disseminates knowledge to the public about Venezuelan fishes and environmental conservation through its exhibits and educational programs. The IZT publishes the scientific journal, *Acta Biologica Venezuelica*, founded in 1951.

Instituto de Zoología Tropical
Universidad Central de Venezuela
Apto 47058
Caracas, 1041-A
VENEZUELA
Web. http://strix.ciens.ucv.ve/~instzool

ORGANIZATION KUYUJANI

Kuyujani is an indigenous organization representing two indigenous peoples of the Caura region, the Ye'kuana and the Sanema who make up 53 communities. It was founded August 8, 1996 and has six areas of interest: Education, Health, Environment and Territory, Human Rights, and Economic and Cultural Development. Kuyujani works to defend the rights and ancestral lands of the indigenous communities ye'kuana and sanema in all aspects; to support and foment dialogue between communities and organizations in the search for integrated development; to encourage the development of courses, seminars or facilities for the technical and scientific capacity of members of the community; to establish and maintain fraternal relations with regional, national and international indigenous organizations with the purpose of uniting criteria and efforts to benefit the rights and interests of our communities; and to establish relations with non-indigenous organizations who work towards the same ends.

Comunidades Santa Maria de Erebato y Boca de Nichare
Barrio Hueco Lindo, Callejón Los Teques
Ciudad Bolívar
VENEZUELA

Acknowledgments

This AquaRAP expedition required the dedication of many people to make it a successful venture. We extend our thanks to the many individuals and institutions providing logistical and scientific support, food services, lodging, meeting facilities and funding.

The AquaRAP team owes its deepest gratitude to the Ye'kuana and Sanema people, and the indigenous organization Kuyujani, for the invitation to conduct this aquatic biodiversity survey in their territory. The continued support and understanding of Capitán Perez and Alberto and Freddy Rodríguez since the inception of this project was critical to the smooth operation of the expedition while also insuring that the information gathered will be applied to community-based environmental conservation.

We would like to thank the Instituto de Zoología Tropical, Universidad Central de Venezuela, for wonderful hospitality, collaborations and scientific support. We are extremely grateful to Mariapia Bevilacqua and ACOANA (Asociación Venezolana para la Conservación de Areas Naturales) for excellent planning, logistical support and scientific expertise. We are grateful to Instituto de Limnología del Orinoco de la Fundación La Salle for providing dormitories and conference facilities at the end of the expedition. We would also like to extend our thanks to Jorge Luís Suárez of Akanan Tours who provided logistical support and who, along with the cook Tenilda Cranes, her assistant Alberto Sarmiento and her husband Jonas Cranes, provided the best fare a field party could ever expect.

We extend special thanks to Dedemai and Simón Caura for welcoming us into Boca de Nichare and for providing excellent guidance. We thank Kuyujani for the use of their infrastructure at El Playon. Thanks are also gratefully extended to the Corporacíon Venezolana de Guyana (CVG) for use of the field station in Entreríos and its indigenous local staff (Jesús and Freddy).

We would like to thank Capitan Raúl Arias of Raul Helicopters for all his courtesy and consideration while flying us from Entreríos to El Playón. The airplane company (Comeravia) pilots and office staff (especially Capitán Andrés Franco and María Isabel de Rivas) did a wonderful job of flying us, along with our equipment and supplies, from Ciudad Bolívar to Entreríos.

For help with all of our other travel and with field collecting we thank the wonderful motoristas and marineros. Motoristas in the Upper Caura were Luís Flores, Wilfredo Flores and Carmelo Castro and in the Lower Caura Miguel Estaba, Wilfredo Flores, José Sosa Silva and Simón Caura. Marineros in the Upper Caura were Rogelio Perez, José Sarmiento, Justino Castro Bonifacio, Lucas González and Juan Núñez (also Kuyujani Coordinator for the expeditions) and in the Lower Caura were Juan Núñez, Nelson Espinoza, Ernesto Sarmiento, Delfín Rivas, Eugenio García and Alberto Sarmiento.

We thank Alberto Rodríguez and Freddy Rodríguez for coordinating with indigenous communities and for assisting with permits for the expedition in their territory. We also thank the following governmental agencies for granting us permits to investigate these incredible areas: Consejo Nacional de Investigaciones Científicas y Tecnológicas (CONICIT); Instituto Nacional de Parques (INPARQUES); Oficina de Asuntos Indígenas del Ministerio de Educación, Cultura y Deportes (DAI); Oficina Nacional de Diversidad Biológica del Ministerio del Ambiente y los Recursos Naturales (MARN); Servicio Autónomo de Pesca del Ministerio de Industria y Comercio (SARPA). We would also like to express our gratitude to Dr. Carlos Genatios, Minister of Science and Technology, and Dr. Héctor Navarro, Minister of Education.

The authors of Chapter 4 thank Professor Rafael Martínez who kindly provided valuable help in the identification of snails, mussels and leeches. The authors of Chapter 6 are grateful to Mary Anne Rogers and Kevin Swagel for help with the processing of fish specimens, and to John Friel and Richard Vari for help with identification of fishes. The authors of Chapter 7 would like to thank CONICIT, the British Council, the Universidad Nacional Experimental de Guayana, University of Birmingham, Carlos Verbin, Claudia Knab-Vispo, Herbario Ovalles, Herbario Nacional de Venezuela, Herbario Regional de Guayana, Herbario de Guanare and Kew Botanical Gardens.

The AquaRAP program and the Caura AquaRAP survey have been made possible by the generous support of the Rufford Foundation. Additional funds for the Caura expedition were generously provided by the Smart Family Foundation, supporting technology and web-broadcasts from the field, and by the Comer Science and Education Foundation. We would like to thank John McCarter (Field Museum), Russ Mittermeier, Peter Seligmann, Anthony Rylands, Jorgen Thomsen and Leeanne Alonso (Conservation International) for continued interest, enthusiasm and support for aquatic conservation. We would like to extend special thanks to Ana Liz Flores and Alfonso Alonso for reviewing this document and to Alberto Rodriguez for the Ye'kuana translations. Joan Goldstein provided invaluable editorial help. As always, we wish to acknowledge the excellent editorial and scientific review of Leeanne Alonso as well as all of her help and guidance throughout the course of the project. The Gordon and Betty Moore Foundation and the Rufford Foundation provided generous funds to publish this report.

Report at a Glance

A BIOLOGICAL ASSESSMENT OF THE AQUATIC ECOSYSTEMS OF THE CAURA RIVER BASIN, BOLÍVAR STATE, VENEZUELA

Expedition Dates

25 November–12 December 2000

Area Description

The Caura watershed, Bolívar State, Venezuela, is a vast expanse of forests and rivers situated upon the Guayana Shield, an ancient geologic formation. The Caura River constitutes a major tributary of the Orinoco River. Inland forests cover approximately 90% of the watershed while the remaining 10% is composed of flooded forests and other non-forest vegetation. Overall, the Caura River Basin hosts 30% of the species registered for Venezuela and 51.3% of Guayanese species, including 88% of the Guayanan endemic plant genera and 28% of the Venezuelan freshwater fish fauna. The extraordinary biological diversity has been attributed to the combination of erosional landscape, the convergence of four Geological Provinces and the marked altitudinal gradient (40–2350 m). In addition to the remarkable diversity and endemism, the forest of this basin is also 85% intact and unaffected by human intervention.

Reason for the Expedition

The principle objective of this AquaRAP survey was to explore and document the aquatic biodiversity of this remote region in cooperation with the Ye'kuana people. The Caura River Basin is part of the Guayana Shield and represents one of the most pristine watersheds in South America. Managing these remaining pristine, and ever more valuable, areas is an increasingly difficult task. This largely pristine area corresponds to the land of the Ye'kuana people, the principle indigenous group. The Ye'kuana are extremely careful in their management and appreciation for their natural resources. The very high environmental quality of this region, especially the area from the Raudal Cinco Mil to the headlands, is due in large measure to the excellent stewardship of the Ye'kuana. Although a water diversion project in the Caura is less imminent now, the region still faces the immediate threat of a major hydroelectric dam at the Salto Pará (waterfalls which divide the region into its Upper and Lower zones and that are sacred to the Ye'kuana) and increased fishing pressure, as well as a rise in tourism and agriculture in the Lower Caura.

Major Results

The Caura River has relatively high fish species richness and high riparian vegetation diversity in comparison to other similar tributaries of the Amazon and Orinoco Basins. While the benthic invertebrates displayed the typical species richness and abundance of inland river systems from the Guayana and Amazon Basins, the decapod team found two undescribed species of palaemonid shrimps, one of which could be endemic to the area. Water quality was typical of a rain driven system draining this type of geology with low ion content and low pH. There were both black and clear water rivers in this system. The highest diversity and abundance of fishes was found in the Cejiato area of the Upper Caura River and the Takoto River in the Lower Caura. The data indicate that each major group of organisms has a specific pattern of distribution throughout the Caura system. Invertebrates were fairly uniformly distributed, while fishes and plants were not. Plants were more diverse above Salto Pará while fishes were more diverse below. Nonetheless, the Upper Caura is more important for plant and fish conservation because the species in the upper sections have narrower distributions.

Number of Species

Plants:	399 species
Fish:	278 species
Aquatic insects:	>87 species
Molluscs:	
gastropods	2 species
bivalves	1 species
Crustaceans:	
isopods	1 species
branchiopod	1 species
decapods	10 species (4 crabs, 6 shrimps)
Annelids:	>3 species

New Records for the Caura River Basin

Fishes:	110 species
Shrimps:	4 species
Crabs:	2 species

New Species Discovered

Shrimps:	*Pseudopalaemon* n. sp.
Fishes:	10 species

Conservation Recommendations

The Caura River represents a pristine region typical of the Guyanas and provides an excellent opportunity for conservation, though the portion of the river below the Cinco Mil rapids has a greater number of actual and potential threats. The most specific recommendations include:

- Stop all development plans that might cause changes in the natural hydrologic cycle of the Caura River Basin including dams, water diversions and major riparian changes.
- Perform a migratory fish study.
- Establish a three-year program on sustainable fishing in the Lower Caura.
- Establish a follow-up monitoring program for the watershed.
- Develop a corridor to protect the flooded forests (up to a 50-year flood cycle).
- Implement educational and public outreach programs about the importance of aquatic ecosystems.
- Prohibit the introduction of exotic aquatic species.
- Restore riparian forests below Cinco Mil rapids.
- Develop management plans for the sustainable harvest of the following plant species: *Ocotea cymbarum, Vochysia venezuelana* (used for making boats), *Acosmium nitens, Geonoma deversa* (used in house construction) and *Heteropsis flexuosa* (used in making baskets and construction).
- Provide immediate protection, as part of a general conservation plan, to the following areas: Raudal Cejiato, Kakada River, the backwaters and flooded lagoons close to Entreríos, the Raudal Suajiditu and the region just above Salto Pará, El Playón and surrounding areas, the Nichare and Tawadu Rivers, the flooded lagoons close to Boca de Nichare and the Takoto River near Raudal Cinco Mil.

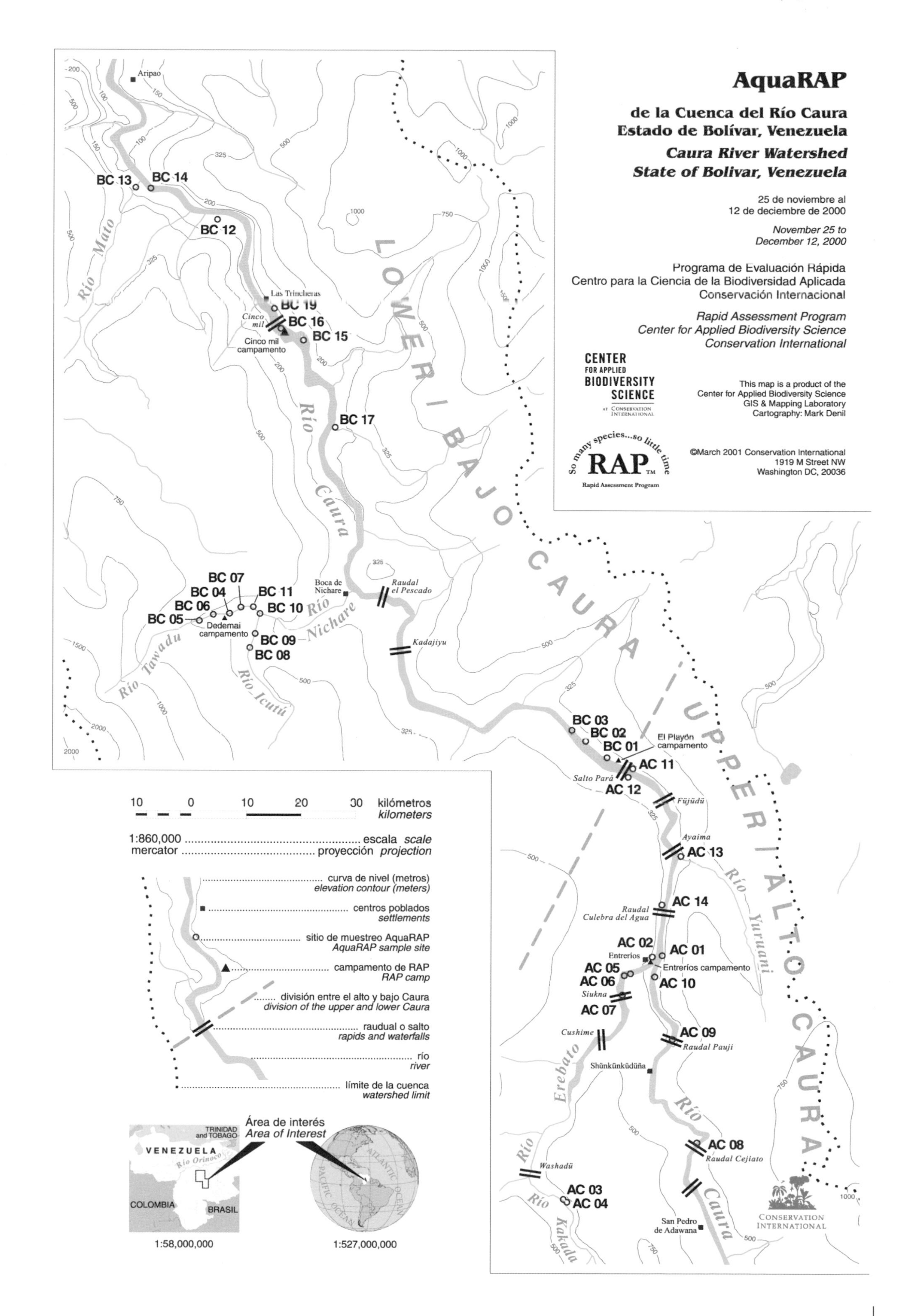

AquaRAP
de la Cuenca del Río Caura
Estado de Bolívar, Venezuela
Caura River Watershed
State of Bolívar, Venezuela
25 de noviembre al
12 de deciembre de 2000
November 25 to
December 12, 2000
Programa de Evaluación Rápida
Centro para la Ciencia de la Biodiversidad Aplicada
Conservación Internacional
Rapid Assessment Program
Center for Applied Biodiversity Science
Conservation International
CENTER FOR APPLIED BIODIVERSITY SCIENCE AT CONSERVATION INTERNATIONAL
This map is a product of the
Center for Applied Biodiversity Science
GIS & Mapping Laboratory
Cartography: Mark Denil
So many species...so little time
RAP
Rapid Assessment Program
©March 2001 Conservation International
1919 M Street NW
Washington DC, 20036
LOWER / BAJO CAURA
UPPER / ALTO CAURA
Aripao
BC 13
BC 14
BC 12
Río Mato
Las Trincheras
BC 19
Cinco mil
BC 16
BC 15
Cinco mil campamento
Río Caura
BC 17
BC 07
BC 04
BC 06
BC 05
BC 11
BC 10
BC 09
BC 08
Dedemai campamento
Boca de Nichare
Raudal el Pescado
Río Nichare
Río Tawadu
Río Icutú
Kadajiyu
BC 03
BC 02
BC 01
El Playón campamento
AC 11
Salto Pará
AC 12
Fújüdü
Ayaima
AC 13
Río Yuruani
Raudal Culebra del Agua
AC 14
AC 02
AC 01
Entrerios
Entrerios campamento
AC 05
AC 06
AC 10
Siukna
AC 07
Cushime
AC 09
Raudal Pauji
Shünkünküdüña
Río Erebato
AC 08
Raudal Cejiato
Washadü
AC 03
AC 04
Río Kakada
San Pedro de Adawana
CONSERVATION INTERNATIONAL
10 0 10 20 30 kilómetros
kilometers
1:860,000 .... escala scale
mercator .... proyección projection
curva de nivel (metros)
elevation contour (meters)
centros poblados
settlements
sitio de muestreo AquaRAP
AquaRAP sample site
campamento de RAP
RAP camp
división entre el alto y bajo Caura
division of the upper and lower Caura
raudual o salto
rapids and waterfalls
río
river
límite de la cuenca
watershed limit
Área de interés
Area of Interest
TRINIDAD and TOBAGO
VENEZUELA
Río Orinoco
COLOMBIA
BRASIL
ATLANTIC OCEAN
PACIFIC OCEAN
1:58,000,000
1:527,000,000

Barry Chernoff

Salto Pará dividing the Upper and Lower sections of the Caura River.
El Salto Pará dividiendo las regiones Superior e Inferior del Río Caura.

Barry Chernoff

The Upper Caura River at Entrerios.
El Río Caura en la región Superior en Entrerios.

Jensen R. Montambault

Genipa spruceana (Rubiaceae) on right side and *Croton cuneatus* (Euphorbiaceae) on left side of the ecologist. These riparian plants are commonly found on depositional or rocky banks, and are highly tolerant of flooding. Their seeds and leaves are eaten by fishes (Knap-Vispo et al. 2002).

Genipa spruceana (Rubiaceae) en el lado derecho y *Croton cuneatus* (Euphorbiaceae) en el izquierdo del ecólogo. Estas plantas ribereñas son comunmente encontradas en bancos deposicionales o rocosos y altamente tolerables a la inundación. Sus hojas y semillas son comidas por los peces (Knap-Vispo et al. 2002).

Barry Chernoff

Helicopter bringing AquaRAP team, equipment, and specimens over the falls at Salto Pará during RAP expedition.

Helicóptero que trae a los participantes, el equipo, y los especímenes de AquaRAP sobre el Salto Pará durante la expedición de AquaRAP.

Barry Chernoff and A. Machado-Allison

The striped headstander, *Anostomus anostomus*, ca. 120 mm standard length, lives in the leaves of members of the Podostemonaceae in the Upper Caura River at Raudal Suajiditu.

El pez, *Anostomus anostomus,* tiene aproximadamente 120 mm de longitud y vive en las hojas de plantas de la familia Podostemonaceae en el Río Caura región Superior en Raudal Suajiditu.

Barry Chernoff and A. Machado-Allison

The cichlid, *Guianacara geayi*, from the Upper Caura River at Raudal Cejiato, approximately 70 mm standard length.

Guianacara geayi, del Río Caura región Superior en Raudal Cejiato, tiene aproximadamente 70 mm de longitud.

Barry Chernoff and A. Machado-Allison

A new species of bloodfin tetra, *Aphyocharax* sp., 50.7 mm standard length, from the Upper Caura River. Discovered during the AquaRAP expedition.

Aphyocharax sp. es una nueva especie de pescado que tiene 50.7 mm de longitud. Encontrado durante la expedición de AquaRAP en el Río Caura región Superior.

Barry Chernoff and A. Machado-Allison

Tetra, *Jupiaba zonata,* 45 mm standard length.

Jupiaba zonata tiene 45 mm de longitud.

Jensen R. Montambault

Candiru, *Vandellia cirrosa*, a small fish that lives in the sandy bottom and leaps into larger fishes gills to suck blood.

Vandellia cirrosa es un pez pequeño que vive en el fondo arenoso y salta en las agallas de peces más grandes para tomar sangre.

Barry Chernoff and A. Machado-Allison

The tentacled armored catfish, *Ancistrus* sp., from the Caura River above Salto Pará, lives among rocks and inside holes in logs.

Ancistrus sp., del Río Caura sobre el Salto Pará, vive entre rocas y dentre y ramas.

Jensen R. Montambault

Apinagia sp. (Podostemaceae), an aquatic species of the vascular river-weed family, is found on rocky substrates in fast currents usually associated with riffles or rapids. These plants are reported in the diet of fishes (Knap-Vispo et al. 2002).

Apinagia sp. (Podostemaceae) es una especie acuática que se encuentra en substratos rocosos en las corrientes con rapidos. Estas plantas son parte de la dieta de varios especies de peces.

Barry Chernoff and A. Machado-Allison

A shrimp, *Macrobrachium brasiliense,* from the Lower Caura River.

El camarón, *Macrobrachium brasiliense,* se encontró en el Río Caura región Inferior.

Barry Chernoff and A. Machado-Allison

A crab, *Fredius stenolobus,* from El Playón, Lower Caura River.

El cangrejo, *Fredius stenolobus*, de El Playón, en el Río Caura región Inferior.

Jensen R. Montambault

This plant is a member of the Family Podostemonacea.

Esta planta es un miembro de la familia Podostemonacea.

Jensen R. Montambault

Caura AquaRAP team.

Equipo del AquaRAP del Caura.

Executive Summary and Overview

INTRODUCTION

The Caura watershed is located in the eastern sector of the Bolívar State, Venezuela. It is a vast expanse of relatively pristine forests and rivers and is a major tributary of the Orinoco River (Rosales and Huber 1996). Evergreen, flooded, gallery and savanna forests cover approximately 90% of the basin (Marín and Chaviel 1996). The basin forms a part of the homelands of the Ye'kuana people who are extremely careful in their management of and appreciation for their natural resources. The very high environmental quality of the Caura River Basin, especially from the Raudal Cinco Mil to the headlands, is due in large measure to the excellent stewardship of the Ye'kuana.

The Caura River Basin is located in the middle region of the Venezuelan Guayana Shield (3°37'–7°47'N and 63°23'–65°35'W). The principal rivers contained within the basin are the Sipao, Nichare, Erebato, and Merewari Rivers on the western margin and the Tigrera, Pablo, Yuruani, Chanaro and Waña Rivers on the eastern margin. The Caura River courses more than 700 km, originating in the highlands of the Guayana Shield at 2,000 meters above mean sea level and crossing several types of physiographic provinces from the rocky complexes of the Guayana Shield in the south to the alluvial planes near its mouth in the Orinoco. The area of the basin is approximately 45,336 km^2, which is 20% of the total surface of Bolívar State or 5% of Venezuela. The Caura River Basin is, thus, the fourth largest basin in Venezuela, preceded in size by the basins of the Apure, Caroní, and Orinoco Rivers (Peña and Huber 1996).

Biological Diversity of the Caura

The biodiversity of the Caura River Basin is not uniformly known across all groups of plants and animals. While much of the terrestrial flora and fauna (Rosales and Huber 1996; Huber and Rosales 1997) is fairly well studied, this is not true for the aquatic fauna (Machado-Allison et al. 1999). The vegetation of the Caura River Basin is very diverse; more than 1180 plant species have been identified from the Basin. The existing data reveal that inland forests cover approximately 90% of the watershed while the remaining 10% of the watershed consists of flooded forests and other non-forest vegetation (CVG-TECMIN 1994; Huber 1996; Marín and Chaviel 1996; Rosales 1996; Aymard et al. 1997; Dezzeo and Briceño 1997; Bevilacqua and Ochoa 2001).

Terrestrial wildlife is known to contain approximately 475 species of birds, 168 species of mammals, 13 amphibian species, and 23 species of reptiles (Bevilacqua and Ochoa 2001). These values represent 30% of the species registered for Venezuela and 51% of Guayanese species. Of the total terrestrial vertebrate species in the Caura River Basin, 5% are considered under threat either nationally or internationally (Bevilacqua and Ochoa 2000). With the addition of the 110 new records for the basin and the 10 new species collected during this AquaRAP survey, the number of freshwater fishes documented with certainty is 278. The 92 species of benthic invertebrates and 12 species of crustaceans comprise the first recorded information for these taxa.

The biogeographic relationships of the flora and the fauna indicate the effect that the waterfall, Salto Pará, has had on the mixing of species. Above the Salto Pará there is a higher degree of endemism than below the falls. For example, in the plants approximately 88% of the Guayana endemic genera are present in the Caura River Basin, and there is a high level of endemism in highland or Tepuy communities (Berry et al. 1995; Huber et al. 1997; Bevilacqua and Ochoa 2001). Above the falls the flora, the aquatic invertebrates and the fishes largely exhibit Guayana Shield and Amazonian affinities. Below the falls these groups include elements from the main stem of the Orinoco River and from the Llanos.

Social and Economic Diversity

The economic and social structures of the human populations in the Caura River Basin are complex and diverse. In the upper and middle areas (above Salto Pará) indigenous cultures such as the Ye'kuana and Sanema (Yanomami) ethnic groups maintain their historic traditions. Seed planting of yucca and plantain in "conucos," hunting, fishing and gathering are fundamental to their domestic and economic activities, such that each member of the population has particular responsibilities according to age and sex (Silva-Monterrey 1997). In the lower part of the basin where human population densities are higher, traditional indigenous elements mix with those of occidental cultures brought by "criollos." In this lower region, the economy incorporates forestry, agriculture, ranching and fisheries, as well as tourist and artisanal activities. The economic relationships in the lower part of the Caura River Basin are more complex. The indigenous ethnic groups use a wide variety of fishes and other aquatic resources for their subsistence.

Beyond the indigenous practices, commercial fisheries for human consumption use a much narrower range of fish species, principally: "cachamas" (*Colossoma macropomum*), "cajaros" (*Phractocephalus hemiliopterus*), "coporos" (*Prochilodus mariae*), "curbinatas" (*Plagioscion squamossissimus*), "laulaos" and "valentones" (*Brachyplatystoma* spp.), "morocotos" (*Piaractus brachypomus*), "Palometas" (*Mylossoma* spp.), "rayaos" (*Pseudoplatystoma* spp.), "sapoaras" (*Semaprochilodus laticeps*) and "sardinatas" (*Pellona castelneana*) (Novoa 1990; Machado-Allison et al.1999). Numerous additional fish species have potential value as ornamental species, an alternative not yet employed but that has the potential to develop into a sustainable industry for indigenous populations in the area. The following species are found in the Caura River and are important and commonly known in the world aquarium trade: "tetras" (*Astyanax, Hemigrammus, Hyphessobrycon, Jupiaba, Moenkhausia*), "palometas" or "silver dollars" (*Metynnis, Myleus, Mylossoma*), "cichlids" (*Aequidens, Apistogramma, Bujurquina, Mesonauta*), "piranhas" (*Pygocentrus, Serrasalmus*) and "headstanders" (*Anostomus, Leporinus*).

Much of the subsistence of the human communities in the Caura River Basin, as well as commercial fisheries in the Orinoco River, depends directly upon the health of the entire watershed. The use of plants by the indigenous communities is high. A total of 358 lowland forest species are known to be used by indigenous peoples (Knap-Vispo 1998) indicating the critical role that these ecosystems play in traditional cultures (Bevilacqua and Ochoa 2001). As a result of our expedition, we found that inundated forests and protected near-shore habitats serve as critical nursery grounds for many species such as the palometa (*Myleus rubripinnis*). Successful recruitment in the Caura River contributes to the approximately 75,000 metric tons of fishes landed at Maripa and Caicara del Orinoco.

Threats to the Region

The Guayana Shield region of Venezuela has endured much environmental alteration from development projects, industry and unregulated fisheries or hunting. The exploitation of strategic minerals such as gold, diamonds and bauxite coupled with the development and construction of one of the largest hydroelectric dam complexes has produced: 1) biodegradation and destruction of extensive green areas in the Caroní and Cuyuní river basins (Machado-Allison et al. 2000); 2) mercury pollution of rivers, forests, wildlife and humans; 3) diminished water quality due to increased sedimentation; and 4) loss of large quantities of potable water for domestic uses (Machado-Allison 1994, 1999; Miranda et al. 1998).

The Caura River Basin is threatened by mining, illegal fishing practices and a proposed dam. Several hundred kilometers of the basin are now being heavily logged close to the Paraguay River Basin. Some mining activities have also been introduced into headwater areas of the basin. At the moment, mining activities are minor in comparison to neighboring basins such as the Caroní and Cuyuní (see Figure 1 of Machado-Allison et al. 2000). A dramatic decrease in fisheries resources in the Lower Caura River is thought to result from illegal commercial fishing in indigenous territory. Another potential threat to the Caura River Basin is a plan to construct a new hydroelectric dam and to divert as much as 75% of the water from the Caura River into the Paragua-Caroní River system. Such a project would not only drastically deplete the amount of water in the Caura River, but also severely alter the hydrological cycle. The health and maintenance of human communities and aquatic and riparian flora and fauna are completely dependent upon the natural hydrological cycle. In the Caura River below Salto Pará, expanding tourism, sportsfishing and agriculture are having a measurable impact, especially below Raudal Cinco Mil. These activities must be studied to allow sustainable use of the ecosystem.

Conservation Opportunities

The Caura River Basin is a large pristine wilderness area. Because of destruction and development in adjacent regions of the northern Guayana Shield (Caroní and Cuyuní Rivers), the Caura River represents an important opportunity to preserve a unique region. The community composition of animals and plants in the Caura River Basin occurs nowhere else in Venezuela or on the Guayana Shield. It is a high diversity area, with many endemic forms and unique communities, such as aquatic plant communities (Podostemonaceae) in rapids and the mid-river rocky-island floras. Protection through education programs, community outreach, and long-term monitoring are highly recommended. Monitoring and establishment of fisheries regulation in the basin is essential to achieve sustainability for its inhabitants.

The AquaRAP Expedition

In response to the projected development of the Caura River Basin, including altering the river course for hydroelectric activity, deforestation, increasing human populations and other threats, a multidisciplinary, multinational team of scientists undertook an Aquatic Rapid Assessment Program (AquaRAP) expedition along the Caura River with the goals of: 1) describing the biological and environmental aspects of the aquatic ecosystems before the implementation of any large-scale modifications; and 2) determining, if possible, the potential changes to the aquatic resources resulting from perceived threats. The expedition took place between 25 November and 12 December 2000 and surveyed the section of the Caura River between the Raudal Cejiato to the south and the Mato River to the north (AquaRAP Map). To facilitate comparisons, the surveyed area was divided into eight regions: Kakada River; Erebato River; Caura River from Entreríos-Raudal Cejiato; Caura River from Entreríos-Salto Pará; Caura River near El Playón; Nichare River; Caura River in the vicinity of Raudal Cinco Mil; and the Mato River (AquaRAP Map). Terrestrial vegetation, aquatic vegetation, physical and chemical water characteristics, plankton, aquatic macroinvertebrates and fishes were assessed. These inventories will be used as a basis for conservation and future research recommendations in the area.

The South American AquaRAP program is a multinational, multidisciplinary program devoted to identifying conservation priorities and sustainable management opportunities for the continent's freshwater ecosystems. AquaRAP's mission is to assess the biological and conservation value of tropical freshwater ecosystems through rapid inventories, and to report the information quickly to local policy-makers, conservationists, scientists and international funding agencies. AquaRAP is a collaborative program managed by Conservation International and the Field Museum.

At the core of AquaRAP is an international steering committee composed of scientists and conservationists from seven countries (Bolivia, Brazil, Ecuador, Paraguay, Perú, Venezuela and the United States). The steering committee oversees the protocols for rapid assessment and the assignment of priority sites for rapid surveys. AquaRAP expeditions, which involve major collaboration with host country scientists, also promote international exchange of information and training opportunities. Information gathered during AquaRAP expeditions is released through Conservation International's *RAP Bulletin of Biological Assessment* designed for local decision-makers, politicians, leaders and conservationists who can set conservation priorities and guide action through funding in the region.

An assessment of the current biodiversity in the Caura River Basin was critical before large scale modifications are made to the region. This part of Venezuela has not been thoroughly sampled, although the flora, birds and mammals are reasonably well known. Additional information was necessary before recognizing management opportunities. The next section summarizes the results of the technical chapters based upon analyses of the data and specimens collected during the AquaRAP expedition. Please see the chapters themselves for more detailed information.

SUMMARY OF RESULTS

Terrestrial and Aquatic Vegetation

A total of 443 samples were taken during the expedition from a variety of riparian and aquatic habitats in the Caura River Basin. The samples contained 399 species of plants with 291 and 185 species from the Upper and Lower Caura, respectively. All of the plant species are included in the list of 1,180 species known from the basin. The investigation reveals that there is a large diversity in the composition of floral communities characteristic of humid climates with low nutrient soils. The variation in community structure comprises a gradient of riparian landscapes determined by the intensity and duration of erosional processes. The diversity of underlying geology as well as the climatic gradient associated with altitude (40–2350 m) also contribute to the exceptional species richness of the Caura River watershed.

The proportion of endemic species is low in the lowlands and in the flooded forest ecosystems of the Caura River Basin. This region is dominated by the palms: *Euterpe precatoria, Attalea maripa, Socratea exhorriza, Genoma baculifera* and *Bactris brongniartii*. Species richness is relatively high in comparison to values for similar forest environments in Amazonia and Guyana. The diversity and species richness of the flooded ecosystems of the Erebato, dominated by *Oenocarpus*, and the middle Caura, dominated by *Mauritiella*, are also similar to values of other riverine corridors on the Guayana Shield. A variable set of unique floral assemblages associated with riverine islands are found in the Erebato River and middle Caura Rivers. These islands have both *terra firme* forests and often dense aquatic forests of Podostemonaceae attached to boulders in rapids.

Water Quality

Forty-three sites within the Upper and Lower regions of the Caura River Basin were sampled for various limnological parameters. The Upper Caura is pristine, comprised of more channelized basins and subject to greater seasonal hydrologic variation than the Lower Caura and as a result far less forest inundation. In this generally low-nutrient tropical river, focus was placed on the analysis of the following parameters: temperature, conductivity, pH, secchi depth, dissolved organic carbon (DOC), alkalinity and sediment.

In general, the waters are slightly acidic and dilute, with very low conductivity and alkalinity. This is typical of a watershed in which the primary input of water is rainwater. These results are congruent with those presented in prior studies and can be attributed to the very ancient geology of the region. The waters were found to be similar for these parameters throughout the Upper and Lower regions of the watershed; however, occasional tributaries were found to vary somewhat from this general theme. DOC (terrestrial input) varied throughout the watershed, and both clearwater and blackwater rivers were sampled.

Overall, water quality was determined to be 'good' for all sites sampled at the time of this study. The Lower region was more disturbed than the Upper region of the watershed. Higher disturbance at the water margins can lead to increased sediment input in the rainy season, decreased terrestrial input of biologically important carbon forms (DOC) and a change of habitat for aquatic organisms. These habitat changes may include changes in water clarity, substrate, river channelization, and the number and diversity of habitats that sustain the current community of aquatic organisms.

Benthic Macroinvertebrates

Diversity of benthic macroinvertebrates in the Caura River was assessed in 25 localities in the Upper as well as in the Lower section of the river. The community of aquatic macroinvertebrates sampled was composed of insects, snails, mussels, oligochaete worms, leeches, turbellarians and crustaceans. Aquatic insects were represented by several orders, of which the most diverse were: Odonata, Diptera, Ephemeroptera, Hemiptera and Trichoptera. The observed species composition is typical of a pristine environment, since it contains a high diversity of Odonata and Trichoptera, which inhabit environments free of pollutants or perturbation. The majority of genera of benthic macroinvertebrates in the watershed were found to be homogeneously distributed. The variation in diversity of benthic faunal groups seems to be related to changes in dissolved oxygen and turbidity of the water. These findings are important because a drastic change in the annual cycle of the water, deforestation or mining would introduce suspended particulate material and increase the conductivity of the water with a subsequent dramatic change in the benthic communities.

Decapod Crustaceans

A total of ten species of decapod crustaceans were found in the surveyed area of the middle Caura River Basin: six species of Palaemonidae (shrimps), one species of Pseudothelphusidae (crabs), and three species of Trichodactylidae (crabs). The region upriver from Salto Pará (Upper Caura) had a lower richness, with five species, than the lower region (Lower Caura) with eight species. The decapod fauna of the Caura River Basin has one possible endemic species, the undescribed shrimp *Pseudopalaemon* sp., while the others are known either from the Llanos region or from the Amazon Basin. The number of decapod species and genera found in the Caura River Basin represents a typical sample of the diversity and abundance of inland river systems from the Guayana and Amazon regions. The shrimp *Macrobrachium brasiliense* was the most frequent and abundant species found in this survey and could be considered as the most typical species of this system. Two undescribed species of palaemonid shrimps were collected. Abundance of decapods was low to moderate, probably reflecting sampling constraints and the overall oligotrophic condition of the habitats. From the decapod community structure, habitat use, and distribution throughout the basin we conclude that the decapod community is at present healthy and will remain so as long as the pristine conditions of the region are preserved.

Fishes

Sixty-five field stations were sampled between Raudal Cejiato and the Kakada River in the south to Raudal Cinco Mil in the north. A total of 278 fish species were identified for the Caura River Basin. The order Characiformes, with 158 species (56.8%) was the most diverse, followed by Siluriformes (74 species, 26.6%), Perciformes (27, 9.7%), Gymnotiformes (9, 3.2%), Clupeiformes (3, 1.1%), Rajiformes and Cyprinodontiformes (2, 0.7%), Beloniformes (1, 0.4%), Pleuronectiformes (1, 0.4%) and Synbranchiformes (1, 0.4%). The Family Characidae with 113 species (40.6 % of the total) was the most species rich. The results add 54 species for the order Characiformes and 39 species of Characidae known from the basin. With respect to other groups the increases were: Siluriformes, 74 vs. 49 previously reported and Perciformes, 27 vs. 12. In total, 110 new records were recorded for the basin. In the Upper areas, the Salto Pará (Upper), Raudal Cejiato and Kakada River (beaches and caño Suajiditu) were the most diverse and abundant with several typical Guayanese forms. The areas below Salto Pará (El Playón), Tawadu and Nichare Rivers and Raudal Cinco Mil (including Takoto River) possessed high diversity and abundance, with numerous species typically found in the Llanos and Orinoco River systems. However, both the upper and lower areas possess their own characteristics in terms of taxonomy, biogeography and conservation importance. A large number of species have potential economic importance as food and as ornamental fishes.

Based upon species richness, diversity and relative abundance of the fishes and the dependence of local human populations upon stocks of fishes, certain areas in the Upper and Lower Caura must be protected and preserved as part of a general conservation plan. These areas include Raudal Cejiato, Kakada River, the backwaters and flooded lagoons close to Entreríos, the Raudal Suajiditu and the region just above Salto Pará, El Playón and surrounding areas, the Nichare and Tawadu Rivers, the flooded lagoons close to Boca de Nichare and the Takoto River near Raudal Cinco Mil.

Ecohydrology and Riparian Vegetation

This chapter combines analyses of ecogeography and the hierarchical study of river systems into an approach of general utility for conservation. We perform a spatial analysis using a Geographical Information System with geo-referenced passport data of target plant taxa to study trends in species diversity along the Caura riparian corridor using the family Leguminosae. We then compare these results with those obtained from different sub-basins of the Amazon and the Orinoco River using selected genera of the tribe Ingeae.

The within-basin and inter-basin analyses of number of species, rarity, commonness and geographical distinctiveness indicated the importance of conserving the lower riparian landscapes on the Caura River. Our results demonstrate the critical importance of conserving the Caura River Basin because not only does the Caura possess a large number of species but also because the species form a unique community originating in a number of different basins and phytogeographical regions. The Caroní River, which has a close floral similarity to the Caura, is highly disturbed. The most critical sections of the Caura River Basin for conservation are the sectors below Salto Pará. The inter-basin analysis showed different regional patterns between selected Ingeae genera that relate to phytogeographical regions and river types. In addition to the Caura River, the Negro River of Brazil and Venezuela is also important for a multinational conservation effort. The ecohydrological approach is demonstrated to be an important conservation tool for analyzing riparian biodiversity and an aid in the selection of reserve areas for *in situ* riparian conservation.

Distribution of Fishes and Patterns of Biodiversity

Null hypotheses concerning random distributions of species with respect to subregions and macrohabitats within the Caura River Basin were tested with data from 97 species of benthic invertebrates, 303 species of riparian plants and 278 species of freshwater fishes. The analysis of the eight subregions split evenly above and below the falls indicated that invertebrates were randomly distributed with respect to subregion. Fishes and plants were not, although the subregional effect in plants was more strongly patterned than that of fishes. Furthermore, fishes were less species rich in the Upper Caura than the Lower Caura. The converse was observed for plants while the invertebrates were almost equally rich. Non-random macrohabitat effects are found in each of the groups with certain commonalities. For example, species of Odonata and Ephemeroptera are found in high oxygen, swift water, rapids habitats and are associated with a large assemblage of fishes that are found in association with dense stands of macrophytes, usually Podostemonaceae. Fish assemblages demonstrate smooth transitions along several macrohabitat gradients (e.g., sand to mud bottoms and shores). At least six macrohabitats are necessary to preserve 82% of the species of fishes.

CONSERVATION AND RESEARCH RECOMMENDATIONS

The basis for recommended conservation and research initiatives can be found in the executive summary and the chapters of the report. Recommendations are not listed in order of priority.

- **Stop all development plans that might cause changes in the natural hydrologic cycle of the Caura River Basin including dams, water diversions and major riparian changes.** Dams or water diversion would upset the hydrologic cycle that is important in tropical rain-dominated systems. The health of the forests, and the successful recruitment and maintenance of the biodiversity, depends entirely upon maintaining the natural hydrologic cycle.

- **Establish a monitoring program on migratory fishes.** Migratory fishes enter the Caura River seasonally and use the flooded forests for breeding and nursery grounds. Critical habitat above and below Salto Pará must be identified to ensure survival of both species and the fisheries.

- **Establish a three-year study program to institute sustainable fishing quotas in the Lower Caura.** Evidence indicates that there is much poaching into Ye'kuana territory and potential overfishing in the Caura River below Salto Pará. The fisheries in this region need to be monitored and evaluated in order to establish management techniques (e.g., quotas, gear type) to ensure sustainable catches for perpetuity.

- **Establish a follow-up monitoring program for the watershed and restore the hydrologic and limnologic monitoring station at Entreríos.** This station would provide a valuable platform for scientists to access and study the region in greater detail and build a long-term database of this tropical rainforest environment. Given that pristine tropical freshwater habitats continue to be degraded at an alarming rate worldwide, preserving the few largely undisturbed regions remaining, including the Caura River Basin, is a priority.

- **Develop a corridor to protect the flooded forests (up to a 50-year flood cycle).** The Caura River Basin presents a unique opportunity to save an endangered resource in Venezuela—the natural Guayanese biome. In accordance with the wishes and customs of the Ye'kuana peoples, a large protective corridor could be established, at least to the level of the 50-year flood cycle, in order to protect the forests and river and the species that use them.

- **Establish a species protection zone plan.**The regions above and below Salto Pará have significantly different species of plants and fishes. In addition, six aquatic macrohabitats (e.g., rapids, river islands, backwater areas, etc.) are necessary to protect the existence of 82% of the species of fishes, including those species used for food and commerce. A plan is needed to determine the amount of area for each of the habitats that is necessary to preserve more than 80% of the animals and plants in the basin. These critical areas should be regarded as core protection zones with minimal human disturbance. Because of the differences in flora and fauna, the core zones must be established in areas above and below Salto Pará.

- **Implement educational and public outreach programs about the importance of aquatic ecosystems.** This needs to be done primarily for communities outside of Ye'kuana territory. Below Ye'kuana territory there is a large degree of degradation both to forest and aquatic habitats. Many communities of people depend upon the basin for their existence and, therefore, education about how protection of the river basin can sustain or enrich their existence will be critical.

- **Prohibit the introduction of exotic aquatic species.** There are sufficient native fishery and aquatic resources available within the Caura River Basin to prohibit the introduction of exotics. The Caura River aquatic fauna is unique within Venezuela and should be protected. There is evidence to show that introduced exotics accelerate the loss of native species and accelerate the degradation of natural habitats.

- **Restore riparian forests below Raudal Cinco Mil rapids.** Regulate and do not permit further agricultural expansion and deforestation within the Caura River Basin. This is especially important in the Lower Caura. This area has been degraded with substantial loss of biodiversity and community structure. This degradation may be coupled with decreased recruitment of fishes, shrimps and crabs. Reforestation techniques are available and would help restore native communities as well as potentially increase fisheries.

- **Develop management plans for the sustainable harvest of the following plant species:** *Ocotea cymbarum, Vochysia venezuelana* (used for making boats), *Acosmium nitens, Geonoma deversa* (used in house construction), and *Heteropsis flexuosa* (used in making baskets and construction). Such plans could help increase economic development of the basin in a sustainable way, as well as help support communities of people currently living in the basin.

LITERATURE CITED

Aymard, G., S. Elcoro, E. Marín and A. Chaviel. 1997. Caracterización estructural y florística en bosques de tierra firme de un sector del bajo Río Caura, Estado Bolívar, Venezuela. *In:* Huber, O. and J. Rosales (eds.). Ecología de la Cuenca del Río Caura. Scientia Guaianae 7:143–169.

Berry, P.E, O. Huber and B.K. Holst. 1995. Floristic Analysis and Phytogeography. *In:* Berry, P. E, B. K. Holst, K. Yatskievych (Ed). Flora of the Venezuelan Guayana, Vol. 1. Introduction. Saint Louis USA: Missouri Botanical Garden. Pp. 161–191.

Bevilacqua, M. and J. Ochoa G. (eds.) 2000. Informe del Componente Vegetación y Valor Biológico. Proyecto Conservación de Ecosistemas Boscosos en la Cuenca del Río Caura, Guayana Venezolana. Caracas, Venezuela: PDVSA-BITOR, PDVSA-PALMAVEN, ACOANA, AUDUBON de Venezuela and Conservation International.

Bevilacqua, M. And J. Ochoa G. 2001. Conservación de las últimas fronteras forestales de la Guayana Venezolana: propuesta de lineamientos para la Cuenca del Río Caura. Interciencia 26(10): 491–497.

CVG-TECMIN. 1994. Informes de avance del Proyecto Inventario de los Recursos Naturales de la Región Guayana. Hojas NB-20: 1, 5, 6, 9, 10, 13 and 14. Ciudad Bolívar, Venezuela: Gerencia de Proyectos Especiales.

Dezzeo, N. and E. Briceño. 1997. La vegetación en la cuenca del Río Chanaro: medio Río Caura. *In:* Huber, O. and J. Rosales (eds.). Ecología de la Cuenca del Río Caura. Scientia Guaianae 7: 365–385.

Huber, O. 1996. Formaciones vegetales no boscosas. *In:* Rosales, J. and O. Huber (eds.). Ecología de la Cuenca del Río Caura. Scientia Guaianae 6: 70–75.

Huber, O. and J. Rosales (eds.). 1997. Ecología de la Cuenca del Río Caura, Venezuela. II Estudios Especiales. Scientia Guaianae No. 7. Caracas, Venezuela: BioGuayana.

Huber O., J. Rosales and P. Berry. 1997. Estudios botánicos en las montañas altas de la cuenca del Río Caura. *In:* Rosales, J. and O. Huber (eds.). Ecología de la Cuenca del Río Caura. Scientia Guaianae 7: 441–468.

Knab-Vispo, C. 1998. A rain forest in the Caura Reserve (Venezuela) and its use by the indigenous Ye'kuana people. PhD Thesis. University of Wisconsin-Madison, USA.

Machado-Allison, A. 1994. Factors affecting fish communities in the flooded plains of Venezuela. Acta Biol. Venez. 14(3):1–20.

Machado-Allison, A. 1999. Cursos de agua, fronteras y conservación. *In:* G. Genatios (ed.). Ciclo Fronteras: Desarrollo Sustentable y Fronteras. Caracas, Venezuela: Com. Estudios Interdisciplinarios, UCV.

Machado-Allison, A., B. Chernoff, C. Silvera, A. Bonilla, H. Lopez-Rojas, C.A. Lasso, F. Provenzano, C. Marcano and D. Machado-Aranda. 1999. Inventario de los peces de la cuenca del Río Caura, Estado Bolivar, Venezuela. Acta Biol. Venez. 19(4):61–72.

Machado-Allison, A., B. Chernoff, R. Royero-Leon, F. Mago-Leccia, J. Velazquez, C. Lasso, H. López-Rojas, A. Bonilla-Rivero, F. Provenzano and C. Silvera. 2000. Ictiofauna de la cuenca del Río Cuyuní en Venezuela. Interciencia 25(1). 13–21.

Miranda, M., A. Blanco-Uribe, L. Hernández, J. Ochoa and E. Yerena. 1998. No todo lo que brilla es oro: hacia un nuevo equilibrio entre conservación y desarrollo en las últimas fronteras forestales de Venezuela. Washington, DC: Inst. Rec. Mundiales (WRI).

Marin, E. and A. Chaviel. 1996. Bosques de tierra firme. *In:* Rosales, J. and O. Huber (eds.). Ecología de la Cuenca del Río Caura. Scientia Guaianae 6: 60–65.

Novoa, D. 1990. El Río Orinoco y sus pesquerías; estado actual, perspectivas futuras y las investigaciones necesarias. *In:* Weibezahn, F., H. Alvarez and W. Lewis (eds.) El Río Orinoco como Ecosistema. Edelca, Fondo Ed. Acta Cienctífica Venezolana, CAVN, USB: 387–406.

Peña, O. and O. Huber. 1996. Características geográficas generales. *In:* Rosales, J. and O. Huber (eds.) Ecología de la Cuenca del Río Caura. Scientia Guaianae 6: 4–10.

Rosales, J. 1996. Vegetación: los bosques ribereños. *In:* Rosales J. and O. Huber (eds.) Ecología de la Cuenca del Río Caura, Venezuela: I. Caracterización general. Scientia Guaianae 6: 66–69.

Rosales, J. and O. Huber (eds.) 1996. Ecología de la Cuenca del Río Caura, Venezuela. I. Caracterización General. Scientia Guaianae 6:1–131.

Silva-Monterrey, N. 1997. La Percepción Ye'kuana del Entorno Natural. En: O. Huber and J. Rosales (eds.) Ecología de la Cuenca del Río Caura. Estudios Especiales Scientia Guaianae, 7: 65–84.

Chapter 1

Introduction to the Caura River Basin, Bolívar State, Venezuela

Antonio Machado-Allison, Barry Chernoff and Mariapia Bevilacqua

INTRODUCTION

The Neotropics still contain hundreds of thousands of square kilometers of pristine forested areas. Much of this pristine region exists in Brazil, Bolivia, Colombia, the Guyanas, Perú and Venezuela. These Neotropical wilderness areas harbor some of the greatest diversity of species and highest biomass of plants, wildlife and freshwater ecosystems on the planet. Today, however, increasing world consumptive demands and increasing human populations are accelerating the exploitation of this once immense reservoir of food, minerals, scenic beauty, energy and biogenetics (SISGRIL 1990; Bucher et al. 1993; Chernoff et al. 1996; Chernoff and Willink 1999; Machado-Allison 1999; Machado-Allison et al. 1999, 2000). Biodiversity studies in South America, especially in the Amazon and Orinoco Basins, are ever more important in order to link economic potential with biological sustainability as a way to reduce actual threats and adverse environmental changes (IUCN 1993; Aguilera and Silva 1997). It is encouraging, however, that recent studies have shown that Venezuela is among those countries possessing high levels of biological diversity (Mittermeier et al. 1997, 1998).

Due to the maintenance of a national conservation policy, Venezuela possesses numerous protected areas. Many receive special protection as National Parks, Forestry Reserves, Natural Monuments, etc. The protected areas are distributed around the country and include Andean highlands ("paramos"), savannas ("llanos"), littoral lagoons, cloud forests and tropical humid forests. All these areas were created following a vision or criteria developed to protect terrestrial habitats and/or wetlands. Few conservation areas are designed to protect river basins, aquatic ecosystems and their adjacent forests.

A significant fraction of Venezuelan ecosystems are currently threatened. The majority of these ecosystems, considered the last forest frontiers of the tropical world, are situated in regions where social and environmental needs conflict (Bevilacqua and Ochoa 2001). The Caura River Basin and other aquatic ecosystems located to the south of the Orinoco River are among these last frontiers.

The Caura watershed is located in the western sector of the Bolívar State, Venezuela. It is a vast expanse of forests and rivers and is a major tributary of the Orinoco River (Rosales and Huber 1996). The first organized expedition into the Caura was directed by Chaffanjon (1889), followed 20 years later by a joint Venezuelan and English expedition directed by André. However, results of the latter expedition were never formally published. In the past century, efforts to study the Caura have been led by scientists from Venezuelan universities and research institutions. Parallel efforts to develop a conservation plan for the Caura River Basin have been made by Venezuelan NGOs, such as ACOANA, Econatura, Fundación La Salle, Jardín Botanico del Orinoco and the Corporación Venezolana de Guyana (CVG). Results of these and other scientific projects in the Caura River Basin have been published in two volumes of *Scientia Guaianae* (Rosales and Huber 1996; Huber and Rosales 1997). Recent publications about the ichthyofauna of the Caura River Basin discuss species diversity, biogeography and conservation (Chernoff et al. 1991; Vari 1995; Balbas and Taphorn 1996; Lasso and Provenzano 1997; Bonilla-Rivero et al. 1999; Machado-Allison et al. 1999).

The Caura River Basin is largely pristine and comprises part of the homelands of the Ye'kuana people. Recent changes to Venezuela's constitution give the Ye'kuana and other indigenous groups rights of determination in their homelands. The very high environmental quality of this region, especially from the Raudal Cinco Mil to the headlands, is due in large measure to the excellent stewardship of the Ye'kuana, who are extremely careful in their management of and appreciation for their natural resources.

HYDROLOGY

The Caura River Basin is located in the middle region of the Venezuelan Guayana Shield (3°37'–7°47'N and 63°23'–65°35'W). The principal rivers contained within the basin are the Sipao, Nichare, Erebato and Merewari Rivers on the western margin and the Tigrera, Pablo, Yuruani, Chanaro and Waña Rivers on the eastern margin. The area of the basin is approximately 45,336 km^2 which is 20% of the total surface of Bolívar State, or 5% of Venezuela. These data place the Caura River Basin as the fourth largest basin in Venezuela, proceeded in size by the basins of the Apure, Caroní and Orinoco Rivers (Peña and Huber 1996).

The Caura River courses more than 700 km, originating in the highlands of the Guayana Shield at 2000 meters above mean sea level. The Caura River crosses several types of physiographic environments from the alluvial planes near the mouth of the Orinoco to the rocky complexes of the Guayana Shield to the south. Evergreen, flooded, gallery and savanna forests cover approximately 90% of the basin (Marín and Chaviel 1996).

The mean annual rainfall varies from 1200 mm toward the mouth of the Orinoco (northerly) to 3000–4000 mm toward the headwaters (southerly). The differences in rainfall correspond to dramatic differences in climatic seasonality between northern and southern portions of the basin. In the north the dry season extends from January through March with a rainy season extending from April through October. In the south there is a short dry season (January-February) and an extensive rainy season throughout the rest of the year. The Caura River contributes close to 3500 m^3/s of water to the Orinoco. This places the Caura River as the second most important affluent of the Guayana Shield margin of the Orinoco (Vargas and Rangel 1996).

Water quality of the Caura River is considered to be good (García 1996). The river has been classified traditionally as a blackwater river following Sioli (1965) due to its apparent brown or tea coloration as well as its low nutrients, low pH and high transparency. However, the river does not fit the true blackwater classification if factors such as certain carbon species and oxygen concentrations (García 1996) and water color after filtration are considered.

Physiographically, the Caura River Basin can be divided into three distinct sections: 1) Lower Caura River, from the confluence with the Orinoco River to the Salto Pará; 2) Middle Caura River, from the Salto Pará to the confluence with the Merewari River and Waña River; and 3) Upper Caura, from the confluence of the Merewari and the Waña Rivers to their headwaters in the Vasade mountain range. The subregions in this report are defined by the geological formation known as the Salto Pará waterfalls. Since we were unable to survey the physiographic Upper Caura due to logistics, in this report we refer to the entire zone above the Salto Pará as "Upper Caura River" and the area below the falls as "Lower Caura River."

GEOLOGY AND GEOMORPHOLOGY

The geology of the Caura River Basin is only partially known. While the mid-elevation and low areas (<1000 m) are well known, the regions corresponding to the Upper Caura are still unexplored from tectonic and petrographic perspectives (Colvee et al. 1990; Rincón and Estanga 1996). In the Caura Basin there are representations of four Geological Provinces (Imataca, Pastora, Cuchivero and Roraima), all belonging to the Guayana Shield formation. These geological provinces, which have been modified by a series of tectonic episodes, contain both Archeozoic and Proterozoic rocks. The first Province (Imataca) corresponds to an Achaean age and is located toward the east of the lower elevational portions of the basin. The Province is characterized by abundance of gneiss, amphibolites, granite intrusions and ferruginous quarzites. The second Province (Pastora) corresponds to the Proterozoic and is located on the center-west area, principally in the Icutú, Tudi and Erebato Rivers. An abundance of granite rocks associated with gneiss is common. The third Province (Cuchivero), formed 2,000–1,700 million years ago (MYA), is the most extensive in the basin. This Province comprises a mixture of complex formations from volcanic origin with acid-volcanic rocks and granite and is located toward the Middle Caura (Entreríos), Mato River, the Upper Caura highlands (Meseta de Jaua) and Merewari River. The fourth Province (Roraima) belongs to the Late Proterozoic and is volcanic with sedimentary volcaniclastic rocks including conglomerates and sandstones. This Province is located close to the Waña River, highlands of Jaua and the Sarizariñama Tepuy, among others. Sediments from pluvial-deltaic origin correspond to the Mesa Formation (Tertiary) that is located in the northern portion of the basin close to Maripa, Aripao and the mouth of the Caura. Finally, alluviums of residual sediments of recent ages (Quaternary) constitute the sediments of the flooded or gallery forests located in the lowlands of the Caura, Nichare and Tawadu Rivers (Colvee et al. 1990; Rincón and Estanga 1996).

The typical Guayana Shield landscapes resulted from the multiple geological and geomorphological processes occurring from the Precambric to the present. The igneous-metamorphic rocks of the Imataca, Pastora and Cuchivero Provinces are related to the development of mountains, plateau, slopes and peneplanes, while the sedimentary rocks of the Roraima Province are more dominant in the high plains and tepuys. Approximately 70% of the area in the basin includes elevated landscapes and high slopes primarily in the south; the lowlands and plains are in the northern section of the basin.

BIODIVERSITY AND BIOLOGICAL VALUE

The biodiversity of the Caura River Basin is not uniformly known. While much of the terrestrial flora and fauna is fairly well studied (Rosales and Huber 1996; Huber and Rosales 1997), this is not true for the aquatic fauna (Machado-Allison et al. 1999). The wildlife of the region, mainly vertebrates, has been studied moderately and is known to contain at least 30 orders with approximately 475 species of birds, 168 species of mammals, 13 amphibian species and 23 species of reptiles (Bevilacqua and Ochoa 2001). These values represent 30% of the species registered for Venezuela and 51.3% of Guayanese species. The orders with greatest taxonomic richness are: birds (Passeriformes, Apodiformes, Falconiformes and Psittaciformes); mammals (Chiroptera, Rodentia and Carnivora); reptiles and amphibians (Squamata and Anura). Of the total terrestrial vertebrate species in the Caura River Basin, 5.2% are considered under threat in the national or international context (Bevilacqua and Ochoa 2001). The survival of these threatened species must be factored into future studies and conservation plans.

The vegetation of the Caura River Basin is very diverse. Approximately 88% of all Guayana endemic genera are present in the Caura River Basin, and there is a high level of endemism in highland or Tepuy communities (Berry et al. 1995; Huber et al. 1997; Bevilacqua and Ochoa 2001). The flora is characteristic of humid nutrient-poor forest ecosystems that have been structured through prolonged erosional processes (Rosales and Huber 1996). The extraordinary biological diversity has been attributed to the combination of erosional landscape, the convergence of four Geological Provinces and the marked altitudinal gradient (40-2,350 m) (Bevilacqua and Ochoa 2001). The existing data reveal that inland forests cover approximately 90% of the watershed while the remaining 10% consists of flooded forests and other non-forest vegetation (CVG-TECMIN 1994; Huber 1996; Marín and Chaviel 1996; Rosales 1996; Aymard et al. 1997; Dezzeo and Briceño 1997; Bevilacqua and Ochoa 2001).

The use of plants by the indigenous communities is high. A total of 358 lowland forest species are known to be used by indigenous peoples (Knab-Vispo et al. 1997) indicating the critical role that these ecosystems play in traditional cultures (Bevilacqua and Ochoa 2001).

Despite what seems to be a reasonable knowledge of the biodiversity of the Caura River Basin, our knowledge has been far from sufficient to manage and protect the natural resources. Using fishes as an example, Mago-Leccia (1970) reported approximately 500 freshwater species from Venezuela mostly found in the Orinoco River Basin, whereas Taphorn et al. (1997) increased that number to almost 1,000. In the Caura River, Balbas and Taphorn (1996) reported 135 species, which was then increased to 191 (Machado-Allison et al. 1999). In this report we raise the total to 278. Despite increasing knowledge, there is no hint that the rate of increase is leveling off as a result of collecting efforts, especially for remote regions (Chernoff and Machado-Allison 1990; Royero et al. 1992; Machado-Allison 1993). Mago-Leccia (1978), Chernoff et al. (1991) and Machado-Allison (1993) have suggested that only 30% of the aquatic flora and fauna of Venezuela is known with certainty.

Our knowledge of the Caura River mirrors that of the Orinoco River. Much recent information has come to light primarily because of numerous collections, expeditions and field efforts commensurate with national and international efforts towards biodiversity and conservation, e.g., in the Atabapo River, in the Orinoco River, at Neblina Tepuy and in the Caura River (Brewer-Carías 1988; Rosales and Huber 1996; Royero et al. 1992; Huber and Rosales 1997; Machado-Allison et al. 2000).

ECONOMY AND SOCIAL STRUCTURE

The economic and social structures of the human populations in the Caura River Basin are complex and diverse. In the areas above Salto Pará, indigenous cultures such as the Ye'kuana and Sanema (Yanomami) maintain their historic traditions. Seed planting of yucca and plantain on small farms or "conucos," hunting, fishing and gathering are fundamental to their domestic and economic activities, such that each member of the population has particular responsibilities according to age and sex (Silva-Monterrey 1997). In the lower part of the basin where population densities are higher, traditional indigenous elements mix with those of occidental cultures brought by non-indigenous Venezuelans or "criollos." In this lower region, the economy incorporates forestry, agriculture, ranching and fisheries, as well as tourist and artisanal activities. The economic relationships in the lower part of the Caura River Basin are more complex than in the upper region.

Given the conservation objectives of this volume, we are interested in cataloguing activities or enterprises that pose potential risks and threats to the ecosystem and biodiversity of the Caura River Basin (e.g. forest exploitation or extraction, pollution of soils and waters and intense fisheries). Most of these activities are concentrated in the lowland floodplains in the northern part of the basin, close to the Orinoco River. Agriculture (tubers, maize and fruit plants) and ranching are major economic activities in the area and have caused loss of forest cover and increased forest fragmentation, pollution of soils and waters by insecticides and erosion near Maripa.

Documenting sustainable activities and enterprises that protect the flora and fauna throughout the basin (Chernoff et al. 1996) is critical. As an example, the Ye'kuana traditions have historical prohibitions to impede over-exploitation of natural resources, in part by utilizing a large variety of food items (Silva-Monterrey 1997). They believe that their food gathering activities must sustain human life while contributing to the maintenance of the natural environment (Silva-Monterrey 1997). Their practice also includes periods of fasting in order to permit resources to be shared among members of the entire community (Silva-Monterrey 1997).

The local indigenous groups use a wide variety of fishes and other aquatic resources for their subsistence. Beyond the indigenous practices, however, commercial fisheries for human consumption use a much narrower range of species, principally: "cachamas" (*Colossoma macropomum*), "cajaros" (*Phractocephalus hemiliopterus*), "coporos" (*Prochilodus mariae*), "curbinatas" (*Plagioscion squamossissimus*), "laulaos" and "valentones" (*Brachyplatystoma* spp.), "morocotos" (*Piaractus brachypomus*), "Palometas" (*Mylossoma* spp.), "rayaos" (*Pseudoplatystoma* spp.), "sapoaras" (*Semaprochilodus laticeps*) and "sardinatas" (*Pellona castelneana*) (Novoa 1990; Machado-Allison et al. 1999). Numerous other species of fishes have potential value as ornamental species, an alternative that is not yet employed but that has the potential to develop into a sustainable industry for indigenous populations in the area. The following species are found in the Caura and are important and commonly known in the world aquarium trade: "tetras" (*Astyanax, Hemigrammus, Hyphessobrycon, Jupiaba, Moenkhausia*), "palometas" or "silver dollars" (*Metynnis, Myleus, Mylossoma*), "cichlids" (*Aequidens, Apistogramma, Bujurquina, Mesonauta*), "piranhas" (*Pygocentrus* and *Serrasalmus*) and "headstanders" (*Anostomus* and *Leporinus*).

Future plans for development of the basin include additional deforestation, alteration of the Caura River for development of hydroelectric power facilities, mining and tourism. Unfortunately, implementation of these proposed development activities will neither enhance regional economies nor increase the quality of life. Data from such projects implemented in other countries or in other parts of Venezuela have shown little or no success (Miranda et al. 1998; Machado-Allison 1999; Machado-Allison et al. 1999, 2000; Bevilacqua and Ochoa 2001).

THREATS

The Guayana Shield region of Venezuela has endured much environmental alteration from development projects, industry and unregulated fisheries and hunting. The exploitation of strategic minerals such as gold, diamonds and bauxite, coupled with development and construction of one of the largest hydroelectric dam complexes of the world has produced: 1) biodegradation and destruction of extensive green areas in the Caroní and Cuyuní River Basins; 2) mercury pollution of rivers, forests, wildlife and humans; 3) diminished water quality due to increased sedimentation; and 4) loss of large quantities of potable water for domestic uses (Machado-Allison 1994, 1999; Miranda et al. 1998).

The Caura River Basin is threatened by mining, illegal fishing practices and a proposed hydroelectric project. In the Caura River Basin, several hundred kilometers are now being heavily logged close to the Paragua River Basin. Some mining activities have also been introduced into headwater areas of the Caura River Basin. At the moment, mining activities are minor in comparison to neighboring basins such as the Caroní and Cuyuní (Machado-Allison et al. 2000: fig. 1). A dramatic decrease in fisheries resources in the Lower Caura River is thought to be due to illegal commercial fishing in indigenous territory. The most important major threat to the Caura River Basin is a plan to construct a new hydroelectric dam and divert as much as 75% of the water from the Caura River into the Paragua-Caroní River system. This will not only drastically deplete the amount of water in the Caura River, but also severely alter the hydrologic cycle. The health and maintenance of human communities and aquatic and riparian flora and fauna are completely dependent upon the natural hydrologic cycle.

OVERALL CONCLUSIONS

The Caura River Basin is a large pristine wilderness area representing an important opportunity to preserve a unique region. The community composition of animals and plants in the Caura River Basin occurs nowhere else in Venezuela or on the Guayana Shield. It is a high diversity area with many endemic forms and unique communities, such as aquatic plant communities (Podostemonaceae) in rapids and the mid-river rocky-island floras. Protection through education programs, community outreach and long-term monitoring are highly recommended. Monitoring and establishment of fisheries regulation in the basin is essential to achieve sustainability for its inhabitants.

LITERATURE CITED

Aguilera, M. and J. Silva. 1997. Especies y diversidad. Interciencia. 22(6): 289–298.

Aymard, G., S. Elcoro, E. Marín and A. Chaviel. 1997. Caracterización estructural y florística en bosques de tierra firme de un sector del bajo Río Caura, Estado Bolívar, Venezuela. *In:* Huber, O. and J. Rosales (eds.). Ecología de la Cuenca del Río Caura. Scientia Guaianae 7:143–169.

Balbas, L. and D. Taphorn. 1996. La Fauna: Peces. *In:* Rosales, J. and O. Huber (eds.). Ecología de la Cuenca del Río Caura. Scientia Guaianae 6: 76–79.

Berry, P.E., O. Huber and B.K. Holst. 1995. Floristic Analysis and Phytogeography. *In:* Berry P. E., B.K. Holst,

K. Yatskievych (eds.). Flora of the Venezuelan Guayana, Vol. 1. Introduction. Saint Louis, USA: Missouri Botanical Garden and Timber Press. Pp. 161–191.

Bevilacqua, M. and J. Ochoa G. 2001. Conservación de las últimas fronteras forestales de la Guayana Venezolana: propuesta dc lineamientos para la Cuenca del Río Caura. Interciencia 26(10): 491–497.

Bonilla-Rivero, A., A. Machado-Allison, B. Chernoff, C. Silvera and H. López-Rojas and C. Lasso. 1999. *Apareiodon orinocensis*, una nueva especie de pez de agua dulce (Pisces: Characiformes: Parodontidae) proveniente de los ríos Caura y Orinoco, Venezuela. Acta Biol. Venez. 19(1): 1–10.

Brewer Carías, C. (ed.). 1988. Cerro de la Neblina. Resultados de la Expedición 1983–1987. Caracas, Venezeula: Academia de Ciencias Físicas, Matemáticas y Naturales.

Bucher, E., A. Bonetto, T. Boyle, P. Canevari, G. Castro, P. Huszar and T. Stone. 1993. Hidrovia: un examen ambiental inicial de la via fluvial Paraguay-Paraná. Humedales para las Americas, publ. 10: 1–74.

Chaffanjon, J. 1889. L'Orinoque et le Caura. *In*: Castellana, M.A. 1986. Relation de voyages exécutés en 1886 et 1887. Hachette et Cie. Paris.: Fund. Cult. Orinoco. 311 pp.

Chernoff, B. and A. Machado-Allison. 1990. Characid fishes of the genus *Ceratobranchia*, with descriptions of new species from Venezuela and Peru. Proc. Acad. Nat. Sci. Philad. 142: 261–290.

Chernoff, B., A. Machado-Allison, and W. Saul. 1991. Morphology variation and biogeography of *Leporinus brunneus* (Pisces: Characiformes: Anostomidae). Ichth. Explor. Freshwaters 1: 295–306.

Chernoff, B., A. Machado-Allison, and N. Menezes. 1996. La conservación de los ambientes acuáticos: una necesidad impostergable. Acta Biol. Venez. 16(2): i–iii.

Chernoff, B. and P. Willink (eds.). 1999. A Biological Assesement of the Aquatic Ecosystems of the Upper Río Orthon Basin, Pando, Bolivia. Bull. Biol. Asses. 15. Washington DC: Conservation International.

Colveé, P., E. Szczerban and S. Talukdar. 1990. Estudios y Consideraciones Geológicas sobre la Cuenca del Río Caura. *In:* Weibezahn, F. H. Alvarez, and W. Lewis (eds). El Río Orinoco como Ecosistema. Edelca, Venezuela: Fondo Ed. Acta Científica Venezolana, CAVN, USB. Pp. 11–44.

CVG-TECMIN. 1994. Informes de avance del Proyecto Inventario de los Recursos Naturales de la Región Guayana. Hojas NB-20: 1, 5, 6, 9, 10, 13 and 14. Ciudad Bolívar, Venezuela: Gerencia de Proyectos Especiales.

Dezzeo, N. and E. Briceño. 1997. La vegetación en la cuenca del Río Chanaro: medio Río Caura. *In:* Huber, O. and J. Rosales (eds.). Ecología de la Cuenca del Río Caura. Scientia Guaianae 7: 365–385.

Garcia, S. 1996. Limnología. *In:* Rosales, J. and O. Huber (eds.). Ecología de la Cuenca del Río Caura. Scientia Guaianae 6: 54–59.

Goulding, M. 1980. The Fishes and the Forest: Explorations in Amazonian Natural History. Berkeley: Univ. Cal. Press. 280 pp.

Huber, O. 1996. Formaciones vegetales no boscosas. *In:* Rosales, J. and O. Huber (eds.). Ecología de la Cuenca del Río Caura. Scientia Guaianae 6: 70–75.

Huber, O. and J. Rosales (eds.). 1997. Ecología de la Cuenca del Río Caura, Venezuela. II Estudios Especiales. Scientia Guaianae 7.

Huber O., J. Rosales and P. Berry. 1997. Estudios botánicos en las montañas altas de la cuenca del Río Caura. *In:* Huber, O. and J. Rosales (eds.). Ecología de la Cuenca del Río Caura. Scientia Guaianae 7: 441–468.

IUCN. 1993. The Convention on Biological Diversity: An explanatory guide (Draft). Bonn: IUCN Environmental Law Centre. (mimeo).

Knap-Vispo, C., J. Rosales and G. Rodríguez. 1997. Observaciones sobre el uso de las plantas por los Ye'kwana en el Bajo Caura. *In:* Huber, O. and J. Rosales (eds.). Ecología de la Cuenca del Río Caura. Estudios Especiales. Scientia Guaianae 7: 215–258.

Lasso, C. and F. Provenzano. 1997. *Chaetostoma vazquezi*, una nueva especie de corroncho del Escudo de Guayana, Estado Bolívar, Venezuela (Siluroidei-Loricariidae) descripción y consideraciones biogeográficas. Mem. Soc. Cienc. Nat. La Salle 57(147): 53–65.

Machado-Allison, A.1993. Los Peces de los Llanos de Venezuela: un ensayo sobre su Historia Natural. (2nda. Edición). Caracas, Venezuela: Consejo de Desarrollo Científico y Humanístico (UCV), Imprenta Universitaria, 121 pp.

Machado-Allison, A. 1994. Factors affecting fish communities in the flooded plains of Venezuela. Acta Biol.Venez. 15(2):59–75.

Machado-Allison, A. 1999. Cursos de agua, fronteras y conservación. *In:* G. Genatios (ed.). Ciclo Fronteras: Desarrollo Sustentable y Fronteras. Caracas: Com. Estudios Interdisciplinarios, UCV. Pp. 61–84.

Machado-Allison, A., B. Chernoff, C. Silvera, A. Bonilla, H. López-Rojas, C. A. Lasso, F. Provenzano, C. Marcano and D. Machado-Aranda. 1999. Inventario de los peces de la cuenca del Río Caura, Estado Bolivar, Venezuela. Acta Biol. Venez. 19:61–72.

Machado-Allison, A., B. Chernoff, R. Royero-Leon, F. Mago-Leccia, J. Velazquez, C. Lasso, H. López-Rojas, A. Bonilla-Rivero, F. Provenzano and C. Silvera. 2000. Ictiofauna de la cuenca del Río Cuyuní en Venezuela. Interciencia 25: 13–21.

Mago-Leccia, F. 1970. Lista de los Peces de Venezuela: incluyendo un estudio preliminar sobre la ictiogeografía del país. Caracas: MAC-ONP. 283 pp.

Mago-Leccia, F. 1978. Los Peces de Agua Dulce del País. Caracas: Cuadernos Lagoven. 35 pp.

Marin, E. and A. Chaviel. 1996. Bosques de Tierra Firme. *In:* Rosales, J. and O. Huber (eds.). Ecología de la Cuenca del Río Caura. Scientia Guaianae 6: 60–65.

Miranda, M., A. Blanco-Uribe, L. Hernández, J. Ochoa and E. Yerena. 1998. No todo lo que brilla es oro: hacia un nuevo equilibrio entre conservación y desarrollo en las últimas fronteras forestales de Venezuela. Washington, DC: Inst. Rec. Mundiales (WRI).

Mittermeier, R.A., P. Robles and C. Goettsch. 1997. Megadiversidad. los paises biologicamente mas ricos del mundo. México: Cemex y Agrupación Sierra Madre, SC.

Mittermeier, R.A., N. Myers, P. Robles and C. Goettsch. 1998. Hotspots. México: Cemex y Agrupación Sierra Madre, SC.

Novoa, D. 1990. El río Orinoco y sus pesquerías; estado actual, perspectivas futuras y las investigaciones necesarias. *In:* Weibezahn, F. H. Alvarez and W. Lewis (eds). El Río Orinoco como Ecosistema. Caracas: Edelca, Fondo Ed. Acta Cienctífica Venezolana, CAVN, USB. Pp. 387–406.

Peña, O. and O. Huber. 1996. Características Geográficas Generales. *In:* Rosales, J. and O. Huber (eds.). Ecología de la Cuenca del Río Caura. Scientia Guaianae 6: 4–10.

Rincón, H. and Y. Estanga. 1996. Geología. *In:* Rosales, J. and O. Huber (eds.). Ecología de la Cuenca del Río Caura. Scientia Guaianae 6: 20–28.

Rosales, J. 1996. Vegetación: los bosques ribereños. *In:* Rosales J. and Huber O. (Eds.) Ecología de la Cuenca del Río Caura, Venezuela: I. Caracterización general. Scientia Guaianae 6: 66–69.

Rosales, J. and O. Huber (eds.). 1996. Ecología de la Cuenca del Río Caura, Venezuela. I. Caracterización General. Scientia Guaianae 6.

Royero, R., A. Machado-Allison, B. Chernoff, and D. Machado. 1992. Los peces del Río Atabapo. Acta Biol. Venez. 14(1): 41–56.

Silva-Monterrey, N. 1997. La Percepción Ye'kwana del Entorno Natural. *In:* O. Huber and J. Rosales (eds.). Ecología de la Cuenca del Río Caura. Scientia Guaianae 7: 65–84.

Sioli, H. 1965. A Limnología e a sua importancia en pesquisas da Amazonia. Amazoniana 1: 11–35.

SISGRIL. 1990. Simposio Internacional sobre los Grandes Ríos Latinoamericanos. Interciencia 15(6): 320–544.

Taphorn, D. R. Royero, A. Machado-Allison, and F. Mago-Leccia. 1997. Lista actualizada de los peces de Agua Dulce de Venezuela. *In:* La Marca, E. (ed.). Vertebrados actuales y fósiles de Venezuela. Mérida, Venezuela: Vol. 1: 55–100.

Vargas, H. and J. Rangel. 1996. Hidrología y Sedimentos. *In:* Rosales, J. and O. Huber (eds.). Ecología de la Cuenca del Río Caura. Scientia Guaianae 6: 48–54.

Vari, R. 1995. The Neotropical Fish Family Ctenoluciidae (Teleostei: Ostaripphysi: Characiformes): Supra and Infrafamilial Phylogenetic Relationships, with a Revisionary Study. Smith Contr. Zoology 654:1–95.

Chapter 2

Riparian Vegetation Communities of the Caura River Basin, Bolívar State, Venezuela

Judith Rosales, Mariapia Bevilacqua, Wilmer Díaz, Rogelio Pérez, Delfín Rivas and Simón Caura

ABSTRACT

A total of 443 samples were taken during the expedition from a variety of riparian and aquatic habitats in the Caura River Basin. The samples contained 399 (291 from the areas above Salto Pará and 185 from Lower Caura) species of plants, all of which are included in the list of 1,180 species known from the basin. The investigation revealed that there is a large diversity in the composition of floral communities, characteristic of humid climates with low nutrient soils. A gradient of riparian landscapes, structured by the intensity and duration of erosional processes, foster this variation in community structure. Both the diversity of the underlying geology and the climatic gradient associated with altitude (40–2,350 m) contribute to the exceptional species richness of the Caura River watershed.

The proportion of endemic species is low in the lowlands and in the flooded forest ecosystems of the Caura River Basin. This region is dominated by palms: *Euterpe precatoria, Attalea maripa, Socratea exhorriza, Genoma baculifera* and *Bactris brongniartii*. The species richness is relatively high in comparison to values for similar forest environments in Amazonia and Guyana. The diversity and species richness of the flooded ecosystems of the Erebato, dominated by *Oenocarpus*, and the Upper Caura, dominated by *Mauritiella*, are also similar to values of other riverine corridors on the Guayana Shield. A variable set of unique floral assemblages associated with riverine islands are found in the Erebato and middle Caura Rivers. These islands have both *terra firme* forests and often, dense aquatic forests of Podostemonaceae attached to boulders in rapids.

INTRODUCTION

The knowledge of florisitic diversity and plant community structure in the Caura River Basin, especially in the riparian forests, is relatively complete. Studies of the flora and plant communities in the basin have been carried out over two extended periods of geographical and scientific exploration (Huber 1996). The first extends from the sixteenth to the nineteenth century during which missionaries and naturalists published important observations characterizing forest ecosystems. The second period began in the twentieth century, during the 1930's, with geographical explorations to the Guayana Shield Region culminating in studies by Williams (1942) and Veillon (1948). More recently, studies within the Caura River Basin have focused upon regional development plans and the conservation of the biological diversity (Steyermark and Brewer-Carias 1976; Lal 1990; CVG-TECMIN 1994; Berry et al. 1995; Briceño 1995; Huber 1995, 1996; Bevilacqua and Ochoa 1996; Marín and Chaviel 1996; Rosales 1996, 2000; Aymard et al. 1997; Briceño et al. 1997; Dezzeo and Briceño 1997; Huber et al. 1997; Knab-Vispo et al. 1997; Rosales et al. 1997; Salas et al.1997; Knab-Vispo 1998; Vispo 2000; Bevilacqua and Ochoa 2001; Rosales et al. 2002a,b; Knab-Vispo et al. in press; Vispo et al. in press).

Approximately 90% of the Caura River Basin is covered by inland and mountain forests, while the remaining area contains riverine flooded forest and savannahs. The basin has high vegetation formation diversity, typical of communities that develop in humid climates, on low-nutrient soils and in a gradient of landscapes structured by continuous and intense erosional processes. The combination of humidity, soils and erosion along with the underlying geological diversity and an altitudinal (40–2,350 m) climatic gradient contribute to the exceptional floristic diversity in the basin.

A large scale perspective (1:2,000,000) permits us to characterize the inland forest (*terra firme*) in four categories: 1) semi-evergreen/deciduous (tropophilus macrothermic) forests, exhibiting marked seasonal changes and usually found in the lowlands in the northern sector of the basin; 2) evergreen humid (ombrophilus macrothermic) forests, distributed primarily on great expanses of lowlands in the middle to high elevations of the basin that have high temperatures and precipitation; 3) evergreen humid premontane (ombrophilus mesothermic) forests, associated with the Guayana Shield landscapes at premontane elevations; and 4) evergreen humid montane (ombrophilus submesothermic) forests, associated with the physiographic landscapes of the Guayana Shield highlands.

In contrast to the *terra firme* forests, the flooded forests cover a relatively small portion of the basin. The flooded forests are distributed in the flood plains and main channels of the Caura and Erebato Rivers. Despite their small extent, the flooded forests play a critical role in the transference of energy and nutrients to the adjacent aquatic and *terra firme* forest ecosystems. Moreover, the flooded forests are reservoirs of great biological diversity, have high landscape value and are resources for the local indigenous populations (Rosales 2000). The proportion of endemic elements is low in the lowlands and in the flooded forest ecosystems. However, the diversity is relatively high and comparable with values registered for similar forest environments in Amazonia and the Guyanas (Klinge et al. 1995; Knab-Vispo 1998; Rosales et al. 1999; Rosales 2000; ter Steege 2000).

The non-forest plant communities are dominated by shrubs and grasses, associated in aquatic communities with rivers and lagoons. The communities are found on alluvial and sandy soils in the savannahs and morichales, located in the lowlands near the Orinoco River. Shrub and grass communities are also found on the tepuys in the Caura River Basin. These are highly valuable floral communities, typical of the Guayana Shield (Berry et al. 1995; Huber 1995; Huber et al. 1997).

Our data show that the flooded ecosystems of the Erebato and Caura River corridors manifest a relatively high diversity of ecological plant assemblages that contribute to the uniqueness of the basin. The diversity in community structure and floristic richness is similar to that found in other riverine tropical corridors (Rosales et al. 2001). The biological and environmental attributes of the Caura River Basin, including the great expanses of primary forests, cultural diversity and potential for sustainable development, classify this bioregion as an important area in the western hemisphere (Bevilacqua and Ochoa 2001).

RESULTS

General description of subregions and field stations

I. Upper Caura River Subregion

This subregion is characterized by abundant rainfall throughout the year, with average precipitation varying from 2,600 mm near Raudal Cejiato to 3,758 mm near Entreríos. The lowest precipitation occurs from January to March, and this study (Nov–Dec) was carried out at the end of the rainy season. The bioclime is macrothermic with an average annual temperature of 25°C and an average altitude of 250 m a.s.l. The riverine landscapes are characteristic of rivers structurally controlled by hydraulic gradients (relative slopes), the sediment load and the age and evolutionary grade of the river in the different sectors. Figures 2.1, 2.2 and 2.3 show examples of typical riverine landscapes and their associated vegetations.

1. Kakada River (AC03, AC04). Tributary of the Erebato River, which drains a sandstone system from the Jaua-Sarisariñama (Roraima Formation) Range. Its waters are strongly acid, black and transparent. The predominant inland landscapes are the old floodplain on quaternary alluviums and the hilly formations of the Cuchivero Province. The riparian landscape, on the other hand, has a gentle slope and is dominated by a recent alluvial plain of depositional origin that extends to the confluence with the Caura. The sampling stations included the flooded plain and the damp forest habitats up to the approximate 50-year flood zone. The flooded margins have differential sandy, limic and clay depositions that modify the topographic gradient. The most elevated positions are dikes and banks with Franco-sandy textures and slopes of 3–4 m. These areas precede depressional, convex plains whose irregular microrelief is formed by a combination of dissectional (erosional cavities) and depositional (marginal buckets) processes. The topography of the riparian zone is formed by alluvial depositions. This zone takes the form of long, narrow, sinuous bars, oriented laterally to the River, with sandy substrates and abundant depositions of organic material coming from the root mat and leaves. On the other hand, at the confluence of Caño Suajiditu, the low area behind the bank is a lateral bar of clay. Just upstream from the confluence there is a series of bars formed by the meander of the river that alternates. The slope of the shore is about 2–3 m high. The bed of the Caño has layers of gravel and sand in addition to accumulations of organic material in different stages of decomposition.

2. Erebato River (AC05, AC06, AC07). Principal tributary of the Caura River that drains parts of the Roraima (Jaua-Sarisariñama Mountain Range), Pastora and Cuchivero (Maigualida and Uasadi Mountain Range) Provinces. The predominant *terra firme* landscapes along the Erebato are old plains and low, frequent hills. An aspect worth noting, the Erebato River is largely confined and structurally controlled by steep banks and rocks with high downstream gradients and many rapids. The River's narrow flood plain suggests that this is a young drainage system. Both riverbanks are steeply carved and there are more erosional habitats than depositional habitats. There are few marshy areas in the riparian zone, however, marshy areas are present in the confluence of caños with the Erebato River forming narrow, sinuous bands that are elevated 2–3 m above the shores. The bottom of the caños present layers of gravel and sand mixed with abundant organic material.

The Erebato River has many rapids and rocky islands. The islands are variable in size and are predominantly elongated downstream. The islands have rocky beaches in addition to lateral and frontal bars that form sandy beaches of rough texture. The hills towards the centers of the islands have flat or convex tops with irregular microrelief.

3. Caura River, Entreríos-Raudal Cejiato (AC01, AC02, AC08, AC09, AC10). Origin in sandstone and igneous systems of the Roraima (Ichún and Roraima Group Formation) Province. The waters are acid, similar to those in the Erebato River. However, the Caura River has higher amounts of suspended material and nutrients. The drainage basin of the Caura River has more tributaries than the Erebato River. The *terra firme* landscape comprises high hills on the western margin of the River and lower hills on the eastern margin. The Caura has a relatively high heterogeneity of habitats due to the development of the floodplain. This indicates that the River basin is relatively mature. There are erosional margins with active slopes. There are also depositional areas in the meanders with bars of sand, clay and organic material. The major forms in the floodplain include: i) levees and elevated banks formed during periods of high waters; ii) zones of swails and troughs that drain off surface waters; iii) depressions that form pools; and iv) lateral depositional bars in the riparian zone. In the caños and tributaries the substrates include exposed rocks and gravels as well as leaves and detritus. There are also island and rapids complexes similar to those described above.

4. Caura River, Entreríos-Salto Pará (AC11, AC12, AC13, AC14). The fundamental characteristics are determined by the confluence of the Caura and Erebato hydrologic systems. In general, waters are acidic, slightly brown and well oxygenated, resembling those in the Upper Caura. This whole area has many rapids and islands with exposed rocky patches and backwater areas. Salto Pará is a mixture of rapids and cataracts that drop 50 m in only 2 km. The floodplains and islands are similar to those described above.

II. Lower Caura Subregion (Figures 2.4 and 2.5)

The Lower Caura subregion manifests a bioclimatic gradient principally due to variation in precipitation which increases from north to south: 1,500–1,900 mm from the Mato River to the mouth of the Caura River; 2,500–3,000 mm in the Nichare River; and 2,970–3,400 mm near Salto Pará. Temperature maintains a macrothermic regime that is similar to that described for the Entreríos area, except that the temperature fluctuates 1–2 degrees higher. The riverine landscapes from Salto Pará to Raudal Cinco Mil are typical of structurally controlled rivers with variations related to altitudinal gradients, sediment load and age or state of evolution. In contrast, meandrous riverine landscapes are found in the Nichare and Icutú Rivers and below Raudal Cinco Mil to the mouth of the Mato River.

1. Caura River, El Playón–Raudal Cinco Mil (BC01, BC02, BC03, BC15, BC16, BC17). This area is characterized by the presence of rocky rapids, turbulent channels and backwaters. Geologically, this area is dominated by the Imataca Province formation with gneiss, anphybolites and ferruginous quarzites in the hills and the Cuchivero Province with igneous and granite rocks in the peneplain. Similarly to the Upper Caura, the dominant riverine landscape is the recent alluvial plain. As in the Upper Caura, the riparian zone also has lateral depositional bars.

2. Nichare, Tawadu and Icutú Rivers (BC04, BC05, BC06, BC07, BC08, BC09, BC10, BC11). In general the waters have lower pH, higher oxygen, less turbidity and conductivity and have lower temperatures than the rest of the Caura River. There are two types of riverine landscapes. The first is the meandering alluvial plain within a valley developed over quaternary and recent alluvia. These plains are located in the Nichare River from the Icutú River to the mouth of the Tawadu River. The variety of terrains includes lateral bars, meandric bars and shore complexes of levees, pools and oxbow lakes. The substrate generally includes clay-lime organic material and organic muds, in addition to aggregations of leaves. The second riverine landscape is found in the Tawadu River. This is a landscape of steep banks carved through hills with granites of the Cuchivero Province. In this area the substrate contains gravel, sand and rocks of different sizes.

3. Caura River from Raudal Cinco Mil to the mouth of the Mato River (BC12, BC13, BC14). The recent alluvial floodplain in this area is the most highly developed of any area that we studied. This area represents the southern limit of flooding in the Lower Caura River due to the damming effect of the Orinoco. The development of alluvial islands is common, as is the development of channel bars and meandric bars in the riparian zone. In the floodplain there are lagoons formed by abandoned river channels or dead arms, oxbow lakes (in the Mato), levees, banks and pools. The floodplain contains quaternary and recent alluvia.

VEGETATION DESCRIPTION

The flooded vegetation can be classified in four types: 1) the herbaceous-shrubby, early successional vegetation of sand bars and backwaters; 2) the flooded forests in margins of caños and rivers; 3) the underbrush-herbaceous vegetation in backwaters; and 4) the rheophylic aquatic vegetation. Each of the four vary in their physiognomy and structure depending upon environmental factors such as flooding periodicity, exposure of rocks and boulders, degree of organic material in alluvial soils, lithological background and topography. The following paragraphs describe each vegetational type:

The first type consists of communities closest to the rivers. These mainly include early successional plants that grow on recent alluvial soils of lateral depositional bars. This community comprises herbaceous and shrubby species, as well as prefruiting plants between 1 to 3 m in height such as: *Psidium* sp., *Mabea* sp., *Miconia* sp., *Vismia* sp., *Croton cuneatus*, *Calycolpus goetheanus, Myrcia splendens, Maytenus guyanensis* and a small, shrubby, white-flowered species of Rubiaceae. These belong to the families Cyperaceae, Graminaeae, Rubiaceae, Labiatae and Gesneriaceae.

On islands, the early successional vegetation borders a narrow, shrubby ecotone in transition to a poorly developed forest generally located in the center of the island (Figures 2.2 and 2.4). The woody elements in the shrubaceous ecotone are constantly being regenerated because of seasonal flooding. This ecotone is related floristically to the flooded forest.

The second type of vegetation is the riparian flooded forest (Figures 2.1 and 2.5). This community occurs on elevated areas, such as banks or levees, as well as in depressions of the floodplains. In general, these are evergreen forests of average height (between 18 and 25 m) and with two strata. The upper stratum has either a continuous or irregular canopy and contains species such as: *Pithecelobium cauliflorum, Macrolobium acaciiefolium, M. angustifolium, Eperua jehnmanii* spp. *sandwichii, Homalium guianense, Caraipa densifolia, Jacaranda copaia, Andira surinamensis, Eschweilera subglandulosa, Catostemma comune, Dialium guianense, Parkia pendula, Micranda minor, Virola surinamensis, Scheflera morototoni, Gustavia coriacea, G. poeppigiana, Tabebuia capitata, Pterocarpus* sp., *Chrysophyllum* sp., *Cupania* sp., *Sterculia* sp., *Lecythis* sp., *Abarema* sp., *Alexa confusa* and *Protium* spp. The lower of the two strata is commonly dense, occupying the zone between 10 and 18 m, with species such as: *Phenakokspermum guyanense, Rinorea flavescens, Swartzia schomburgkii, Cassipourea guianensis, Amphirrox latifolia, Psychotria* spp., *Erythroxylum* sp., *Diplasia* sp., *Alexa* sp., *Mabea* sp., *Chrysophyllum* sp., *Inga* sp., *Sterculia* sp., *Lecythis* sp., *Abarema* sp., *Alexa* sp., *Protium* sp., *Calycolpus* sp., *Myrcia* sp., *Ficus* sp. and *Zigia* sp. Occasionally, there is a third, lower stratum dominated by a few species with scattered individuals at heights of only 6 to 10 m. A number of species commonly grow above the canopy at heights from 26 to 30 m, including: *Parkia pendula, Micranda minor, Catostenma commune, Ceiba pentandra* and *Tabebuia* sp.

The forest floor is an active area of forest regeneration and while occasionally sparse, generally varies in density from medium to high. The forest floor contains many juvenile trees, grasses, shrubs and prefruiting species, such as: *Piper* sp., *Amphirrox* sp., *Costus* sp., *Heliconia* sp., *Renealmia* sp., *Tabernamontana* sp., *Eugenia* sp., *Miconia* sp., *Psychotria* sp., *Erythroxylum* sp. and *Diplasia* sp. Colonies of cyperaceans are localized in open areas in the forest, as are small patches of musaceans, principally *Phenakospermum guyanense*, reaching heights of 8 m.

Palm trees are frequent and characteristic of flooded forests, occurring in all strata. The most frequently encountered palms are *Euterpe precatoria, Attalea maripa, Astrocaryum gynacanthum, Socratea exorrhiza* and *Desmoncus* sp. Occasionally, we found colonies of *Geonoma deversa* and *Bactris brongniarti*. The species *Oenocarpus bacaba* and *O. bataua* are common in the flooded forests of the Erebato River.

In response to an increase of relative humidity, there can be abundant mosses, lichens, araceans and epiphytes such as bromeliads and orchids. Ferns also are common and can be either free-living or epiphytic. Lianas (vines) and reeds are found in areas with natural perturbations, such as in forest gaps resulting from fallen trees. In areas with poor drainage, the lianas and reeds often form dense interlaced communities.

Flooded forests are shorter and less floristically complex in poor drainage conditions, in long periods of flooding and in soils of depositional-residual origin (Figures 2.1 and 2.5). In the confluence of caños, or in depressed areas, there are patches of low forests with variable canopies reaching 8 to 15 m. On the other hand, the plant coverage of the forest floor is moderate to sparse. Leaves accumulating on the forest floor are an important source of nutrients for flooded forests.

The third type of flooded vegetation is underbrush, dominated ecologically by *Inga vera*, and found in depositional margins and beaches of backwaters and meandric curves. Below Raudal Cinco Mil the same habitats are occupied by other species of *Inga*, frequently along with *Alibertia latifolia* and *Coccoloba obtusifolia*.

This type of vegetation can grow to 2 m in height and can be extremely dense, forming an intricate network of branches and strong trunks. Some reeds, vines and lianas of the families Apocynaceae, Vitaceae, Leguminosae, Convolvulaceae and Malvaceae co-occur where the flooded forest makes contact with the underbrush, forming mosaic "reed forests."

In an area between the Nichare and Tawadu Rivers we made a transect of this habitat. The area was an exposed depositional bar forming a beach approximately 30 m wide. The substrate of the exposed beach was soft mud due to the high percentage of organic material and clay. The underbrush was totally exposed and contained *Solanum* sp. as the dominant species. This community was ecologically equivalent to the *Inga vera* underbrush. We noted an early-successional herbaceous community in front of the underbrush. This herbaceous community had species of the families Cyperaceae, Graminaea, Rubiaceae, Labiatae and Gesneriaceae, in addition to *Montrichardia arborescens*, the tallest element, which tolerates the cyclical flooding. The location of these communities preceding the underbrush depends upon depositional dynamics, beach exposure and extention of water interface. In transition toward the riparian forest there was an ecotone

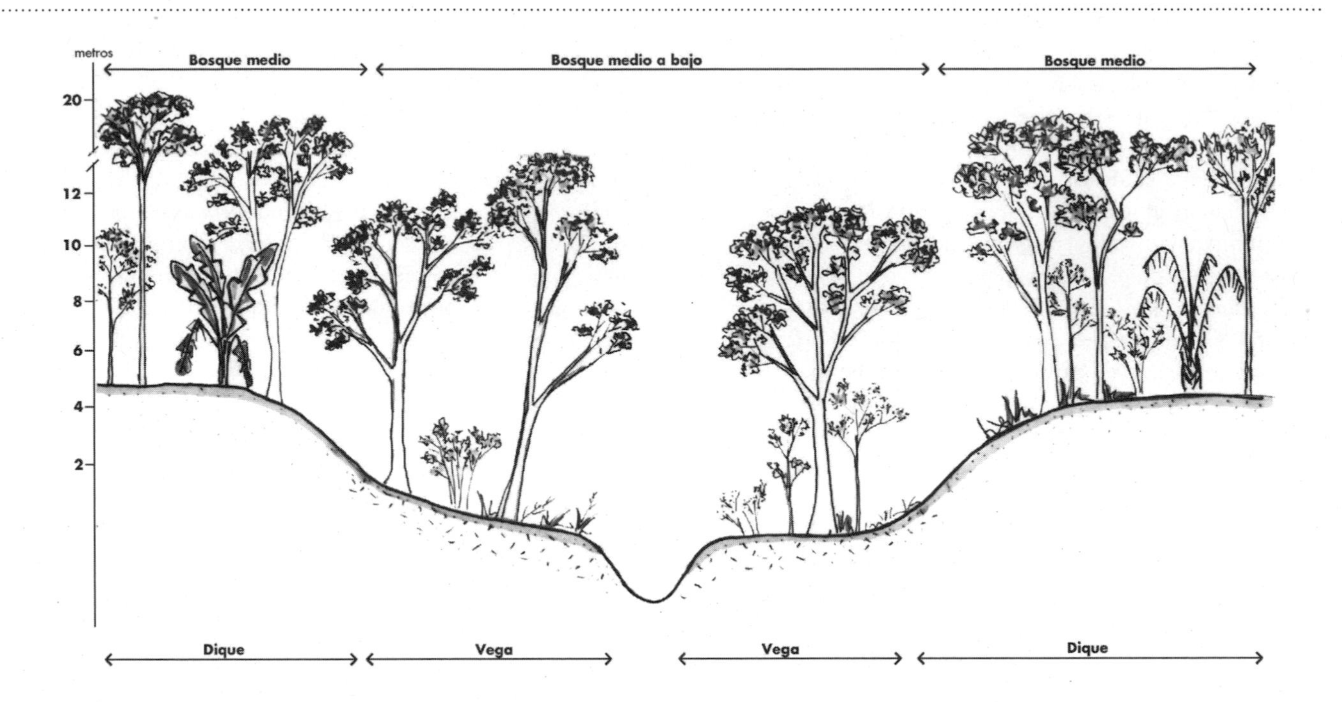

Figure 2.1. Diagram of the medium-height flooded forest at the junction of the Caño Wididi and the Erebato River (Georeference Point AC06). The soil in the riparian zone contains fine, organic material, while that in the levee is coarse. Bosque medio means medium-height forest; bosque medio o bajo means medium to low forest; dique means levee; and vega means riparian zone.

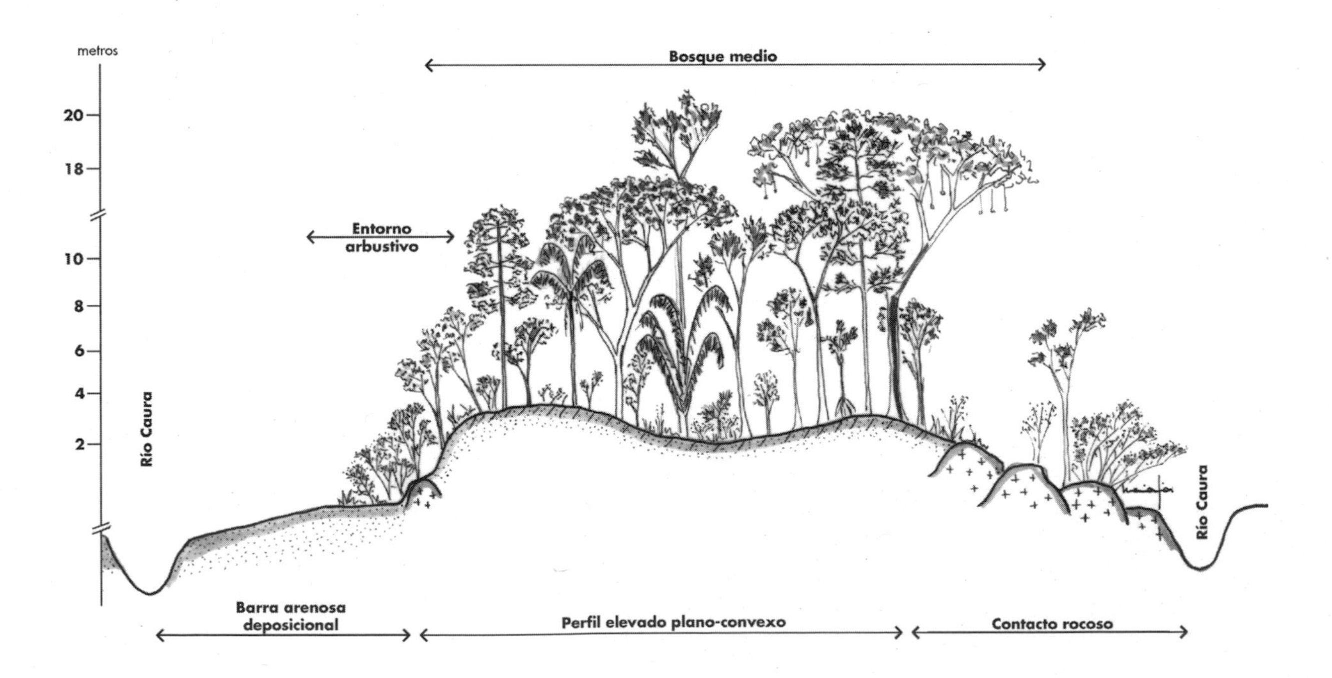

Figure 2.2. Topographic diagram of the transect across the medium-height forest of an island surrounded by rapids in the Caura River (Georeference Point AC09). Bosque medio means medium height forest; barra arenosa deposicional means depositional sandy bank; perfil elevado plano-convexo is the profile of the elevated convex plain at the top of the island; and contacto rocoso means rocky contact.

Symbols: crosses–exposed rock; stippled points–sand; diagonal lines–transported organic material.

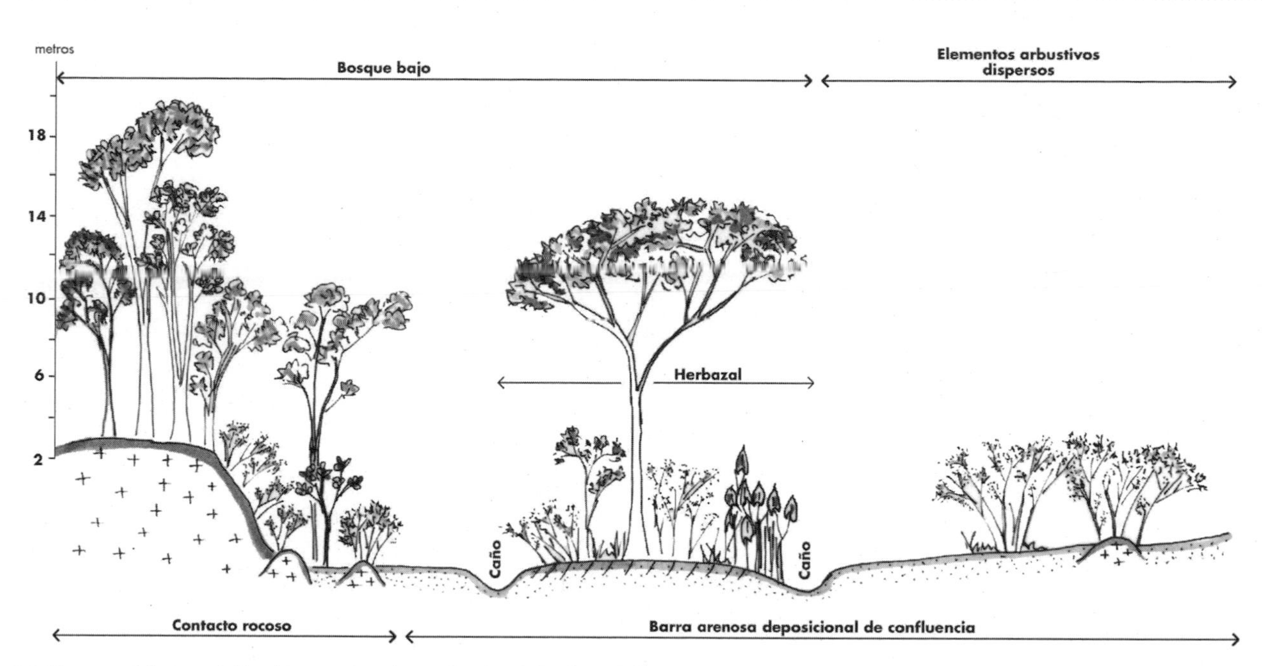

Figure 2.3. Diagram of the vegetational transect at the confluence of the Caño Cejiato with the Caura River. (Georeference Point AC08.) Bosque bajo means low forest; elementos arbustivos dispersos means dispersed shrubby elements; herbazal means an area containing herbs; contacto rocoso means rocky contact; and barra arenosa deposicional de confluencia means sandy depositional bar in the confluence.

Symbols: crosses–exposed rock; diagonal lines–transported organic material; and stipples–sand.

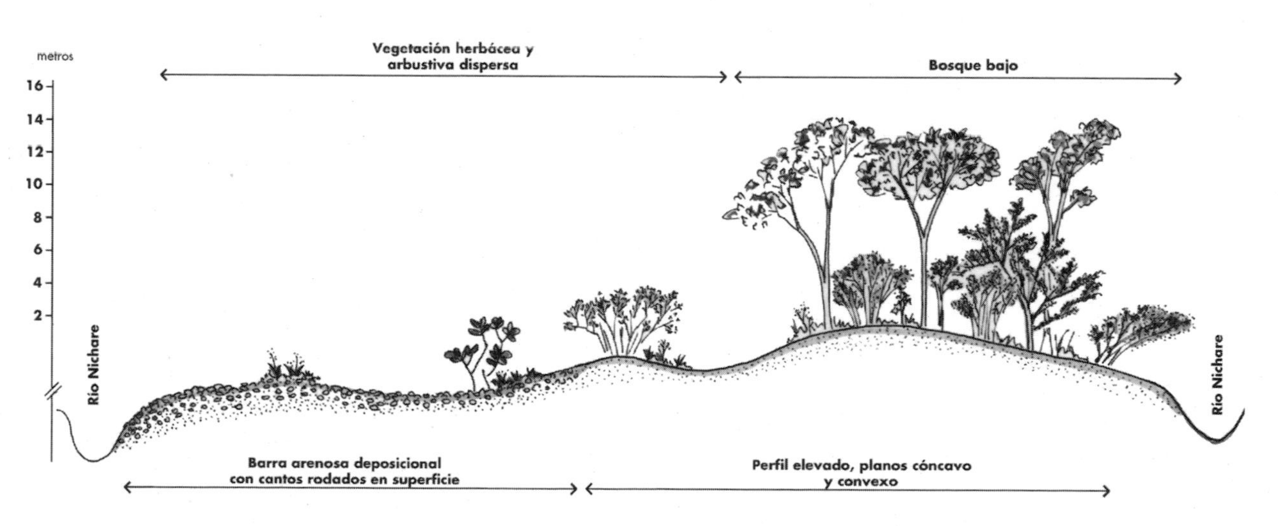

Figure 2.4. Topographic diagram of the vegetational transect of an island in the Nichare River near the mouth of the Tawadu River (Georeference point BC10). Vegetación herbácea y arbustiva dispersa means herbaceous vegetation and dispersed herbs; bosque bajo means low forest; barra arenosa depsicional con cantos rodados en superficie means sandy depositional bar with many round stones on the surface; perfil elevado, planos cóncavo y convexo means elevated profile of the concave and convex plains. The stipple points indicate sand.

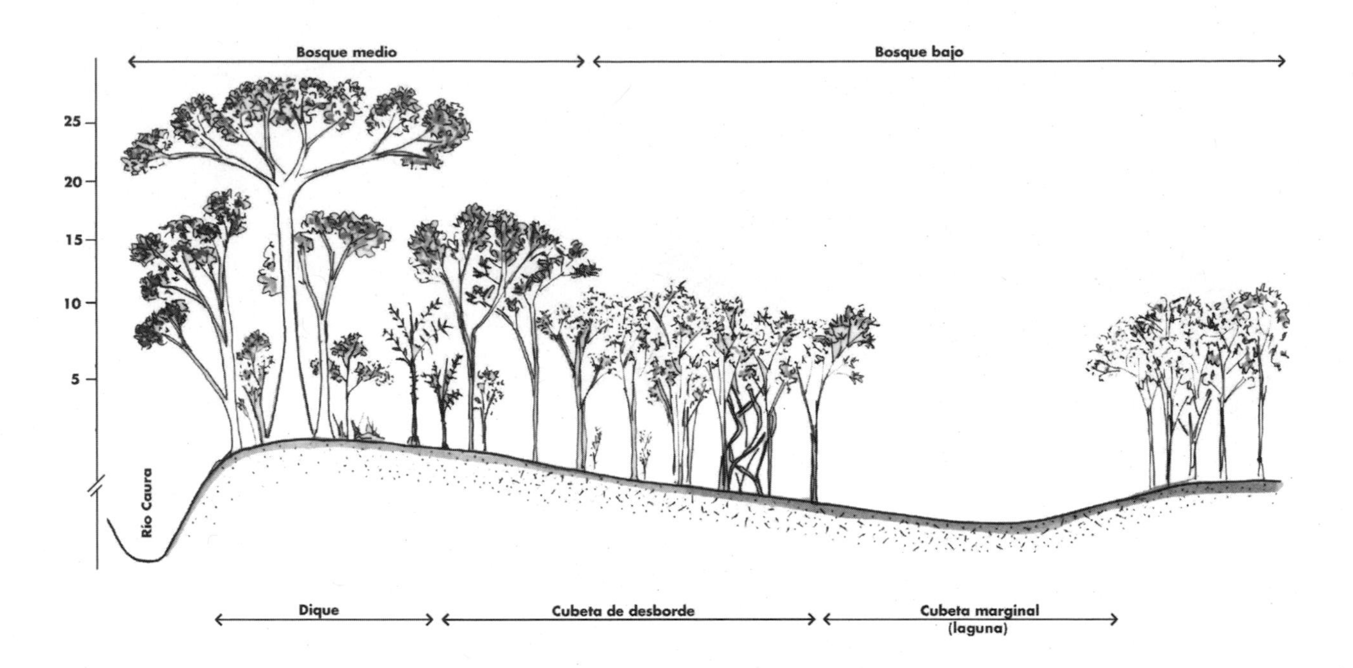

Figure 2.5. Topographic transect of medium to low riverine forests of the Río Guiare, tributatry of the Río Caura (Georeference point BC12). Bosque medio means medium height forest; bosque bajo is low forest; dique means river levee; cubeta de desborde is the backslope of the levee; cubeta marginal (laguna) indicates the lowest margin of the backslope in which water accumulates forming a lagoon. The soils of the levee proper are of clay, whereas the remaining soils contain clay and fine organic deposits.

of woody plants such as *Genipa* sp. *ruceana, Cecropia peltata, Myrcia* sp., *Corton* sp. and *Macrolobium acaciiefolium.*

The fourth type of vegetation is the strictly aquatic vegetation. Aquatic plants are relatively scarce in the main channel of the Caura and Erebato Rivers. Basically, the aquatic vegetation comprises reophilic macrophytes of the Family Podostemonaceae, most often found in rapids. We collected a total of five species; the highest diversity was found in the region from Entreríos to Salto Pará. While few species were present, these macrophytes are important as environmental indicators of water quality because of their fragility. The macrophytic communities also provide critical sustenance and refuge for aquatic organisms.

DISCUSSION

We collected a total of 443 samples from the riparian zone, flood plain, islands and rapids from which we identified 399 species (Appendix 2). We identified 291 species from areas above Salto Pará and 185 from the Lower Caura.
In general, the flora of the Caura compares well with the literature for Neotropical forests, principally containing the families: Leguminosae (*sensu lato*), Lauraceae, Anonaceae, Rubiaceae, Moraceae, Myristicaceae, Sapotaceae, Meliaceae, Euphorbiaceae, Chrysobalanaceae and Melastomataceae. Phytogeographically, the flora of the Upper Caura is principally related to the Guayanan and Amazonian floristic provinces. On the other hand, the Lower Caura contains elements from these provinces as well as from the Llanos Province.

The gamma or regional vegetation diversity of the Caura River Basin is moderately high. The flooded forests have elements from Amazonian igapos (*terra firme* streams with blackwater), as well as common species of the Amazonian varzea (flooded blackwater forests). These Amazonian elements become more frequent below Salto Pará to the confluence of the Caura and Orinoco Rivers.

In general, the palms *Euterpe precatoria, Attalea maripa, Socratea exhorriza, Geonoma baculifera* and *Bactris brongniartii* are characteristic floral elements of flooded forests of the Caura River. Whereas the genus *Oenocarpus* is common in the flooded forests of the Erebato River, the genus *Mauritiella* dominates flooded areas with sandy soils in the Upper Caura.

Eperua jehnmanii spp. *sandwichii* is arboreal and very common in the forests of the Upper Caura though not present in the Lower Caura. On the other hand, *Ocotea cymbarum* and *Campsiandra laurifolia* are only found in the Lower Caura.

In the Lower Caura below Raudal Cinco Mil we documented a new floristic communitiy that is distinct from those observed above the rapids. The new community contains the species *Campsiandra laurifolia, Piranhea trifoliata, Homalium racemosum, Mabea nítida, Gustavia augusta, Ruprechtia tenuiflora, Symmeria paniculata* and *Simira aristiguieta.* Whereas the analogous community above the rapids contains: *Simira aristiguietae, Homalium racemosum, Gustavia augusta, Mabea nitida, Luelea candida, Ruprechtia tenuiflora, Macrolobium angustifolium, M. acaciaefolium, Montrichardia arborescens, Podostemonaceae* sp. 1, *Inga vera, Homalium guianense, Caraipa densifoli* and *Croton cuneatus.*

The Podostemonaceae are common and diverse in rapids such as Culebra de Agua, Perro de Agua and Dimoshi rapids. We collected five species (*Apinagia ruppioides, A. Staheliana, Mourera fluviatilis, Rhyncochlays* sp. and *Rhyncholacis* sp.) and possibly a new genus. The macrophyte communities of the Upper Caura are more diverse than those of other rivers of the Venezuelan Guayana Shield.

Riverine Rocky Communities: Ecosystems of High Conservation Value

Data obtained during the AquaRAP expedition demonstrate that there is a relatively high frequency of rocky habitats with unique ecological assemblages in the Erebato River, and less so in the Caura River.

These Rivers have exceptional landscape value because of the floral communities associated with rocky substrates, particularly the rheophyllic communities of macrophytes (Podostemonaceae) and the woody vegetation of the islands.

The continual transformation of riverine plant communities is dependent upon the intensity of perturbations (inundation, sedimentation, erosion) that derive from the dynamic cycle of flooding. Moreover, the low-nutrient soils and the exposed, non-sedimentary rocks help determine the origin and ecological heterogeneity of the vegetational assemblages. The mosaic of vegetation resulting from the above process, includes: i) the presence of herbaceous and shrubby communities in different degrees of cover; and ii) flooded pioneer forests in different stages of successional sequence. These pioneer forests range from low underbrush to a medium-height, late-successional forest and also have abundant climbing and woody vine species. The structure and richness are, in some cases, comparable with those of riverine and *terra firme* forests.

The small islands, situated among the rapids, have patches of low, dispersed forest due to the exposed rocky surface, year-long high humidity and recurring impacts of the hydrologic cycle. In these forests there are important assemblages of epiphytes and mosses, and many woody plants have twisted trunks and small coarse leaves. The physiognomies of these forests are reminiscent of transitional cloud forests or of highland vegetation (>2000 m asl). All of these characteristics confer high biological and landscape values to these islands.

CONSERVATION OPPORTUNITIES

The riparian forests in the Caura River Basin, as illustrated in this chapter, provide the following environmental services: fixation and sequestration of carbon; refuge for biodiversity; conservation of soils; production of water; climatic regulation; and the preservation of food resources to sustain an expanding local human population as well as wildlife (Rosales and Huber 1996; Silva 1996, 1997; Rosales et al. 1997; Knab-Vispo 1998; Bevilacqua and Ochoa 2000; Centeno 2000; Vispo et al. in press). Recent conservation and development initiatives proposed for the region highlight the current and potential use of these services.

Conclusions and Recommendations

- The rocky islands in the Erebato River in the Upper Caura River Basin are ecologically special and require protection.

- Protection should be afforded to the rapids of the Upper Caura River Basin that are covered with the macrophytes of the family Podostemonaceae. These habitats provide refuge for many animal groups that form unique assemblages in the macrophytes.

- Maintenance of the natural hydrologic cycle that produces the intense dynamics within the riparian vegetational communities should be guaranteed. Construction of dams or water diversion projects will endanger riparian plant communities.

- Further agricultural expansion and deforestation within the Caura River Basin should not be permitted and current activities should be regulated. This is especially important in the Lower Caura.

LITERATURE CITED

Aymard, G., S. Elcoro, E. Marín and A. Chaviel. 1997. Caracterización estructural y florística en bosques de tierra firme de un sector del bajo Río Caura, Estado Bolívar, Venezuela. *In:* Huber, O. and J. Rosales (eds.). Ecología de la Cuenca del Río Caura. Scientia Guaianae 7:143–169.

Berry, P. E, O. Huber and B. K. Holst. 1995. Floristic Analysis and Phytogeography. *In:* Berry, P. E., B. K. Holst, K. Yatskievych (eds.). Flora of the Venezuelan Guayana, Vol. 1. Introduction. Saint Louis, USA: Missouri Botanical Garden and Timber Press. Pp. 161–191.

Bevilacqua, M., and J. Ochoa G. 1996. Áreas Bajo Régimen de Administración Especial (ABRAE). *In:* Rosales, J., and O. Huber (eds.). Ecología de la Cuenca del Río Caura. Scientia Guaianae 6: 106–112.

Bevilacqua, M., and J. Ochoa G. (eds.). 2000. Informe del componente Vegetación y Valor Biológico. Proyecto Conservación de Ecosistemas Boscosos en la Cuenca del Río Caura, Guayana Venezolana. Caracas, Venezuela: PDVSA-BITOR, PDVSA-PALMAVEN, ACOANA, AUDUBON de Venezuela and Conservation International.

Bevilacqua, M., and J. Ochoa G. 2001. Conservación de las últimas fronteras forestales de la Guayana Venezolana: propuesta de lineamientos para la Cuenca del Río Caura. Interciencia 26: 491–497.

Briceño, E., L. Valvas and J. A. Blanco. 1997. Bosques ribereños del Bajo Río Caura: vegetación, suelo y fauna. *In:* Huber, O., and J. Rosales (eds.). Ecología de la cuenca del Río Caura. Scientia Guaianae 7: 259–289.

Briceño, J. A. 1995. Análisis Fitosociológico de los bosques ribereños del río Caura en el Sector Cejiato–Entrerios, Distrito Aripao del Estado Bolívar. Informe de Pasantía. Universidad de los Andes, Facultad de Ciencias Forestales, Mérida. Mimeografiado.

Centeno, J. C. 2000. Compensación de las emisiones de carbono provenientes del consumo de orimulsión: Viabilidad económica y política. Informe del Proyecto Conservación de Ecosistemas Boscosos en la Cuenca del Río Caura, Guayana Venezolana. Caracas, Venezuela: PDVSA-BITOR, CI, ACOANA, AUDUBON de Venezuela and PDVSA-PALMAVEN.

CVG-TECMIN. 1994. Informes de avance del Proyecto Inventario de los Recursos Naturales de la Región Guayana. Hojas NB-20: 1, 5, 6, 9, 10, 13 and 14. Ciudad Bolívar, Venezuela: Gerencia de Proyectos Especiales.

Dezzeo, N., and E. Briceño. 1997. La vegetación en la cuenca del Río Chanaro: medio Río Caura. *In:* Huber, O., and J. Rosales (eds.). Ecología de la Cuenca del Río Caura. Scientia Guaianae 7: 365–385.

Huber, O. 1995. Vegetation. *In:* Berry P. E., B.K. Holst, K. Yatskievych (eds.). Flora of the Venezuelan Guayana, Vol. 1. Introduction. Saint Louis, USA: Missouri Botanical Garden and Timber Press. 97–160.

Huber, O. 1996. Formaciones vegetales no boscosas. *In:* Rosales, J., and O. Huber (eds.). Ecología de la Cuenca del Río Caura. Scientia Guaianae 6: 70–75.

Huber, O., J. Rosales and P. Berry. 1997. Estudios botánicos en las montañas altas de la cuenca del Río Caura. *In:* Huber, O., and J. Rosales (eds.). Ecología de la Cuenca del Río Caura. Scientia Guaianae 7: 441–468.

Klinge, H., J. Adis and M. Worbes. 1995. The vegetation of seasonal varzea forest in the lower Solimoes River, Brazilian Amazonia. Acta Amazonica 25:201–220.

Knab-Vispo, C. 1998. A rain forest in the Caura Reserve (Venezuela) and its use by the indigenous ye´kwana people. PhD Thesis. University of Wisconsin-Madison USA.

Knab-Vispo, C., J. Rosales and G. Rodríguez. 1997. Observaciones sobre el uso de plantas por los ye´kwana en el bajo río Caura. *In:* Huber, O., and J. Rosales (eds.). Ecología de la Cuenca del Río Caura. Scientia Guaianae 7: 211–257.

Knab-Vispo, C., J. Rosales, P. Berry, G. Rodríguez, Salas I. Goldstein, W. Díaz and G. Aymard. In press. Annotated floristic checklist of the riparian corridor of the lower and middle Río Caura with comments on animal use. Scientia Guaianae 13.

Lal, J. R. 1990. Estudios Fitosociológicos de varios tipos de bosque en la Reserva Forestal El Caura. Estado Bolívar. Informe de Pasantía, Facultad de Ciencias Forestales. Universidad de los Andes. Mérida. Mimeografiado.

Marín, E., and A. Chaviel. 1996. La vegetación: bosques de tierra firme. *In:* Rosales, J., and O. Huber (eds.). Ecología de la Cuenca del Río Caura. Scientia Guaianae 6: 60–65.

Rosales, J. 1996. Vegetación: los bosques ribereños. *In:* Rosales, J., and O. Huber (eds.). Ecología de la Cuenca del Río Caura. Scientia Guaianae 6: 66–69.

Rosales, J. 2000. An ecohydrological approach for riparian forest biodiversity conservation in large tropical river. PhD Thesis. School of Geography and Environmental Sciences, The University of Birmingham, Inglaterra.

Rosales, J., and O. Huber (eds.).1996. Ecología de la Cuenca del Río Caura, Venezuela. I. Caracterización General. Scientia Guaianae 6.

Rosales J., C. Knab-Vispo and G. Rodríguez. 1997. Bosques ribereños del bajo Río Caura entre el Salto Pará y los Raudales de Cinco Mil: su clasificación e importancia en la cultura ye´kwana. *In:* Huber, O., and J. Rosales (eds.). Ecología de la Cuenca del Río Caura. Scientia Guaianae 7:171–214.

Rosales, J., G. Petts and J. Salo. 1999. Riparian flooded forests of the Orinoco and Amazon River Basins: a comparative review. Biodiversity and Conservation 8: 551–586.

Rosales, J., G. Petts and C. Knab-Vispo. 2001. Ecological gradients in riparian forests of the lower Caura River, Venezuela. Plant Ecology 152: 101–118.

Rosales, J., G. Petts, C. Knab-Vispo, J. A. Blanco, A. Briceño, E. Briceño, R. Chacón, B. Duarte, U. Idrogo, L. Rada, B. Ramos, H. Rangel and H. Vargas. 2002a. Ecohydrological assessment of the riparian corridor of the Caura River in the Venezuelan Guiana Shield. Scientia Guaianae 13. (in press).

Rosales, J., C. Vispo, N. Dezzeo, L. Blanco, C. Knab-Vispo, N. Gonzalez, C. Bradley, D. Gilvear, G. Escalante, N. Chacon and G. Petts. 2002b. Riparian forests ecohydrology in the Orinoco River Basin. *In:* McClain, M. (ed.). The Ecohydrology of South American Rivers and Wetlands. UNESCO IHP Ecohydrology. Ecohydrology Programme. (in press).

Salas, L., P. E. Berry and I. Goldstein. 1997. Composición y estructura de una comunidad de árboles grandes en el valle del Río Tabaro, Venezuela: una muestra de 18.75 ha. *In:* Rosales, J., and O. Huber (eds.). Ecología de la Cuenca del Río Caura. Scientia Guaianae 7: 291–308.

Silva, M. N. 1996. Etnografía de la Cuenca del Caura. *In:* Rosales, J., and O. Huber (eds.). Ecología de la Cuenca del Río Caura. Scientia Guaianae 6: 98–105.

Silva, M. N. 1997. La percepción Ye'kwana del entorno natural. *In:* Rosales, J., and O. Huber (eds.). Ecología de la Cuenca del Río Caura. Scientia Guaianae 7: 65–84.

Steyermark, J., and C. Brewer-Carías.1976. La vegetación de la cima del macizo de Jaua. Boletín Sociedad Venezolana de Ciencias Naturales 22: 179–405.

ter Steege, H. 2000. Plant diversity in Guyana, with recommendations for a National Protected Area Strategy. Tropenbos Series 18:1–220.

Veillon, J. P. 1948. Cuenca del bajo y medio Caura. Estado Bolívar. Mapa Forestal. Caracas, Venezuela: Departamento de Divulgación Agropecuaria, Ministerio de Agricultura y Cría.

Vispo, C. 2000. Uso criollo actual de la fauna y su contexto histórico en el bajo Caura. Memorias Sociedad de Ciencias Naturales La Salle 149:115–144.

Vispo, C., J. Rosales and C. Knab-Vispo. In press. Ideas on a conservation strategy for the Caura's riparian ecosystem. Scientia Guaianae 12.

Williams, L. 1942. Exploraciones Botánicas en la Guayana Venezolana. I. El medio y bajo Caura. Caracas Venezuela: Servicio Botánico, Ministerio de Agricultura y Cría.

Chapter 3

A Limnological Analysis of the Caura River Basin, Bolívar State, Venezuela

Karen J. Riseng and John S. Sparks

ABSTRACT

Forty-three sites within the Upper and Lower regions of the Caura River Basin, state of Bolívar, Venezuela, were sampled for various limnological parameters. In this generally low-nutrient, tropical river, focus was placed on the analysis of the following parameters: temperature, conductivity, pH, secchi depth, dissolved organic carbon (DOC), alkalinity and sediment. The Upper Caura is pristine, comprised of more channelized basins and subject to greater seasonal hydrologic variation than the Lower Caura and, as a result, far less forest inundation. Herein we report on the chemical aspects of the waters sampled, whereas the biotic components of the system are covered by other groups in this volume.

In general, the waters in this region were found to be slightly acidic and dilute, with very low conductivity and alkalinity, typical of a rain-dominated system. These results are congruent with those presented in prior studies and can be attributed to the very ancient geology of the region. The waters were found to be similar for these parameters throughout the Upper and Lower segments of the watershed; however, occasional tributaries varied somewhat from this general theme. DOC (terrestrial input) varied throughout the watershed, in clearwater and blackwater rivers.

Overall, water quality was determined to be 'good' for all sites sampled at the time of this study. The Lower section of the Caura basin was more disturbed than the Upper region of the watershed. Higher disturbance at the water margins can lead to increased sediment input in the rainy season and decreased terrestrial input of biologically important carbon forms (DOC), thereby causing a change of habitat for aquatic organisms. These habitat changes may include changes in water clarity, substrate, river channelization and the number and diversity of habitats that sustain the current community of aquatic organisms.

Our conservation recommendation is stasis. Dams or water diversion would upset the hydrologic cycle that is important in tropical rain-dominated systems. We also recommend the creation of a biological monitoring station along the main channel of the Caura, possibly at Salto Pára. This station would provide a valuable platform for scientists to access and study this region in greater detail and build a long-term database for this tropical rainforest environment. Given that pristine tropical freshwater habitats continue to be degraded at an alarming rate worldwide, preserving the few largely undisturbed regions remaining, including the Caura River watershed, must be a priority for conservation management protocols.

INTRODUCTION

The AquaRAP team was dispatched to the Caura River to assess the value of this area for future conservation effort. The Caura River Basin is a relatively pristine region in Venezuela, and there is great potential for future exploitation (commercial fishing, mining and hydro-electric power) of this important natural resource. The goals of this limnological survey were: 1) to assess general water conditions and quality throughout the watershed, 2) to examine limnological trends along the river and its tributaries and 3) to provide conservation recommendations for the watershed.

The Caura River watershed is 45.3 km^2 and is part of the larger Orinoco River drainage basin (1,000,000 km^2), South American's third largest basin (AquaRAP Map). The Orinoco possesses the third largest discharge in the world at 1100 km^3/year (Paolini et al. 1987), with maximums in mid-August and early September (Depetris and Paolini 1991). Most of the tributaries of the Caura River watershed run over the Guayana Shield. Rivers draining the Guayana Shield may be characterized as more acidic, lower in conductivity and lower in total suspended solids than rivers draining the Andes or Plains (Depetris and Paolini 1991). The headwaters of the Caura River originate in the southern portion of the state of Bolívar, near Venezuela's border with Brazil. The watershed is bordered by the Pakaraima Mountains to the south and the Sierra Maigualida to the west. Although there are tributaries originating at greater than 2,000 m, the vast majority of the watershed lies at less than 500 m. Many sections of rapids are present throughout the Caura Basin and its tributaries, and major falls, Salto Pará, separate the Upper and Lower regions of the basin. Above the falls at Entreríos, the major confluence with the Caura is the Erebato River (~10.5 km^2 watershed). The Yuruani River (2.8 km^2) also joins the Caura above the falls. Below Salto Pará, the Nichare (4.0 km^2) and the Mato (2.6 km^2) Rivers comprise the major tributaries of the Caura.

Seasonality

The Caura is a tropical, rain-dominated system with strong seasonal cycles. Rain-dominated systems are those in which rain is the main source of water input into the system. Weather in the region is typical of a moist tropical system, warm (mean annual air temperature range 25.7–27.7°C) with average annual humidity of 75% at Maripa (Lower Caura) and 83–85% at Entreríos and Salto Pará (Upper Caura) (Vargas and Rangel 1997a). The rainy season extends from approximately May to October with total annual precipitation of 1980 mm (Maripa), 2970 mm (Salto Pará) and 3260 mm (Entreríos) (Vargas and Rangel 1996a). Maximum discharge is in July and August. The Caura may experience a 10-fold variation in annual discharge, ca. 7.5 m^3/s maximum at the mouth (Lewis et al. 1987). These large fluctuations in discharge translate into dramatic annual changes in water level, for example 7 m in 20-year mean annual water level fluctuations at Maripa (Vargas and Rangel 1996b), and large forest inundation in areas where the landscape is favorably inclined. The Lower Caura, especially below Raudal Cinco Mil, has greater forest inundation (Rosales et al. 2001) than upstream sites and thus will have a strong interaction with the terrestrial system.

Link with terrestrial system

The Caura system contains both 'clearwater' and 'blackwater' tributaries and flows over an extremely old geologic formation (Guayana Shield). The color in blackwater rivers is associated with humic and fulvic acids that accumulate from the decomposition of organic material. These compounds can leach out of terrestrial material that accumulates in the river and into the aquatic environment, and are evident by higher levels of dissolved organic carbon (DOC). Once in the system, the DOC is an important source of carbon for microbial production. DOC can be produced by the breakdown of material within the stream, such as macrophytes, or from material originating outside the stream, such as leaf litter. For a review of the transformation of DOC in streams see Allan (1995). Dissolved organic carbon (DOC) is the more dominant form in rivers draining the Guayana Shield, whereas dissolved inorganic carbon is the dominant form in the Andes and Plains draining rivers of the Orinoco watershed (Depetris and Paolini 1991). Floodplain lakes contribute DOC to the river system (e.g. middle Orinoco (Castillo 2000)) when they overflow into the river basin during the rainy season. Amount of forest inundation, river morphology and riparian vegetation all contribute significantly in the interaction between terrestrial and aquatic systems.

An overview of the Caura River Basin ecology, including limnology, can be found in *Scientia Guaianae* no. 6 and 7 (García 1996; Ibañez and Lara 1997) and results of a two-year water chemistry study in the Lower Caura have been published by Lewis (1986), Lewis et al. (1986) and Lewis et al. (1987). Herein we report on the chemical aspects of the waters sampled, while the biotic components of the system are covered elsewhere in this volume. In this generally low-nutrient, tropical river (Lewis 1986; García 1996), focus was placed on analyzing the following parameters: temperature, conductivity, pH, secchi depth, dissolved organic carbon (DOC), alkalinity and sediment.

METHODS

A Horiba water quality meter was used to measure conductivity (µS/cm), dissolved oxygen (mg/l) and water temperature (deg C). The pH was measured using an Orion 250 portable meter. Transparency was measured using an 8" secchi disk. Water turbidity was recorded as low, medium or high. Additionally, the percentage of shade on the water was recorded when sun was directly overhead.

All water samples were collected from approximately 10 cm below the surface. After collection, the raw water was filtered through a Whatman GF/F filter with nominal pore size of 0.7 µm. The color of the filtered water was recorded as either clear, very slightly brown (vsb), slightly brown (sb), light brown (lb) or brown (b). Filters were analyzed for particulates and the filtrate was analyzed for alkalinity and DOC. Sediment matter was measured using preweighed GF/F 47 mm filters that were dried at 40°C for 48 hours after use. The dried filters were reweighed on a Metler Balance. The weight difference was divided by the volume filtered for the particulate matter greater than the nominal pore size of the filter (0.7µM) per liter of sample.

Filtrate for the DOC analysis was preserved with 1ul HCl per ml sample and stored in a borosilicate glass scintillation vial. Filtrate analyzed for alkalinity was stored in a 60 ml HDPE bottle. The filtered water was analyzed for DOC and alkalinity in the laboratory of George Kling at the University of Michigan. DOC was measured on a Shimadzu 5000 TOC analyzer. Alkalinity was measured by direct titration to the inflection point using a Radiometer autotitrating system with 0.25N or 0.1N HCl as titrant.

RESULTS AND DISCUSSION

The main channel of the Caura River progressively increases in width moving downstream. There are several notable and unique habitats within the basin, in addition to the main river channel, including lagoons, varsea lakes and numerous smaller tributaries comprising both clear and blackwater rivers. Increased external perturbation is evident downstream, especially near the confluence with the Orinoco River.

Table 3.1. Average of temperature, pH, conductivity (Cond), dissolved oxygen (D.O.), alkalinity (Alk), dissolved organic carbon (DOC) and sediment by category. Note that in some cases data are missing for a particular site, for a complete list see Appendices 3–5. The lake in the Lower Caura is completely inundated at high water (determined by botany group).

	Number of Sites Averaged	Temp (°C)	pH	Cond (µS/cm)	D.O. (mg/l)	Alk (µeq/L)	DOC (µM)	Sediment (mg/L)
Average all sites	43	24.9	5.60	13.0	6.7	105	266	11.0
Range		22.7–26.8	4.33–5.98	6–41	0.9–9.1	20–375	98–915	1.3–21.9
Standard Deviation		1.1	0.34	7.4	1.4	74	149	6.4
Upper Caura Watershed (above Salto Pará)								
All sites	21	24.9	5.60	9.9	6.6	71	223	11.1
Rivers								
Upper Caura	8	26.0	5.78	10.7	7.0	78	275	12.5
Kakada	1	23.9	5.04	6.0	6.4	20	233	4.2
Erebato	3	25.0	5.78	9.3	6.7	68	191	7.4
Yuruani	1	25.4	5.68	12.0	6.4	94	278	21.8
Small streams	5	24.0	5.39	9.0	6.5	68	193	13.7
Caños	3	23.9	5.46	10.3	4.9	79	130	6.8
Lower Caura Watershed (below Salto Pará)								
All sites	22	26.2	5.69	10.3	8.0	73	272	11.4
Rivers								
Lower Caura	4	24.9	5.59	16.0	6.9	129	294	10.9
Nichare	2	24.6	5.61	12.0	6.9	98	187	17.3
Tawadu	5	23.2	5.76	9.6	8.1	61	352	1.3
Takoto	1	25.4	5.99	41.0	6.8	375	183	16.2
Icutú	1	24.8	5.87	14.0	6.8	126	184	13.8
Mato	1	25.9	5.82	20.0	6.2	165	528	21.9
Caños	4	24.8	5.80	21.5	7.3	174	199	9.5
Lagoons	2	26.0	4.90	12.0	4.1	87	272	5.1
Varzea Lake	1	26.6	4.34	19.0	0.9	86	915	10.6
Spring	1			35.0		280	200	

Throughout the watershed, a total of 43 sites were sampled corresponding to 31 georeference points/localities. Elevation in the region surveyed ranged from 500 m in the Upper Caura to less than 100 m in the Lower Caura. A complete list of results for water chemistry analyses is presented in Appendices 3, 4 and 5 and is summarized below (Table 3.1).

For all sites sampled, the water was dilute, a result most likely attributable to regional geology (ancient geologic formation, i.e., Guayana Shield) and the time of the year the samples were taken. In this survey all samples were collected in November-December, immediately following the end of the wet season. On average, waters sampled in the Upper Caura Basin were approximately 2 degrees centigrade cooler than those sampled throughout the Lower Caura.. Conductivity was generally very low, with few exceptions. The water was slightly acidic, more so in lagoons, and generally well oxygenated, approaching saturation below large rapids and falls. Lagoonal areas were much less well oxygenated, and completely isolated lagoons (i.e., no inflow or outflow at the time of collection) were extremely poorly oxygenated at this time of the year. The smaller tributaries and caños generally had lower turbidity and higher transparency.

In the Upper Caura Basin (above Salto Pará), 21 limnology sites were sampled within 14 georeference localities. Overall, the water was moderately acidic, with a pH ranging from 5.0 to 5.9. Conductivity was exceedingly low, ranging from 6 to 12 µS. Water in the Upper Caura was generally well oxygenated, however saturation was only found to occur immediately downstream from rapids or waterfalls. Noteworthy sites in the Upper Caura Basin, in terms of habitat uniqueness, include both the Kakada River (blackwater) and the Yuruani River (green water, very productive algal community).

In the region of the Lower Caura (Salto Pará), 22 limnology sites were sampled within 17 georeference localities. Based on the parameters examined, the water in this region is similar to that of the Upper Caura, except that on average water temperatures are approximately 2°C for the range of habitats sampled in the Lower Caura. Sites sampled included several along the main channel of the Caura River, completely isolated and semi-isolated lagoons and backwaters, and a number of both large and small tributaries. Both right and left bank tributaries were sampled, as well as both clear and blackwater systems. The water in the region was moderately acidic to acidic, with pH ranging from 4.4–6.0. Conductivity was generally low, although in some tributaries moderate readings were obtained (e.g., Takoto River), and ranged from 9–41 µS/cm. Temperature of waters in this region ranged from 23–26°C, with the blackwater Tawadu River standing out as the coldest in the region (<23°C). The Lower Caura Basin encompasses more varied aquatic systems than the Upper Caura Basin, including an increasing number of lagoonal and floodplain regions.

Similar values to those documented in previous studies of the Caura River Basin were recorded in this survey (Lewis 1986; Lewis et al. 1987; García 1996; Ibañez and Lara 1997). Several interesting limnological trends and observations in DOC, pH, conductivity, temperature and alkalinity were noted. Lewis (1986) reported DOCs of 417–1417 µM at the mouth of the Caura. The maximum DOC (915 µM) reading in this study was recorded from a seasonal lake within an island. The botany group determined that this lake is completely connected to the main river in the rainy season. As in Orinoco floodplain lakes (Castillo 2000), the DOC will likely become more concentrated prior to the lake being flushed into the Caura when water levels rise. This demonstrates an important source of organic carbon for the aquatic system. Readings for conductivity, pH and alkalinity were similar to values published in prior investigations (García 1996; Ibañez and Lara 1997). In this study, pH was found to increase continuously as one moves downstream, starting at 5.5 and ending at 7.2. Conductivity overall was very low (10–15µS/cm), but higher values were found in a small stream near Salto Pará (35µS/cm) and near the Takoto River (41µS/cm). The Takoto River flows over a different geological formation, the Imataca Province, than do rivers in other parts of the basin. The Imataca Province is characterized by the abundance of gneiss, amphibolites, granite intrusions and ferruginous quarzites. Water temperature in the Caura River was generally found to increase as one moves downstream. Temperature was also lower in rapids and in small tributaries with sufficient volume to have an effect on the temperature in the main channel of the Caura River (such as the Nichare and Erebato Rivers). While overall alkalinity ranged from 20–375µeq/L in the watershed, in general it ranged from 60–90µeq/L, which is considered very low. The higher alkalinity values were recorded from caños and small tributaries.

CONCLUSION AND CONSERVATION RECOMMENDATIONS

Based on the limnological parameters examined, we find the entire area sampled in this AquaRAP to be relatively intact. The Upper Caura is a pristine region, whereas the Lower Caura is generally pristine, although several areas were somewhat disturbed and habitat degradations were noted. Deforestation in lowland Amazonian streams has been shown to alter soil and water chemistry, including dissolved and particulate nutrients (Neill et al. 2001). The regions sampled in this survey currently experience little impact from deforestation, erosion and other outside influences that could negatively affect water quality in the Basin. Conservation efforts directed at maintaining the future stability of this river system should include:

- Maintenance of the riparian zone throughout the watershed, a natural buffer for the aquatic system.

- Maintenance of the natural hydrologic cycle, given that seasonal forest inundation is clearly an important source of carbon, and likely other nutrients fundamental to the aquatic organisms inhabiting these regions. Dams and water diversion schemes could seriously alter this annual cycle.

- Preservation of the Upper Basin.Preventing environmental degradation throughout the Upper Caura Basin is paramount to maintaining the quality of aquatic habitats throughout the entire watershed.

- Establishing a scientific monitoring station along the Caura, possibly near Salto Pára, for access to the Upper Caura. Disturbances to the hydrologic cycle, including any dams or water diversion projects along this river, pose a great threat to the ecosystem.

LITERATURE CITED

Allan, J. D. 1995. Stream Ecology: Structure and function of running waters. London England: Chapman and Hall.

Castillo, M. M. 2000. Influence of hydrological seasonality on bacterioplankton in two neotropical floodplain lakes. Hydrobiologia 437: 57–69.

Depetris, P. J., and J. E. Paolini. 1991. Biogeochemical Aspects of South American Rivers: The Paraná and the Orinoco. *In:* Degens, E. T., S. Kempe, J.E. Richey (eds.). Biogeochemistry of Major World Rivers. Scope 42: 105–122.

García, S. 1996. Limnología. *In:* Rosales, J. and O. Huber (eds.). Ecología de la Cuenca del Río Caura. Scientia Guaianae 6: 54–59.

Ibañez, A. M., and J. I. Lara. 1997. Algunos aspectos fisicoquimicos y biologicos de las aguas del Río Caura (Venezuela), en su parte media. *In:* Huber, O. and J. Rosales (eds.). Ecología de la Cuenca del Río Caura. Scientia Guaianae 6: 34–39.

Lewis, W. J., S. K. Hamilton, S. L. Jones and D.D. Runnels. 1987. Major element chemistry, weathering, and element yields for the Caura River drainage, Venezuela. Biogeochemistry 4: 159–181.

Lewis, W. M. 1986. Nitrogen and phosphorus runoff losses from a nutrient-poor tropical moist forest. Ecology 67(5): 1275–1282.

Lewis, W. M., J. F. Saunders, S. N. Levine and F. H. Weibezahn. 1986. Organic carbon in the Caura River, Venezuela. Limnol. Oceanogr. 31(3): 653–656.

Neill, C., L. A. Deegan, S. M. Thomas and C. C. Cerri. 2001. Deforestation for pasture alters nitrogen and phosphorus in small Amazonian streams. Ecological Applications 11(6): 1817–1828.

Paolini, J. E., R. Hevia and R. Herrera. 1987. Transport of carbon and minerals in the Orinoco and Caroní Rivers during the years 1983–84. *In:* Degens, E. T., S. Kempe, Gan Weibin (eds.). Biogeochemistry of Major World Rivers. Scope 64: 325–38.

Rosales, J., G. Petts and C. Knap-Vispo. 2001. Ecological gradients within the riparian forests of the lower Caura River, Venezuela. Plant Ecology 152: 101–118.

Vargas, H., and J. Rangel. 1996a. Clima: Comportamiento de las variables. *In:* Rosales, J. and O. Huber (eds.). Ecología de la Cuenca del Río Caura. Scientia Guaianae 6: 34–39.

Vargas, H., and J. Rangel. 1996b. Hidrología y Sedimentos. *In:* Rosales, J. and O. Huber (eds.). Ecología de la Cuenca del Río Caura. Scientia Guaianae 6: 48–53.

Chapter 4

Diversity of Benthic Macroinvertebrates from the Caura River Basin, Bolívar State, Venezuela

José Vicente García and Guido Pereira

ABSTRACT

Diversity of benthic macroinvertebrates in the Caura River was assessed at 25 localities throughout the Upper and Lower sections of the river. The community of aquatic macroinvertebrates was composed of insects, snails, mussels, oligochaete worms, leeches, turbellarians and crustaceans. Aquatic insects were represented by several orders of which the most diverse were: Odonata, Diptera, Ephemeroptera, Hemiptera and Trichoptera. The observed species composition is typical of a pristine environment, since it contains a high diversity of Odonata and Trichoptera, which are inhabitants of environments free of pollutants or perturbation. The majority of genera of benthic macroinvertebrates in the watershed were found to be homogeneously distributed. The variation in diversity of benthic faunal groups seems to be related to changes in dissolved oxygen and turbidity of the water. These findings are important because a drastic change in the annual cycle of the water, deforestation or mining would introduce suspended particulate material and increase conductivity causing a subsequent dramatic change in the benthic communities.

INTRODUCTION

The Caura River watershed comprises a vast region of relatively pristine forests and rivers. This watershed is one of the principal tributaries of the Orinoco River in Venezuela. In this region, the terrestrial flora and fauna is quite well known, while of the aquatic fauna only the fishes are fairly well known (Machado-Allison et al. 1999, 2003). Invertebrate groups such as crabs, shrimps and aquatic insects are totally unknown, despite their importance in the ecology of aquatic ecosystems. Benthic organisms are important for the transfer of energy through trophic levels in aquatic systems. Many benthic invertebrates incorporate allochtonous material while transforming and transfering autochtonous material within the system (Wiggins 1927; Wallace and Merrit 1980). Some groups such as aquatic insects feed on unicellular algae, bacteria, fungi, vascular plants and leaf detritus, zooplankton, other small invertebrates and small fishes (McCafferty 1981), while other groups such as mussels, snails and some insects are suspension feeders of seston. They also comprise the main food items for young stages of fishes (Lowe-McConnell 1975; Machado-Allison 1999), crabs, shrimps and birds (Epler 1995). Additionally, the species composition of benthic communities is often used to assess water quality since they are important indicators of the level of perturbation or pollution in aquatic systems (Wiggins 1927; Hynes 1970; Epler 1995).

This work represents a rapid survey of the biodiversity of the benthic macro-invertebrate fauna in which we evaluate the degree of degradation. To the extent possible, we compare our results with information from large, nearby rivers of the region such as the Orinoco and Caroní Rivers. Lastly, we evaluate the importance of these groups in the establishment of conservation strategies for this region.

METHODS

During the survey we sampled 25 georeferenced localities: 11 in the Upper Caura and 14 in the Lower Caura. These localities and the physicochemical variables are shown in Appendices 3, 4 and 5. We used a variety of methods including: a Surber sampler; an Eckman grab, filtration through graded sieves, visual search method and several kinds of nets including a d-frame hand net, small aquaria nets and a minnow net. In addition, collection of submerged leaves, rocks and wood also produced samples. All organisms were fixed in the field in a 70% ethanol solution. In the field, sediments were fixed in a 5% formalin solution and then transferred to 70% ethanol. This technique was not useful to preserve turbellarians, because of tissue disintegration during the transportation to laboratory. All biological samples were ultimately preserved in 90% ethanol solution. For identification, a stereoscopic microscope, a compound microscope and conventional keys to level of genera were used for the majority of groups (Richardson 1905; Johannsen 1937a, b; Van Deer Kuyp 1950; Needham and Westfall 1955; Hilsenhoff 1970; Peters 1971; Bryce and Hobart 1972; Benedetto 1974; Flint 1974; Edmunds et al. 1976; Hulbert et al. 1981a, b; McCafferty 1981; Limongi 1983; Stehr 1987; Kensley and Schotte 1989; Daigle 1991; Epler 1995, 1996; Milligan 1997). For midge larvae (Chironomidae), we followed the methods described by Bryce and Hobart (1972), boiling the heads of larvae in a KOH solution for 3 to 5 minutes and then mounting them on slides with polyvinyl-lactophenol. This medium was also used to observe oligochaete setae. A Principal Component Analysis based on the number of genera of insects and class for other groups, and some physicochemical variables such as dissolved oxygen, turbidity and conductivity for each geo-referenced locality, were performed using MVSP program version 3.1 (Kovach 1998), in order to determine the existence of some distributional patterns and the affinity of the groups to these physicochemical variables. Additionally, the total richness and the Shannon-Wiener diversity index were calculated for each locality.

RESULTS

The community of benthic macroinvertebrates of the Caura River watershed is composed of aquatic insects, snails, mussels, oligochaete worms, leeches, turbellarians and crustaceans. The main macrohabitats are: aquatic rooted vegetation, marginal vegetation, rocky beds, submerged woods, quiet zones with leaf beds, sand beaches, muddy bottom beaches, marginal pools and pools on large rocks. This great heterogeneity of habitats was found in the main channel of the river as well as in tributaries.

Aquatic insects are represented by the following orders: Odonata, Ephemeroptera, Hemiptera, Trichoptera, Coleoptera, Diptera, Plecoptera, Neuroptera, Lepidoptera and Collembola. Aquatic mollusks showed a low diversity and are represented by two abundant species of *Pomacea* and *Limnopumus* (Gastropoda, Ampullariidae), two very abundant species of *Doryssa* (Gastropoda, Melaniidae) and one species of *Eupera* (Bivalvia, Sphaeriidae) with a low abundance. Crustaceans were represented by one species of isopod (*Exocorallana berbicensis*) of the family Corallanidae, and one species of clam shrimp (*Cyclestheria hislopi*) of the family Cyclestheriidae, both found only in the Takoto River in the Lower Caura. Oligochaetes were represented by species of the family Lumbriculidae, Enchytraeidae and an undetermined family. The leeches comprised one species, *Haementeria tuberculifera* (Glossiphoniidae) and an undetermined species. Aquatic turbellarians (Plathyhelminthes) were

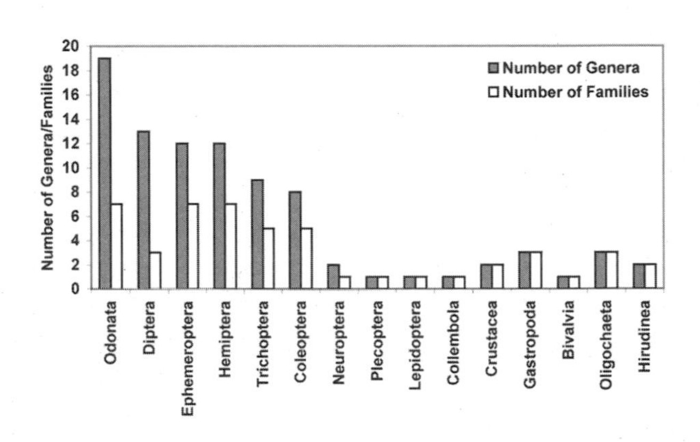

Figure 4.1. The number of families and genera in each order found in the Caura River Watershed.

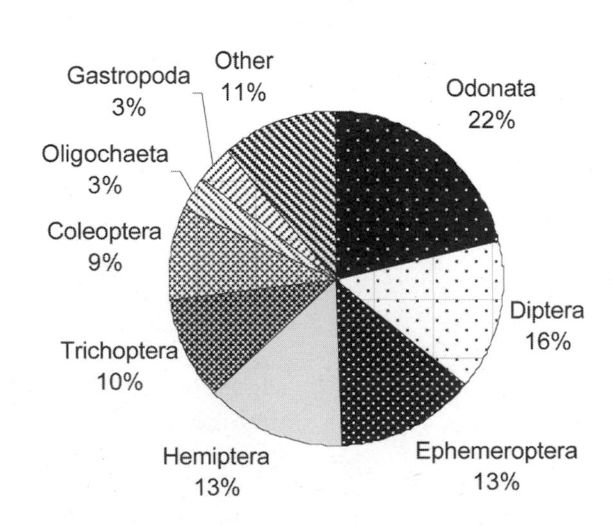

Figure 4.2. The percentage of total genera in each order or class found in the Caura River Basin.

observed in the Upper and Lower Caura. A complete list of classes, families and genera are shown in Appendix 6.

Among aquatic insects, the Odonata (dragonflies and damselflies) are represented by seven families with at least 19 genera (Figure 4.1) of which one genus was found only in the Upper Caura and two only in the Lower Caura. Odonata were found in the majority of sampled habitats. Ephemeroptera (mayflies) were represented by seven families and 12 genera (Figure 4.1), three of them found only in the Lower Caura and one only in the Upper Caura. The genera *Thraulodes* and *Ulmeritus* are particularly abundant and distributed in the whole watershed. The Hemiptera (water bugs) are also distributed in the whole watershed and were represented by 12 genera and seven families (Figure 4.1). Trichoptera (caddisflies) were represented by nine genera and five families, and were moderately abundant. Water beetles (Coleoptera) of at least eight genera and five families were distributed in the whole watershed (Figure 4.1). The genus *Megadytes,* when present, was particularly abundant. The Diptera were abundant and represented by species of 13 genera and three families, the family Chironomidae being the most diverse. The Plecoptera (stoneflies) were represented by one species of the genus *Anacroneuria*, which was abundant. Neuropterans of the family Corydalidae were found mainly in the Upper Caura, while Lepidopterans of the family Pyralidae were distributed in the whole watershed. One species of *Willowsia* (Colembolla, Entomobryidae) was found only in La Ceiba lagoon. In order of importance, Odonata represented 22% of all genera, followed by Diptera with 16%, Ephemeroptera and Hemiptera with 13% each and Trichoptera and Coleoptera with 10 and 9%, respectively (Figure 4.2).

Two Principal Component Analyses were performed. In the first, the number of genera in each locality was included in order to infer if distributional patterns existed. In the second analysis, physicochemical variables and the number of genera in each locality were included to check if geographic variation among taxa is associated with variation in physicochemical variables such as dissolved oxygen, turbidity and conductivity. Because the pH values are low and display low variablity among localities, pH was not included in the analysis. The variable loadings from both analyses are shown in Tables 4.1 and 4.2.

Table 4.1. Eigenvalues and percentage of variation explained by first three principal components axes using number of genera found in each georeferenced site in the first PCA analysis.

	Principal components axes		
	Axis 1	Axis 2	Axis 3
Eigenvalues	9.01	4.46	3.05
Percentage	41.39	20.48	14.00
Cumulative percentage	41.39	61.87	75.87
PCA variable loadings			
Odonata	0.84	-0.15	-0.25
Ephemeroptera	0.47	0.36	0.42
Hemiptera	-0.07	-0.59	0.64
Trichoptera	0.14	0.18	0.47
Coleoptera	0.08	-0.15	0.07
Diptera	-0.17	0.64	0.13
Plecoptera	0.10	0.09	0.05
Neuroptera.	0.04	0.00	0.09
Lepidoptera	0.02	0.10	0.09
Crustacea	0.01	-0.06	0.01
Gastropoda	-0.02	0.06	0.23
Bivalvia	-0.04	-0.02	0.06
Oligochaeta	0.01	0.08	0.18
Hirudinea	0.03	0.01	0.04

Table 4.2. Eigenvalues and percentage of variation explained by first three principal components axes using number of genera and physicochemical parameters for each georeferenced site in the second PCA analysis.

	Principal components axes		
	Axis 1	Axis 2	Axis 3
Eigenvalues	52.39	10.45	6.71
Percentage	65.32	13.03	8.37
Cumulative percentage	65.30	78.33	86.70
PCA variable loadings			
Odonata	-0.01	0.71	-0.44
Ephemeroptera	-0.01	0.47	0.05
Hemiptera	-0.01	-0.17	-0.39
Trichoptera	0.05	0.10	-0.01
Coleoptera	-0.04	0.04	-0.15
Diptera	-0.05	-0.02	0.53
Plecoptera	0.00	0.10	0.02
Neuroptera	-0.02	0.03	-0.02
Lepidoptera	-0.01	0.02	0.06
Crustacea	0.00	0.00	-0.05
Gastropoda	-0.01	-0.04	-0.03
Bivalvia	0.03	-0.04	0.02
Oligochaeta	-0.01	0.03	0.00
Hirudinea	0.01	0.03	0.00
Conductivity	0.99	-0.01	0.02
Dissolved oxygen	0.10	0.33	0.18
Turbidity	0.05	-0.33	-0.55

The analysis based upon taxa only shows that the first three principal components comprise about 76% of the total variability. The results (Figure 4.3) indicate that the distribution in the majority of orders is homogeneous. However, 5 localities (AC06, AC09, AC14, BC05 and BC08) show high correlations to Odonata and Ephemeroptera, two localities (AC05, AC11) are highly correlated to Diptera, and about four localities (BC12, BC13, BC15 and BC17) are negativly correlated with the previous localities, but positively correlated with each other because of the presence of Hemiptera.

In the analysis with taxa and environmental variables, the first three components comprise about 84% of the total variability. This analysis (Figure 4.4) shows a high positive correlation between Odonata, Ephemeroptera and dissolved oxygen, and high negative correlation between these groups, turbidity and conductivity. Other benthic faunal groups were not related to any of the three physicochemical variables. All groups, however, are inversely related to the degree of conductivity (Figure 4.4).

DISCUSSION

The majority of genera of benthic macroinvertebrates in the Caura River are distributed in the Upper as well as in the Lower Caura. They inhabit a great diversity of macrohabitats largely repeated along the river in the principal channel as well as in tributaries, rapids and lagoons. Because PCA analyses do not detect associations between the majority of geo-referenced localities and the faunal benthic components, we concluded that the community is homogeneous with respect to region. Only four orders of aquatic insects (Odonata, Ephemeroptera, Hemiptera and Diptera) show some affinity with some localities.

The greatest diversity of Odonata and Ephemeroptera occur in the caño Wididikenü (Erebato River), Paují and Culebra de Agua rapids (Caura River), Dimoshi rapid (Tawadu River) and Icutú River. The zones with high diversity of Diptera are: Perro rapid (Erebato River) and Fiaka island (Caura River). For Hemiptera, the greatest diversity occurs in La Ceiba lagoon (Mato River) and Takoto River, both located in the Lower Caura.

The snails of the genus *Doryssa*, very abundant in the Upper and Lower Caura, have been previously reported in the upper Orinoco and upper Siapa River, Amazonas State, inhabiting rocky beds with acid waters (pH = 5.7) (Martínez and Royero 1995). Snails of the genus *Pomacea*, abundant in the Lower Caura, are indicative of the benthic fauna of the lower Orinoco River.

The Caura River has almost uniformly low pH and conductivity. The variation in diversity of benthic faunal groups, therefore, seems to be related to changes in dissolved oxygen and turbidity of the water. These findings are important because a drastic change in the annual cycle of the water, deforestation or mining would introduce suspended particulate material and increase the conductivity of the water with a subsequent dramatic change in the benthic communities.

The observed species composition is typical of a pristine environment, since it contains a high diversity of Odonata and Trichoptera, which are inhabitants of environments free of pollutants or perturbation (Wiggins 1927; Daigle 1991,1992). Coleopterans of the family Elmidae, which cannot tolerate pollution by wetting agents such as soaps and detergents (Epler 1996) were found inhabiting the watershed. The same is true for some genera of Ephemeroptera and Plecoptera (Hilsenhoff 1970; Edmunds et al. 1976). The orders Odonata, Ephemeroptera and Trichoptera are the major components of the benthic macro-invertebrate fauna, which suggests that the Caura River has high water quality.

The majority of localities have similar Shannon-Wiener diversity indices. However, some localities that have high values, including: Raudal Perro (AC05), Raudal Cejiato (AC08), Raudal Pauji (AC09), region above Salto Pará (AC12), Raudal Culebra de Agua (AC14), Raudal Tajañano (BC06), caño above Raudal Cinco Mil (BC15) and Takoto River (BC17). Thus, these localities should be included in the conservation strategy for the watershed.

The Odonata and Ephemeroptera contain relatively fragile species and their presence indicates high environmental quality of aquatic ecosystems (degree of pristineness). The areas with the highest diversity of these groups are: caño Wididikenü (AC06), Raudal Pauji (AC09), Raudal Culebra de Agua (AC14), Raudal Dimoshi (BC05) and Icutú River (BC08). We consider these areas priority for conservation within the Caura River Basin.

Finally, there are very few works on diversity in other nearby rivers such as the Orinoco and Caroní Rivers, but based on the works of Vásquez et al. (1990) and Marrero (2000), we conclude that the diversity of the Caura is higher than that of the Caroní River and similar to that of the Orinoco River.

CONSERVATION CONCLUSIONS

The observed species composition is typical of a pristine environment because it contains a high diversity of taxa normally associated with environments free of perturbation.

The orders Odonata, Ephemeroptera and Trichoptera are the major components, in order of importance, which indicates that the watershed has high water quality.

Although eight localities have high Shannon-Wiener diversity indices, conservation priority should be given to those areas that have high diversity of Odonata and Ephemeroptera: caño Wididikenü, Raudal Pauji, Raudal Culebra de Agua, Raudal Dimoshi and Icutú River. Odonata and Ephemeroptera are important environmental indicators of pristine environments.

The variation in diversity of benthic faunal groups seems to be related to changes in dissolved oxygen and turbidity of

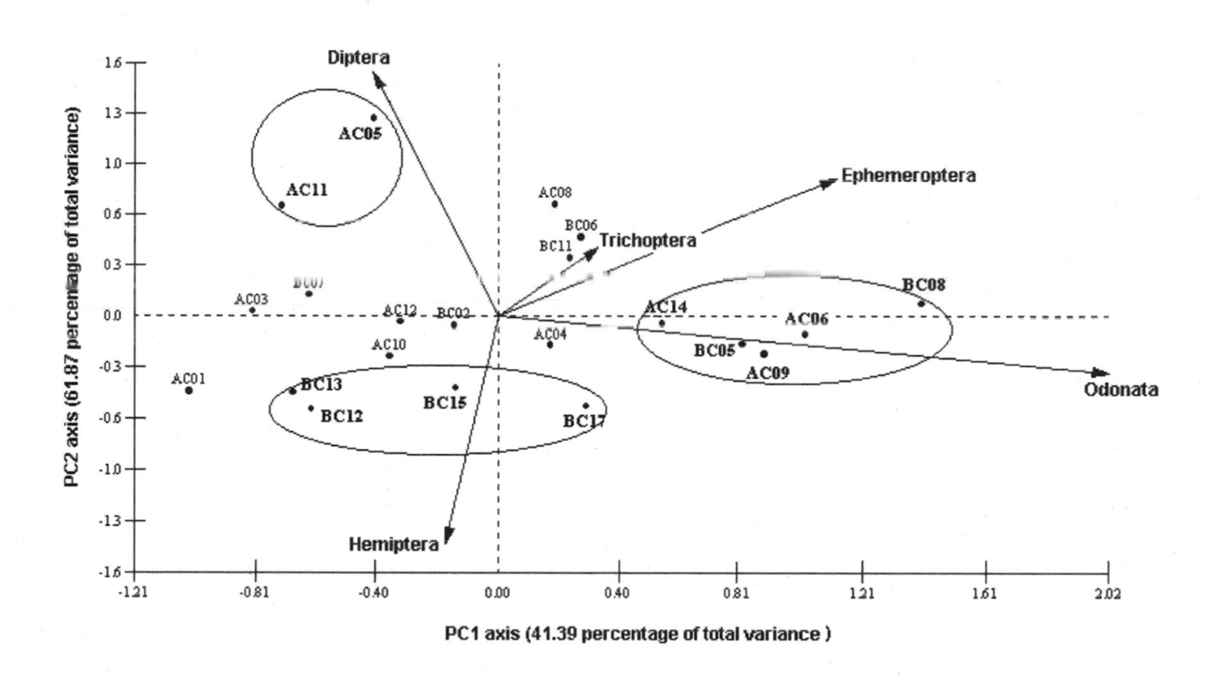

Figure 4.3. Scatter diagram of variables on axes 1 vs. 2 of PCA based upon number of genera at each georeference site.

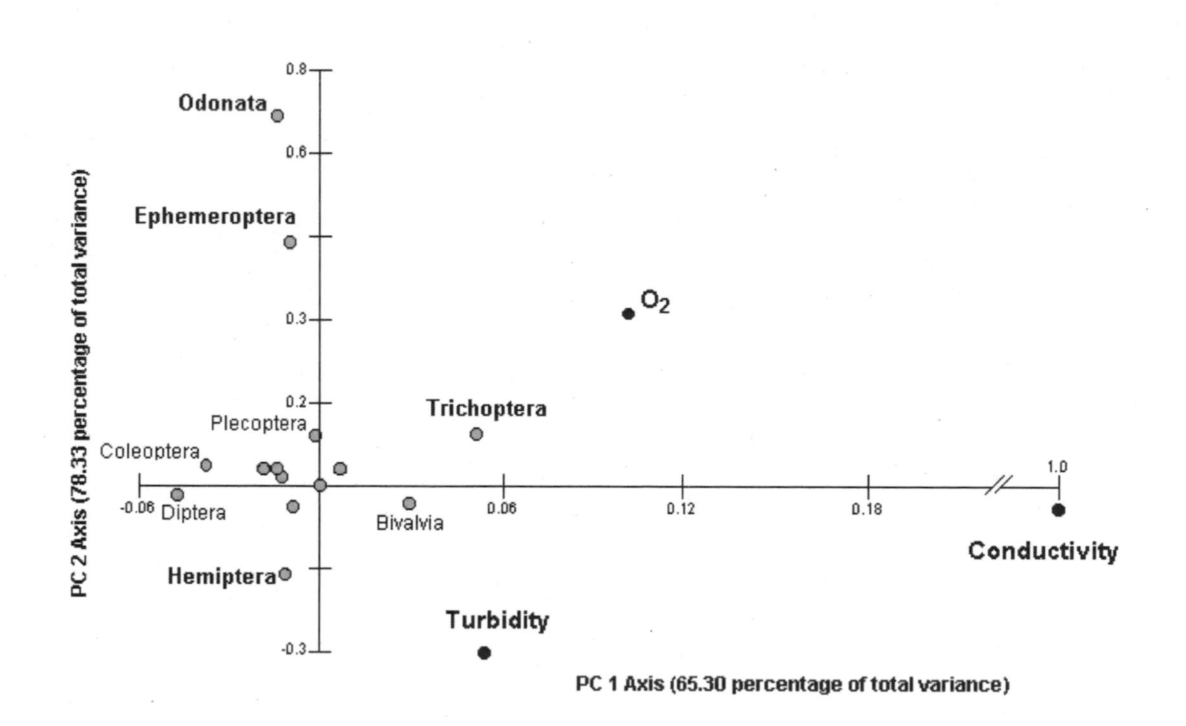

Figure 4.4. Scatter diagram of scores on axes 1 vs. 2 of PCA, using number of genera and physicochemical variables of each georeferenced site.

the water. In consequence, a change in the annual cycle of the water, deforestation or mining would generate a dramatic change in the benthic macro-invertebrate community of the watershed.

LITERATURE CITED

Benedetto, L. 1974. Clave para la determinación de los plecópteros suramericanos. Studies on the Neotropical Fauna 9: 141–170.

Bryce, D., and A. Hobart. 1972. The biology and identification of the larvae of the Chironomidae (Diptera). Entomologist's Gazette 23: 175–215.

Daigle, J. J. 1991. Florida Damselflies (Zygoptera): A species key to the aquatic larval stages. Department of Environmental Regulation. Florida State. Technical Series 11(1).

Daigle, J. J. 1992. Florida Dragonflies (Anisoptera): A species key to the aquatic larval stages. Department of Environmental Regulation. Florida State. Technical Series 12(1).

Edmonson, W. T. (ed.) 1959. Fresh-Water Biology. Second Edition. New York USA: John Wiley and Sons, Inc.

Edmunds, G. F., S. L. Jensen and L. Berner. 1976. The mayflies of North and Central America. Minneapolis: University of Minnesota Press.

Epler, J. H. 1995. Identification manual for the larval Chironomidae (Diptera) of Florida. Final Report DEP Contract Number WM579. Tallahassee USA: Department of Environmental Protection.

Epler, J. H. 1996. Identification manual for the water beetles of Florida (Coleoptera: Cryopidae, Dytiscidae, Elmidae, Gyrinidae, Haliplidae, Hydraenidae, Hidrophilidae, Noteridae, Psephenidae, Ptilodactyllidae, Scirtidae). Final Report DEP Contract Number WM621. Tallahassee USA: Department of Environmental Protection.

Flint, O. S. 1974. Studies of Neotropical Caddisflies, XVII: The Genus *Smicridea* from North and Central America (Trichoptera: Hydropsichidae). Smithsonian Contributions to Zoology 167. Washington DC. USA: Smithsonian Institution Press.

Hilsenhoff, W. L. 1970. Key to genera of Wisconsin Plecoptera (stonefly) nymphs Ephemeroptera (mayfly) nymps and Trichoptera (caddisfly) larvae. Research Report 67. Madison, USA: Department of Natural Resources.

Hulbert, S. H., G. Rodríguez and N. W. Dos Santos (eds.). 1981. Aquatic Biota of Tropical South America. Part 1. Arthropoda. San Diego USA: San Diego University Press.

Hulbert, S. H., G. Rodríguez and N. W. Dos Santos (eds.). 1981. Aquatic Biota of Tropical South America. Part 2. Anarthropoda. San Diego USA: San Diego University Press.

Hynes, H. B. 1970. The ecology of running waters. Toronto Canada: University of Toronto Press.

Johannsen, O. A. 1937. Aquatic Diptera part III. Chironomidae: Subfamilies Tanypodinae, Diamesinae, and Orthocladiinae. New York: Cornell University Experiment Station Memoir 205.

Johannsen, O. A. 1937. Aquatic Diptera part IV. Chironomidae: Subfamily Chironominae. New York : Cornell University Experiment Station Memoir 210.

Kensley, B., and M. Schotte. 1989. Guide to the marine isopod crustaceans of the Caribbean. Washington DC, USA: Smithsonian Institution Press.

Kovach, W. L. 1998. MVSP, a multivariate statistical package for Windows. Version 3.0. Pentraeth, England: Kovach Computing Services.

Limongi, J. 1983. Estudio morfo-taxonomico de náyades en algunas especies de Odonata (Insecta) en Venezuela. Caracas Venezuela: Trabajo de Grado, Universidad Central de Venezuela.

Lowe-McConnell. 1975. Fish communities in tropical freshwaters. Chapter 3, Equatorial forest rivers: ecological conditions and fish communities. New York USA: Logman.

Machado-Allison, A. 1992. Larval ecology of fishes. *In:* Hamlett, W. (ed.). Reproductive Biology of South American Vertebrates. New York: Springer-Verlag. Pp. 45–59.

Machado-Allison, A., B. Chernoff, C. Silvera, A. Bonilla, H. López-Rojas, C. A. Lasso, F. Provenzano, C. Marcano and D. Machado-Aranda. 1999. Inventario de los peces de la cuenca del Río Caura, Estado Bolívar, Venezuela. Acta Biol. Venez. 19:61–72.

Machado-Allison, A., B. Chernoff, F. Provenzano, P. Willink, A. Marcano, P. Petry and B. Sidlauskas. 2003. Inventory, relative abundance, diversity and importance of fishes in the Caura River Basin. *In:* Chernoff, B., A. Machado-Allison, K. Riseng, and J. R. Montambault (eds.), A Biological Assessment of the Aquatic Ecosystems of the Caura River Basin, Bolívar State, Venezeula. RAP Bulletin of Biological Assessment, No. 28. Washington DC. USA: Conservation International. Pp. 64–74.

Marrero, C. 2000. Biomonitoreo de poblaciones de insectos acuáticos bentónicos en ríos de la cuenca del Río Caroní, para detectar bioacumulación de subproductos mercuriales. Final Report. Project number 96001791. Caracas, Venezuela: CONICIT.

Martínez, R., and R. Royero. 1995. Contribución al conocimiento de *Diplodon (Diplodon) granosus granosus* Brugiere (Bivalvia: Hyriidae) y *Doryssa hohenackeri happleri* Vernhout (Gastropoda: Melaniidae) en el alto Siapa (Departamento de Río Negro), Estado Amazonas, Venezuela. Acta Biológica Venezuelica 16: 79–84.

McCafferty, W. P. 1981. Aquatic Entomology. Boston USA: Jones and Bartlett Publishers.

Merrit, R. W. and K. W. Cummins. 1978. An introduction to the aquatic insects of North America. Dubuque: Kendall-Hunt Publishing Co.

Milligan, M. R. 1997. Identification manual for the aquatic Oligochaeta of Florida. Volume 1. Freshwater Oligochaetes. Final Report DEP Contract Number WM550. Tallahassee: Department of Environmental Protection.

Needham, J. G., and M. J. Westfall. 1955. A manual of the Dragonflies of North America, including the Greater Antilles and provinces of Mexican border. Berkely: University of California Press.

Peters, W. L. 1971. A revision of the Leptophlebiidae of the West Indies (Ephemeroptera). Smithsonian Contributions to Zoology 62. Washington DC: Smithsonian Institution Press.

Richardson, H. 1905. Isopods of North America. Bulletin No. 54. Washington DC: United States National Museum.

Stehr, F. W. (ed.) 1987. Inmature Insects. Vol. 1. Dubuque: Kendall-Hunt Publishing Co.

Van Deer Kuyp. 1950. Mosquitoes of the Netherlands Antilles and their hygienic importance. Studies on the Fauna of Curacao and other Caribbean Islands 23: 37–114.

Vásquez, E., L. Sánchez, L. E. Pérez and L. Blanco. 1990. Estudios hidrobiológicos y piscicultura en algunos cuerpos de agua (ríos, lagunas y embalses) en la cuenca baja del Río Orinoco. *In*: Weibezahn, F. H., H. Alvarez and W. Lewis (Eds.). El Río Orinoco como ecosistema. Results of the Symposia: Ecosistema Orinoco: conocimiento actual y necesidades de estudios futuros. XXXVI Convención anual AsoVAC Caracas. 430p.

Wallace, J. B., and W. Merrit. 1980. Filter-feeding ecology of aquatic insects. Annual Review of Entomology 25: 103–132.

Waltz, R. D., and W. P. McCafferty. 1979. Freshwater springtails (Hexapoda, Collembola) of North America. Purdue University Agricultural Experiment Station Research Bulletin 960. West Lafayette, Indiana.

Wiggins, G. B. 1927. Larvae of the North American caddisfly genera (Trichoptera). Toronto: University of Toronto Press.

Chapter 5

Survey of Decapod Crustaceans in the Caura River Basin, Bolívar State, Venezuela: Species Richness, Habitat, Zoogeographical Aspects and Conservation Implications

Célio Magalhães and Guido Pereira

ABSTRACT

A total of ten species of decapod crustaceans were found in the surveyed area of the middle Caura River Basin: six species of Palaemonidae (shrimps), one species of Pseudothelphusidae (crabs) and three species of Trichodactylidae (crabs). The region upriver from Salto Pará (Upper Caura) had a lower richness, with five species, than the lower region (Lower Caura) with eight species. The decapod fauna of the Caura River Basin has one possible endemic species, the undescribed shrimp *Pseudopalaemon* sp., while the others are either known from the Llanos region or from the Amazon Basin. The number of decapod species and genera found in the Caura River Basin represents a typical sample of the diversity and abundance of inland river systems from the Guayana and Amazon regions. The shrimp *Macrobrachium brasiliense* was the most frequent and abundant species found in this survey and could be considered the most typical species of this system. Two undescribed species of palaemonid shrimps were collected. Abundance of decapods was low to moderate, probably reflecting sampling constraints and the overall oligotrophic condition of the habitats. From the decapod community structure, habitat use and distribution throughout the basin we conclude that the decapod community is at present healthy and will remain that way as long as the pristine conditions of the region are preserved. We recommend the Nichare River subregion as a potential area for conservation of the decapod crustacean fauna of the Caura River.

INTRODUCTION

The Venezuelan freshwater decapods are fairly well known due to the many contributions of Rodríguez (1980, 1982a, b, 1992, and others) and Pereira (1985, 1986, 1991). However, there are only a few faunistic studies concerning decapods of a particular region or hydrographic basin of the country. Until now, such studies have only been made for the Paria peninsula (López and Pereira 1994) and the Orinoco River Delta (López and Pereira 1996, 1998).

The decapod faunal composition of the Caura River system is known from sporadic collections. Such collections identify that five species occur in this basin: the shrimps *Macrobrachium brasiliense* (Heller), *Palaemonetes carteri* Gordon, an undescribed species of *Macrobrachium* (Rodríguez 1982b), the trichodactylid crab *Valdivia serrata* (Rodríguez 1992), and a species of pseudothelphusid crab, *Fredius stenolobus*, described by Rodríguez and Suárez (1994). We expect this number to increase as the basin is better explored. An AquaRAP biological assessment of the middle Caura River Basin in November–December 2000 offered the opportunity for such an exploration. In this chapter, we report our findings about species richness, habitat and longitudinal distribution of the decapod fauna along the study area. Conservation implications concerning this particular group are also included.

METHODS

The decapod fauna of the Caura River Basin was assessed based on material from 55 collecting stations of 31 georeference points ranging from latitudes 05°29,563′N to 07°11,890′N (Map), following the AquaRAP Sampling Protocol (Chernoff and Willink 2000). The collections were made from 25 November to 10 December 2000. The explored area of the middle Caura River system was arbitrarily divided into two large regions by the presence of a large waterfall, the Salto Pará. The region upriver from the waterfall (Upper Caura) includes the Caura, Erebato, Kakada and Yuruani Rivers, and ranges from Salto Pará as far upstream as Raudal Cejiato (Caura River) and Kakada River. The region downstream from Salto Pará (Lower Caura) comprises the Caura, Nichare, Tawadu, Takoto and Mato Rivers.

We attempted to sample all habitats found in a georeference station for decapods. These habitats consisted of river, forest creek, rapids, pond and lagoon. Within each habitat, as many microhabitats as possible were sampled. Microhabitats were classified as overhanging vegetation, rocks, Podostemonaceae (aquatic rooted vegetation), sandy beach, submerged leaf litter and woody debris. Collecting gear consisted of dip nets, 2 m and 5 m seine nets, traps and, eventually, bare hands. Some sampling constraints (such as diurnal collecting time while crustaceans have nocturnal habitats, short collecting time, difficulties using collecting gear in rocky and woody substrates) were certainly impediments to a better survey. Sampling time was 1–3 hours. The survey was strictly qualitative and no attempts were made to standardize the efforts. Analysis of species composition similarities among subregions of approximately compatible scale was performed with UPGMA cluster analysis using Jaccard's similarity coefficient applied to presence/absence data (Ludwig and Reynolds 1988). Six subregions were recognized: Erebato (Erebato River and Kakada River); Caura (Caura River upstream from Salto Pará); Playón (Caura River and forest streams in the Salto Pará area); Nichare (Nichare, Tawadu and Icutú Rivers); Cincomil (Caura and Takoto Rivers, around Raudal Cinco Mil); and Caumato (Mato and Caura Rivers downstream from Raudal Cinco Mil). Also, Principal Component Analysis (PCA) was done to summarize information into fewer variables and to determine which variables contribute to the high variance among localities. The data include species abundance and physicochemical data (dissolved oxygen, turbidity, conductivity, bottom type, pH and temperature) as variables for each locality sampled using the computer program MVSP ver. 3.1 (Kovach 1998).

Specimens were preserved in 70% ethanol and identifications were completed using the descriptions of Rodríguez (1980, 1982a, b, 1992), Rodríguez and Suárez (1994), Pereira (1986) and Magalhães and Türkay (1996a, b). Specimens were deposited in the crustacean collections of the Museo de Biología de la Universidad Central de Venezuela (Caracas - Venezuela) and the Instituto Nacional de Pesquisas da Amazônia (Manaus - Brazil).

RESULTS AND DISCUSSION

Species richness and longitudinal distribution

The decapod fauna collected by the AquaRAP expedition consisted of ten species of shrimps and crabs, from three families and seven genera (Table 5.1, Appendix 11). All shrimps belong to the family Palaemonidae. One species of crab is a Pseudothelphusidae and three species are Trichodactylidae. The specific composition of the decapod fauna varied between the Upper and Lower Caura regions. In the Upper Caura only five species were found: the shrimps *Macrobrachium brasiliense* and *Macrobrachium* sp. 1, and the crabs *Poppiana dentata*, *Valdivia serrata* and *Fredius stenolobus*. In the Lower Caura region, a larger number of shrimp species were found (*Macrobrachium amazonicum*, *M. brasiliense*, *Palaemonetes carteri*, *P. mercedae* and *Pseudopalaemon* n. sp.) while only one additional species of trichodactylid crab was found (*Forsteria venezuelensis*). In the Lower Caura region, the species richness becomes higher towards

Table 5.1. List of species of decapod crustaceans collected by the AquaRAP expedition to the Caura River Basin (Bolívar State, Venezuela), November–December 2000, according to the regions surveyed.

Taxa	Upper Caura	Lower Caura	Previously recorded from basin	New record for basin
Palaemonidae (shrimps)				
Macrobrachium amazonicum		X		X
Macrobrachium brasiliense	X	X	X	
Macrobrachium sp. 1	X			X
Palaemonetes carteri		X	X	
Palaemonetes mercedae		X		X
Pseudopalaemon n. sp.		X		X
Pseudothelphusidae (crabs)				
Fredius stenolobus	X	X	X	
Trichodactylidae (crabs)				
Forsteria venezuelensis		X		X
Poppiana dentata	X	X		X
Valdivia serrata	X	X	X	

the lowland part of the basin. Except for *Macrobrachium* sp. 1, all species occurring in Upper Caura are also present downstream of the falls, and other species, typical to lowland areas, begin to appear. This is the case for the shrimp species *M. amazonicum* and *P. carteri* and the crab species *F. venezuelensis*.

The shrimp *M. brasiliense* was the most frequent taxon in the collections. A precise abundance is not available as sampling efforts were not homogenous. However, it seems that abundance of all taxa was usually low, except for a few collecting stations where the shrimp *M. brasiliense* appeared with moderate profusion. Abundance of crabs was usually low regardless of the region. However, this evaluation is certainly compromised by the sampling difficulties in collecting these nocturnal animals.

Diversity of the decapod fauna within the Caura River Basin can be considered moderate. Although there are not many systematic surveys for larger tributaries of the Orinoco and Amazon Basins, the number of decapod species and genera found in the Caura River Basin represents a typical sample of the diversity and abundance of inland river systems from the Guayana and Amazon regions. Similar surveys conducted in the Tahuamanu/Manuripe River system (secondary tributaries of the Madeira River) in Bolivia (Magalhães 1999), and in the Pastaza River in Ecuador/Peru, also rendered ten species for each region, although with different taxonomic composition. The species accumulation curve (Figure 5.1) suggests that the number of species found should not be much higher than ten, as the curve is reaching an asymptote after 13 days of sampling. The curve also indicates the higher species richness in the lower part of the basin: the graph displays an asymptote between the third and seventh days of sampling, which represent the survey time in the Upper Caura. After that period, the curve increases as long as the expedition proceeded towards the lower areas of the basin.

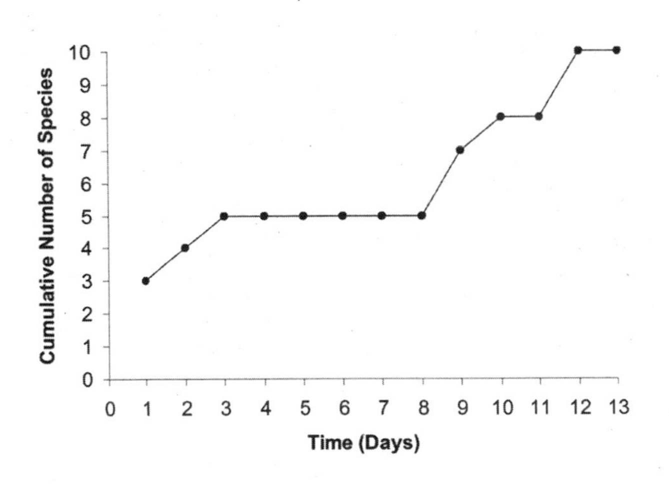

Figure 5.1. Species accumulation curve for decapod crustaceans collected in the Caura River Basin (State of Bolívar, Venezuela) during the AquaRAP expedition, November–December 2000.

The Upper Caura region has fewer species then the Lower Caura. The area just downstream from Salto Pará (El Playón) also has a low number of species, and presents the same species as the Upper Caura River; both subregions are very similar to the Erebato subregion (Table 5.2, Figure 5.2). These subregions have a similar pattern of habitat availability for decapods, namely river, rapids and forest creeks, which could account for such similarity of species.

The subregion comprising the Nichare River sub-basin and Caura River downstream from the mouth of Nichare River is characterized by the appearance of lowland species. The similarities in the species composition between these subregions increases as new species appeared towards the lower course of the river. The Nichare River subregion presents the higher number of species (seven) and was the only place in which the new species of the palaemonid shrimp, *Pseudopalaemon* n. sp., was found. In addition, this subregion showed a faunal composition very similar to most of the other subregions (Figure 5.2), which makes it a potential area for directing eventual conservation efforts.

The knowledge of the decapod species composition of the Caura River Basin has been greatly increased with the results of the AquaRAP expedition, now totaling eleven species. In

Table 5.2. Jaccard binary similarity coefficients of regions within Caura River Basin for the decapod crustacean fauna collected during the AquaRAP expedition, November–December 2000.

	Caura	Erebato	Playón	Nichare	Cinco Mil	Caumato
Caura	1.000					
Erebato	0.800	1.000				
Playón	1.000	0.800	1.000			
Nichare	0.571	0.500	0.571	1.000		
Cinco Mil	0.667	0.571	0.667	0.625	1.000	
Caumato	0.286	0.250	0.286	0.333	0.571	1.000

addition to the five species (three palaemonid shrimps, one pseudothelphusid crab and one trichodactylid crab) already known (Rodríguez 1982b; Rodríguez and Suárez 1994), the expedition recorded four species of shrimps and two of crabs new to the basin (see Table 5.1). *Macrobrachium* sp., an undescribed species recorded by Rodríguez (1982b) in the Tauca River (a tributary of the Lower Caura) was not collected during our expedition.

Habitat

The main habitats explored were the forest streams ("caños"), main river channel, rapids and islands. In the Upper Caura region, the decapods occurred mostly in the small, shadowed forest streams with sandy and rocky bottom preferentially, but were also found in rapids or rocky islands along the main channel of the rivers. The pseudothelphusid crabs were usually found associated with rocky substrates in the main channel or in the forest streams; in the latter, they can also inhabit holes in submerged dead tree trucks. The juvenile shrimps and trichodactylid crabs were commonly found among the submerged leaf litter and wood debris while the adults occurred mainly in cryptic microhabitats, such as holes in submerged tree trucks. Table 5.3 summarizes the occurrence of species according to habitat, microhabitat and bottom type.

Habitats occupied by these crustaceans were very similar and found all over the region. Absence of species in some stations could well be due to sampling constraints rather than other reasons. *F. stenolobus* was not collected in the Kakada River stations in spite of the presence of rocky substrates in this area. However, short time and diurnal collecting may have prevented us from collecting this species in those stations.

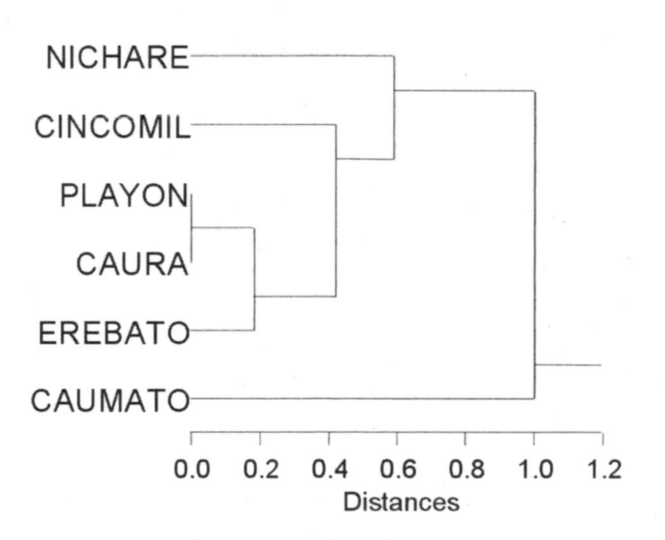

Figure 5.2. UPGMA cluster analysis of regions within Caura River Basin concerning the decapod crustacean fauna collected during the AquaRAP expedition, November–December 2000. (Euclidean Distances based on Jaccard binary similarity coefficients; number of observations: 10.)

Habitat preferences in the Lower Caura resemble those verified in the Upper Caura. However, environments such as sandy and sandy-muddy beaches, rocky pools with marginal vegetation, lakes and lagoons were more frequently found in the lower region, particularly in the subregion downstream from the Nichare River subregion. These habitats were usually inhabited by the lowland species, for instance: *Macrobrachium amazonicum* occurred in the sandy-muddy beaches, lake and rocky pool with marginal vegetation, while *Palaemonetes carteri* appeared in a lagoon. In general, abundance was low, but it was moderate for *M. brasiliense*, and, perhaps, for *Poppiana dentata* which occurred in forest streams with sandy bottoms and much wood debris.

Analyses

Principal component analysis shows that 92.1% of the variance is explained by the first three principal components (Table 5.4). Because of the high amount of variance described by a few principal components, the analysis provides a reasonable description of the correlation structure among biotic and abiotic variables for the Upper and Lower Caura regions. Component 1 explains 69.9% of total variation; however this component is characterized by a single variable with a high eigenvalue, *M. brasiliense* with 0.99, while the rest of the variables possess values of 0% or near 0%. *M. brasiliense* is the most common and widespread species in the system. It is present in almost every locality and as such it can be considered the most typical and representative species of the Caura River system.

Principal component 2 explain 12.7% of the total variation; it is characterized by the fact that some species have high and positive values of variation, such as *M. amazonicum, M. brasiliensis, Palaemonetes mercedae* and *Pseudopalaemon* n. sp., while another group of species possess variation values that range from 0 or near 0 positives and negatives such as *Macrobrachium* sp. 1, *Palaemonetes carteri, Poppiana dentata, Valdivia serrata* and *Forsteria venezuelensis*. Finally, of all physicochemical variables only temperature shows a high positive load. This axis could be interpreted as the community or biological component. It takes into account the species that are more widely distributed and possess relatively high abundance, also species more restricted with a low abundance and probably the interactions among them. Finally, temperature seems to be a key factor that explains distribution.

Component 3 explain 9.4% of total variation; a high and positive eigenvalue for physicochemical variables (pH, dissolved oxygen, temperature, turbidity) and negative for bottom type. Following a similar reasoning as for the previous component, this could be interpreted as the abiotic component reflecting environmental heterogeneity along the Upper and Lower Caura regions. When localities are plotted against axes 2 and 3, a general representation of the biotic and abiotic components of the system is drawn (Figure

Table 5.3. Decapod crustacean distribution according to aquatic habitats, microhabitats and botton types sampled in the Caura River Basin (Bolívar State,Venezuela) during the AquaRAP expedition, November–December 2000.

	Habitat					Microhabitat						Bottom				
	Forest creek	Lagoon	Pond	Rapids	River	Overhanging vegetation	Podeoste-monaceae	Rocks	Sandy beach	Submerged leaf litter	Wood debris	Clay	Gravel	Muddy	Rocky	Sandy
Palaemonidae (shrimps)																
Macrobrachium amazonicum		x	x		x	x			x	x		x			x	x
Macrobrachium brasiliense	x	x	x		x		x	x	x	x	x	x	x	x	x	x
Macrobrachium sp. 1	x									x			x	x	x	x
Palaemonetes carteri		x								x		x				
Palaemonetes mercedae			x		x	x					x	x			x	
Pseudopalaemon n. sp.	x				x					x	x		x	x		x
Pseudothelphusidae (crabs)																
Fredius stenolobus	x			x	x			x		x	x		x		x	x
Trichodactylidae (crabs)																
Forsteria venezuelensis	x				x				x	x	x		x	x		x
Poppiana dentata	x	x			x					x	x	x		x		x
Valdivia serrata	x		x		x					x	x	x	x	x	x	x

Table 5.4. Eigenvalues and percentage of variation explained by first three principal components axes using abundance of species and physicochemical variables in each geo-referenced station surveyed during the AquaRAP expedition to the Caura River Basin, Venezuela, November–December 2000.

	Principal components axes		
	Axis 1	Axis 2	Axis 3
Eingenvalues	460.196	83.288	62.082
Percentage	69.99	12.67	9.44
Cumulative percentage	69.99	82.66	92.10
PCA variable loadings			
Macrobrachium brasiliense	0.99	0.41	0.04
Macrobrachium amazonicum	-0.04	0.86	-0.39
Macrobrachium sp.	-0.01	0.00	0.00
Palaemonete carteri	0.00	0.00	0.00
Palaemonetes mercedae	0.00	0.27	-0.12
Pseudopalaemon n. sp.	0.00	0.14	-0.06
Poppiana dentata	0.08	-0.06	-0.19
Valdivia serrata	0.04	-0.03	-0.04
Forsteria venezuelensis	0.04	0.00	0.00
Bottom type	-0.02	0.03	-0.11
pH	0.01	0.08	0.17
Dissolved Oxygen	0.01	0.06	0.27
Temperature	-0.01	0.39	0.82
Turbidity	0.00	0.08	0.12

5.3). It can be seen that physicochemical variables (except bottom type) lay above the origin in the right quadrant with temperature showing the highest value. Three species located in the lower right quadrant show progressively higher values from *Pseudopalaemon* n. sp., then *Palaemonetes mercedae* to *Macrobrachium amazonicum* (highest value). Thus, together with *M. brasiliense*, these three species play a significant role in this ecosystem. *M. amazonicum* appears to have some relationship with water temperature and turbidity; field observations agree with this statement.

In conclusion the PCA analysis establishes that *M. brasiliense* is the most typical species of the system studied, and that components 2 and 3 may reflect community characteristics of Caura River system.

Zoogeographical aspects

The decapod fauna of the Caura River Basin consists of Amazonian and Orinocoan elements. Except possibly for *Pseudopalaemon* n. sp., no decapod species is endemic to the basin. The *Pseudopalaemon* n. sp. could be endemic to the Caura River Basin, but such statement is premature as palaemonid shrimps usually have wider distribution. The occurrence of this species in other river systems of the Guayana region may be recorded in the future by more intensive collections. The shrimp *M. brasiliense* has a very wide distribution throughout South American river basins (Holthuis 1952; Rodríguez 1981; Coelho and Ramos-Porto 1985) and is one of the few palaemonid species found at altitudes above 300 m. The other species, *Macrobrachium* sp. 1, which is being described (G. Pereira unpublished data) also seems to be a highplain river species, as it was first collected in the upper reaches of the Caroní River Basin. Its occurrence in the Caura River Basin suggests that it could have a wider distribution in rivers of the Guayana Shield. The shrimps *M. amazonicum* and *Palaemonetes carteri* are widespread species. The former is a very common species all over the Orinoco, Amazon, Paraguay and Parana River lowlands (Holthuis 1952; Rodríguez 1980, 1981, 1982b; Coelho and Ramos-Porto 1985; López and Pereira 1996, 1998; Ramos-Porto and Coelho 1998), but has never been recorded in the Caura River before this expedition. The latter species is also present in Guyana, Suriname, French Guiana and other areas of the Amazon and Orinoco Basins (Holthuis 1952; Rodríguez 1981; Coelho and Ramos-Porto 1985; López and Pereira 1996, 1998; Ramos-Porto and Coelho 1998) and was recorded in the lower reaches of the Caura River by Rodríguez (1982b).

Three crab species found in the Caura River have Orinocoan and Amazonian areas of distribution. *V. serrata* is a trichodactylid crab widely occurring in both river basins and its presence in the Lower Caura River was previously known (Rodríguez, 1992). *P. dentata* was known to be distributed in a narrow strip near the coast of northern South America, from Venezuela to French Guiana (Rodríguez 1992), and *F. stenolobus* was restricted to the Caura River Basin (Rodríguez and Suárez 1994). However, recent collections indicate that the latter two species also occur in other rivers of the Venezuelan Guayana and the Amazonian Basin (J. Meri and C. Magalhães, unpublished data). The occurrence of *P. dentata* in the Caura River, particularly in the Upper Caura, suggests that its distribution also encompasses other southern tributaries of the Orinoco River.

The trichodactylid crab *F. venezuelensis* is endemic to the Orinoco River Basin (Rodríguez 1980, 1992) but its occurrence in the Caura River was verified for the first time during the AquaRAP expedition. This seems to be a typical lowland species and the absence of records in the Upper Caura was expected.

CONSERVATION IMPLICATIONS

The habitats occupied by these crustaceans along the surveyed area were in good condition and the usually low abundance of specimens can be due to the overall oligotrophic situation of such areas. Possible threats to the decapod fauna include deforestation of the riparian vegetation and silting of forest creek beds. However, this was not noticed in the evalu-

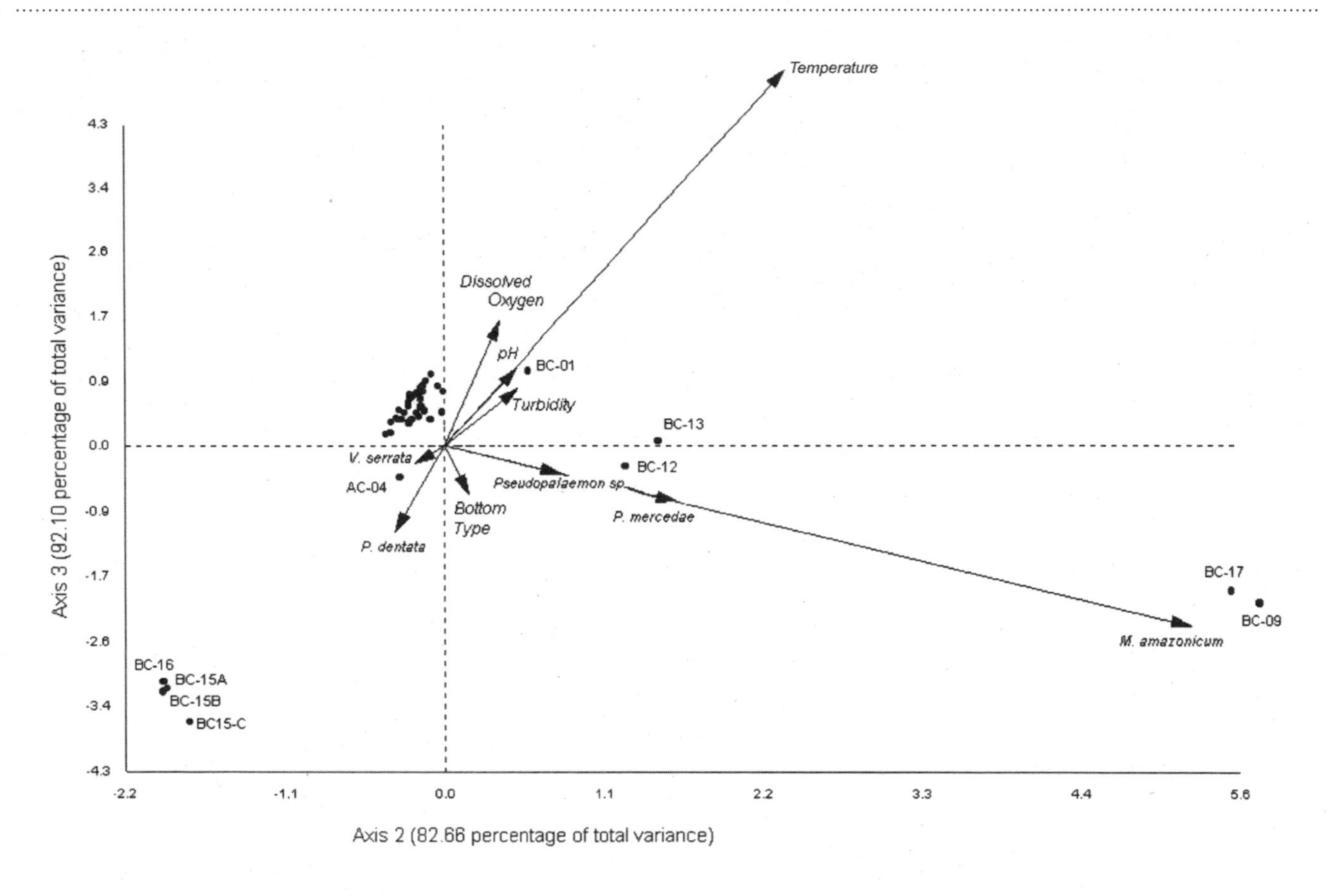

Figure 5.3. Scatter diagram of variables on axes 2 vs. 3 of PCA using abundance of species and physicochemical variables for each georeferenced station surveyed during the AquaRAP to the Caura River Basin.

ated area in such a way that could constitute a current threat to the decapod community. Although the crustacean species are not commercially exploited in this region, they play an important role in the ecological processes of the environment and are sensitive to the impacts of deforestation.

The Nichare River subregion is an important area for the conservation of decapods in the Caura River. In addition to the higher number of species, there is also a possibly endemic species of palaemonid shrimp. The faunal composition of the Nichare subregion is very similar to most of the other subregions (Figure 5.2). Conservation of this area would eventually contribute to the preservation of most of the shrimp and crab species of the Caura River Basin.

CONSERVATION AND RESEARCH RECOMMENDATIONS

Conservation recommendations from the standpoint of decapod crustaceans follow the general recommendations for this region. Specific recommendations concerning crustaceans are:

- Develop a detailed census of the small tributaries in order to pinpoint those that could better act as refuge for reproduction and maintenance of shrimp and crab populations and then propose specific conservation measures.
- Create a detailed and extensive inventory of all groups of macro and microcrustaceans in order to evaluate more precisely the specific composition of the crustacean community along the basin.
- We recommend the Nichare River subregion as the most appropriate area for conservation of the decapod crustacean fauna of the Caura River.

LITERATURE CITED

Chernoff, B., and P. W. Willink. 2000. AquaRAP sampling protocol/Protocolo de amaostragem do AquaRAP. *In*: Willink, P.W. et al (eds.). A biological assessment of the aquatic ecosystems of the Pantanal, Mato Grosso do

Sul, Brasil. RAP Bulletin of Biological Assessment 18: 241–242. (Appendix 1)

Coelho, P. A., and M. Ramos-Porto. 1985. Camarões de água doce do Brasil: Distribuição geográfica. Revista brasileira de Zoologia 2(6): 405–410.

Holthuis, L. B. 1952. A general revision of the Palaemonidae (Crustacea Decapoda Natantia) of the Americas. II. The subfamily Palaemonidae. Occasional Papers, Allan Hancock Foundation 12:1–396.

Kovach, W. L. 1998. A Multi-Variate Statistical Package for Windows. (ver. 3.1). Anglesey, Wales: Kovach Computing Services.

López, B., and G. Pereira. 1994. Contribución al conocimiento de los crustaceos y moluscos de la Peninsula de Paria / Parte I: Crustacea: Decapoda. Memoria de la Sociedad de Ciencias Naturales La Salle 54(141): 51–75.

López, B., and G. Pereira. 1996. Inventario de los crustaceos decapodos de las zonas alta y media del Delta del Rio Orinoco, Venezuela. Acta Biologica Venezuelica 16(3): 45–64.

López, B., and G. Pereira. 1998. Actualización del inventario de crustáceos decápodos del Delta del Orinoco. Pp. 76–85. *In*: Sánchez, J. L. L., I. I. S. Cuadra and M. D. Martínez (eds.). El Rio Orinoco. Aprovechamiento Sustentable. Caracas, Venezuela: UCV. (Memorias de las Primeras Jornadas Venezolanas de Investigacion sobre el Rio Orinoco.)

Ludwig, J. A., and J. F. Reynolds. 1988. Statistical Ecology. A Primer on Methods and Computing. New York USA: John Wiley and Sons.

Magalhães, C. 1999. Diversity and abundance of decapods crustaceans in the rio Tahuamanu and rio Manuripi basins. *In*: Chernoff, B. and P. W. Willink (eds.). A biological assessment of the aquatic ecossystems of the Upper Río Orthon basin, Pando, Bolivia. Bulletin of Biological Assessment 15. Washington DC: Conservation International. Pp. 35–38, Appendix 5.

Magalhães, C., and M. Türkay. 1996a. Taxonomy of the Neotropical freshwater crab family Trichodactylidae I. The generic system with description of some new genera (Crustacea: Decapoda: Brachyura). Senckenbergiana biologica 75(1/2): 63–95.

Magalhães, C., and M. Türkay. 1996b. Taxonomy of the Neotropical freshwater crab family Trichodactylidae II. The genera *Forsteria*, *Melocarcinus*, *Sylviocarcinus*, and *Zilchiopsis* (Crustacea: Decapoda: Brachyura). Senckenbergiana biologica 75(1/2): 97–130.

Pereira, G. 1985. Freshwater shrimps from Venezuela III: *Macrobrachium quelchi* De Man and *Euryrhynchus pemoni* n. sp. (Crustacea: Decapoda: Palaemonidae). Proceedings of the Biological Society of Washington 3: 615–621.

Pereira, G. 1986. Freshwater shrimps from Venezuela I: Seven new species of Palaemoninae (Crustacea: Decapoda: Palaemonidae). Proceedings of the Biological Society of Washington 99(2): 198–213.

Pereira, G. 1991. Camarones de agua dulce de Venezuela II: Nuevas adiciones en las familias Atyidae y Palaemonidae (Crustacea, Decapoda, Caridea). Acta Biologica Venezuelica 13(1-2): 75–88.

Ramos-Porto, M., and P. A. Coelho. 1998. Malacostraca–Eucarida. Caridea (Alpheoidea excluded). *In*: Young, P. S. (ed.), Catalogue of Crustacea of Brazil. Rio de Janeiro, Museu Nacional. p. 325–350. (Série Livros n. 6)

Rodríguez, G. 1980. Crustaceos Decapodos de Venezuela. Caracas, Venezuela: IVIC.

Rodríguez, G. 1981. Decapoda. *In*: Aquatic Biota of Tropical South America, Part 1: Arthropoda. Hurlbert, S. H., G. Rodríguez and N. D. Santos (eds.). San Diego State USA: San Diego University. Pp. 41–51.

Rodríguez, G. 1982a. Les crabes d'eau douce d'Amerique. Famille des Pseudothelphusidae. Collection Faune Tropicale, 22. Paris, France: Editions Office de la Recherche Scientifique et Technique Outre-mer (ORSTOM).

Rodríguez, G. 1982b. Fresh-water shrimps (Crustacea, Decapoda, Natantia) of the Orinoco basin and the Venezuelan Guayana. Journal of Crustacan Biology 2(3): 378–391.

Rodríguez, G. 1992. The Freshwater Crabs of America. Family Trichodactylidae and Supplement to the Family Pseudothelphusidae. Collection Faune Tropicale, 31. Paris, France: Editions Office de la Recherche Scientifique et Technique Outre-mer (ORSTOM).

Rodríguez, G., and H. Suárez. 1994. *Fredius stenolobus*, a new species of freshwater crab (Crustacea: Decapoda: Pseudothelphusidae) from the Venezuelan Guayana. Proceedings of the Biological Society of Washington 107: 132–136.

Chapter 6

Inventory, Relative Abundance and Importance of Fishes in the Caura River Basin, Bolívar State, Venezuela

Antonio Machado-Allison, Barry Chernoff, Francisco Provenzano, Philip W. Willink, Alberto Marcano, Paulo Petry, Brian Sidlauskas and Tracy Jones

ABSTRACT

Over 21 days (Nov.–Dec. 2000) we surveyed 65 field stations between Raudal Cejiato and Kakada River in the south to Raudal Cinco Mil in the north of the Caura River Basin. A total of 278 species were identified. The order Characiformes with 158 species (56.8%) was the most diverse, followed by Siluriformes (74, 26.6%), Perciformes (27, 9.7%), Gymnotiformes (9, 3.2%), Clupeiformes (3, 1.1%), Cyprinidontiformes (2, 0.7%), Rajiformes (2, 0.7%), Beloniformes (1, 0.4%), Pleuronectiformes (1, 0.4%) and Synbranchiformes (1, 0.4%). The Family Characidae with 113 species (40.7% of the total) was the most specious. The results add 54 species for the order Characiformes and 39 species of Characidae to previous lists from the basin. For catfishes and electric fishes (Siluriformes, Gymnotiformes) we increased the known species to 72 from 49. While for Perciformes, 27 species are now confirmed, increased from the previous 12. Overall 110 new records were collected for the basin. In the Upper Caura, Salto Pará, Raudal Cejiato and Kakada River were the richest and most diverse with several typical Guayanese forms. In the Lower Caura, El Playón, Tawadu-Nichare Rivers and Raudal Cinco Mil, including the Takoto River, possessed high diversity and richness, with numerous species typical of the Llanos and Orinoco River. However, both areas possess their own characteristics in terms of taxonomy, biogeography and conservation importance. A large number of species have potential economic importance as food and ornamental fishes. Only a few species are captured and used by indigenous human populations.

Based upon species richness, diversity and relative abundance of the fishes and the dependence of local human populations upon stocks of fishes, certain areas in the Upper and Lower Caura must be protected and preserved as part of a general conservation plan. These areas include Raudal Cejiato, Kakada River, the backwaters and flooded lagoons close to Entreríos, the Raudal Suajiditu and the region just above Salto Pará, El Playón and surrounding areas, the Nichare and Tawadu Rivers, the flooded lagoons close to Boca de Nichare and the Takoto River near Raudal Cinco Mil.

INTRODUCTION

Venezuela, like other tropical American countries, possesses extensive pristine areas covering several hundred thousand square kilometers. These areas, shared by several countries (Brazil, Bolivia, Colombia, Ecuador, Guyana, Perú and Venezuela), have diverse and rich concentrations of wildlife and biomass. Today, the region is threatened by development and rapid exploitation of its immense reservoir of prime food, mineral, scenic and energy resources (SISGRIL 1990; Bucher et al. 1993; Machado-Allison 1994, 1999; Chernoff et al. 1996, 1999; Machado-Allison et al. 1999a, b).

The Guayana Region (Bolívar State) in Venezuela is not immune to these processes and risks. The exploitation of minerals such as gold, diamonds and bauxite, together with the development of one of the most extensive hydroelectric systems in the world, has resulted in:

1) biodegradation and destruction of extensive forest areas in the Caroní and Cuyuní Basins; 2) pollution of water, wildlife and humans by mercurial residues; 3) increased sedimentation; and 4) decrease in water quality and reduction of water volume in numerous Venezuelan rivers (Machado-Allison 1994). The Caura River Basin has recently been considered for diversion of waters to the Caroní Basin, in order to compensate for the waterpower that the hydroelectric dams lost due to anthropogenic activities in the headwaters of this river.

Knowledge of freshwater fishes of Venezuela, particularly of the Guayana Shield, has been enhanced due to the recent attention to biodiversity and to numerous collecting expeditions. For example, Mago-Leccia (1970) reported a total of 500 freshwater species for Venezuela only three decades ago. Recently, Taphorn et al. (1997) increased that number to 1,065 species, not including the 119 marine species that occasionally enter rivers. However, there have been projections indicating that this number could easily be increased and reach as high as 1,200 freshwater species (Chernoff and Machado-Allison 1990; Machado-Allison et al. 1993b). Furthermore, Mago-Leccia (1978), Chernoff et al. (1991) and Machado-Allison (1993) have suggested that only 30% of Venezuelan freshwater species have been precisely identified, indicating that many more species could be endemic to the Orinoco or Caura Basins, especially in areas such as the Venezuelan Guayana and Amazonas. Recent publications include discussions on the diversity of species or biogeographical aspects of the ichthyofauna of the Caura River Basin (Chernoff et al. 1991; Buckup 1993; Balbas and Taphorn 1996; Lasso and Provenzano 1997; Bonilla-Rivero et al. 1999; Machado-Allison et al. 1999b).

The main objective of this chapter is to present an inventory of the fish species of the Caura River collected during the expedition and to provide observations on the diversity, relative abundance, economic importance and biogeographic relationships of the fishes. This knowledge is essential for making decisions that could affect the aquatic ecosystems.

MATERIALS AND METHODS

Study Area

The Caura River Basin is located in the central region of the Venezuelan Guayana Shield (3°37'–7°47'N and 63°23' and 65°35'W). Its borders are: to the northeast the Aro River; to the east and southeast the Paragua River Basin; to the south and southwest the Uraricoaera and Avaris Basins in Brazil; to the southwest and west the Ventuary River; and to the north and northwest the Orinoco and Cuchivero Rivers, respectively. The main tributaries are the Sipao, Nichare, Erebato and Merewari Rivers to the west, and the Tigrera, Pablo, Yuruani, Chanaro and Waña Rivers to the east. The approximate surface area is 45.336 km^2 (20% of the total area of Bolívar State and 5% of the country), which places it third in size behind only the Apure and Caroní Basins (Peña and Huber 1996). Physiographic characteristics, hydrology, climate and other information are described in the Introduction to this volume (Machado-Allison et al. 2003). (See also Peña and Huber 1996; Rosales and Huber 1996; and Huber and Rosales 1997.)

Ichthyological Stations

Sixty-five fish collections were made in different macrohabitats from the Kakada River (Erebato) and Cejiato Rapids (Caura) in the south, to Raudal Cinco Mil in the north, including the Nichare and Tawadu Rivers (Figure 6.1, Appendix 7). The criteria for choosing the stations generally followed suggested AquaRAP protocols (Chernoff and Willink 2000). The region was arbitrarily divided into two

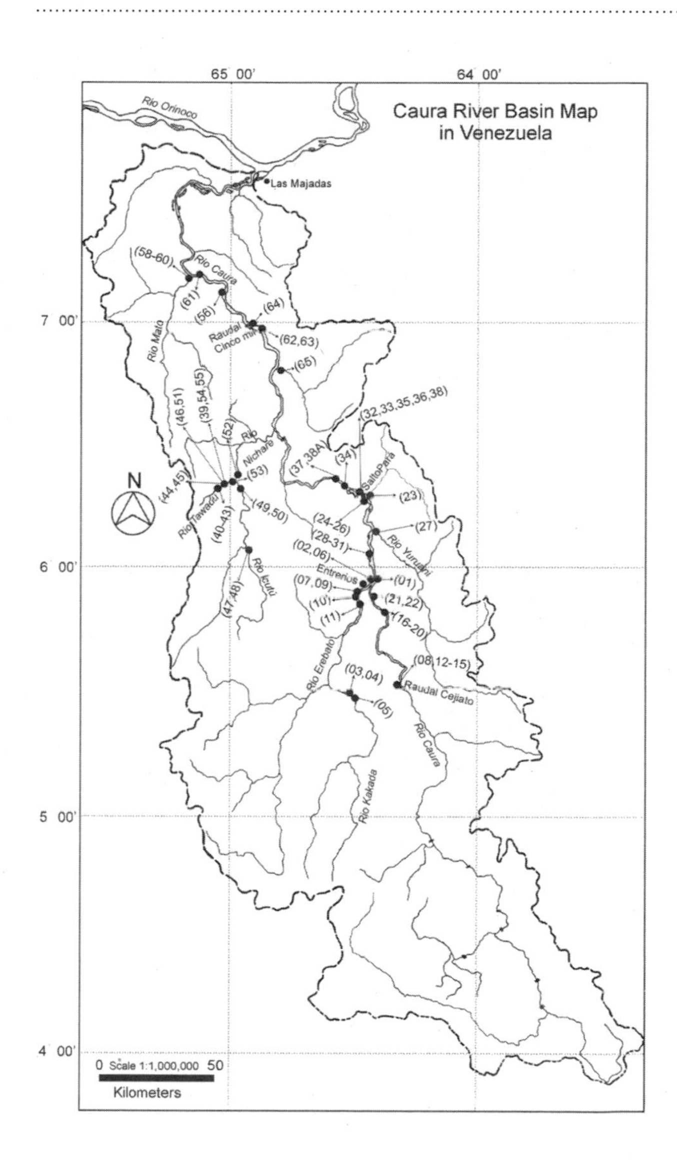

Figure 6.1. Map of the Caura River Basin, Bolívar State, Venezuela. The dark circles show ichthyological collecting stations. The numbers are the ICT collection numbers given in Appendix 7, along with georeference numbers and exact coordinates.

subregions identified as Upper Caura (AC) and Lower Caura (BC). The waterfall, Salto Pará, was the border between the regions. The Upper Caura was further subdivided into 14 ecoregions or georeference points (AC01-AC14) and the Lower Caura was divided in 17 ecoregions or georeference points (BC01-BC17). These ecoregions are described in Appendix 1.

The Collecting Stations corresponding to fish collections were identified as (ICT-xxx). We made 31 collections in the Upper Caura (ICT-01 to ICT-31) and 34 collections in the Lower Caura (ICT-32 to ICT-65). Coordinates were obtained using a calibrated GPS. For each station we recorded ecological variables and a description of the habitat that included physionomic and floristic description, soils, shores, bottom types, habitats, water classification, velocity, pH, transparency, conductivity, color and temperature (Appendices 1, 3, 4, 5).

Fishes were collected with beach seines (several mesh and sizes), experimental gillnets, cast nets, minnow traps, 3 m trawl and hand nets. Specimens were fixed in buffered formalin solution (10%) and later stored in ethyl alcohol (70%). Fishes were shipped to the Field Museum in Chicago, where they were sorted and identified. Afterwards, samples were divided and distributed to the Museo de Biología (MBUCV) in Caracas, Field Museum (Chicago, USA), Museu de Zoologia, Universidade de São Paulo (São Paulo, Brazil), and Museo de Historia Natural, Universidad Nacional Mayor de San Marcos (Lima, Perú). Some larger specimens were skeletonized for anatomical studies and are stored in the MBUCV.

Identifications were made in a careful but relatively rapid fashion. General works such as Eigenmann (1918–1928), Eigenmann and Myers (1929) and Gery (1977) were used, but preference was given to revisionary articles such as Nijsen and Isbruker (1980), Vari (1992,1995), Buckup (1993), Machado-Allison et al. (1993a), Mago-Leccia (1994), Machado-Allison and Fink (1996) and Lasso and Machado-Allison (2000) and original descriptions such as Lasso and Provenzano (1997), Bonilla-Rivero et al. (1999), Chernoff and Machado-Allison (1999) and many others. Species were identified to the lowest level possible (usually genus or species). However, in some cases identification was impossible. We chose a conservative approach and did not include all of the taxa that we collected, and eliminated from our analysis those taxa whose identifications were ambiguous or unknown on our list. Some species, such as *Ancistrus* sp. A, possess enough characters to separate them from congeners previously described and reported in the literature. These forms are potentially new species.

RESULTS AND DISCUSSION

Diversity and distribution: general

A total of 19,266 specimens belonging to 278 species were captured and identified from the expedition. The previous works of Balbas and Taphorn (1996) and Machado-Allison et al. (1999a) reported 130 and 191 species, respectively, and included areas near the Orinoco River (Lower Caura) that were not sampled in this study. Our results increase the number of previously known species by more than 30%. This demonstrates the importance of our expedition because the study area only represents a modest subsection of the whole basin. The numbers of species can be put into a larger context by comparing them with inventories published from nearby basins (Table 6.1).

As was established in other studies of river systems (e.g., Río Tahuamanu, Bolivia, Chernoff et al. 1999), it is difficult to state with any certainty the degree of endemism in the Caura River system. Our lack of knowledge on the actual distributions of species in the majority of the aquatic systems in South America, and in particular those of the Guayana

Table 6.1. Number of species recorded for several rivers and lagoons in the Guayana Shield and Orinoco Basin in Venezuela.

River/System	Basin	# Species	Reference
Atabapo	Orinoco	169	Royero et al. 1986
Caroni	Orinoco	120	Lasso 1989; Lasso et al. 1991
Caroni (Rio Claro)	Orinoco	81	Taphorn and Garcia 1991
Caroni (Lagoons)	Orinoco	54	Rodriguez and Lewis 1990
Caura	Orinoco	130(450?)	Balbas and Taphorn 1996
Caura	Orinoco	191	Machado-Allison et al. 1999b
Cuyuni	Essequibo	136	Machado-Allison et al. 2000
Suapure	Orinoco	140	Lasso 1992
Llanos	Apure	226	Machado-Allison et al. 1993b
Caura	Orinoco	278	This study

Shield, make this task difficult. However, it appears that the Caura is a system with high richness and variety of species, an exceptional number of new records and the high possibility of discovering new species. Using a conservative approach, we documented 110 species not previously recorded for the Caura (Appendix 8). Furthermore, we are convinced that at least 10 new species are present in the genera *Apareiodon, Aphyocharax, Astyanax, Bryconops, Harttia, Imparfinis, Moenkhausia* and *Paravandellia,* of which several are in the process of being described (Chernoff et al. 2003).

We have no doubt that a more detailed study would yield more knowledge of the fish fauna in the Caura. We base this conclusion on the species accumulation curve (Figure 6.2). After 21 days of sampling the accumulation rate of new species for the expedition did not decrease or even reach an asymptote. The slope actually increases in the lower parts of the river where the complexity of habitats and diversity is higher. As an example, we collected 60 species in the last collection in the Takoto River, from which at least 10 additional species were added to the list of the expedition.

Biological and Economic Importance

The fishes captured during this expedition (Appendix 8) belong to a large variety of trophic and functional ecological groups, for example: predators—*Acestrorhynchus, Hoplias, Serrasalmus, Hydrolicus, Crenicichla*; herbivores—*Piaractus, Leporinus*; detritivores—*Bunocephalus, Cyphocharax, Curimata, Prochilodus, Semaprochilodus*; planktivores—*Anchoviella*; parasites—*Exodon, Paravandellia, Vandellia*; and insectivores—*Astyanax, Jupiaba, Moenkhausia, Hyphessobrycon, Hemigrammus and Knodus.* This last group depends almost entirely upon an allotochtonus resource from the forest. There were also species that were found living in patches of macrophytic vegetation attached to large rocks and boulders in rapids. The fishes mostly belonged to rheophilic groups, such as *Anostomus, Ancistrus, Characidium, Leporinus, Melanocharacidium* and *Parodon,* though there were some important exceptions that are mentioned below.

Worth noting is that the species in the Caura River contain many migratory species, such as: *Piaractus, Prochilodus, Curimata and Semaprochilodus*. While species of this group are broadly distributed throughout the Orinoco River Basin, they ascend the Caura River and its tributaries.

We also collected many commercially important species such as: *Piaractus, Prochilodus, Ageneiosus, Hoplias, Hydrolicus, Myleus, Plagioscion* and ornamentals: *Apistogramma, Geophagus, Guianacara, Hemigrammus, Jupiaba, Moenkhausia, Pyrrhulina, Ramirezella, Rivulus, Satanoperca* and *Xenagoniates*. Finally, others are of taxonomic interest, particularly members of the genera: *Ancistrus, Astyanax, Apareiodon, Aphyocharax, Bryconops, Doras, Harttia, Hypostomus, Moenkhausia, Pimelodella* and *Satanoperca*, which have potentially new species and will help document biogeographic relationships among adjacent basins.

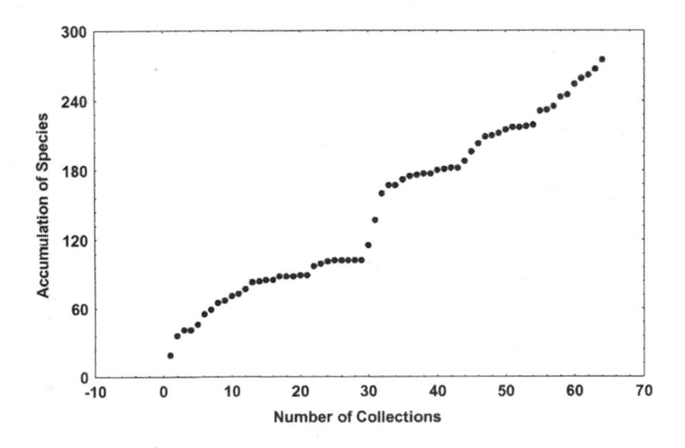

Figure 6.2. Species accumulation curve for fishes collected in the Caura River, Bolivar state, Venezuela during AquaRAP Expedition November–December 2000.

Species richness

An analysis of the ichthyofauna shows that the 278 species represent 10 orders and 36 families. We found the highest richness in the order Characiformes with 158 species (56.8% of the total), followed by Siluriformes (74, 26.6%), Perciformes (27, 9.7%), Gymnotiformes (9, 3.2%), Clupeiformes (3, 1.1%), Cyprinodontiformes (2, 0.7%), Rajiformes (2, 0.7%), Beloniformes (1, 0.4%), Pleuronectiformes (1, 0.4%) and Synbranchiformes (1, 0.4%). The family Characidae was the best represented in this study, with 113 species or approximately 41% of the total. Our results increased by 54 species within Characiformes and 39 species within Characidae the number of previously known taxa for the basin. With regards to other groups, 74 species of Siluriformes were identified compared to 49 species previously; and for Perciformes 27 vs. 12 previously reported.

COMPARISON BETWEEN THE REGIONS (UPPER AND LOWER)

As established previously, the study area was divided into two regions, using Salto Pará as the border between them. In the Upper region we captured a total of 4,659 (24.1%) specimens, representing 103 (37%) species, while in the Lower we collected 14,607 (75.9%) individuals, representing 226 (81.3%) species. There were 52 species (18.3%) that were only found in the Upper region while 175 species (62.6%) were only found in the Lower region. A total of 51 (18.7%) species were shared between the two areas.

There are several factors that could explain the differences in species richness and relative abundance between these two areas. Possibly the most important is the Salto Pará that physically forms a natural barrier for the dispersal of fishes as species entering the Caura River from the Orinoco River cannot ascend Salto Pará. Other factors such as habitat heterogeneity, food availability, stream gradient and the degree of oligotrophy of the environments could also be factors promoting differences. The waters above Salto Pará are generally transparent and acidic with low conductivity (Garcia 1996). Only in those environments in which there was accumulation of organic material such as detritus or leaf litter (e.g., beaches at Raudal Cejiato) did we find a moderate abundance and high diversity. Similar conditions in the Lower region gave quite similar results (Garcia 1996), permitting us to conclude that ecology plays an important role in the distribution and abundance of the species today (see below).

Table 6.2. Summary of the Field Stations, number of species, number of specimens and habitats (h) sampled during the AquaRAP Expedition (Nov–Dec 2000). Bw Backwater; C Caño; R Rapids; Pa Sandy Beach; Pr Rocky Beach; Pm Muddy Beach; L Lagoon; Ia Sandy Island; Ir Rocky Island; T tributary (stream).

UPPER CAURA				LOWER CAURA			
Stations	Species	Specimens	h	Stations	Species	Specimens	h
ICT-01	18	220	C	ICT-32/33	37	519	Pa
ICT-02	25	139	R	ICT-34	46	720	C
ICT-03	12	151	Pa	ICT-35	31	344	Ir
ICT-04	9	31	R	ICT-36	13	238	Pr
ICT-05	22	447	C	ICT-37	24	504	C
ICT-06	13	41	C	ICT-38	10	107	Ir
ICT-07	16	493	R	ICT-38a	2	10	C
ICT-08	20	220	Pa	ICT-39	14	164	Pr
ICT-09	15	38	C	ICT-40	4	90	R
ICT-10	29	392	C	ICT-41	15	246	R
ICT-11	18	113	C	ICT-42	5	112	C
ICT-12	28	236	Pa	ICT-43	2	2	R
ICT-13	25	159	T	ICT-44	8	284	Ir,R
ICT-14	9	20	T	ICT-45	15	58	C,R
ICT-15	7	28	R,Pr	ICT-46	12	84	L
ICT-16	9	178	Pa	ICT-47	35	563	Pa
ICT-17	12	50	R	ICT-48	41	693	C
ICT-18	11	14	Pr	ICT-49	33	848	Pa
ICT-19	9	52	Pa	ICT-50	22	326	Bw
ICT-20	16	201	Bw	ICT-51	50	1977	L
ICT-21	--	--		ICT-52	44	740	Ir
ICT-22	5	15	C	ICT-53	29	1100	Pm
ICT-23	32	429	R	ICT-54	22	304	Pm
ICT-24	--	--		ICT-55	9	39	C
ICT-25	30	359	Pr	ICT-56	36	811	L
ICT-26	13	222	Bw	ICT-58	6	15	Pm
ICT-27	14	62	Bw	ICT-59	46	590	Pm
ICT-28	16	174	P	ICT-60	21	234	Pm
ICT-29	16	61	R	ICT-61	20	201	Pa
ICT-30	9	20	T	ICT-62	43	1288	Ia
ICT-31	6	13	R	ICT-63	36	433	C
				ICT-64	25	281	R,Pr
				ICT-65	62	780	R,Pa

Table 6.2 (Appendix 7) is a summary of the total number of species captured at each field station and their relative abundance. The stations that contributed most to richness and abundance in the Upper Caura were ICT-23, ICT-25, ICT-10, ICT-12, ICT-02, ICT-13, ICT-05 and ICT-07. The first two correspond to the area above Salto Pará that is characterized by a mixture of rocky/sandy beaches with abundant aquatic plants (Podostemonaceae) and rapids. The stations ICT-12 and ICT-13 correspond to beaches and the mouth of a tributary in Raudal Cejiato. ICT-02 is a backwater close to Entreríos and ICT-10 is a backwater in the Erebato River. Finally, the stations ICT-03 (beach) in the Kakada River and ICT-05, Caño Suajiditu in the same river, are important. It should also be noted that in backwaters and on beaches of Raudal Cejiato and Suajiditu there were abundant juveniles of several species, which suggests that the areas act as nurseries for fishes in the reproductive season. In summary, the richness and abundance of the Upper section was highest in three major areas: i) the area near Salto Pará; ii) the area near Raudal Cejiato; and iii) the area near Kakada River, especially beaches and Caño Suajiditu.

The stations that contributed the most in the Lower Caura in terms of diversity and relative abundance were ICT-65, ICT-51, ICT-59, ICT-34, ICT-52, ICT-62, ICT-48, ICT-32-33, ICT-63, ICT-47 and ICT-35. The richest station was that in the Takoto River (ICT-65), an exceptionally heterogeneous area with rapids, pools, sandy beaches and backwaters. This area of the Takoto River deserves special attention for future conservation. Habitats in the vicinity of Raudal Cinco Mil (ICT-62 and 63) also had high richness and moderate abundance. At these stations we collected a number of species that we found nowhere else in the basin. The stations ICT-32, 33, 34 and 35 were from the region immediately below Salto Pará, including El Playón and a caño. Finally, ICT-47, 48, 51 and 52 correspond to diverse zones in the Tawadu and Nichare Rivers including beaches and lagoons. All these areas must be included in future conservation plans.

Special attention needs to be given to the lagoon (ICT-51) in the Nichare River Basin due to its importance to the local people. The lagoon is commonly used by fishermen and the presence of bigger fishes leads us to recommend its protection as a fishery reserve.

Biogeographic aspects

Few studies have presented information on the distribution and relationships of the freshwater fishes of Venezuela. Mago-Leccia (1970, 1978) was the first to construct several biogeographic zones, in which he places the Caura River Basin within the Guayana Shield Province. Later, Chernoff et al. (1991), Lasso et al. (1991) and Lasso and Provenzano (1997) among others constructed biogeographic hypotheses regarding the Orinoco-Amazonas and Guayanese faunas. However, further understanding of the biogeography must await detailed phylogenetic studies on the major groups of fishes of the region.

The data obtained in this study indicate that the majority of species located in the Upper Caura are apparently elements of Guayanese faunas present in adjacent basins such as the Caroní and Cuyuní in Venezuela and Guyana. Species such as *Bryconops* cf. *colaroja*, *Harttia* sp., *Chaetostoma vasquezi, Lebiasina uruyensis, Aequidens chimantanus* and *Guianacara geayi* support this hypothesis.

In the Lower Caura, we recognize the components of two distinct biogeographic regions: i) the Lower Orinoco River and Llanos; and ii) the Upper Orinoco and Río Negro-Casiquiare (Amazon). Of the Lower Orinoco–Llanos faunas are: *Acestrorhynchus falcatus, A. microlepis, Anchoviella jamesi, Aphyocharax alburnus, Boulengerella lucia, Brycon bicolor, Bujurquina mariae, Eingenmannia virescens, Gymnotus carapo, Hypostomus plecostomus, Microphilipnus ternetzi, Orinocodoras eigenmanni, Ochmacanthus alternus, O. orinoco, Pellona castelneana, Potamorhina altamazonica, Potamotrygon dorbigny, Prochilodus mariae, Pygocentrus cariba, Semaprochilodus laticeps, Raphiodon vulpinnus, Sorubim lima, Sternopygus macrurus* and *Triportheus albus.* Of the Upper Orinoco–Amazon faunas are: *Micoschemobrycon casiquiare, Serrasalmus* sp., *Anostomus anostomus, A. ternetzi, Bryconops alburnoides, Cynodon gibbus, Hydrolicus armatus, Crenicichla lenticulata, Exodon paradoxus* and *Hyphessobrycon serape* among others.

Economic and Social Importance

Goulding (1979, 1980), Goulding et al. (1988) and Peres and Terborgh (1995) document not only the importance of rivers structuring human settlements throughout Amazonia, but also the increasing dependence of humans on aquatic resources for sustenance. The Ye'kuana and the Sanema peoples live in the basin and many of their domestic and commercial activities, as well as their myths and religious beliefs, are intimately associated with the river, its flora and its fauna. The conservation and sustainable use of the basin's resources is very important for the culture and social traditions of these people.

Many of the species found in the Caura River possess high value as food or as part of an ornamental fisheries. Little attention has been placed on the Upper Caura due principally to natural barriers such as Salto Pará but due also to the expressed wishes of the Ye'kuana and Sanema people.

In the Lower Caura and its confluence with the Orinoco River and flooded lagoons, there is an intense commercial fisheries. Some of the most important species in this fishery are: "cachamas" (*Colossoma macropomum*), "morocotos" (*Piaractus brachypomus*), "sapoaras" (*Semaprochilodus laticeps*), "coporos" (*Prochilodus mariae*), "palometas" (*Mylossoma duriventre*), "caribe colorado" (*Pygocentrus cariba*), "valentones," "dorados" and "laolaos" (*Brachyplatystoma filamentosus, B. rosseauxi* and *B. vaillanti*), "rayaos" (*Pseudoplatystoma fasciatum* and *P. tigrinum*), "cajaro" (*Phractocephalus hemiliopterus*), "curbinata" (*Plagioscion squamossissimus*) (Novoa and Rámos 1978; Novoa 1982). Despite the importance of the fishery, there are no reliable data on size of the catch, prices, and the level of commercial activity. Moreover, critical biological information from which to develop sustainable fisheries is lacking, such as growth, migrations, reproductive cycles, minimal age of spawning, fecundity, ecology, etc.

The fishery pressure in the Upper Caura is distributed over many kinds of species with the intent of obtaining animal protein for the indigenous local populations. Although to a lesser degree than in the lower areas, we are aware that extensive use of rotenone or "barbasco" in caños can cause considerable damage to the natural resources. This activity has been denounced at several meetings and workshops with the community and the indigenous groups are aware of the potential environmental damage (ACOANA, pers. comm.). On the other hand, hook and line fishing is very common, targeting mainly the capture of big species such as "aimaras" and "guavinas" (*Hoplias macrophthalmus* and *H. malabaricus*), "pacus" (*Myleus rubripinnis, M. asterias* and *M. torquatus*), "bagres" (*Ageneiousus* sp.), "guitarrillos" (*Pseudodoras* sp. and *Doras* sp.), "caribes" (*Serrasalmus rhombeus*) and "payaras" (*Hydrolicus armatus* and *H. tatauaia*). There are no data on the utilization of nets in the Upper Caura.

We have obtained information through the internet that species such as "pavones" (*Cichla orinocensis, C. temensis* and *C. monoculus*), "sardinatas" (*Pellona castelneana*) and "payaras" (*Hydrolicus armatus* and *H. tatauaia*) are advertized in special tourist brochures or websites promoting sports fishing and tourism in the Lower Caura. This requires careful study because no data are available on the quantity of sport-fishes being caught. However, several of the advertisements describe daily catch values which could prove harmful to fish populations if true.

Finally, it is important to note the potential for the development of an ornamental fishery (i.e., one that provides ornamental species for the pet trade) or fish culture in the Caura River. Numerous species captured in the Upper, as well as the Lower, portions of the basin possess high international commercial value. Species such as *Ammocryptocharax elegans, Ancistrus* sp., *Anostomus anostomus, A. ternetzi, Aphyocharax alburnus, A. erythrurus, Apistogramma iniridae, Brachychalcinus opercularis, Bryconops giacopini, Bujurquina mariae, Carnegiella strigatta, Caenotropus labyrinthicus, Chaetostoma vasquezi, Chalceus microlepidotus, Corydoras blochii, C. boehlkei, C.bondi, C. osteocarus, Eigenmannia virescens, Exodon paradoxus, Farlowella vittata, Guianacara geayi, Hyphessobrycon bentosi, H. serpae, Jupiaba zonata, Leporinus arcus, L. brunneus, L. grandti, L. maculatus, Melanocharacidium dispiloma, M. nigrum, Moenkhausia collettii, M. copei, M. lepidura, M. oligolepis, Myleus rubripinnis, M. asterias, Nannostomus erythrurus, Ramirezella newboldi, Poptella longipinnis, Potamotrygon schoederi, Pygocentrus cariba, Pyrrhulina brevis, Rineloricaria fallax, Tatia galaxias, T. romani* and *Xenagoniates bondi* can be found in pet or aquarium stores. Scientifically regulated, this activity could provide an ecologically sound means for increasing incomes of the local indigenous populations.

In summary, data on the fisheries of this area (Lower and Upper) are either lacking or not currently reliable. This makes the establishment of programs that will permit sustainable use of the natural resources very difficult. Our discussion presented in this paper is intended to stimulate research into and the gathering of reliable information on the exploited aquatic resources of the Caura River Basin in order to guarantee the protection of biodiversity through sustainable use.

Critical habitats

We identified a number of critical habitats where protection is required for the continual survival of freshwater fishes and maintenance of the spectacular biodiversity. These are the same habitats that support the reproduction and growth of the species which are economically valuable either as food or as ornamentals. The habitats are described in detail in other chapters of this volume. They fall into three groups: i) flooded areas, including forest, savannahs and lagoons; ii) rapids; and iii) caños and beaches associated with backwaters.

The flooded forests, including associated lagoons and flooded savannas, comprise the most critical and highly endangered areas, especially in the Lower Caura, as has been demonstrated in other studies (Welcomme 1979; Goulding 1980; Lowe-McConnell 1987; Chernoff et al. 1991; Machado-Allison 1993, 1994). The flooded areas provide nursery grounds (permanent or temporary) for perhaps more than 60% of the species present. Moreover, many species such as "cachamas," "morocotos," "palometas," "pacus" and "palambras" (*Colossoma, Mylossoma, Myleus, Piaractus* and *Brycon*) feed on fruits and seeds dropped by plants into the water or when flooding allows fishes to exploit the forest floor (Goulding 1980; Machado-Allison 1982, 1993). The phenology of the forest and its fauna (including fishes) is intimately tied to the flooding cycle. Maintenance of this coevolved association among the many elements inhabiting the flooded forests is critical for the survival of the entire ecosystem, including the community of fishes.

As has been noted for other river basins such as the Río Madeira, Brazil (Goulding 1979) and the Tahuamanu-Manuripi, Bolivia (Chernoff et al. 1999), the Caura and Erebato Rivers possess a relatively narrow floodplain. This is especially true in the Upper areas. The narrow floodplain has little area to serve as a buffer between logging and ranching activities and the critical flooded zones. Near the mouth of the Caura River the flooded forest is wider. However, logging and other human activities are more intense, promoting domestic, industrial and agricultural pollution.

Another important habitat within the basin is the rapids in the areas above Salto Pará and in the Nichare and Tawadu Rivers. These rapids contain extensive coverage by aquatic plants of the Family Podostemonaceae. The fish communities that live among the rocks and these macrophytes are unique to the Caura Basin as far as we can determine. The fishes intimately associated with these ecosystems use the macrophytes for protection, for food and as a nursery. The degree of coverage by the macrophytes may provide shelter from the currents because some species, such as *Crenicichla* or *Jupiaba*, are quiet water species, not usually found in the fast currents of rapids. Reduction of the natural water levels in these habitats will drastically damage these unique communities.

Lastly, it is worth noting a few specialized habitats that are important for fishes. The Upper Caura and Erebato Rivers are characterized by having low nutrient levels, low conductivities, and low dissolved or suspended organic particulates in the water column. Thus, the higher accumulations of fish species and biomass occur near deposits of organic debris and detritus, usually in protected areas such as backwaters, oxbows, lagoons or protected beaches. This situation was noted for Raudal Cejiato, Tawadu and Suajiditu. These areas serve as reproductive, nursery and shelter areas. As with the rapids, the functionality of these zones depends on maintenance of the natural hydrologic cycle.

In the Lower Caura, this phenomenon occurs in areas near El Playon and in the Nichare and Tawadu Rivers. These regions have beaches in backwaters or lagoon habitats where accumulated organic material, mostly leaf detritus, supports rich fish communities. Many of these communities are fished by local people for food.

Conservation Implications

Our data and that gathered from other aquatic ecosystems allows us to conclude that conservation of aquatic biodiversity is one of the most important and difficult challenges facing humans today (Chernoff et al. 1996; Machado-Allison et al. 1999a). Our relatively poor knowledge of the use of aquatic ecosystems, including the use of water, by humans for development and exploitation is one of the principal obstacles to effective management and conservation (IUCN 1993; Gleick 1998).

Aquatic ecosystems are poorly known in comparison to others such as rain forests. We lack knowledge about the taxonomy of aquatic species, their phylogenetic and biogeographic relationships (Böhlke et al. 1978; Mago-Leccia 1978; Chernoff et al. 1991; Mago-Leccia 1994), and their ecologies and life histories (Goulding 1979; Lundberg et al. 1979; Winemiller 1989; Machado-Allison 1992, 1993; Menezes and Vazzoler 1992). Almost nothing is known about interactions among organisms and with their surrounding physical environment. What is known is limited to a small sample of habitats or climatic circumstances (Lowe-McConnell 1964, 1969, 1987; Goulding 1980; Machado-Allison 1993).

This study, based on diversity and relative abundance of fishes and the relationship of fishes with local populations, demonstrates that certain areas in the Upper and Lower Caura must be protected and preserved as part of a general conservation plan. These areas include Raudal Cejiato, Kakada River, the backwaters and flooded lagoons close to Entreríos, the Raudal Suajiditu and the region just above Salto Pará, El Playón and surrounding areas, the Nichare and Tawadu Rivers, the flooded lagoons close to Boca de Nichare and the Takoto River near Raudal Cinco Mil.

We stress that any development project that will change the seasonal hydrologic or climatic regime will adversely affect the natural environment. The hardest hit would be the shallow and fast water habitats such as rapids and riffles where an important and special microhabitat is currently structured by the presence of species of aquatic plants of the family Podostemonaceae.

Prior to making decisions that will irreversibly affect aquatic ecosystems (e.g., diverting, channelization or damming) such as those proposed for the Caura and adjacent basins, the development of Hidrovia (Bucher et al. 1993) that proposes to connect the three major South American basins, or the regional national plan called "Eje Orinoco-Apure," it is necessary to have reliable ecological, economic and social information (SISGRIL 1990; Bucher et al. 1993; Machado-Allison 1994; Aguilera and Silva 1997; Machado-Allison 1999). It is imperative to have an appreciation of the ecological complexities of aquatic ecosystems and life histories of aquatic organisms both to be aware of the damage that development could cause and to be able to study alternative plans.

The information that we have presented about fish diversity and aquatic ecosystems within the Caura River Basin, in terms of economic potential and importance for human populations, argues strongly for the development of immediate conservation plans. Conservation efforts in the Caura River Basin can provide a leadership role towards conservation of freshwater ecosystems in Venezuela. Large-scale conservation efforts are necessary immediately from the perspectives of society and of science. The aquatic ecosystems of Venezuela are critical natural resources that sustain both human and animal populations. They help provide food security and clean water, while also preserving wildlife. The high quality of human life in Venezuela is integrally tied to the maintenance and preservation of these tropical environments.

CONCLUSIONS AND CONSERVATION RECOMMENDATIONS

The region studied in the Caura River Basin, Bolívar State, Venezuela is a potential hotspot for the biodiversity of freshwater fishes. We discovered and identified 278 species in the region. This represents close to 28% of the total freshwater species known for Venezuela.

The known diversity in the basin will increase with increased sampling. Numerous areas were not sampled such as higher altitude areas of the Caura and Erebato Rivers. Nor was the Lower Caura close to the Orinoco sampled. The number of species was still increasing at a considerable rate at the end of the expedition.

The ichthyofauna possesses an interesting assemblage of species with biogeographic associations with Guayanese faunas in the Upper Caura, and elements from the Orinocan and Amazonian faunas in the Lower Caura.

Many species are economically valuable as food or ornamental fishes. Fishes are currently used for subsistence, mainly in Upper areas. The development of a fisheries based on ornamental species could be an economic alternative for the local indigenous populations. Also, we recommend the development of fisheries for local consumption and the careful regulation of catches for exportation to other areas of the country.

Fishery studies in the Lower Caura are necessary and urgent. Ecological data to determine what levels of harvest are sustainable is of prime importance.

Zones of critical habitats such as rapids, flooded forest and lagoons and backwater areas should be protected. Rather than designating a National Park, we suggest an integrated-use plan with some habitats receiving full protection. Protection must be given to the narrow floodplain, as well as to Raudal Cejiato, Kakada River, the backwaters and flooded lagoons close to Entreríos, the Raudal Suajiditu and the region just above Salto Pará, El Playón and surrounding areas, the Nichare and Tawadu Rivers, the flooded lagoons close to Boca de Nichare and the Takoto River near Raudal Cinco Mil.

Programs should be developed to work with local people in order to exchange information to explain the relationships between maintenance of habitats and the biodiversity of fishes. Educational programs should be developed, enabling local residents and fisherman to participate in monitoring fish stocks and habitats. These programs should promote identification of fishes and how to recognize new or unusual species that should be brought to the attention of scientists.

LITERATURE CITED

Aguilera, M., and J. Silva. 1997. Especies y diversidad. Interciencia. 22(6): 289–298.

Balbas, L., and D. Taphorn. 1996. La Fauna: Peces. *In:* Rosales, J. and O. Huber (eds.). Ecología de la Cuenca del Río Caura. Scientia Guaianae 6: 76–79

Böhlke, J., S. Weitzman and N. Menezes. 1978. The status of systematic studies of South American fresh water fishes. Acta Amazonica 8: 657–677.

Bonilla-Rivero, A., A. Machado-Allison, B. Chernoff, C. Silvera, H. López-Rojas and C. Lasso. 1999. *Apareiodon orinocensis,* una nueva especie de pez de agua dulce (Pisces: Characiformes: Parodontidae) proveniente de los ríos Caura y Orinoco, Venezuela. Acta Biol. Venez. 19(1): 1–10.

Bucher, E., A. Bonetto, T. Boyle, P. Canevari, G. Castro, P. Huszar and T. Stone. 1993. Hidrovía. Un examen ambiental inicial de la vía fluvial Paraguay-Paraná. Humedales para las Américas, Publ. 10:1–10.

Buckup, P. 1993. Review of the characidiin fishes (Teleostei: Characiformes), with descriptions of four new genera and ten new species. Ichth. Explor. Freshwaters 4(2): 97–154.

Chernoff, B., and A. Machado-Allison. 1990. Characid fish of the genus *Ceratobranchia* with description of new species from Venezuela and Peru. Proc. Acad. Nat. Sci. Phil. 142:261–290.

Chernoff, B., and A. Machado-Allison. 1999. *Bryconops colaroja* and *B. colanegra,* two new species from the Cuyuni and Caroní drainages of South America. Ichth. Explor. Freshwaters. 10(4): 355–370.

Chernoff, B., and P. W. Willink. 2000. AquaRAP sampling protocol. *In:* Willink, P. W., B. Chernoff, L. E. Alonso, J. R. Montambault and R. Lourival (eds.). A Biological Assessment of the Aquatic Ecosystems of the Pantanal, Mato Grosso do Sul, Brasil. RAP Bull. Biol. Asessment 18. Washington, DC: Conservation International. Pp. 241–242.

Chernoff, B., A. Machado-Allison and W. Saul. 1991. Morphology variation and biogeography of *Leporinus brunneus* (Pisces: Characiformes: Anostomidae). Ichth. Explor. Freshwaters 1(4): 295–306.

Chernoff, B., A. Machado-Allison and N. Menezes. 1996. La conservación de los ambientes acuáticos: una necesidad impostergable. Acta Biol. Venez. 16 (2): i–iii.

Chernoff, B., P. Willink, J. Sarmiento, S. Barrera, A. Machado-Allison, N. Menezes and H. Ortega. 1999. Fishes of the rios Tahuamanu, Manuripi and Nareuda, Dpto. Pando, Bolivia: Diversity, Distribution, Critical Habitats and Economic Value. *In:* Chernoff, B. and P. Willink (eds.). A Biological Assessment of the Aquatic Ecosystems of the Upper Río Orthon Basin, Pando, Bolivia. Bull. Biological Assessment 15. Washington, DC: Conservation International. Pp. 39–46.

Chernoff, B., A. Machado-Allison, F. Provenzano, P. Willink and P. Petry. 2003. *Bryconops* n. sp., a new species from the Rio Caura Basin of Venezuela (Characiformes, Teleosteii). Ichthyological Explorations of Freshwaters 13(4).

Eigenmann, C. 1918–1928. The American Characidae (I–IV). Mem Mus. Comp. Zool. 43.

Eigenmann, C., and G. Myers. 1929. The American Characidae (V). Mem Mus. Comp. Zool. 43:429–574

García, S. 1996. Limnología. *In:* Rosales, J. and O. Huber (eds.). Ecología de la Cuenca del Río Caura. Scientia Guaianae 6: 54–59.

Gery, J. 1977. The Characoids of the World. Neptune City USA: TFH Publications.

Gleick, P. 1998. The World's Water: The Biennial Report on Freshwater Resources. Washington, DC, USA: Island Press.

Goulding, M. 1979. Ecologia da Pesca do Rio Madeira. Cons. Nac. Des. Cient. e Tec., INPA.

Goulding, M. 1980. The Fishes and the Forest: Explorations in Amazonian Natural History. Berkeley: Univ. Cal. Press.

Goulding, M., M. L. Carvalho and E. G. Ferreira. 1988. Río Negro: rich life in poor water: Amazonian diversity and foodchain ecology as seen through fish communities. The Hague, Netherlands: SPB Academic Publ.

Huber, O., and J. Rosales (eds.). 1997. Ecología de la Cuenca del Río Caura, Venezuela. II Estudios Especiales. Scientia Guaianae 7: 1–473.

IUCN. 1993. The Convention on Biological Diversity: An explanatory guide (Draft). Bonn Germany: IUCN Environmental Law Centre. (mimeo).

Lasso, C. 1989. Los peces de la Gran Sabana, Alto Caroní, Venezuela. *Mem. Soc. Cienc. Nat. La Salle*, 49–50 (131–134): 208–285.

Lasso, C. 1992. Composición y aspectos ecológicos de la ictiofauna del Bajo Suapure, serranía Los Pijiguaos (Escudo de Guayana), Venezuela. Mem. Soc. Cienc. Nat. La Salle 52(138):5–54.

Lasso, C., and F. Provenzano. 1997. *Chaetostoma vazquezi*, una nueva especie de corroncho del Escudo de Guayana, Estado Bolívar, Venezuela (Siluroidei-Loricariidae) descripción y consideraciones biogeográficas. Mem. Soc. Cienc. Nat. La Salle 57(147): 53–65.

Lasso, C., and A. Machado-Allison. 2000. Sinopsis sobre las especies de la Familia Cichlidae en la Cuenca del Orinoco. Caracas Venezuela: Conicit.

Lasso, C., A. Machado-Allison and R. Perez. 1991. Consideraciones zoogeográficas de los peces de la Gran Sabana (Alto Caroní) Venezuela, y sus relaciones con las cuencas vecinas. Memoria, Soc. Cien. Nat. La Salle IL and L: 21.

Lowe-McConnell, R. 1964. The fishes of the Rupununi Savanna District of British Guyana. Pt.1. Grouping of fish species and effects of the seasonal cycles on the fish. Journ. Linn. Soc. (Zool.) 45:103–144.

Lowe-McConnell, R. 1969. Some factors affecting fish populations in amazonian waters. Atas do simposio sobre a Biota Amazonica 7:177–186.

Lowe-McConnell, R. 1987. Ecological Studies in Tropical Fish Communities. New York USA: Cambridge Univ. Press.

Lundberg, J., J. Baskin and F. Mago-Leccia. 1979. A preliminary report on the first cooperative U.S. - Venezuelan ichthyological expedition to the Orinoco River. 14 p. (Mimeo).

Machado-Allison, A. 1982. Estudios sobre la subfamilia Serrasalminae (Teleostei-Characidae). Parte I. Estudio comparado de los juveniles de las "cachamas" de Venezuela (Géneros *Colossoma* y *Piaractus*). Acta Biol. Venez. 11(3): 1–102.

Machado-Allison, A. 1992. Larval ecology of fishes of the Orinoco Basin. In: W. Hamlett (ed.). Reproductive Biology of South American Vertebrates. Springer Verlag. Pp. 45–59.

Machado-Allison, A. 1993. Los Peces del Llano de Venezuela: un ensayo sobre su Historia Natural. (2nda. Edición). Caracas Venezuela: Consejo de Desarrollo Científico y Humanístico (UCV), Imprenta Universitaria.

Machado-Allison, A. 1994. Factors affecting fish communities in the flooded plains of Venezuela. Acta Biol.Venez. 15(2):59–75.

Machado-Allison, A. 1999. Cursos de agua, fronteras y conservación. *In:* Genatios, G. (ed.) Ciclo Fronteras: Desarrollo Sustentable y Fronteras. Caracas, Venezuela: Com. Estudios Interdisciplinarios, UCV. Pp. 61–84.

Machado-Allison, A. and W. Fink. 1996. Los Peces Caribes de Venezuela. Caracas Venezuela: Univ. Central de Venezuela-Conicit.

Machado-Allison, A., B. Chernoff, P. Buckup and R. Royero. 1993a. Las especies del género *Bryconops* Kner, 1859 en Venezuela (Teleostei-Characiformes). Acta Biol. Venez. 14(3):1–20

Machado-Allison, A., F. Mago-Leccia, O. Castillo, R. Royero, C. Marrero, C. Lasso and F. Provenzano. 1993b. Lista de especies de peces reportadas en diferentes cuerpos de agua de los bajos llanos de Venezuela. *In:* Machado-Alllison, A. (ed.) Los Peces del Llano de Venezuela: un ensayo sobre su Historia Natural. (2nda. Edición). Caracas, Venezuela: Consejo de Desarrollo Científico y Humanístico (UCV), Imprenta Universitaria. Pp. 129–143.

Machado-Allison, A., J. Sarmiento, P. W. Willink, B. Chernoff, N. Menezes, H. Ortega, S. Barrera and T. Bert. 1999a. Diversity and abundance of fishes and habitats in the Rio Tahuamanu and Rio Manuripi Basins (Bolivia). Acta Biol.Venez. 19(1): 17–50.

Machado-Allison, A., B. Chernoff, C. Silvera, A. Bonilla, H. Lopez-Rojas, C. A. Lasso, F. Provenzano, C. Marcano and D. Machado-Aranda. 1999b. Inventario de los peces de la cuenca del Río Caura, Estado Bolivar, Venezuela. Acta Biol. Venez. 19 (4):61–72.

Machado-Allison, A., B. Chernoff, R. Royero-León, F. Mago-Leccia, J. Velazquez, C. Lasso, H. López-Rójas, A. Bonilla-Rivero, F. Provenzano and C. Silvera. 2000. Ictiofauna de la cuenca del Río Cuyuní en Venezuela. Interciencia 25(1):13–21.

Machado-Allison, A., B. Chernoff and M. Bevilacqua. 2003. Introduction to the Caura River Basin, Bolívar State, Venezuela. *In:* Chernoff, B., A. Machado-Allison, K. Riseng, and J. R. Montambault (eds.). A Biological Assessment of the Aquatic Ecosystems of the Caura River Basin, Bolívar State, Venezeula. RAP Bulletin of Biological Assessment, No. 28. Washington DC, USA: Conservation International. Pp. 28–33.

Mago-Leccia, F. 1970. Lista de los Peces de Venezuela. Caracas, Venezuela: Minist. Agric. y Cría, Ofic. Nac. Pesca.

Mago-Leccia, F. 1978. Los Peces de agua dulce de Venezuela. Caracas, Venezuela. Cuadernos Lagoven, Caracas.

Mago-Leccia, F. 1994. Electric Fishes of the Continental Waters of America. Biblioteca Acad. Ciec. Fis. Mat. y Nat. XXIX: 206 pp.

Menezes, N., and P. Vazzoler. 1992. Reproductive characteristics of Characiformes. *In:* Hamlett, W. (ed.). Reproductive Biology of South American Vertebrates. Springer-Verlag. Pp. 60–70.

Nijsen, H., and J. Isbruker. 1980. A review of the genus *Corydoras* Lacepede, 1803 (Pisces, Siluriformes, Callichthyidae) Bijdragen tot de Dierkunde 50(1):190–220.

Novoa, D. 1982. Los recursos pesqueros del Río Orinoco y su Explotación.Caracas, Venezuela: Corp. Venez. Guayana (CVG).

Novoa, D., and F. Ramos. 1978. Las Pesquerías Comerciales del Río Orinoco. Caracas, Venezuela. Corp. Venez. Guayana.

Peña, O., and O. Huber. 1996. Características Geográficas Generales. *In:* Rosales, J. and O. Huber (eds.). Ecología de la Cuenca del Río Caura. Scientia Guaianae 6: 4–10.

Peres, C. A., and J. W. Terborgh. 1995. Amazonian nature reserves: an analysis of the defensibility status of existing conservation units and design criteria for the future. Conservation Biology 9:34–45.

Rodríguez, A., and W. Lewis. 1990. Diversity and species composition of fish communities of Orinoco floodplain lakes. Nat. Geograp. Res. 6(3): 319–328.

Rosales, J., and O. Huber (eds.). 1996. Ecología de la Cuenca del Río Caura, Venezuela. I. Caracterización General. Scientia Guaianae 6: 1–131.

Royero, R., A. Machado-Allison, B. Chernoff and D. Machado-Aranda. 1986. Peces del Río Atabapo. Territorio Federal Amazonas, Venezuela. Acta Biol. Venez. 14(1):41–55.

SISGRIL. 1990. Simposio Internacional sobre Grandes Rios Latinoamericanos. Interciencia 15(6): 1–193.

Taphorn and J. Garcia. 1991. El río Claro y sus peces, con consideraciones de los impactos ambientales de las presas sobre la ictiofauna del Bajo Caroní. Biollania, 8: 1–15.

Taphorn, D., R. Royero, A. Machado-Allison and F. Mago-Leccia. 1997. Lista actualizada de los peces de agua dulce de Venezuela. *In*: La marca E (ed.). Vertebrados actuales y fósiles de Venezuela, Vol. 1. Venezuela: Merida. Pp. 55–100.

Vari, R. 1992. Systematics of the Neotropical Characiform Genus *Cyphocharax* Fowler (Pisces:Ostariophysi). Smith.Contr. Zool. 529:1–137.

Vari, R. 1995. The Neotropical Fish Family Ctenoluciidae (Teleostei:Ostariophysi: Characiformes: supra and infrafamilial phylogenetic relationships with a revisionary study. Smith. Contr. Zool. 654:1–97.

Welcomme, R. 1979. Fisheries Ecology of Floodplain Rivers. London UK: Logman.

Winemiller, K. 1989. Pattern of variation in life story among South American fishes in seasonal environments. Oecologia 81:225–241.

Chapter 7

Ecohydrological and Ecohydrographical Methodologies applied to Conservation of Riparian Vegetation: the Caura River as an example

Judith Rosales, Nigel Maxted, Lourdes Rico-Arce and Geoffrey Petts

ABSTRACT

We combine analyses of ecogeography and the hierarchical study of river systems into an approach of general utility for conservation. We perform a spatial analysis using a Geographical Information System with geo-referenced passport data of target plant taxa to study trends in species diversity along the Caura riparian corridor using the family Leguminosae. We then compare these results with those obtained from different sub-basins of the Amazon and the Orinoco Rivers using selected genera of the tribe Ingeae.

The within-basin and inter-basin analyses of number of species, rarity, commonness and geographical distinctiveness indicated the importance of conserving the lower riparian landscapes on the Caura River. Our results demonstrate the critical importance of conserving the Caura River Basin because not only does the Caura possess a large number of species but also because the species form a unique community originating in a number of different basins and phytogeographical regions. The Caroní River, which has a high floral similarity to the Caura, is highly disturbed. The most critical sections of the Caura River Basin for conservation are the sectors below Salto Pará. The inter-basin analysis showed different regional patterns between selected Ingeae genera that relate to phytogeographical regions and river types. In addition to the Caura River, the Río Negro of Brazil and Venezuela is also important for a multinational conservation effort. The ecohydrological approach is demonstrated to be an important conservation tool for analyzing riparian biodiversity. The ecohydrological approach also is appropriate to aid in the selection of reserve areas for *in situ* riparian conservation.

INTRODUCTION

As defined in the Convention on Biological Diversity (UNEP 2000), *in situ* conservation means: i) the conservation of ecosystems and natural habitats; ii) the maintenance or recovery of viable populations in their natural surroundings; and, in the case of domesticated or cultivated species, iii) recovery in the surroundings where they have developed their distinctive properties. Establishing a network of protected areas is proposed in Article 8 of the Convention, and is seen as vital for conservation of the world's natural and cultural resources. The values of protected areas range from preserving natural habitats and associated flora and fauna, to maintaining environmental stability of surrounding regions.

Indicators used for selecting reserves for *in situ* conservation include species richness, biological distinctiveness, current conservation status and economic value (Dinerstein et al. 1995; Maxted et al. 1997b). The data required for assessing these values, however, is not always easy to evaluate. Therefore, some authors (e.g., Prance 1997) have emphasized that conservation protocols should preserve as many species of plants, particularly endemic species, as possible.

In this context, an ecogeographical approach proposed by Maxted et al. (1997b) for *in situ* genetic conservation could also be useful for evaluating conservation priorities in riparian forest ecosystems. Ecogeography is defined by Maxted et al. (1995) as a synthesis of ecologic, geographic and taxonomic information for a particular taxon. This information includes various aspects of biology, such as taxonomy, ecology, distribution, phenology, reproductive biology, genetic diversity and seed storage behavior. Selecting target taxa is important in tropical regions since in many areas the floristic composition is unknown and it is impractical to gather information from all taxa. Selection of target taxa is based on objective, scientifically repeatable and economic principles related to the perceived value of species in the communities targeted for conservation. After selecting a target taxon for conservation, a survey is conducted to gather data for an ecogeographic analysis (Maxted et al. 1997c). The results can be used to formulate conservation priorities.

In riparian ecosystems, an ecogeographic approach (Maxted et al. 1995, 1997a) can be linked to a hierarchical categorization of rivers (Petts and Amoros 1996). The riparian corridor (river channel and the adjacent land affected by the characteristics of the riparian habitat) is evaluated over a nested set of progressively more limited spatial scales:

- drainage basin;
- functional sectors or stream segments, defined by hydrological and geomorphological criteria;
- functional 'ecological' units associated with specific landforms and characterized by typical plant communities indicative of localized habitat conditions.

The exact terminology is of secondary importance and indeed varies in other hierarchical strategies where 'reach', 'section' or 'sector' are often used with similar meaning (Curry and Slater 1986; Frissel et al. 1986; Gregory et al. 1991). The key characteristic of linking the fluvial hydrosystem approach with the ecogeographical approach is that the location of any given plant species along a forested riparian corridor can be described according to a set of hierarchical descriptors of the riparian environment. Each community can thereby be linked to possible influencing factors on an array of scales.

This paper aims to test the utility of this linked ecohydrological approach with data from the Caura River Basin in order to determine conservation priorities. A feasibility analysis using this mixed approach for the South American bioregions was undertaken by contrasting previous results from extensive plot inventories of tree species and from environmental data in the Caura River Basin (Rosales 2000).

METHODS

Ecogeographic Survey of the Fluvial Hydrosystems

The proposed methodology follows a spatial approach. A Geographical Information System is used to geo-reference data from different sources. Using digitized versions of riparian networks of the selected river basins, a spatial analysis of patterns of species diversity and environmental variation held in specific databases is conducted. The software used is (i) ARCView 3.1, (ii) EXCEL and ACCESS for handling of the databases and (iii) SPSS for the statistical calculations.

The functional hydrosystem approach (Petts and Amoros 1996) is used for the spatial analysis within a basin in the Caura River (a blackwater tributary of the Orinoco River draining the Guayana Shield, Rosales et al. 2002b). The basin level comparison includes tributaries of the Orinoco and Amazon Rivers depending upon the water classification as reviewed by Rosales et al. (1999).

In the Caura River, the riparian corridor was classified by Rosales et al. (2003) in functional units, sectors and landscapes. The riparian corridor indicates a stretch of the river, its banks and the land close by, including the river and river channel together with their associated wildlife and the adjacent riparian ecosystem (following Angold et al. 1995). Functional units represent different geomorphic features such as backswamps, point-bars, levees and terraces. Functional sectors are defined as stream segments differentiated by spatial discontinuities in two hydrological variables—the estimated discharge, which increases after the confluence of each tributary, and the slope of the valley, which is related to the presence of major rapids or waterfalls. A functional riparian landscape is a set of riparian functional sectors spatially connected, and related to similar dominant fluvial processes, vegetation types and patterns of land use (Rosales et al. 2003). It collates functional sectors which are spatially contiguous and connected and share similar functional sectors and sets of functional units.

Functional sectors and functional riparian landscapes (for more detailed explanation see Rosales et al. 2002) are the basic elements for a GIS model for the within-basin analysis, whereas complete riparian corridors are used for the between-basin analysis. An investigation of the taxonomic variation in target taxa in relation to changes in the riparian sectors and landscapes (within-basin analysis) and different riparian corridors (between-basin analysis) is conducted using the methodology proposed by Maxted et al. (1997b) for *in situ* genetic conservation:

- Phase 1 includes: (i) delimiting target areas, (ii) selecting target taxa, (iii) identifying taxon information, and (iv) constructing databases.
- Phase 2 includes: (i) collecting ecogeographic data, (ii) selecting representative specimens, (iii) verifying data, (iv) analyzing geographical, ecological and taxonomic data, and (v) assessing current conservation status.
- Phase 3 includes: (i) ecogeographic synthesis and (ii) identifying conservation priorities.

I. Within basin analysis

a) Phase 1. Selecting target area and target taxon

The target area for the within-basin analysis is the Caura River. The family Leguminosae was selected as the target taxon for the following reasons:

1. Leguminosae was found to be the most common and species rich family of tree species in the riparian forests of Amazon and Orinoco Basins (Rosales et al. 1999; Rosales et al. 2002).
2. Although the species of Leguminosae do not always have the highest individual abundances, they do occupy all functional units present in the riparian corridors. Some of the legume species are indicators for certain functional units (e.g. in bed-rock constrained sectors: *Macrolobium angustifolium* in backswamps and confluence swamps or *Inga vera* in bars and point-bars) and could also be considered as keystone species given the high biomass and contribution to productivity and food chains.
3. Many riparian species of Leguminosae are used by the local human populations and are important economically at local and regional levels in the Caura River Basin and adjacent riparian areas along the Orinoco (Rosales 1990; Knab-Vispo et al. 1997; Rosales et al. 1997).
4. Many legume species are functionally important because of their association with nitrogen-fixing bacteria.

b) Phase 2. Collecting ecogeographic data

All legume specimens collected in the Caura River Basin were surveyed in the following herbaria:

- Herbario Nacional de Venezuela (VEN)
- Herbario Ovalles–Facultad de Farmacia, Universidad Central de Venezuela (MYF)
- Herbario Universidad Nacional Experimental Ezequiel Zamora (PORT)
- Herbario Regional de Guayana, Jardín Botanico del Orinoco (GUYN)
- Royal Botanic Gardens, Kew (K)

Information was also taken from herbarium collections at the following Internet locations:

- TROPICOS database, Missouri Botanical Garden (http://mobot.mobot.org/W3T/Search/vast.html)
- The Vascular Plants Type Catalogue at the New York Botanical Garden, where many Legume-type specimens collected in the Neotropics are deposited (http://www.nybg.org/bsci/hcol/vasc/tflow.html)

The following variables were recorded in a database: subfamily, species, collector, collection number, habitat, site, latitude and longitude. This database was linked as points of a theme-database-coverage layer in a GIS project using ARCView. The system also used vector layers of a previously digitized map of the Caura River Basin shown in Figure 7.1 (based on Rosales et al. 2003). The layers have: (i) the main channel divided in sectors between major tributaries' confluence or presence of major rapids or waterfalls, and (ii) tributaries and main channels. The main channel layer is subdivided in 15 functional sectors, which were classified in Rosales et al. (2002) as homogeneous functional riparian landscapes, according to a series of associated environmental variables. The associated table manages a database with the following variables: functional sector, estimated discharge, slope, width, unit stream power and sinuosity.

Furthermore, the following functional landscapes along the Caura were described by Rosales et al. (2003):

- Sectors 1 to 3 are located in the lower Caura between Cinco Mil Rapids and the mouth of the Caura. It is influenced by a backwater effect where the Orinoco flood pulse controls the hydrological variations in flooding characteristics and biochemistry of the Caura floodplain. These sectors are characterized by an unconfined to bedrock confined valley, and single thread sinuous to meandering channel patterns. The climate is in the transition from humid to dry macrothermic in altitudes ranging from 20 to 50 m asl, and the geology of the valley is dominated by La Mesa Formation. These sectors are characterized by the presence of the following geomorphic units: confluence swamps, backswamps, abandoned channels, oxbow lakes, bank benches, levees, point-bar systems, alluvial islands and lateral sand bars.

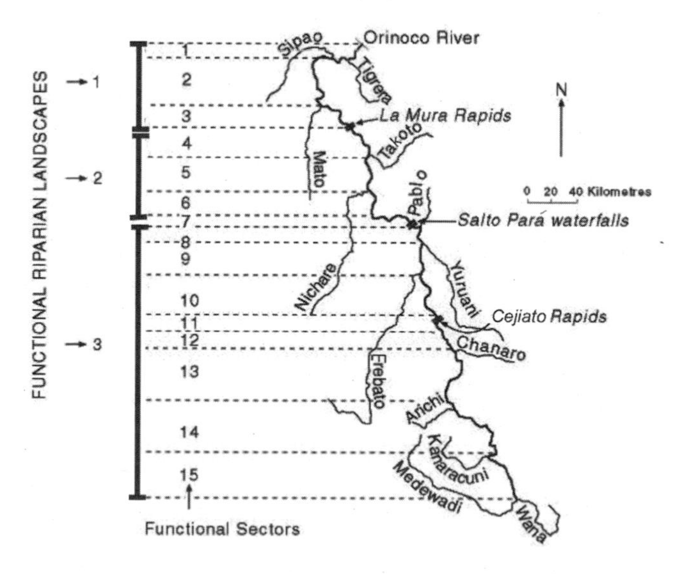

Figure 7.1. Digitized version of the Caura River Basin showing the location sectors and riparian landscapes.

- Sectors 4 to 6 are located in the lower Caura, between Cinco Mil Rapids and Salto Pará. They present a bedrock-confined valley, multithread-anabranching and single thread channel patterns. The climate varies from macrothermic humid to transition humid to dry, altitudes range from 60 to 100 m asl, and the geology of the valley is dominated by the Cuchivero geological group. It has the following geomorphic units: i) erosional and in-channel units (rocky margins, bank benches, and lateral sandbars), and ii) vertical accretion and overbank deposits plus in-channel units (levees, backswamps, confluence swamps, slack-water areas and point-bars).
- Sectors 7 to 15 are located in the Upper Caura, between Salto Pará waterfalls and the confluence of the Kanaracuni tributary. The river presents bedrock confined valley and multithread-anabranching and single thread channel patterns. The climate is humid macrothermic, the altitudes range between 200 to 500 m asl, and the geology of the valley is dominated by the Cuchivero. It presents the following geomorphic units: i) erosional and in-channel units (rocky margins, bank benches and lateral sandbars) and ii) vertical accretion and overbank deposits plus in-channel units (levees, backswamps, confluence swamps, slack-water areas and point-bars).

c) **Phase 3. Analysis of the ecogeographic data and establishment of conservation priorities**

Queries in the GIS allowed the spatial searching of the legume database along the three main areas of environmental variation occurring between the riparian landscapes. The selected records from the databases were further summarized as the total number of species per functional sector and functional riparian landscape considering the subfamilies Caesalpinioideae, Mimosoideae and Papilionoideae. Also summarized were members of the tribe Ingeae, and *Inga* spp. within the Mimosoideae.

Values of functional sectors and functional riparian landscapes for conservation are given the following criteria: the numerical contribution to biodiversity, the distinctiveness in terms of species composition (rarity, uniqueness) but also the presence and abundance of selected target species. Some of the criteria for targeting species for biodiversity conservation (Spellerberg 1996; Maxted et al. 1997d) are used to preliminarily test the feasibility of the methodology:

- Presence in the sectors of possible keystone species, *Inga vera* and *Macrolobium angustifolium,* due to their dominance in the functional habitats, bars and swamps, respectively (Rosales et al. 2001; Rosales et al. 2002).
- Presence and abundance of the species *Acosmium nitens* and *Eperua jenmanii* ssp. *sandwithii*, which are of socio-economic use for local human populations (Knab-Vispo et al. 1997; Rosales et al. 1997).
- Presence and abundance of locally endangered species, e.g. *Acosmium nitens* which are used frequently for house construction in this region.

II. Between basins analysis

a) **Phase 1. Selecting target area and target taxa**

The selected target areas were the Orinoco and Amazon River basins (see Rosales et al. 1999) and the target taxon was the Ingeae, family Leguminosae, subfamily Mimosoideae that includes six genera: *Macrosamanea, Zygia, Hydrochorea, Albizia, Abarema* and *Cedrelinga.*

The Ingeae was selected because it has a high percentage of species associated with riparian habitats in the Neotropics (Guinet and Rico-Arce 1988; Rico-Arce 1987, 2000; Barneby and Grimes 1996, 1997).

b) **Phase 2. collecting ecogeographic data**

For this task, the presence of riparian species for a selected number of rivers was recorded from the most recent publication presenting ecogeographical data and map distributions in America for genera of the Ingeae excluding the *Inga* genus (Barneby and Grimes 1996, 1997).

c) **Phase 3. Analyzing ecogeographic data and establishing conservation priorities**

The rivers selected for the study are classified into the following types:

1. white-water rivers transporting Andean waters rich in nutrient-sediments
    - Central and lower Amazon (between the Solimões and the Tapajos)
    - Lower Madeira River
    - Japurá River

2. black or clear oligotrophic-water rivers draining the Guayana Shield and lowlands of the Amazon
    - Caura River
    - Caroní River
    - Upper Orinoco-Casiquiare
    - Negro River

After the selection of the sample of species per river, data was tabulated in terms of presence-absence per river and three different analyses were carried out:

1. Richness as total number of species

2. Percentages of rarity and commonness as a percentage of species per river classified as very rare or very

common, respectively, according to four ranges of relative rarity:

- Very rare—occurring in only one river
- Rare—occurring in two to three rivers
- Common—occurring in four to five rivers
- Very common—occurring in more than six rivers

3. Geographical distinctiveness is the separation obtained from the different river basins using TWINSPAN analysis (Gauch 1979) and a Canonical Correspondence Analysis ordination using CANOCO (ter Braak 1988) with presence-absence of species per river basin. Ordination techniques are useful to study environmental gradients at different scales (Greig-Smith 1983) and can reveal distributional patterns related to environmental and/or evolutionary distinctiveness.

A final assessment is conducted from an analysis of species richness, rarity and commonness as well as distinctiveness using species distributions as indicators for conservation.

RESULTS

I. Analysis within Caura Basin

Species richness and rarity

Five hundred thirty eight collections of woody legume species were surveyed in the lowlands of the whole Caura River Basin. Although this number reflects a low collection effort in comparison to the size of the basin (45,336 km^2), it is the best available at present. As has been noted by Huber (1995), most of the collection effort in the Guayana Shield of Venezuelan has been concentrated along the rivers. Because of this, the 341 collections (Appendix 9) that were included in a 5 km wide buffered area along the sectors in approximately 500 km of the riparian corridor of the Caura are considered highly representative for this preliminary study. A total richness of 110 woody legume species is summarized in Table 7.1. This table also presents the results of

Table 7.1. Summary of species richness along the riparian corridor of the Caura River. L-1, L-2 and L-3 represent the functional riparian landscapes, the numbers below represent the functional sectors.

		L-1			L-2				L-3		
Species	**1**	**2**	**3**	**4**	**5**	**6**	**7**	**10**	**11**	**12**	**13**
Total number of species per sector (110)	14	24	9	26	48	19	34	15	14	14	15
Total species per riparian landscape		34			74				62		
Total collections per sector	23	36	15	39	65	26	58	21	21	17	20
Total collections per landscape		74			130				137		
Ratio total species/total collections per sector	0.6	0.7	0.6	0.7	0.7	0.7	0.6	0.7	0.7	0.8	0.18
Length of the corridor (km)		134			127				189		
Ratio number of collections/ length of corridor		0.55			1.02				0.72		
Ratio total species/total collections per landscape		0.46			0.57				0.45		
Ratio total species/length of corridor		0.25			0.58				0.33		
Ratio number species/number collections/Length of the corridor x 100		0.34			0.45				0.24		
Number of species unique to the landscape and percentage of unique species per number of species per length of the corridor		9 20%			28 32%				25 22%		

species richness, number of collections (related to collection effort) and the ratio of species richness/number of collections per sector and grouped by riparian landscapes. The sector analysis indicates that the number of species per sector is dependent on collection effort, having a strong positive correlation to the number of collections (R^2 = 0.96). The correlation is lower at the riparian landscape level (R^2 = 0.84) given the increase in the probability of collecting the same species in different landscapes. The landscape level was considered more appropriate for the analysis because it consititutes a functional level of the riparian ecosystem. There was not a significant correlation between length of the river channel for each sector and the number of collections or number of species. A similar result was found at the landscape level. In all the relations, the ratio of the number of collections to the length of the corridor and the ratio of the number of species to the length of the corridor seems to be higher in the intermediate landscape at Cinco Mil – Salto Pará. This riparian landscape has been the most intensively sampled (Williams 1942; Knab-Vispo 1998; Rosales 2000).

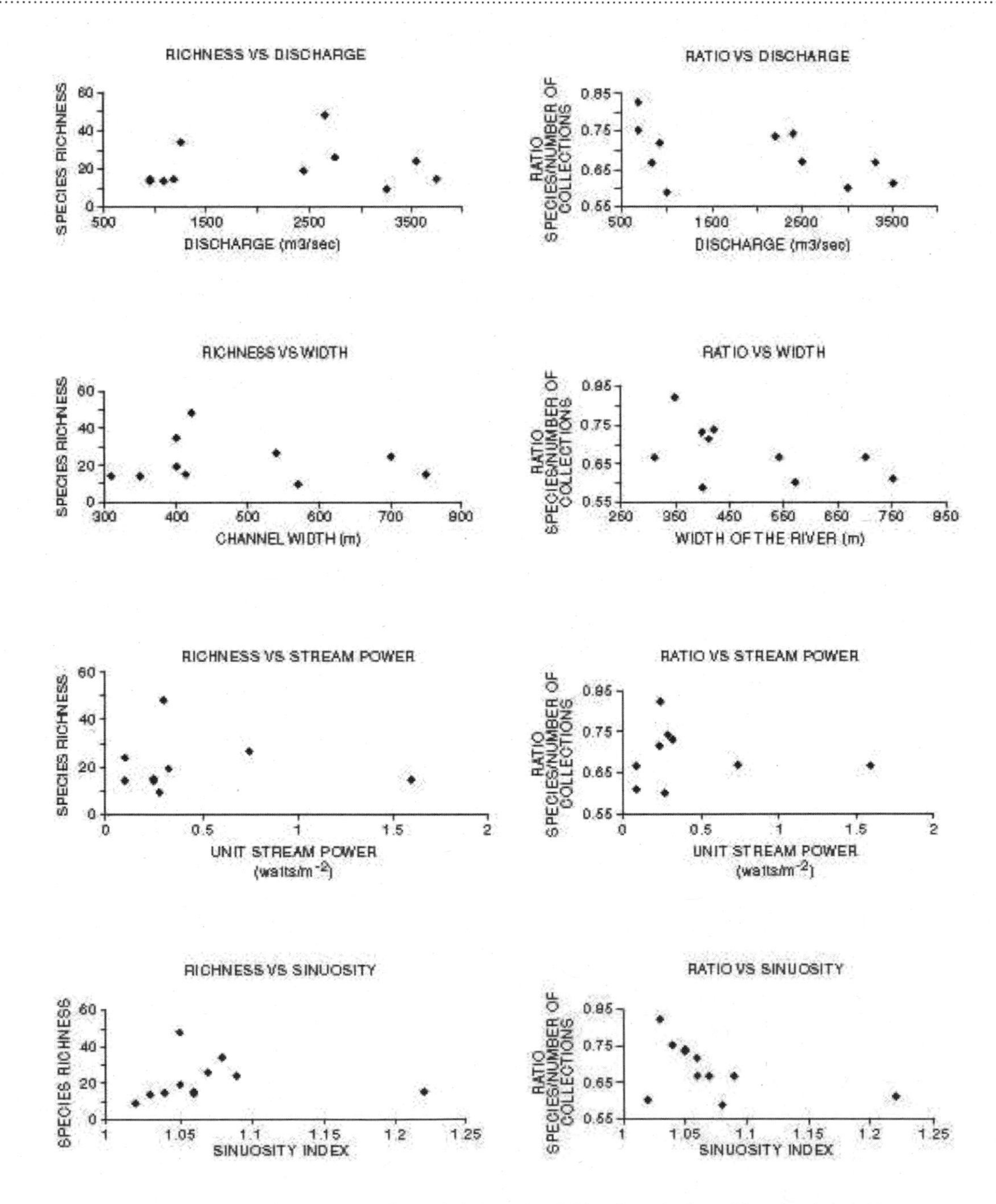

Figure 7.2. Relationships of legume species diversity and different hydrological variables along the Caura River riparian corridor.

Figure 7.2 shows the relations between some hydrological measurements at the sector level (estimated discharge, channel width, unit stream power and sinuosity index) and two diversity measurements (species richness and species richness/number of collections). The trends indicate that sectors having intermediate values in the hydrological variables are richer in species. However, the ratio species richness/number of collections gives negative trends. Although these results are preliminary, they are critical for future studies.

When evaluated by landscape, the number of species was higher at the intermediate landscape (Cinco Mil-Salto Pará). Figure 7.3 shows that this pattern also holds true for different subfamilies, excepting the Papilionoideae. Within the Mimosoideae, the same pattern is observed in the genus *Inga*, whereas in the Ingeae, lower richness is found in the lower riparian landscape (Mouth-Cinco Mil). The ratio species richness/number of collections however resulted in relatively higher values at sectors upstream but when evaluated by landscape, it was higher in the intermediate landscape (Table 7.1).

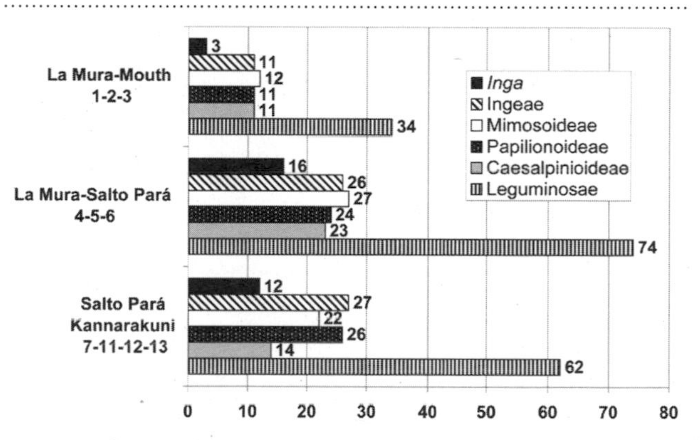

Figure 7.3. Species richness for the family Leguminosae and the genus *Inga* in different riparian landscapes along the Caura River riparian corridor. La Mura = Raudal Cinco Mil.

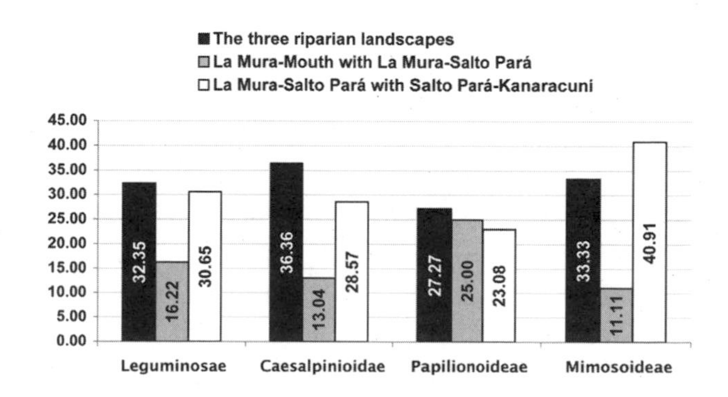

Figure 7.4. Percentages of similarity in species composition among the different riparian landscapes of the Caura River riparian corridor.

The percentage of similarity between the three groups of riparian sectors is shown in Figure 7.4. The three riparian landscapes have a similar percentage of common species. The comparison of Cinco Mil-Mouth with Cinco Mil-Salto Pará resulted in the lowest similarity, while the comparison of Cinco Mil-Salto Pará with Salto Pará-Kanaracuni resulted in the highest. The subfamilies Caesalpinioideae and Mimosoideae account for the highest contrasts if compared with Papilionoideae. In terms of relative rarity, a higher number of relatively unique species was found in the Cinco Mil-Mouth landscape.

Target species distribution

The distribution within the Caura River Basin of locally-important species for conservation (*Acosmium nitens* and *Eperua jenmanii* ssp. *sandwithii*) is shown in Figure 7.5. If *Acosmium nitens* is targeted for conservation, only the Lower Caura, sectors one to four, contains this typical Igapó species (Prance 1979; Kubitski 1989), indicating a low recurrence of its habitat. *Eperua jehnmanii* ssp. *sandwithii*, an economic species for Ye'kwana populations only occurs in parts of the Upper Caura. This species is present, but not dominant, in levees of the Upper Caura (Briceño 1995). It also occurs in upland forests, giving a wider range of opportunities if targeted for conservation. *Inga vera*, a species indicator of bars and point bars, and *Macrolobium angustifolium*, a species indicative of swamps, are widely distributed in the upstream sectors. However, these species occur with low frequency in the Erebato River, the major tributary of the Caura, dominated by high stream power and poorly developed depositional features.

Following this approach with the data in Appendix 9, it is possible to evaluate the distribution observed within each sector of other target species, including those used by human populations (Knab-Vispo et al. 1997):

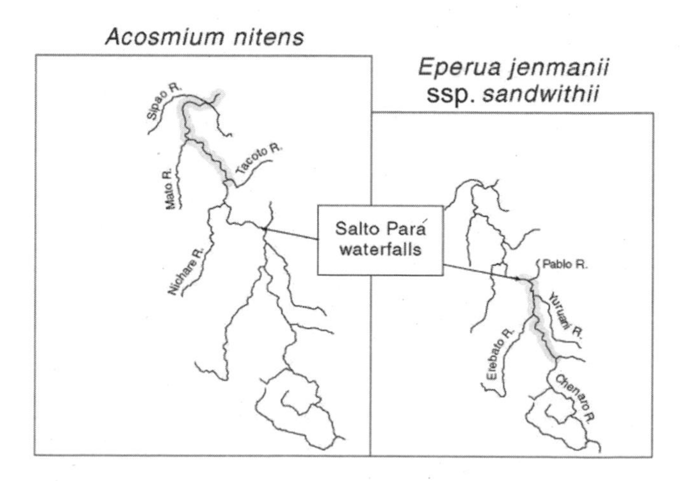

Figure 7.5. Distribution of two legume species along the Caura River riparian corridor.

Directly

1. In construction of dugout canoes: *Andira surinamensis, Sclerolobium guianense, Macrolobium multijugum*;
2. In house construction: *Etaballia dubia, Cassia moschata, Dialium guianense, Centrolobium paraense, Swartzia arborescens, Acosmium nitens, Sclerolobium guianense*;
3. As food: *Dialium guianense, Hymenaea courbaril, Dypterix punctata, Inga alba, Inga bourgoni, Inga coruscans, Inga crocephala, Inga densiflora, Inga dumosa, Inga edulis, Inga laurina, Inga leiocalycina, Inga nobilis, Inga oerstediana, Inga pilosula, Inga splendens, Inga thibaudiana* and *Inga umbellifera.*

Indirectly:

1. Eaten by game or fish: *Campsiandra laurifolia, Dialium guianense, Macrolobium angustifolium, Macrolobium acaciaefolium, Macrolobium multijugum, Sclerolobium chrysophyllum, Sclerolobium guianense, Senna silvestris* var. *silvestris, Dalbergia glauca, Dalbergia hygrophylla, Dypterix odorata, Dypterix punctata, Swartzia leptopetala, Abarema jupunba, Hydrochorea corymbosa, Inga vera* and others referred to in 3 (above), *Newtonia suaveolens, Parkia pendula, Zygia cataractae, Zygia latifolia* and *Zygia unifoliolata.*

II. Between basin analysis

A list of 51 species of Ingeae (Appendix 10) recorded from distribution maps (Barneby and Grimes 1996, 1997) is ordered according to a classification using TWINSPAN (Figure 7.6). All of these species are found in riparian habitats. A summary of the number of species, rarity and commonness percentages is given in Table 7.2. It highlights a lower number of species in the Caroní and Caura Rivers, highest rarity values for the Negro, Madeira and Japura Rivers and highest commonness for the Caura, Caroní and Madeira Rivers.

In terms of geographical distinctiveness, Figure 7.6 shows the results of the TWINSPAN classification. A differentiation between Amazonian and Guayanan phytogeographical regions is observed in the graph. In the Amazonian region it separates the blackwater Negro River as a distinct group from the whitewater Japura, Madeira and Amazon rivers, whereas the blackwater rivers of the Guayana Region seem to be differentiated into western and eastern components of the Guayana Shield as predicted by Huber (1995).

The Canonical Correspondence Analysis reflects three major axes of variation accounting for 70% of the variance in species distributions (eigenvalue =1.146). The first two axes explain 53% of the variance. Figure 7.7 shows the ecogeographical distinctiveness among the rivers. That distinctiveness probably represents the evolution of the Amazon-Orinoco basins. The Negro and Upper Orinoco-Casiquiare would be the more recent lowlands draining podzolic white-sand soils that are separated from older Shield regions where the Caura, Caroní (Guayana Shield), and Madeira (Brazilian Shield) are located. The close relation of the Japura with the Negro and Casiquiare might be related to previous phytogeographic links. Currently, a tributary of the Caqueta, after joining the Japura, also drains podzolic white-sand soils. (Duivenvoorden 1995).

DISCUSSION

Conservation priorities for the Caura River Basin

The results of the ecogeographic data analysis are similar to those from complete tree species surveys of three riparian landscapes with 74, 0.1 ha plots (Rosales 2000; Rosales

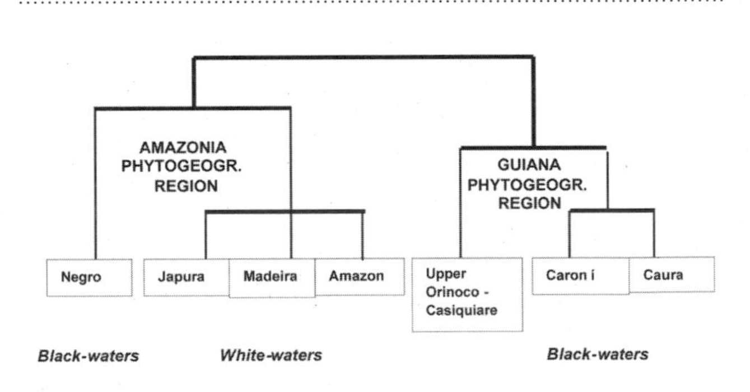

Figure 7.6. Classification of riparian corridors within the Orinoco and Amazon River Basins according to a TWINSPAN analysis of presence-absence of species in selected genera of the Ingeae.

Table 7.2. Summary of conservation indicators for different riparian corridors in the Orinoco and Amazon River Basins, using results from the Ingeae database.

	Negro	Upper Orinoco-Casiquiare	Amazon	Madeira	Japura	Caroni	Caura
Species Number	37	27	27	21	20	13	9
Rarity Percentage	11	7	4	10	10	0	0
Commonness Percentage	16	22	22	29	20	38	56

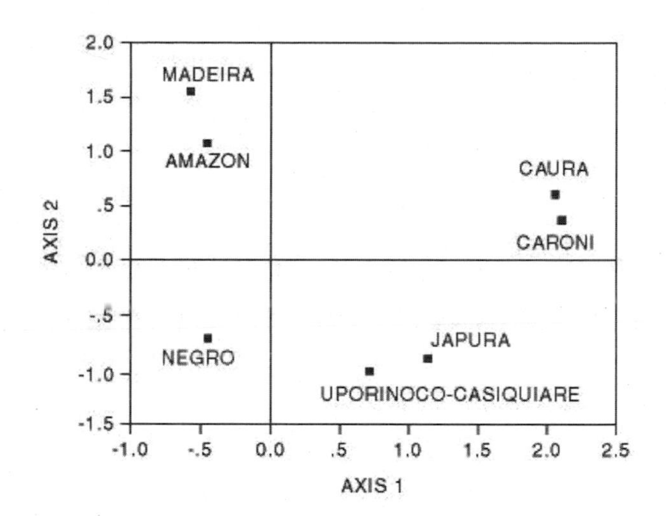

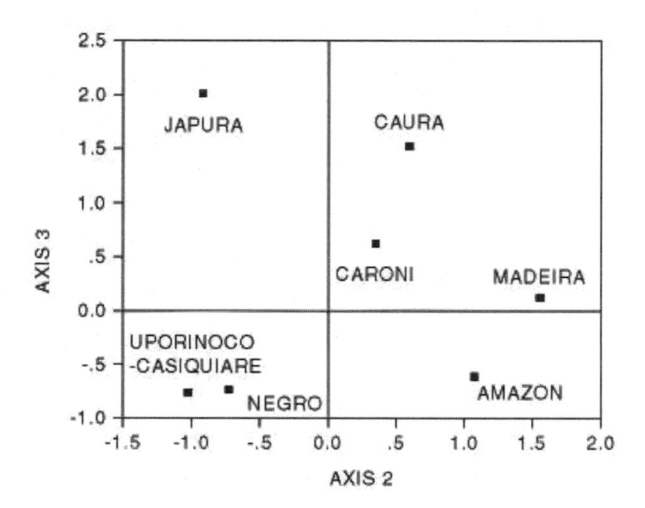

Figure 7.7. Results of a Canonical Correspondence Analysis of riparian corridors in the Orinoco and Amazon River Basins using presence-absence of species in selected genera of the Ingeae.

et al. 2002b.). These authors gave total species numbers of 56, 191, and 180 in 2.3, 2.8 and 2.3 has respectively, with species densities of 24.4, 68.2 and 78.3 species/ha. The results also show that the Lower Caura riparian landscape, representing the confluence zone under the backwater effect of the Orinoco (Rosales et al. 2001), has lower diversity but is more unique, at the basin level, than those associated with bed-rock constrained channels. The latter have the highest values of diversity, but more widely distributed taxa.

In terms of value for conservation, the results indicate that from the three landscapes along the Caura's riparian corridor, Mouth-Cinco Mil and Cinco Mil-Salto Pará should have the highest priority given that together they would preserve a higher number of species richness and species that are relatively rare within the basin. Although the riparian landscape upstream of Salto Pará has a higher number of species, it also has a high similarity with the Cinco Mil-Salto Pará section.

Conservation strategies should also use the degree of landscape level recurrence, in a manner similar to that presented by Walker (1995) for species. The riparian landscape close to the mouth of the Caura River has relatively low diversity of plants but the landscape is unique within the basin. Certain functional units within the Caura River have the potential to disappear, depending upon their vulnerability to potential environmental threats, such as the Caura-Paragua diversion (Rosales et al. 2001). Some of the species targeted as having economic importance for local human populations depend on the recurrence of riparian landscapes that contain particular habitats. For example, the frequency of occurrence of the keystone species for swamps and channel bars, *Macrolobium multijugum* and *Inga vera,* respectively, depends upon the recurrence of sectors having low stream power. The perpetuation of low stream power sectors cannot be guaranteed, even in the Erebato River, if the Caura-Paragua diversion is constructed. Furthermore, the increasing urbanization along the Orinoco, including the Lower Caura, has the potential to increase the pressure on *Acosmium nitens*, if this species continues to be used for house construction; its endangered status should increase. Experiments in sustainable use of potential economically valuable species and in habitat restoration should be considered in order to preserve the uniqueness of the lower riparian landscape of the Caura River; levels of species richness will require monitoring.

Between Basin Evaluation of Conservation Priorities in the Orinoco and the Amazon Basins

The high percentage of commonness obtained in the Caura River is very important for its conservation status even though it does not have high rarity or species richness. Given that the Caroní River basin is already highly disturbed, preserving the Caura has the potential to protect a large number of the species representative of Guayana Shield and Amazonian regions. The Negro River region is characterized by high endemism and diversity, along with the Upper Orinoco-Casiquiare (Prance 1982; Kubitzki 1989; Pires-O'Brien and O'Brien 1995). Although limited to a small number of rivers, our analysis also identifies the effects of the underlying geological diversity, which includes fertility, river dynamics and the orogeny of the river basin. Caroní and Caura River Basins form a distinctive, northern group of the eastern Guayana Phytogeographical Region (Huber 1995), whereas the Upper Orinoco-Casiquiare and Negro River are in the western Guayana Province. The latter are similar to the Japurá due to phytogeographical affinities. The Amazon and Madeira rivers form another distinctive group. In terms of riparian vegetation, the constrained channels of the Caroní and Caura seem to be intermediates between the highly fertile and dynamic varzea flooded forests of the Madeira, Amazon and Japurá, and the highly oligotrophic Igapó forests of the Negro and the Upper Orinoco-Casiquiare. The ecohydrographic analysis of the Ingeae suggests that the Negro River in Brazil and Venezuela is also a high priority for conservation, due to the high number of species, high rarity and high distinctiveness. Because the Orinoco and Amazon watersheds span different countries, a regional, multinational strategy for the conservation of the riparian corridors is needed. The ecohydrographical approach presented herein is useful in order to develop conservation strategies for riparian biodiversity.

The Importance of the Methodology for the Selection of In Situ Reserves for Riparian Biodiversity Conservation

These results demonstrate the value of the ecohydrological approach as a tool for rapid assessments and for the selection and *in situ* conservation of riparian biodiversity. Within the Caura River Basin, analysis of the data from the AquaRAP expedition gave similar results to those obtained using extensive plot analyses. Additionally, the between-basin study identifies regional patterns reported by phytogeographers. The ecohydrological methodology also offers an avenue for integrating genetic diversity. The dynamics and history of connectivity and fragmentation in different riparian sectors can be associated with levels of genetic diversity. For example, intermediate disturbances to the rates of channel dynamics were used to predict where higher levels of heterozygosity should be found in three riparian species of legumes along the lower Solimões (Hill et al. 1978). Although limited in terms of the quantity and quality of information that can be found in herbaria (which often lack precise localities) or in the literature, our approach could be useful for planning and conducting rapid assessments such as Aquatic Gap (Haverland and Muir 1998) or AquaRAP.

LITERATURE CITED

Angold, P., A. Gurnell and P. Edwards. 1995. Information from river corridor surveys. Water and Environmental Management 9: 489–498.

Barneby, R., and J. Grimes. 1996. Silk Tree, Guanacaste, Monkey's Earring: A generic system for the synandrous Mimosaceae of the Americas. Part I: *Abarema, Albizia,* and allies. Memories of the New York Botanical Garden 74: 1–292.

Barneby, R., and J. Grimes. 1997. Silk tree, Guanacaste, Monkey's Earring: A generic system for the synandrous Mimosaceae of the Americas. Part II. *Pithecellobium, Cojoba,* and *Zygia.* Memories of the New York Botanical Garden 74: 1–292.

Briceño, A. 1995. Analisis fitosociologico de los bosques ribereños del Río Caura en el Sector Cejiato-Entrerios. Distrito Aripao del Edo. Bolívar. Forest Engineer Thesis, Universidad de los Andes, Mérida, Venezuela.

Curry, P., and F. Slater. 1986. A classification of river corridor vegetation from four catchments in Wales. Journal of Biogeography 13: 119–132.

Dinerstein, E., D. Olson, D. Graham, A. Webster, S. Primm, M. Bookbinder and G. Ledec. 1995. A conservation assessment of the terrestrial ecoregions of Latin America and the Caribbean. Washington, DC, USA: The World Bank, published in association with The World Wildlife Fund.

Duivenvoorden, J. 1995. Tree species composition and rainforest-environment relationships in the Middle Caqueta area, Colombia, NW Amazonia. Vegetatio 120, 91–113.

Frissel, C., W. Wiss, C. Warren and M. Huxley. 1986. A hierarchical framework for stream classification: viewing streams in a watershed context. Environmental Management 10: 199–214.

Gauch, H. 1979. TWINSPAN a FORTRAN program for two-way indicator species analysis and classification. New York USA: Cornell University Press.

Gregory, S., F. Swanson, W. Arthur McKee and K. Cummins. 1991. An ecosystem perspective of riparian zones: focus on links between land and water. BioScience 41: 540–551.

Greig-Smith, P. 1983. Quantitative Plant Ecology. Oxford, England: Blackwell Scientific Publications.

Guinet, Ph., and L. Rico-Arce. 1988. Pollen characters in the genera *Zygia, Marmaroxylon* and *Cojoba* (Leguminosae, Mimosoideae, Ingeae) a comparison with related genera. Pollen and Spores 30: 313–328.

Haverland, P., and T. Muir. 1998. Aquatic GAP—current status and next steps. GAP Annual Meetings - Available at http://www.gap.uidaho.edu/GAP.

Hill, R., G. Prance, S. Mori, W. Steward, D. Shimabukuru and J. Bernardi. 1978. Estudo eletroforetico da dinamica de variacao genetica em tres taxa ribeirinhos ao longo do rio Solimoes, America do Sul. Acta Amazonica 8: 183–199.

Huber, O. 1995. Vegetation. *In:* Berry, P., B. Holst and K. Yatskievych (eds.). Flora of the Venezuelan Guayana. Vol. 1. Introduction. Oregon USA: Missouri Botanical Garden and Timber Press. Pp. 97–160.

Knab-Vispo, C. 1998. A rain forest in the Caura Reserve (Venezuela) and its use by the indigenous ye'kwana people. PhD Thesis. University of Wisconsin-Madison USA.

Knab-Vispo, C., J. Rosales and G. Rodríguez. 1997. Observaciones sobre el uso de plantas por los ye'kwana en el bajo río Caura. *In:* Rosales, J. and O. Huber (eds.). Ecología de la Cuenca del Río Caura. Scientia Guaianae 7: 211–257.

Kubitzki, K. 1989. The ecogeographical differentiation of Amazonian inundation forests. Plant Systematics and Evolution 162: 285–304.

Maxted, N., M. van Slageren and J. Riham. 1995. Ecogeographic surveys. *In*: Guarino, L., V. Ramanatha and R. Reid (eds.). Collecting Plant Genetics Diversity. Wallingford: CAB International. Pp. 255–285.

Maxted, N., B. Ford-Lloyd and J. Hawkes (eds.). 1997a. Plant Genetic Conservation, the *in situ* approach. London UK: Chapman and Hall.

Maxted, N., B. Ford-Lloyd and J. Hawkes. 1997b. Complementary conservation strategies. *In:* Maxted, N., B. Ford-

Lloyd and J. Hawkes (eds.). Plant Genetic Conservation, the *in situ* approach. London UK: Chapman and Hall. Pp. 15–39.

Maxted, N., L. Guarino and M. Dullo. 1997c. Management and monitoring. *In:* Maxted, N., B. Ford-Lloyd and J. Hawkes (eds.). Plant Genetic Conservation, the *in situ* approach. London UK: Chapman and Hall. Pp. 144–159.

Maxted, N., J. Hawkes, L. Guarino and M. Sawkins. 1997d. The selection of plant conservation targets. Generic Resources Crop Evolution 7, 1–12.

Petts, G., and C. Amoros. 1996a. Fluvial Hydrosystems: a management perspective. *In:* Petts, G. and C. Amoros (eds.). Fluvial Hydrosystems. London UK: Chapman and Hall. Pp. 263–278.

Pires-O'Brien, M., and O. O'Brien. 1995. Ecologia e modelamento de florestas tropicais. Ministerio da Educacao e do Desporto, Facultade de Ciencias Agrarias do Para. Belem.

Prance, G. 1979. Notes on the vegetation of Amazonia III. The terminology of Amazonian forest types subject to inundation. Brittonia 31: 26–38.

Prance, G. (ed.). 1982. Biological diversification in the Tropics. New York USA: Columbia University Press.

Prance, G. 1997. The conservation of botanical diversity. *In:* Maxted, N., B. Ford-Lloyd and J. Hawkes (eds.). Plant Genetic Conservation, the *in situ* approach. London UK: Chapman and Hall. Pp. 1–14.

Rico-Arce, L. 1987. Generic patterns in the tribe Ingeae with emphasis in Zygia-Caulathon. Bulletin IGSM 15: 51–69.

Rico-Arce, L. 2000. New combinations in Mimosaceae. Novon 9: 554–556.

Rosales, J. 1990. Análisis florístico-estructural y algunas relaciones ecológicas en un bosque estacionalmente inundable en la boca del Río Mapire, Edo. Anzoategui. MSc. Thesis. Instituto Venezolano de Investigaciones Científicas, Venezuela.

Rosales, J. 2000. An ecohydrological approach for riparian forest biodiversity conservation in large tropical rivers. PhD Thesis, University of Birmingham, UK.

Rosales, J. and O. Huber (ed.). 1996. Ecología de la cuenca del Río Caura. I. Caracterización general. Scientia Guaianae 6. Ediciones Tamandua, Caracas.

Rosales, J., M. Bevilacqua, W. Díaz, R. Pérez, D. Rivas and S. Caura. 2003. Riparian vegetation communities of the Caura River. *In:* Chernoff, B., A. Machado-Allison, K. Riseng, and J. R. Montambault (eds.). A Biological Assessment of the Aquatic Ecosystems of the Caura River Basin, Bolivar State, Venezeula. RAP Bulletin of Biological Assessment, No. 28. Washington, DC USA: Conservation International. Pp. 34–43.

Rosales, J., C. Knab-Vispo and G. Rodríguez. 1997. Los bosques ribereños del bajo Caura entre el Salto Pará y Los Raudales de Cinco Mil: su clasificación e importancia en la cultura Ye'kwana. *In:* Huber, O. and J. Rosales (eds.). Ecología de la Cuenca del Río Caura. Scientia Guaianae 7: 171–213.

Rosales, J., G. Petts and J. Salo. 1999. Riparian flooded forests of the Orinoco and Amazon River basins: a comparative review. Biodiversity and Conservation 8: 551–586.

Rosales, J., G. Petts and C. Knab-Vispo. 2001. Ecological gradients in riparian forests of the lower Caura River, Venezuela. Plant Ecology 152 (1): 101–118.

Rosales, J., G. Petts, C. Knab-Vispo, J. A. Blanco, A. Briceño, F. Briceño, R. Chacón, B. Duarte, U. Idrogo, L. Rada, B. Ramos, J. Rangel and H. Vargas. 2002. Ecohydrological Assessment of the Riparian Corridor of the Caura River in the Venezuelan Guayana Shield. *In:* Vispo, C. and C. Knab-Vispo (eds.). Plants and Vertebrates of the Caura´s Riparian Corridor: Their Biology, Use and Conservation. Scientia Guaianae 13: in press.

Spellerburg, I. 1996. Conserving biological diversity. *In:* Spellerburg, I. (ed.). Conservation Biology. Edinburg: Longman. Pp. 25–35.

ter Braak, C. 1988. CANOCO—a FORTRAN Program for Canonical Community Ordination by [Partial] [Detrended] [Canonical] Correspondence Analysis, Principal Components and Redundancy Analysis. Wageningen: Agricultural Mathematics Group.

UNEP. 2000. Documents of the Convention of Biological Diversity and the Conferencies of the Parties COP (1992–2000). Secretariat of the Convention on Biological Diversity. http//www/biodiv.org/.

Walker, B. 1995. Conserving biological diversity through ecosystem resilience. Conservation Biology 9, 747–752.

Williams, L. 1942. Exploraciones Botánicas en la Guayana Venezolana. I. El medio y bajo Caura. Servicio Botánico-Ministerio de Agricultura y Cría.

Chapter 8

The Distribution of Fishes and Patterns of Biodiversity in the Caura River Basin, Bolívar State, Venezuela

Barry Chernoff, Antonio Machado-Allison, Philip W. Willink, Francisco Provenzano-Rizzi, Paulo Petry, José Vicente García, Guido Pereira, Judith Rosales, Mariapia Bevilacqua and Wilmer Díaz

ABSTRACT

We test null hypotheses concerning random species distributions with respect to subregions and macrohabitats within the Caura River Basin with data from 97 species of benthic invertebrates, 399 species of riparian plants and 278 species of freshwater fishes. The analysis of the eight subregions split evenly above and below the falls indicated that invertebrates were randomly distributed with respect to subregion. Fishes and plants were not random, and the subregional effect in plants was more strongly patterned than that of fishes. Furthermore, fishes were less species rich in the Upper Caura than the Lower Caura. The converse was observed for plants while the invertebrates were almost equally rich. Non-random macrohabitat effects are found in each of the groups with certain commonalities. For example, species of Odonata and Ephemeroptera are found in high oxygen, swift water and rapids habitats and are associated with a large assemblage of fishes and dense stands of macrophytes, usually Podostemonaceae. Fish assemblages demonstrate smooth transitions along several macrohabitat gradients (e.g., sand to mud bottoms and shores). At least six macrohabitats are necessary to preserve 82% of the species of fishes.

INTRODUCTION

The Caura River watershed is a relatively large pristine region and is home to thousands of species of plants and animals. Of the aquatic and flood zone organisms surveyed during the AquaRAP expedition to the Caura River, more than 90 species of benthic invertebrates, 399 species of plants and 278 species of fishes were collected. In addition to the quantity of new information about species distributions, and new species occurrences, it is critical from a conservation perspective to test hypotheses about the distributions of animals and plants within the basin.

A pattern of heterogeneous flora and fauna distribution within the Caura River subregions and macrohabitats would have important ramifications for conservation recommendations. For example, if species were homogeneously distributed then a core conservation area could be established that might effectively protect the vast majority of the species. However, as the distribution of the species either among subregions or among macrohabitats becomes increasingly distinct and patchy, then a single core area, apart from the entire region, may not provide the desired level of protection. Chernoff and Willink (2000) and Chernoff et al. (1999, 2001a) demonstrated how we can use information on the relative heterogeneity of distributions among sub-regions or among macrohabitats to predict possible faunal changes in response to specific environmental threats and that such analyses can and should be carried out within the framework of a rapid assessment program.

This paper will begin with tests of two null hypotheses critical to freshwater fish conservation in the Caura River Basin as follows: that the fishes are randomly distributed among i) eight subregions; and ii) 20 macrohabitats. We will then compare the results from fishes to those for benthic invertebrates and the plants.

METHODS

The collections from the Caura River were divided into two principal regions—above (Upper Caura) and below Salto Pará (Lower Caura). Eight geographic subregions were then designated (Map). The four regions in the Upper Caura are: Kakada, Erebato, Entrerríos-Cejiato, and Entreríos-Salto Pará. In the Lower Caura the four regions are: El Playón, Nichare, Cinco Mil, and Mato.

At each collecting locality a number of ecological variables, such as bottom type, shore type, habitat type, etc. described in Appendices 1 and 3 were recorded. We were able to categorize each station into a principal macrohabitat type. Twenty principal macrohabitat types were identified. There were no true lakes, water bodies with endorheic drainage basins; instead lakes refer to lagoons with either a small connection to a river or temporally isolated from the river. At each locality, we evaluated whether aquatic grasses were present, whether there was riparian forest and if there was flooded vegetation. Flooded vegetation included mats of vegetation that were attached to rocks in rapids (e.g., Podomostemacea) or floating (e.g., *Eichhornia*).

In order to determine if the number of collections per region or per macrohabitat was affecting the estimates of species richness, we calculated a linear regression, pooled among groups. The regression line is logically forced through the origin (e.g., Chernoff and Willink 2000). Because the analysis of variance (ANOVA) of the regression was significant, testing the slope against a null hypothesis of zero, an analysis of covariance (ANCOVA) was performed to see if other hypothesized effects (i.e., headwaters vs. lowlands) were significant. In ANCOVA, the qualitative group variable (i.e., elevational group) is entered as the independent variable, the number of species is the dependent variable and the number of collections serves as the covariate. If the F-statistic of the ANCOVA is significant, two further tests must be carried out to determine if the difference is attributable to mean differences of the independent variable. The first tests the null hypothesis that the within-group variances are equal. The second tests the null hypothesis of homogeneity of within-group slopes. If one fails to reject both null hypotheses, then the F-statistic significance is attributable to the differences indicated by the independent variable.

Chernoff et al. (1999, 2000) selected Simpson's Index of Similarity, S_s, as the most consistent with data collected during rapid inventories or with point source data. Simpson's Index uses the following table format to calculate the similarity between two lists or samples of species:

		Sample 1	
		1	0
Sample 2	1	*a*	*b*
	0	*c*	*d*

where, ***a*** is the number of positive matches or species present in both samples, ***b*** is the number of species present in sample 2 and absent from sample 1, ***c*** is the converse of ***b***, and ***d*** is the number of negative matches or species absent from both localities. Simpson's index of similarity, $S_s = a/(a+b)$, where $b < c$, or $S_s = a/n_s$ where n_s is the number of species present in the smaller of two lists. The denominator of the index eliminates interpretation of the negatives—absent species. The 0's in the matrices are really coding artifacts or place holders for missing data.

In order to interpret the observed similarity of two samples, both of which are drawn from a fixed larger universe (e.g., the set of all species captured in the Caura), we undertake a four-step procedure. In **step 1**, we calculate S_s by reducing via rarefaction the number of species in the larger sample to equal the number of species in the smaller sample, n_s. This rarefaction and calculation of S_s is iterated 500 times. From the 500 simulations a mean similarity, S'_s, is calculated and reported in tables of similarity. This procedure is repeated to calculate an S'_s for each pair of samples in the analysis.

Interpreting the significances of the mean similarities among the samples requires simulations across the range of number of species found in the samples of subregions, water classes, and macrohabitats. In **step 2**, we simulate 200 random pairs of samples by bootstrapping with replacement from the set of all species captured during the expedition with the constraint that each random sample contains a fixed number of species for a given point in this range. For each random pair of the 200 we calculate their Simpson's Similarity. These 200 random similarities approximate a normal distribution from which we calculate a mean and standard deviation due to random causes; henceforth called mean random similarity, S^*_n, where n refers to the number of species present in the sample.

Random similarity distributions were generated at intervals of 10 species in order to estimate S^*_n and its standard deviation for samples containing between 20 and 140 species. This range of random list-sizes encompasses the actual number of species observed in subregions and in macrohabitats. In **step 3**, the means and standard deviations are plotted against number of species present in a sample. As the number of species present in a sample increases the observed similarity due to random effects also increases but the variance decreases.

In **step 4**, we compare the observed mean similarity, S'_s, calculated from rarefaction (step 1) to the predicted value of S^*_n and its standard deviation. Using a 2-tailed parametric approach, we calculate the probability of obtaining the observed similarity at random from the number of standard deviations that the observed similarity was either above or below the mean of the bootstrap random distribution. This probability was obtained by interpolation of the values presented in Rohlf and Sokal (1995: table A). The significance of the probability values was adjusted with the sequential Bonferoni technique (Rice 1989) because each sample is involved in multiple comparisons. The sequential Bonferoni procedure is conservative, making it harder to reject a null hypothesis. We selected the P=0.01 level as our criterion for rejection of a null hypothesis. The value, 0.01, was divided by the number of off-diagonal comparisons present in the upper or lower triangle of the matrix of observed similarities. This new result is used as the criterion to evaluate the null hypothesis that $S'_s = S^*_n$. For example, in the lower triangle of Table 8.1 there are 28 similarities. In order to reject the null hypothesis of this two-tailed test, S'_s must be more than 3.5 standard deviations above or below S^*_n so that P<0.0002.

If S'_s is found to be significantly different from S^*_n, then we reject the null hypothesis and conclude that the observed *similarity* is not due to random effects. However, if S'_s falls within the random effects, or if S'_s is greater than the random mean, we fail to reject the null hypothesis concerning the *samples*—that the two *samples* are equal. In the former case we conclude that the two lists are drawn homogeneously from a larger distribution. In the latter case, we conclude that the similarity is due to biological dependence or correlation, such as nested subsets. That is, one population forms the source population for another. If S'_s is significantly less than S^*_n, we reject the null hypotheses for equality of similarity and for equality of the samples. We can then search for biological or environmental reasons for the dissimilarities.

If we discover that similarities are not random, we can investigate whether the pattern of species presences in relation to environmental variables is non-random. The measure of matrix disorder as proposed by Atmar and Patterson (1993) calculates the entropy of a matrix as measured by temperature. Temperature measures the deviation from complete order (0°) to complete disorder (100°) in which the cells of a matrix are analogous to the positions of gas molecules in a rectangular container. After the container has been maximally packed to fill the upper left corner (by convention), the distribution of empty and filled cells determines the degree of disorder in the species distributions and corresponds to a temperature that would produce the degree of disorder (Atmar and Patterson 1993). To test whether the temperature could be obtained due to random effects, 500 Monte Carlo simulations of randomly determined matrices of the same geometry were calculated. The significance of observed temperatures is ascertained in relation to the variance of the simulated distributions. Because we are not using this procedure to test specifically whether a non-random pattern can be ascribed to either nested subsets or clinal turnover the modifications proposed by Brualdi and Sanderson (1999) are not required. Software to calculate matrix disorder is available from Atmar and Patterson at the following internet site: http://www.fieldmuseum.org

Relationships among regions and macrohabitats are summarized in branching diagrams. Two types of procedures were used both involving parsimony criteria. The first is a minimum "evolution" network calculated from a distance matrix as 1 minus the means of the rarefied Simpson's Index (=1- S'_s). The second method uses the presence of species as shared characters in A Camin-Sokal Parsimony Analysis (Sneath and Sokal 1973) does not permit reversals, hence, shared absences are not taken as characters. Paup* 4.0b was used for these analyses.

RESULTS

Regions

A total of 67 collecting stations were made during the expedition resulting in the capture of 278 presumptive fish species. A total of 31 collections within 14 georeference areas were made above the falls and 35 collections within 17 georeference areas were made below the falls.

The apparent species richness was much greater in the Lower Caura than the Upper Caura with 226 and 103 fish species captured, respectively. The species richness was not even among regions (Figure 8.1). The number of fish species per region ranged from 26 to 120. However, we found almost twice as many fish species per region in the lowlands (mean = 104.8) than we found above the falls (mean = 54.5). There is, however, a strong effect of collecting effort on the number of species captured (Figure 8.1). The pooled regression analysis demonstrates that species richness is a significant, linear function of the number of collections made for each georeference area with a non-zero slope ($F_{1,6}$= 26.2, P<.002). The results of ANCOVA correct for the effects of

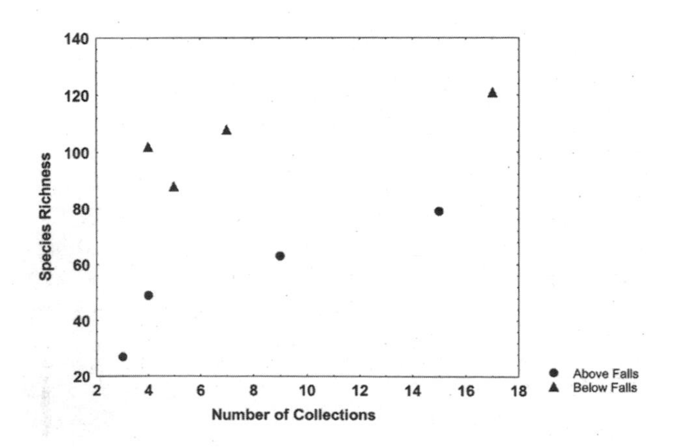

Figure 8.1. Species richness as a function of number of collections for subregions of the Caura River taken above (circles) and below (triangles) the Salto Pará.

collecting effort and reject the null hypothesis of equality of mean species richness above and below the falls ($F_{1,5}$=45.4, P<.002). The result reflects the differences in means because we could not reject null hypotheses of homogeneity of variances or of slopes (P>.24). Thus, for eight collections per region, the predicted and 95% confidence value in the Upper Caura is 55.4 ± 21 species as compared to 104.3 ± 20 species in the Lower Caura.

The disparity in fish species richness between headwaters and lowlands (n=157) is apparent from the total list of species (Appendix 8). As we proceed up river systems, above large falls and into the piedmont, we expect to find fewer species and many with more restricted ranges than species in the unimpeded lowlands (Lowe-McConnell 1987). Therefore, in the Upper Caura we predict that our samples should contain: i) species with broad elevational distributions, shared both by headwater and lowland species; and ii) species with narrower elevational preferences, not found in the lowlands, especially those with mud bottoms. The Lower Caura shares 52 fish species, from which we observe their similarity, S'_s, to be 50.5%. The Upper Caura is a higher gradient environment, with many rocks, rapids and sandy bottoms and narrower flooded margins (Machado et al. 2003). Of the 51 species captured only above the falls, many are representative of more widespread taxa (e.g., *Myleus asterias, Cyphocharax* cf. *festivus, Gymnotus carapo, Crenicichla alta, Plagioscion* cf. *auratus*). However, a large number of the species found uniquely in the upper section are characteristic of the environments described. For example, those characteristic of piedmont, rather than lowland habitats include: *Ancistrus* spp., *Cetopsorhamdia* cf. *picklei, Chaetostoma vasquezi, Harttia* sp., *Hypostomus* cf. *ventromaculatus, Rineloricaria fallax, Apareiodon* sp., *Leporinus arcus, L.* cf. *granti, Melanocharacidium melanopteron* and *Guianacara* sp. In addition there were a number of species usually associated with sandy bottoms that tend not to occur in the broad lowland, mud-bottom, flood plains of the Orinoco River: *Corydoras* spp., *Imparfinis* sp. B., *Characidium* spp., *Knodus* spp. and *Geophagus* sp.

We should also expect the true headwater species to have relatively narrow ranges because headwater regions are relatively isolated from one another, acting much like islands (Lowe-McConnell 1999). Of the 39 species collected only in the headwaters, only seven species were found to inhabit three or more of the four regions (Table 8.1). Although a few species appearing to have narrow or spotty distributions are artifactual (e.g., *Myleus* spp., *Pimelodus* cf. *ornatus*), the majority are not. The result is that more than 80% of the species found only in the upper regions were captured in a single region. Although continued sampling would undoubtedly increase the distribution of the collected taxa (Alroy 1992; Chernoff and Willink 2000), as well as add more fish species to the known list, we doubt seriously whether the majority of the species will be ubiquitously distributed. From this we conclude that the regions above the falls contain a combination of species with broad elevation tolerances as well as those preferring more upland, headwater habitats.

The lowland fish fauna was very rich, containing 226 species. The fishes were not distributed as broadly in the Lower Caura as we would have expected (e.g., in the Pantanal; Chernoff and Willink 2000) only 26% were collected in three or four of the lowland regions. And only 28 species (12.5%) were collected in each of the four lowland regions. There were 174 fish species collected only in the lowlands. The lowland-only fauna contained species that are distributed in the Caura River, other Guayana Shield rivers, the Negro River (e.g., *Microschemobrycon* spp., *Leporinus brunneus, Ammocryptocharax elegans, Anostomus ternetzi, Serrasalmus* sp., *Crenicichla* cf. *lenticulata, C.* cf. *wallacei*) and those that are typical of the main flooded areas of the lower Orinoco River (e.g., *Anchoviella* spp., *Pellona castelneana, Pygocentrus cariba, Aphyocharax erythrurus, Triportheus albus, Sorubim lima, Pimelodus blochii, Bujurquina mariae, Achirus* sp.). Thirty-seven of the lowland only species (=16.8%) were distributed in three or four of the regions (Table 8.2) comprising an assemblage of many ornamental or diminutive species (e.g., *Anostomus ternetzi, Microschemobrycon callops, Ramirezella newboldi, Parvandellia* sp., *Apistogramma* cf. *indirae, Paravandellia* sp.). But there were also large species (> 200 mm SL, e.g., *Hydrolycus tatauaia* or *Hypostomus* cf. *plecostomus*) and a diverse set of trophic specialists, from mud-eating and herbivorous (e.g., *Curimata incompta*) to piscivorous (e.g., *Serrasalmus* sp.).

The pattern of similarities, S'_s (Table 8.3) demonstrates the effects of the Salto Pará on the structure of fish communities in the Caura River. Among the regions above the falls, the coefficients are significantly different from random and are positive. This means that the regions in the Upper Caura are positively correlated or biologically dependent upon each other (Chernoff et al. 1999; Chernoff and Willink 2000). Good evidence of this can be found in the number of species shared between regions relative to those that are found in only one region. Although the four regions above the falls share large percentages of their species overall, the numbers of species shared uniquely among these regions is exceptionally low—less than four and modally zero. This result indicates that species do not seem to be segregating or partitioning the areas above the falls differentially; there is no evidence of species turnover or transition boundaries. This result is consistent with the regions having nested subset relations to each other (see below).

Our rarefaction and simulation analyses of similarities demonstrate that out of 16 similarity coefficients for regions separated by the Salto Pará, only a single coefficient is significantly different from mean random similarity (Table 8.3). The species found above and below the falls come almost entirely from the set of species that are found in five or more of the eight regions (Table 8.4). That is, essentially those species that are ubiquitous. In one case, the similarity of Entreríos-Cejiato to the Mato River is significant (P<.01) and is negative—much lower than expected due to random processes. Negative relationships indicate displacement or strong regional sorting of taxa (Chernoff et al. 1999). Indeed the Mato River sample is the most downstream and includes many species from the Orinoco River that penetrate only partway up the Caura River (e.g., the piranha, *Pygocentrus cariba*, the catfish, *Xyliphius* cf. *melanopterus*).

In the Lower Caura, four of the six similarity coefficients are significantly different from random (P<.01) and are positive. The two random coefficients compare the Mato River with the Playón and Nichare subregions. These are the two subregions in the Lower Caura River that are most distant from the Mato River subregion; the Cinco Mil subregion is positively and significantly correlated with Mato River. The significant similarity between Mato River and Cinco Mil is due to the Orinoco River elements that largely characterize Mato River extending only as far upriver as the rapids at Cinco Mil.

The overall distribution matrix is marginally significantly different from random. The observed matrix temperature (49.93) was 1.82 standard deviations below the mean of 500 Monte Carlo simulations (P=.03). This is entirely due to the distribution of widely distributed taxa shared by the lowland regions. When we analyzed the subregions in the Upper Caura by themselves, highly significant results were obtained. The observed matrix temperature (24.09) was more than 3.38 standard deviations cooler than the mean of 500 Monte Carlo simulations (P<.0004). Given the data on similarities from above, the upper subregions are consistent with a pattern of nested subsets. The most interpretable patterns within the distributional data comprise the following: i) that there is a significant faunal turnover due to the Salto Pará; ii) the subregions of the Upper Caura are structured as nested subsets; and iii) in the Lower Caura there is more of a disjunction at the lower end due to the incursion of Orinoco River fauna. These conclusions are well summarized by the results of the tree structure using Camin-Sokal Parsimony (Figure 8.2). The retention index, which measures the amount of information due to shared taxa, is 0.706. With the exception of the Mato River being the most disparate of the lowland group, the pattern among Cinco Mil, Nichare

Table 8.1. Species of fishes found only in the Upper Caura River above the Salto Pará.

One or Two Regions (n=39)	
Acestrorhynchus	cf. *apurensis*
Aequidens	sp.
Ageneiosus	sp.
Ancistrus	sp. A
Ancistrus	sp. B
Anostomus	*anostomus*
Apareiodon	sp.
Brachychalcinus	*orbicularis*
Cetopsorhamdia	cf. *picklei*
Chaetostoma	*vasquezi*
Characidae	sp. A
Characinae	sp. A
Corydoras	cf. *osteocarus*
Creagrutus	sp.
Ctenobrycon	*spilurus?*
Cyphocharax	cf. *festivus*
Doras?	sp.
Farlowella	*oxyrryncha*
Geophagus	sp.
Guianacara	cf. *geayi*
Gymnotus	*carapo*
Harttia	sp.
Hemigrammus	sp. B
One or Two Regions (n=39) (continued)	
Hemiodus	cf. *unimaculatus*
Hemiodus	*goeldii*
Hypostomus	cf. *ventromaculatus*
Imparfinis	sp. B
Jupiaba	cf. *zonata*
Jupiaba	sp. B
Knodus	sp. C
Melanocharacidium	*melanopteron*
Moenkhausia	cf. *grandisquamis*
Moenkhausia	cf. *miangi*
Moenkhausia	sp. B
Myleus	*asterias*
Myleus	*torquatus*
Phenacogaster	sp. B
Pimelodus	cf. *ornatus*
Plagioscion	cf. *auratus*
Three or More Regions (n=7)	
Aphyocharax	sp.
Bryconops	sp. A
Corydoras	*boehlkei*
Crenicichla	*saxatilis*
Guianacara	*geayi*
Knodus	cf. *victoriae*
Rineloricaria	*fallax*

Table 8.2. Fishes captured only in three or four regions of the Lower Caura River, below the Salto Pará (n=37).

Ancistrus	sp. C
Anostomus	*ternetzi*
Aphyocharax	*alburnus*
Apistogramma	sp. A
Astyanax	sp.
Brycon	*pesu*
Bryconamericus	cf. *cismontanus*
Characidae	sp. B
Corydoras	cf. *bondi*
Creagrutus	cf. *maxillaris*
Curimata	*incompta*
Cyphocharax	*oenas*
Farlowella	*vittata*
Hemigrammus	cf. *tridens*
Hemiodus	*unimaculatus*
Hydrolycus	*tatauaia*
Hyphessobrycon	*minimus*
Hypostomus	cf. *plecostomus*
Ancistrus	sp. C
Anostomus	*ternetzi*
Jupiaba	*polylepis*
Knodus	sp. B
Leporinus	cf. *maculatus*
Microschemobrycon	*callops*
Microschemobrycon	*casiquiare*
Microschemobrycon	*melanotus*
Moenkhausia	cf. *lepidura* D
Moenkhausia	*copei*
Ochmacanthus	*alternus*
Paravandellia	sp.
Pimelodella	cf. *cruxenti*
Pimelodella	cf. *megalops*
Ramirezella	*newboldi*
Rineloricaria	sp.
Serrasalmus	sp. A
Steindachnerina	*pupula*
Tetragonopterus	*chalceus*
Triportheus	*albus*
Vandellia	*sanguinea*

Table 8.3. Number of species shared (upper triangle) and mean Simpson's similarity coefficients, S's (lower triangle) among regions within Caura River. Similarity coefficients shown in bold are significantly different from random similarity (P<0.001). The shaded cells are comparisons among regions below the falls. Abbreviations: Ent-Cejiato–Caura River between Entreríos and the Raudal Cejiato; Ent-SP–Caura River between Entreríos and Salto Pará; Cinco Mil–Raudal Cinco Mil; Mato–Río Mato; n–number of species; u–number of unique species; %u–percentage of unique species.

	Kakada	Erebato	Ent-Cejiato	Ent-SP	Playon	Nichare	Cinco Mil	Mato
Kakada	1	25	25	18	11	12	7	4
Erebato	**50.67**	1	39	36	21	22	16	9
Ent-Cejiato	**32.33**	**49.52**	1	43	27	27	21	12
Ent-SP	**28.61**	**57.16**	**54.47**	1	24	24	22	15
Playon	9.92	19.49	25.04	-22.18	1	66	57	38
Nichare	9.92	18.3	22.42	-19.88	**54.53**	1	61	45
Cinco Mil	-6.91	-15.82	-20.62	-21.66	**52.88**	**50.38**	1	43
Mato	-4.44	-10.27	**-13.65**	-17.03	35.23	37.3	**42.1**	1
n	27	49	79	63	108	121	102	88
u	0	1	19	6	21	27	22	30
%u	0	2.04	24.05	9.52	19.44	22.31	21.57	34.09

and Playón subregions was random and, therefore, is not to be interpreted. In the subregional group from Upper Caura there is a principal river grouping extending from just above the falls to Cejiato; the samples from the Erebato River and Kakada River share sequentially fewer species with the mainstem group. This pattern is confirmed by the results of the principal coordinates analysis (Figure 8.3), in which the first two principal coordinates (64% of the variance) array the subregions congruent with their geographical relationships—the groups separated by the falls, and the Mato River is most disparate.

Table 8.4. Species of freshwater fishes commonly found in the Caura River. Common is defined as having been captured in five or more of the eight regions.

Aequidens	cf. *chimantanus*
Astyanax	*integer*
Bryconops	cf. *colaroja*
Characidium	sp. A
Cyphocharax	*festivus*
Cyphocharax	sp.
Hoplias	*macrophthalmus*
Hypostomus	sp. B
Jupiaba	*atypindi*
Jupiaba	cf. *atypindi*
Jupiaba	cf. *polylepis*
Jupiaba	*zonata*
Moenkhausia	cf. *lepidura* A
Moenkhausia	cf. *lepidura* B
Moenkhausia	cf. *lepidura* C
Moenkhausia	cf. *lepidura* E
Moenkhausia	*collettii*
Moenkhausia	*grandisquamis*
Moenkhausia	*oligolepis*
Phenacogaster	sp. A
Pimelodella	sp. B
Pimelodella	sp. C
Poptella	*longipinnis*
Pseudocheirodon	sp.
Satanoperca	sp. A
Serrasalmus	*rhombeus*
Synbranchus	*marmoratus*
Tetragonopterus	sp.

Macrohabitats

The distribution of fish species was found to be non-randomly distributed with respect to the sample of 20 macrohabitats. All of the coefficients were significantly different from random ($P<.005$) and the matrix was significantly more ordered (cooler) than expected at random ($P<.0001$).

The non-random associations due to habitats are evident in the Camin-Sokal Parsimony analysis (Figure 8.4). The

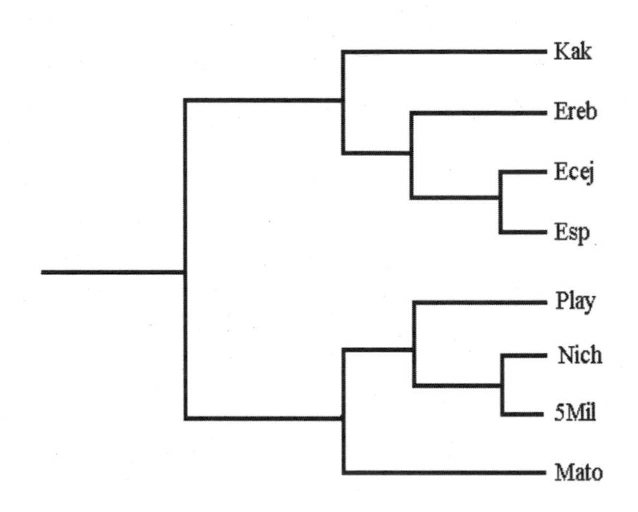

Figure 8.2. Camin-Sokal parsimony analysis of Simpson's similarities among subregions in the Caura River for fishes. Abbreviations: Kak–Kakada River; Ereb–Erebato River; Ecej–Caura River from Entrerios to Cejiato; Esp–Caura River from Entrerios to Salto Pará; Play–El Playón; Nich–Nichare River; 5Mil–Raudal Cinco Mil; and Mato–Mato River.

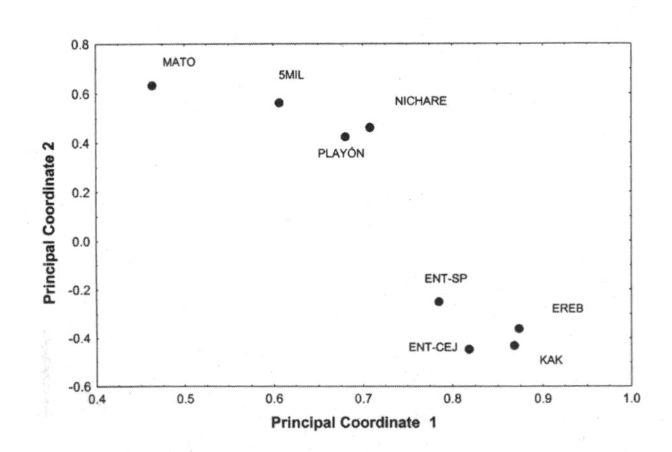

Figure 8.3. Principal coordinates analysis of Simpson's similarity matrix among subregions of the Caura River. Abbreviations: Kak–Kakada River; Ereb–Erebato River; Ent-Cej–Caura River from Entrerios to Cejiato; Ent–SP–Caura River from Entrerios to Salto Pará; Play–El Playón; Nichare–Nichare River; 5Mil–Raudal Cinco Mil; and Mato–Mato River.

retention index is 0.783. In the analysis, many like habitats are grouped together. For example, on the right hand side is a cluster containing muddy bottoms and shore, logs, leaves and detritus along with forested shores. A somewhat different community is found in big river habitats with sand, rocks, and beaches. The upper Caura River is well characterized by many island habitats that share many species in common with the main channel.

The assemblage of species inhabiting grasses, macrophytes and rapids is a major discovery of our expedition. In the upper Caura River, there are many rapids or areas with swift water over large boulders and rocks, including over the Salto Pará. In these swift water habitats we found up to seven species of the vascular plant family Podostemonaceae. There were 130 species of fishes living among the macrophytes overall, and 120 species in macrophytes and rapids.

The principal coordinates analysis (Figure 8.5) better demonstrates the transitions among habitat groups. For example, the transition between mud to sand in riverine communities occurs in the upper left side of the graph. There is also a smooth transition from sand bottoms and beaches through rocky substrates to islands, and rapids with macrophytes. From center to top right (Figure 8.5) represents a transition among lower velocity habitats (e.g., caños) to lakes and areas with still waters with floating vegetation.

Given the diversity of and non-random patterning of species among macrohabitats, we ask how many macrohabitats are required to protect the majority of the species. The accumulation curve of the percent of total species over the number of macrohabitats is shown in Figure 8.6. The curve was calculated from polynomial regression in which all coefficients are significantly different from zero. The equation of the regression is:

$$Y = 50.9 + 24.2X - 9.7X^2 + 1.8X^3 - 0.2X^4 + .004X^5 + e$$

where Y is the cumulative percentage of species, X is the number of macrohabitats and e is the error term. The greatest return of cumulative percentage to the number of macrohabitats is determined from the inflection point (setting the second derivative to zero). The inflection point is 6.0, which corresponds to 82% of the species (Figure 8.6). Thus, if adequate protection can be given to 6 macrohabitats,

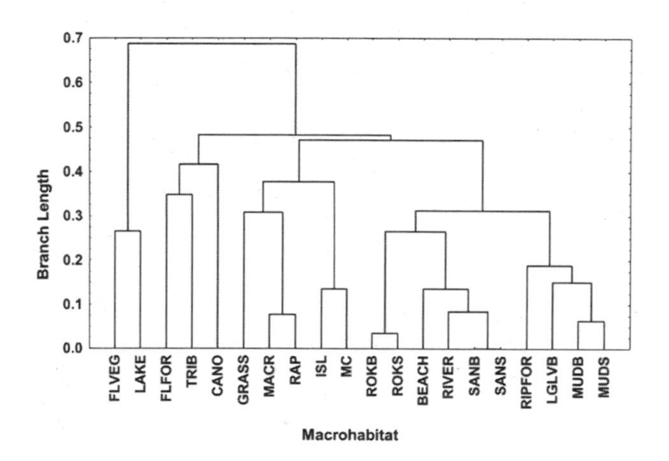

Figure 8.4. Camin-Sokal Parsimony analysis of 20 macrohabitats in the Caura River. Abbreviations: B–bottom; Flveg–floating vegetation; Flfor–flooded forest; Cano–caño; Isl–island; Lglv–logs and leaves; Macr–macrophytes; MC–main channel; Rap–rapids; Ripfor–riparian forest; Rok–rocky; San–sandy, S–shore; Trib–tributary.

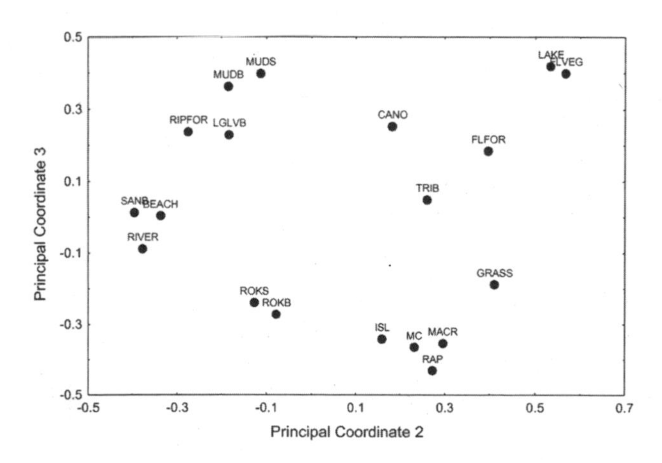

Figure 8.5. Principal coordinates analysis of Simpson's similarity coefficients among 20 macrohabitats in the Caura River. Abbreviations: B–bottom; Cano–caño; Flfor–flooded forest; Flveg–floating vegetation; Isl–island; Lglv–logs and leaves; Macr–marophytes; Rap–rapids; Rok–rocky; S–shore; San–sandy; Trib–tributary.

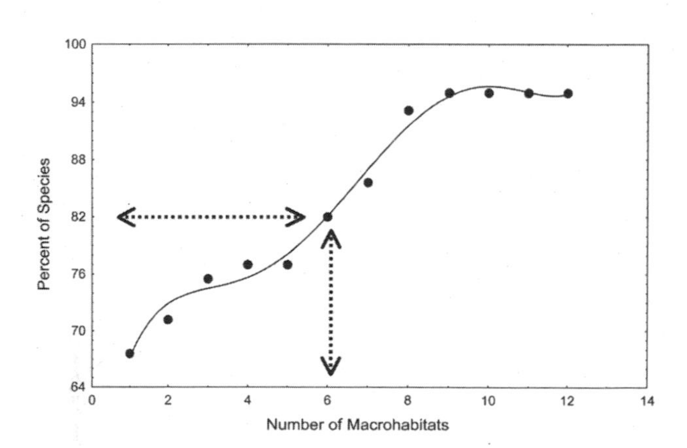

Figure 8.6. Cumulative percentage of species plotted against the number of macrohabitats. The actual values are shown as dots; the polynomial regression line is shown. The arrows indicate the point of inflection for the regression equation, indicating the maximum percentage of species that can be preserved with the fewest macrohabitats.

the large majority of species can be protected. This group includes 228 species with all of the commercially important species found in Appendix 2.1.

DISCUSSION

The distribution of fishes in the Caura River show strong subregional and macrohabitat effects, and are largely non-random and non-homogeneous. The subregional effects that are non-random are due to separation by the Salto Pará. These results accord well with our studies in the Tahuamanu-Manuripe rivers, Bolivia, the southern Pantanal, Brazil, and the Paraguay River in Paraguay, as well as the studies on the Jau River, Brazil (Forsberg et al. 2001), lowland Amazon floodplain (Cox Fernandes 1995; P. Petry, pers. comm.) and the Napo River, Ecuador (Ibarra and Stewart 1989). These studies disagree with the general statements of Lowe-McConnell (1987) and Goulding et al. (1988) who claim that distributions of freshwater fishes in lowland habitats are largely random. The non-random, non-homogeneous distribution have important implications for the conservation of the ichthyofauna.

The regions in the Upper Caura have significantly fewer species of fishes than the regions below the Salto Pará. This is because species from the Orinoco River penetrate the Caura River to varying degrees. The species richness of the plants overall is greater above the river than below (291 vs 185, respectively). However, there were more vascular aquatic macrophytes and grasses above the falls than below (Rosales et al. 2003a). This difference in species richness associated with elevational change in the fish and the plant data is congruent with the results reported from the southern Pantanal (Chernoff and Willink 2000). The benthic invertebrate data (Appendix 6) do not show an appreciable difference with 70 species in the Lower Caura and 75 species in the Upper Caura.

The distribution of benthic invertebrate species differs from those of the fishes in that the invertebrates do not show a geographic pattern of similarities among the subregions (Garcia and Pereira 2003). We have reanalyzed their data to correspond to the subregions presented herein and in the botanical analyses. The matrix correlation of subregional similarities between the fish and benthic invertebrate data sets is not significant ($r = 0.32$, $P>.05$). The results of Camin-Sokal Parsimony analysis and principal coordinates analysis (Figures 8.7, 8.8) show that although the below-falls (Salto Pará) samples from El Playón and the Mato River are outliers to the remaining subregions, Cinco Mil and Nichare share significant numbers of species with the above-fall subregions. This is in distinct contrast to the fish and botanical distributions. The fishes show a strong upstream-downstream component (Figures 8.2, 8.3) but are not well structured above the falls. The plants, on the other hand are well structured with regards to subregions (Rosales et al. 2003a,b). The plants show a marked discontinuity at the falls, but subregions such as the Kakada and Erebato also represent floral turnovers. The distribution of the fishes is intermediate between the unstructured invertebrate data and the well-structured plant distributions.

Macro-habitat effects are significant, non-random and critical to properly understanding the distributions of plants,

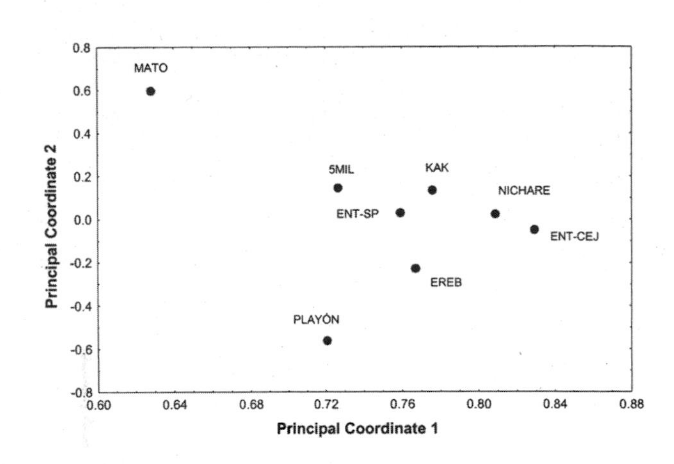

Figure 8.7. Principal coordinates analysis among Jaccard's similarity matrix for subregions of the Caura River for aquatic benthic invertbrates. Abbreviations: Ent-Cej–Caura River from Entrerios to Cejiato; Ent-SP–Caura River from Entrerios to Salto Pará; Ereb–Erebato River; Kak–Kakada River; Mato–Mato River; Nichare–Nichare River; Play–El Playón; 5Mil–Raudal Cinco Mil.

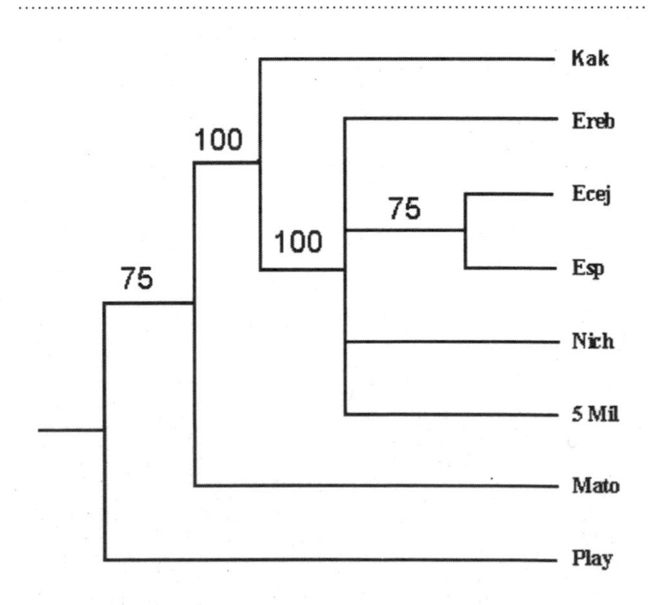

Figure 8.8. Majority rule consensus tree of Camin-Sokal Parsimony analysis of subregions in the Caura River for benthic aquatic invertebrates. Ecej–Caura River from Entrerios to Cejiato; Ereb–Erebato River; Esp–Caura River from Entrerios to Salto Pará; Kak–Kakada River; Mato–Mato River; Nich–Nichare River; Play– El Playón; 5Mil–Raudal Cinco Mil.

invertebrates and fishes within the Caura River watershed (see above, García and Pereira 2003; Rosales et al. 2003a.). It is difficult to precisely compare the macrohabitats among these groups. However, Garcia and Periera (2003) do demonstrate the correlations between physicochemical characteristics of habitats (such as dissolved oxygen) and that species of Odonata and Ephemeroptera, which were characteristic of swift water and rapids. The fishes are partitioning the environment relative to bottom type, forest structure, water current and the presence of macrophytes or other vegetation. Smooth transitions in faunal composition are evident (Figure 8.5) based upon these variables. Interestingly, many of the patterns of fish assemblages correspond to landscape changes in forests. As noted by Rosales et al. (2003a), the islands, particularly above the falls are unique habitats with characteristic assemblages of plants. These island habitats are unique for the fishes, as well, with rocky outcrops, lush stands of Podostemonaceae and rapids or swift currents.

Developing any conservation strategy for the Caura watershed must take into account both the subregional and macrohabitat effects upon species distributions. For example, 37 out of 278 species of fishes were found in three of four subregions, and only 28 species are found in five of the eight subregions and regarded as common. Different subregions and macrohabitats are critical as nursery areas or for particular life stages. Commercially important species, such as the palometa, *Myleus rubripinnis,* were found in backwater areas. Diminution of water quality through deforestation, runoff or pollution due to mining will kill off the aquatic macrophytes, invertebrates and ultimately the fishes—130 species of fishes were found closely associated with aquatic macrophytes. The effects of the number of macrohabitats on number of fish species (Figure 8.6) demonstrated that if six or more macrohabitats could be preserved in sufficient quantity, then more than 82% of the known fish fauna of the Caura River could be saved. It is particularly critical to incorporate the Lower Caura, especially from Raudal Cinco Mil downstream, because of the increasing effects of human pressures on the river.

LITERATURE CITED

Alroy, J. 1992. Conjunction among taxonomic distributions and the Miocene mammalian biochronology of the Great Plains. Paleobiology 18:326–343.

Atmar, W., and B. D. Patterson. 1993. The measure of order and disorder in the distribution of species found in fragmented habitat. Oecologia 96:373–382.

Brualdi, R. A., and J. G. Sanderson. 1999. Nested species subsets, gaps and discrepancy. Oecologia 119:256–264.

Chernoff, B., and P. W. Willink. 2000. Biodiversity patterns within the Pantanal, Mato Grosso do Sul, Brasil. *In:* Willink, P. W., B. Chernoff, L. E. Alonso, J. R. Montambault and R. Lourival (eds.). A Biological Assessment of the Aquatic Ecosystems of the Pantanal, Mato Grosso do Sul, Brasil. RAP Bulletin of Biological Assessment 18. Washington, DC: Conservation International. Pp. 103–106.

Chernoff, B., P. W. Willink, J. Sarmiento, A. Machado-Allison, N. Menezes and H. Ortega. 1999. Geographic and macrohabitat partitioning of fishes in Tahuamanu-Manuripi region, Upper Río Orthon basin, Bolivia: conservation recommendations. *In:* Chernoff, B. and P.W. Willink (eds.). A Biological Assessment of Aquatic Ecosystems of the Upper Río Orthon Basin, Pando, Bolivia. RAP Bulletin of Biological Assessment 15. Washington, DC: Conservation International. Pp. 51–68.

Chernoff, B., P. W. Willink, M. Toledo-Piza, J. Sarmiento, M. Medina and D. Mandelburger. 2001a. Testing hypotheses of geographic and habitat partitioning of fishes in the Río Paraguay, Paraguay. *In:* Chernoff, B., P. W. Willink and J. R. Montambault (eds.). A Biological Assessment of the Aquatic Ecosystems of the Río Paraguay Basin, Alto Paraguay, Paraguay. RAP Bulletin of Biological Assessment 19. Washington, DC: Conservation International. Pp. 82–101.

Chernoff, B., P. W. Willink, A. Machado-Allison, M. Fatima Mereles, C. Magalhaes, F. A. R. Barbosa, M. Callisto Faria Pereira and M. Toledo-Piza. 2001b. Congruence of diversity patterns among fishes, invertebrates and aquatic plants within the Río Paraguay Basin, Paraguay. *In:* Chernoff, B., P.W. Willink and J.R. Montambault (eds.). A Biological Assessment of the Aquatic Ecosystems of the Río Paraguay Basin, Alto Paraguay, Paraguay. RAP Bulletin of Biological Assessment 19. Washington, DC: Conservation International. Pp. 102–114.

Cox Fernandes, C. 1995. Diversity, distribution and community structure of electric fishes (Gymnotiformes) in the channels of the Amazon River system, Brasil. Ph.D. Dissertation. Duke University, Durham.

Forsberg, B., J. G. D. Castro, E. Cargnin-Ferreira and A. Rosenqvist. 2001. The structure and function of the Negro River Ecosystem: insights from the Jau Project. *In:* Chao, N. L., P. Petry, G. Prang, L. Sonneschein and M. Tlusty (eds). Conservation and Management of Ornamental Fish Resources of the Río Negro Basin, Amazonia, Brazil. Manaus Brazil: Universidade do Amazonas. Pp. 125–144.

García, J. V., and G. Pereira. 2003. Diversity of benthic macroinvertebrates from the Caura River watershed, Venezuela. *In:* Chernoff, B., A. Machado-Allison, K. Riseng, and J. R. Montambault (eds). A Biological Assessment of the Aquatic Ecosystems of the Caura River Basin, Bolívar State, Venezuela. RAP Bulletin of Biological Assessment 28. Washington, DC: Conservation International. Pp. 49–55.

Goulding, M., M. L. Carvalho and E. G. Ferreira. 1988. Rio Negro: rich life in poor water: Amazonian diversity and foodchain ecology as seen through fish communities. The Hague Netherlands: SPB Academic Publishing.

Ibarra, M., and D. J. Stewart. 1989. Longitudinal zonation of sandy beach fishes in the Napo River basin, eastern Ecuador. Copeia 1989:364–381.

Lowe-McConnell, R. H. 1987. Ecological studies in tropical fish communities. New York USA: Cambridge University Press.

Machado-Allison, A., B. Chernoff, F. Provenzano, P. Willink, A. Marcano, P. Petry and B. Sidlauskas. 2003. Inventory, relative abundance, diversity and importance of fishes in the Caura River Basin. *In:* Chernoff, B., A. Machado-Allison, K. Riseng, and J. R. Montambault (eds). A Biological Assessment of the Aquatic Ecosystems of the Caura River Basin, Bolívar State, Venezuela. RAP Bulletin of Biological Assessment 28. Washington DC: Conservation International. Pp. 64–74.

Rice, W. R. 1989. Analyzing tables of statistical tests. Evolution 43: 223–225.

Rohlf, F. J. and R. R. Sokal. 1995. Statistical tables. New York USA: W. H. Freeman and Co.

Rosales, J., M. Bevilacqua, W. Díaz, R. Pérez, D. Rivas and S. Caura. 2003a. Riparian Vegetation Communities of the Caura River Basin, Bolívar State, Venezuela. *In:* Chernoff, B., A. Machado-Allison, K. Riseng, and J. R. Montambault (eds). A Biological Assessment of the Aquatic Ecosystems of the Caura River Basin, Bolívar State, Venezuela. RAP Bulletin of Biological Assessment 28. Washington DC: Conservation International. Pp. 34–43.

Rosales, J., N. Maxted, L. Rico-Arce and G. Petts. 2003b. Ecohydrological and ecohydrographical methodologies of riparian vegetation applied to conservation: the Caura River as an example. *In:* Chernoff, B., A. Machado-Allison, K. Riseng, and J. R. Montambault (eds). A Biological Assessment of the Aquatic Ecosystems of the Caura River Basin, Bolívar State, Venezuela. RAP Bulletin of Biological Assessment 28. Washington DC: Conservation International. Pp. 75–85.

Sneath, P. H. A., and R. R. Sokal. 1973. Numerical Taxonomy. San Francisco USA: W. H. Freeman and Co.

Glossary

Aquaculture—Raising fish in enclosed areas, usually with the intent of selling the fish.

Arboreal—Pertaining to trees.

Basin—See *watershed.*

Benthic—Of or pertaining to the bottom of a river, lake, or other body of water.

Biodiversity—Description of the number of species, their abundance, and the degree of difference among species.

Carnivorous—Organisms that feed on animals.

Caño—Tributary in which water can flow in both directions depending on whether the water is rising, due to rain or floods, or falling.

Cretaceous—A time period of the earth's history extending from about 145 million years ago to about 65 million years ago.

Dead arm—An arm of a river extending into the forest in an old river channel.

Endemic—Found only in a given area, and nowhere else.

Endorheic—Waters flowing into an enclosed watershed.

Erosion—The act of water washing away soil.

Floating meadows—Large aggregations of floating vegetation.

Fluvial—Pertaining to rivers or streams.

Herbivores—Organisms that feed on plants.

Heterogeneity—Degree of difference among items.

Hydrological—Pertaining to water.

Insectivores—Organisms that feed on insects.

Inundation—Flood.

Laguna—Lagoon.

Landscape—Pertaining to the structure of an area at a coarser level than that of site, at least at the level of a georeference area.

Lentic—Pertaining to still water, as in lakes and ponds. See *lotic.*

Liana—Vine.

Littoral—The aquatic zone extending from the beach to the maximum depth at which light can support the growth of plants.

Lotic—Pertaining to flowing water; as in rivers and streams. See *lentic.*

Macrophyte—A non-microscopic aquatic plant.

Madrevieja—Dead arm.

Miocene—A time period of the earth's history extending from about 25 million years ago to about 5 million years ago.

Omnivores—Organisms that feed on a variety of food types, both plant and animal.

Periphyton—Algae attached to rocks, logs, and other underwater substrates.

Physiognomy—The overall appearance or constituency of an area.

Piscivores—Organisms that feed on fishes.

Pleistocene—A time period of the earth's history extending from about 2 million years ago to about 100,000 years ago.

Precambrian—A time period of the earth's history extending from the origin of the earth to about 580 million years ago.

Quaternary—A time period of the earth's history extending from about 2 million years ago to the present.

Riachuelo—Small stream, forest stream.

Río—River.

Riparian—Found along the edge of a river. Often used in the context of vegetation.

Savanna—Grasslands.

Seine—A mesh net, often used to catch fish and other larger aquatic organisms.

Siltation—The deposition of fine sediment (know as silt), oftentimes covering existing structures.

Stagnant—Still water (that is often foul).

Substrate—The soil found between the roots of plants or at the bottom of lakes and rivers.

Watershed—A region drained by a particular river and its associated streams. Also known as a basin.

Tuna Medewadi, Wenesuweda de´wö

TAMEDÖ YEICHÜ

Tuna Medewadi na shi weja´katojo dü´se wodiwa nonoodü de´wö yeichü, Wenesuweda, annawööne Escudo Guayanés de´wö(3°37'–7°47'N y 63°23'–65°35'W). Na ajo´jo chuuta tumakudojunu yeichüüdö tunakomo maja yaawö dinñaku´kwaka yekenkajöötüdü (Rosales y Huber 1996). Edo chuuta na tüwü seneedö yeichü, tukunnö´ato maja, woi aiño maja naadenha yaawö chü´tadü 90% je medeewadi nonoodü de´wö (Marin y Chaviel 1996). Edö naadü na yaawö ye´kwana nonoodü weichü,sotto eda´chönnamo tünwanno iyö nono de´wö naadü. Inñataje naadü tüwü chuuta medewadi nonoodü de´wö, muuda soodü jokonno yujuudaka tüdüjoone, tünwanno ye´kwanaakomo kenñe.

Medewadi motadükoomo ajo´jokoomo na yaawö Sipawo, Wünküyaadi, Dedewatö, Medewadi maja shi womontojo dü´se, yotonno yaawö Tikededa, Jadde, Yuduwani, Sawaadu, Wanña maja shi weja´katoojo dü´se. Medewadi na 700 kilimetros je, neja´ka yaawö Escudo Guayanés kawö yeichü de´wö, 2000 metudu je dama jonno kajunñadödö: nija´dö´a önnene nono weichü de´kökö, önnene tooja´jano jonno yujuudawö akanajaato tooja´jemjünü dinña tüdüjoone dinñaku´kwaka yeekekadünña. Medewadi nonoodü na yaawö yü´seeto´jüdü 45.336 Km2 je, 20% je Wodiwa nonoodü jökö yeichü, 5% je tameedöödö wenesuweda nonoodü jökö yeichü; edö tuna nonoodü ajo´jo yeichü na aaköcheinña tooje Apuude, Kadooni, Dinñaku´jeje yeichü (Peña y Huber 1996).

ÖNNEENE MEDEWAADI NONOODÜ DE´WONKOMO YEICHÜ

Önnene Medewadi nonoodü de´wö naadü ajichajeene owaanökö´da na. Ooje woowanoma´jö yeichame chuuta jökö chuutakankomo jökö maja (Rosales y Huber 1996, 1997), na´kwakankomo jökömmaane yööje wö´duto´me´dana (Machado-Allison et al. 1999).

Medewadi nonoodü de´wö na önnene chuuta, 1180 e´joye chuuta yeichü edantaajö. Jajeeda jökö ajoijajö na chuuta 90% na Medewadi nonoodü omomjünü de´wö, 10% kene yaawö chuuta tukunnö´ato yö´jejemaja chuuta ya´wakukaajö (CVG-TECMIN 1994; Huber 1996; Marin y Chaviel 1996; Rosales 1996; Aard et al 1997; Dezzeo y Briceño 1997; Bevilacqua y Ochoa 2001).

Chuutakankomo Medewaadinñankomo töne nato 475 je tadinñaamo, 168 je odookoja´komo, 13 kütooja´komo, 23 je makasana okoyuja´komo yeichomjökö (Bevilacqua y Ochoa 2001). Edo nadü yawö na 30% je chuutakankomo edanta´komo Wenesuweda de'wö yeichü yotonno yaawö 51% je Guayana de´wö yeichü. Tameedöödö ye´jö´ja´komo weichü Medewaadinña naatoodü, 5% je naato amonche´da yaatamedükomo eetö Wenesuweda de´wö aneja nono de´wö maja yaawö (Bevilacqua y Ochoa 2000). Edö AquaRap küntöjö´a´todawö küneedantoicho 113 je eduwa´komo, 10 je eduwa´komo ju´jokomo weichü, mödö aka naato na´kwakankomo weichö tuna saayu´jemjünü akankomo naato yaawö yojodüa´komo 278 je. 92 je se´jömjünükomo naatodü 12 je yaawö wayakani ja´komo mödööna yaawö awa´deenato mödööwa´kö yeeja´kaajö.

Chuuta chutakankomo maja Medewaadi nonoodü de´wö naatodü na yaawö Jaada soodü e´nei na yöökamooajö Medewadinña naatoodü. Jaada soodü de´wonno maadödö na yaawö inñankomo jooje inñawoono we´yeje´da, mödö nene´ju´jö´a yaawö chuuta ke, tüwü na yaawö yü´seeto 88% je Guayana de´wono denha na yöötö mödö töneejeene na yaawö ojejene jüü de´kökö (Berry et al. 1995; Huber et al. 1997; Bevilacqua y Ochoa 2001). Yööjemmaja ö´düjai weiño nadenña yaawö chuuta, se´jömjünükomo na´kwakankomo kudaada maja u´joye naatodenha Escudo Guayanés yotonno Amazonas de´wonkomo denha tünwanno, anejanaadö´ja nadenha jü´waye, dinñaku´kwakaano yotonno yanonñano maja tüweiye.

ÖNNENE SOTTO WEICHÜKOMO YEICHO´KOMO MAJA

Sotto weicho'komo yeicho´komo maja Medewadinñankomo, na ooje önnene. Medewadi ü´joye ju´wayeedö maja jaada de´wö naato ye´kwaanakomo Ye´kwana Sanama maja, tünwanno nato yaawö yöötödö nadöato tüwotunnoichomo. Kiyeede yotonno jaduudu ñaatü´tödü, wesenünnö, o´tödü, chööjüdüdü amukudu na chadawajuichomo yeicho´komo maja. Mödööje yaawö chööwaadönñe ñanno sotto tadawajuichomo na yaawö inchomoodü woije yanwaje yeichü o wodije yeichü woije (Silva-Monterrey 1997).

Sotto chaadawajoichomo jökö yeichü Medewaadi jü´waye na önnene. Inña na yaawö wanna´köjeene sotto iyö yaawö ye´kwana weichü noojodüa yaawö yadanawi weichü akö. Mödööje yaawö chadawaajuichomo nadenha yaawö iye akötödü, natü ñaatü´tödü, oküünü inñejenkadü, o´tödü, Waademanö jökö wetadawa´kajoonö, yotonno yaawö tükaajüdü o akaajötüdü jeiñemma.

Ye´kwanakomo najöiyaato tümenka´da kudaka yö´jöje yawö anejakomo na´kwakankomo totükomooje. Yaanc yööje´da yaawö ye´kwanakomo weichükomo nünhe´da, yadanawi kudaka najöiya sotto otükomooje tümenka inñatakomoodenña: kachamas *(Colossoma macropomum)*, kajaadu *(Phractocephalus hemiliopterus)*, ya´koto *(Prochilodus mariae)*, Akujja *(Plagioscion squamossissimus)*, Laulaos yotonno muukudi (*Brachyplatystoma* spp.), muudu (*Piaractus brachypomus*), jadumeeta (*Mylossoma* spp.), kudidi (*Pseudoplatystoma* spp.), sajuwaada (*Semaprochilodus laticeps*), yotonno sadidinata (*Pellona castelneana*) (Machado-Allison et al. 1999; Novoa 1990).

Oje kudaka nato yaawö tujunna´komo ökünüje tüdüüamo (yadanawi wö) mödö na yaawö tadawajuje eijaicho tajoojodenha tüwü yaawö ye´kwanakomo tadawajuije tüwatamemjünüje eijaicho mane tüwü yaawö. Kanno ñanno yaawö kudaka okünüje tüdüüamo medewaadinña edanta´komo: tetras (*Astyanax, Hemigramus, Hyphessobrycon, Jupiaba, Moenkhausia*), jadumeta o silver dollars (*Metynnis, Myleus, Mylossoma*), cichlids (*Aequidens, Apistogramma, Bujurquina, Mesonauta*), Ka´shai (*Pygocentrus, Serrasalmus*) yö´jöje yaawö headstanders (*Anostomus, Leporinus*).

Jata weicho´komoje na yaawö medewadinñankomo, dinñaku´kwaichomo maja yaawö, aónmakudö´da naato yaawö tünna´kwadükomo´kwai ñonoodu de´wö naadü maja yaawö. Iye nako´aato ye´kwanaakomo tüweicho´komooje, 358 je iye weichü owanökö na yaawö ye´kwanaakomo tünaakö´e yeichü (Knap-Vispo 1998). AquaRap nedantödü´kene yaawö chuuta tünnadüüato na´kwakankomo weicho´komo maja yaawö tuna e´jichokono na yaawö önnene kudaka wennejenkato´komooje naadünane jadumeta (*Myleus rubripinnis*). Enmenkaajo na yaawö akene ke´ja aneja´kwaichomo kudaka wüta´komo medewaadi´chai ni´ya´ta yaawö yü´seto´jüdü 75,000 tonedada métrica je Madijanña Kaikadaña maja ajöiya´komo.

MEDEWAADI MAKUDÖNEI WEEJÜDÜ YAAWÖ

Escudo Guayanés wenesuweda de´wö na tünkone´madü yeichame jata tü´tajötüdü tüwennejenkato´me, amode´notojo, na´kwakano ajöichü o wesenünnö inñammödö chü´tammeküdü´jo´da tüdüdü nejoodüjoodü. Uudu, widiki, Bauxita u´kadü yoojodüajö yaawö tuna a´dudu jadö nejodüja yaawö: 1) Chuutakankomo yotonno chuuta wö´jajoodü ooje kadoninña yotonno yaawö kuyuweninña maja (Machado-Allison et al. 2000); 2) Tuna yemmadü, chuuta, ose, sotto maja yaawö asoke ke; 3) Nossaje tuna wööyemmadü tumuuke; 4) Tuna inñatadü tönhemmü´je´da yö´düdü yaawö (Machado-Allison 1994, 1999; Miranda et al. 1998).

Medewaadi kone´manei weejüdü na yaawö uudu, widiki u´kadü ke, inñammödö na´kwakano ajoichü jajeda´je´da tüweichame, tuna inña´düdü maja yaawö. Uudu, widiki u´kadü na oojeje tünonñato yaawö, jadawa akö amonche´da ööjökö yeijökö medewaadi, inña ooje uudu widikimaja u´kadü na (Machado-Allison et al. 2000). Yotonno yaawö, nöneadenha yaawö ömje´sato oje kudaka´je´da yeichü eduwa medewaadi jü´waye ye´kwanakomo nonoodü de´wö ajöichü töjemajomuje yeijököjenñema inñammödö jajedaje´da tüweichame.

Naadenha yaawö aku´sana tü´tajöötüdü Medewaadi inña´dudu jökö 75% tuna adudu jökönchödö Medewaadi jonno Jadawa´kwaka kadoninña tüdüjone, yotonno yaawö Jaada inña´düdü. Mödoö aku´sana tü´tajöötüdü naadü tuna medewaadi se´kö yödüdümma jünü na yaawö, tuna waawüyümüdü nichone´madenha yaawö. Sotto jataadükomo, na´kwakankomo, chuuta, tunajakokonkomo weichü naato yaawö tuwoije tuna wawüyümudüke. Yotonno nadenha yaawo, Jaada jonno ju´wakaadö wö´kaaje´da ooje´kömma no´düa wönwenaanö jökö wetadawa´kajoonö, o´tödü enno´janködö, tumjune´da natü ñaatü´tödü maja yaawö, mödööna yaawö jojejeene Muuda sodünwawonno jü´wakaadö. Edö tadawaju naadü yaawö wowanoomanö eijaiña yaawö ajichajeene wetadawakajoono wetö nono de´wö.

SOOMATOJO EIJAICHO

Medewaadi nonoodü na yaawö ajo´jo chuuta ye´wö. Chuuta wekone´maduke jata tadawajui wennejenkadüüke nortenñano wadödoö Escudo Guayanés de´wö yeichü (Kadoninña Kuyuwininña maja), edö Medewaadi na yaawö tüsomamje yaadödö yeijökö inñatajeedö. Chuutakankomo Chuuta maja yaawö Medewaadi de´wono naato yaawö a´ke aninñaja eetö wenesuweda de´wö yeichü a´ke maja yaawö Escudo Guayana de´wö. Na yaawö tüwü onnene ye´wonkomo, yotonkomo tötü tünwanno, kanno na´kwakankomo (podostemonaceae) soodü jökö yeichukomo tünwanno yotonno chuuta yantadü de´wonomma tüwü. Modooje yeijökö yaawö tüdüdü eijanña wowanomanö ñanno yöötö jata nadükomo, mödö yaawö Medewaadi eda´choto´me ma tuweiyemüje. Yojemmaja yaawö na´kwakano ajöchü aje´kadü tujunne nadenha chonekadö maja yaawö iyö yeichojo ataame´da yeichükomo wetö yotonkomo weichojoje

AQUARAP WÜTÖJÖTÖ´JÜDÜ

Aku´sana tu´tajötüdü watannö´nö jökö Medewaadinña, tuna adöödü medewaadi jadawa wadödö, iye akötödü, sotto wejamüdükomo, yotonno yaawö aneja kone´manei weejüdü ekwojötüdü jokonchödö, künojodücho tawanojo´na´komo wanna aninñajankomo jadö kuntonto yaawö tuna enmenka tamjö´ne (AquaRap), edö tüdü jökönchödö yaawö: 1) Ekammajotüdü jökönchödö na´kwakankomo weichükomo yeichü´komo awiyakoko nadü maja yaawö iyö de´wö natoodü tumjune´da ñaakudödü owajo; 2) Eneedü, na´kwakankomo eijaicha naicho iyö yeicho´komo choone´ma´jökö.

Küntöjö´a´to yaawö 25 noviembre yeichü 12 diciembre jona tüdüjoone 2000 wedu yeichü, mödö künnücho yaawö sejiyato sodü jonno matu kankoi tüdüjoone. Ashicha enmenkadü jökönchödö, künö´datokajötüi yaawö aaduwawö amojato´kwakö töneemü: 1) Ka´kada; Dedewatö; 3) Medewaadi dedewatö kanonno sejiyato sodü jonane; 4) Medewaadi dedewato kanonno jaada sodü jonane; 5) Jaadanwakökö; 6) Wünküyade; Muuda sodüchökö; Matu maja. Künötömmenkai yaawö chuuta weichü mawoono nakwakano maja, tuna weichü yakaano maja yaawö, tunanwakonño, se´jömjünükomo na´kwakankomo yotonno yaawö kudaka maja. Edö tüdüajo yaawö, kone´manei weejüdü yeijökö, no´dua´de yaawö medewadi somato´me wowanomatojojene eijaiña aneedawö ke wö´dütojoje.

Programa suramerica wadödö AquaRap tüwü na yaawö iyö nono de´wono aninñajano akö aatantai tawanojo´na´komo maja yaawö soomatojo e´se´totoojo edantödü jökö maja yeichü tüwü ataame´da na´kwakankomo öse chuuta maja yeichükomo wetö ta´kwiti´yemjünü kwaka naatodü America Latina de´wö. AquaRap tadaawajui na yaawö oneejadü iyö tuna´kwawö naatodü yeicho´komo soomatojo maja, tamjö´ne ñe´ku´tödukomoai, ekammajotudu weto yawo kajichanakomowo, jajeeda chonekannamowo, eda´chödüjökö nichü´tajö´aatodü wönñe , científico komowö, jüdata u´namowo maja edo nunhato jokono. AquaRap na yaawö we´wa´tönö jokono ne´se´ta yaawö Coservación Internacional yotonno Field Museum.

AquaRap aka na yaawö tüweiye comité internacional mödö niju´jö´ta yaawö científicokomo eda´chö´sa´komo aköamojato´to´kwakö nono de´wö (Blivia, Brasil, Ecuador, Paraguya, Perú, Venezuela yotonno yaawö Estados Unidos). Edö comité nadü nemmenka yaawö jajeda tamjö´ne ö´düjai yeichü tujunne yeichü maja yaawö enmenkadü jökö yeichü. AquaRap wütöjötüdawö chöjadönñe ichödükomo na yaawö yeichü iyö nonode´wonkomo científicokomo jadö, towanokomo ekammajö´anködö ñanno aninñajankomo we´a´komojadö iyö wadadödö wowanomatojoje eijaicho mödödenha. Yööje wütöjö´nawö yötunnöi watamu´kajö sotto owanökönñe nö´döa yaawö jajeda (Boletin de evaluación Biológica) Conservación Internacional nüdüdü, iyö chonekajö na yaawö kajichanakomo, poditikokomo, eda´sö´cha´komo wadodonoje, tüdüjai natodü eda´chödü jökö yeichü yotonno jüdata iyöjokono tüdüdü jökönchödö yaawö.

Edö nono naadü wenesuweda de´wono chönünhejene owanökö´dana, joduje nö´düa yaawö enmenkadü iyö tuna nonodü de´wonkomo (biodiversidad) choone´madü owajo . Chuuta, tadinñamo, odookoja´komo jökönñe woowanoma´jö tüweiye yeichame, tujunne na yaawö ooje´köjeene iyö nono jökö chönünhejene e´se´tödü wetö yaawö.

CHÖWADÖNÑE SA´DIMINCHAJÖ´AJÖ

Chuuta weichü mawoono na´kwakaano maja

Medewaadinña küntöjö´a´todawö künajoicho yaawö 443 je töneejomü öönünhatojünü yeicho´komo awoono. Iyö töneemü ajöiyajö aka 302 je chuuta weichü, tamedödö tüwü yaawö 1180 je chötükomo ajoijaajö owanökö naadü medewaadinñano aka yeeja´kajö tüwü.

Iyö wowanoma´jö kümja´kai yaawö önneene chuuta weichü, tuna´janonñano nono de´wono nunhedenña nono yaawö chuuta ewanshiñü´je´da. Önnene chuutakomo weichü tunajakökö ooje yo´kadüke tuna jedü. Öönunhe´da ta´ne yeichü jejechö kawö nono weichüke (40–2350 m) mödöje yeichü mödö tamedödö ni´ya´tadenha yaawö önnene na yaawö medewaadi nonodü de´wö.

Nono akanajaato tünnadüato maja de´wö nato yaawö se´kö aatantawö yeichükomo yootonkomo. Yöötö na jooje yaawö wasai, waju ja´komo: *Euterpe precatoria, Attalea maripa, Socratea exhorriza, Genoma baculifera,* yotonno yaawö *Bactris brongniartii.* Ooje´kö önnene yeichü na yaawö Amazoníanñano chutakankomo Guyananñano maja e´joye´kö iyönünhato denha yeichame chuuta weichü. Tüwomonhato nono dedewatö´kwainño na yaawö ooje na yaawö *Oenocarpus,* medewadi´chai´chene yaawö, oojena *Mauritiella,* mödö na yaawö aninñajano tuna´kwainño Escudo Guayana de´wono. Önnene chuuta ejüüdü na yaawö yantadukomo de´wö dedewatö´kwai yotonno yaawö medewaadi´chai jaada to´na tüdüjone. Iyö yantadukomo na yaawö chuuta tüweiye mawoono na´kwakano maja yaawö, anedawö ooje yaawö Podostemonaceae soodü jakökö yeichü tüwü yaawö.

Iñataje tuna weichü

43 je chööwütu künatammüi tuna töneemüje makiñaai ü´joyeno yootonno jü´wayeno maja yaawö. Medewaadi ü´joye na yaawö yaadödü tümaakudö´da, Iñña na yaawö ooje´kö tuna ñootadukomo yootonno yaawö tuna weichü woije tüwü na yaawö jü´wayeno nünhe´da, mödööje yaawö ooje inñadü´dana yaawö chuuta. Ajo´jo yeichüna yaawö tuna medewaadi weichü ooje awansi´je´da yaka iyö tuna kunötömmenkai edö jökö yeichü: Ta´ne o tukuna´se yeichü, tukuna´sato o ta´nato tünado yeichü, sü´je yeichü, tumuuke yeichü maja yotonno anejakomo maja yaawö.

Ajo´jo yeichü, iyö tuna na yaawo tüda´yeiche (ácida), se´kö ta´nato o tukuna´sato tünadö yeichü tüwü na yaawö. Yeichü mödöjena yaawö tuna konojojato weejüdü. Edö wowanoma´jö weja´kaajö na yaawö awa´deeto wowanoma´jö nünhe maja yeeja´kaajö mödöje na yaawö chü´tadü jena´do´jö nono madono yeijökö. Önünhe küneja´kai yaawö u´joyaano akö jü´wayeeno, yaane, yantai ñootadükomo anijanadö´ja yaawö.

Tumuuke yeichü jököödö´ja, inñataje küneeja´kai tamedödö ajöiyajö küna´ja´dü. Jü´waye tumuuke´kö küneeja´kai u´joyaano e´joye´kö, tuna jedü tükone´ma yeijökööjenñemma kunnöjö awö taku´ne´kö na yaawö, mödö nichamjiyakaja yaawö na´kwakankomo weicho´komo.

Se´jömjünüükomo tunanwakonchomo

25 je jaatadü künootonejai yaawö ñanno se´jömjünükomo tunanwakoichomo önnene yeichükomo. Ñanno weichükomo küneeja´kai yaawö wanna memuuja´komo, mutuuja´komo, diichö, suuduja´komo, ködödöi, wayakanija´komo, ojemma yotonno yaawö. Kanno wechükomo yeichü mödö yaawö tümakudööjünü nossajemjünü tuna aka yeichükomo tünwanno. Künödantöi yaawö tamedöiche tuna wadadödö nato tünwanno se´jömjünükomo tunanwakonchomo. Öönünhe´da önnene yeichükomo tunanwakonchomo na yaawö chü´tadü okisijeno woije tumuuke tuna weichü woijemmaja jenñemma. Mödööje yöödantödü yeichü tujunne na yaawö ne´kö´se ke´ja tuna weichü chamjiyaka´jökö jayeedömma, chuuta akö´a´jökö, o uudu u´kwadüje wetadawakajonö tüdüüa´jökö aneja nüta yaawö tuna aka, chamjiyakajoone yaawö ñanno tunanwakonchomo weicho´komo.

Wayaakani ja´komo

10 je künödantoicho yaawö wayaakani weichukomo enmenka´jüdü dü´tö medewadinña: Toniamojato je suudu weichükomo (Palaemonidae), toni yaawö wayaakani weichü (Pseudothelphusidae), aduwawö yaawö wayakanijato maja (Trichodactylidae). Medewadi ü´joye na yaawö ooje´damma kanno weichükomo jaatodenhamma, jü´waye´kene yaawö ooje´kö aduwawö amojatoto´kwakö. Inñano töötüjanñone aninñaja a´ke tüwü yaawö na suudu, wayaakanija´komo´kene tüweiye natodenha yaawö Llanonña yotonno yaawö Amazonanña. Kanno weichükomo künödantoichu yaawö yeichükomo mödö Escudo de Guayana yotonno yaawö Amazonico de´wö tuna naadü´kwawonkomo kanno. Suudu jadasideedu küna´jaakö yaawö ojejene yöödantödü enmenkadaawö. Kanno natodü yaawö eduwa ajicha nato tünwanno mödöjedenña naato´de tükone´ma´da medewaadi weichü´kö wadaadödö.

Kudaka

65 je jaatadü künö´düi ajoicho´komo sejiyato soodü jonno ka´kadanña tüdüjoone yotonno yaawö muuda soodü jona tüdüjoone jü´waye. Tamedödö yojodüüa´komo yaawö 278 je kudaka weichükomo künödantöi yaawö medewaadi´chawonkomo. Carasiformes weichükomo (orden) küneeja´kaicho yaawö 158 je, 74 je siluriformes, 27 je Perciformes, 9 je Gymnotiformes, 3 je Clupeiformes, 1 je yaawö Beloniformes, Pleuronectiformes, Synbranchiformes. Characidae weichükomo (familia) 113 je. Künödantoicho yaawö owaano de´wö yeichü jonno 45 je Characiformes weichükomo (orden) yootonno yaawö 31 je Characidae. Anejakomo´kene yaawö Siluriformes 49 je küna´ja´to owaano 74 je na eduuwa, Perciformes 12 je yeichü jonno 27 je eduuwa. Tamedöödö yoojodüajö na yaawö eduuwa´komo yöödanta´komo medewadi´chawonkomo 110 je. Tuweiye naato yaawö tönööamoje tüweiyamo ökünüje tüweiyamommaja.

Tunajakokono Chuuta

Medewadinñano chuuta tunajakokono enmenka´jüdü chuuta weichü, yootonoojünü chuuta, iyö nunhato chuuta weichü tuna wadadödö, nono weichü ene´ju´nei maja yaawö küneeja´kai yaawö tujunne eda´chödü nonojoya´komo chuuta tunajakokonkomo. Mödöje yeichü edantotojo yaawö Ingeae weichü jökö nonoweichü woijato wa´tadü tüwüüyeijökö. Chone´nadiyü´janojanñone na yaawö jaadanwawonno ju´wakadö.

Kudaka wöökamodükomo

Kudaka wöökamoa´komo künotonejai yaawö 97 je se´jömjünükomo tunanwakonchomo weichü weja´kajöke, 303 je tunajakokono chuuta yotonno yaawö 278 je kudaka weichükomo tuna ta´kwiti´yemjünü´kwakankom o. 8 jaatadü tömmenkamü künö´datokai öönünhe jaada de´wonno maadödö jaadanwawonno jü´wakadö maja mödö künekammai yaawö: 1) se´jömjünükomo nato chöwadadödömma, yööje´da´kene yaawö kudaka chuutammaja yaawö yööje´da yöökamoa´komo; 2) nono weichü e´nei töneejene küna´jaakö yaawö chuuta jökö; 3) medewaadi ü´joye oojejene´da kudaka weichükomo küneeja´kai yaawö jü´wayeenoje´da, chuuta´kene yaawö yööje´da, se´jömjünükomo´kene yaawö önünhe.

EDÖÖJE YEICHOJONA YAAWÖ EDA´CHOTOJO YOTONNO WOWANOMATOJOJE MAJA

Eda´chotojo eijaicho yaawö yotonno wowanomatojoje edö yennajö jökö yeeja´kajö; edö jonno yö´mennajö nadü tujunnejeene´da na yaawö.

- **Edennamjüdü tamedöödö aku´sana tü´tajötüdü tüwoije tuna medewaadi weichü wawüyümüdü chamjiyakajai naadü, iñña´dudu, aneja´kwaka tuna adöödü, tuna wütotojo chamjiyakadü.** Tuna iñña´dudu aneja´kwaka adöödümmaja yaawö anijana dö´ja tüdüjai na yaawö tuna wawüyümüdü konojo jökö yeichü. Chuuta weichü yotonno yaawö yöötöödö ichödükomo yeichükomo maja yaawö iyö de´wö nudöiñe naatodü naato yaawö yöötöödö tuna wawüyümüdü tüwoije yeichüke.
- **Tüdüdü wetadawa´kajotojo kudaka wütödükomo ene´mato´komo.** Kudaka wütödükomo nomonhato medewaadi´chaka yotonno yaawö chuuta chunnö´ajö nüdüato yaawö tũwe´moichato´komoje. Yeicho´komo ekone´majaicho jaada de´wonno ü´jonaadö iññawonno jü´wakaado maja ötödantöjai na yaawö o´jajo´da yeichükomo wetö yaawö.
- **Tüdüüdü wetadawa´kajootojo aduwaawö wedu to´kwa´kö kudaka ajoicho´komo choonekatojo jökö jaadanwawonno yeichü.** Töneena yaawö takade´da o´tödü ye´kwanakomo nonoodü de´wö yotonno yaawö oojedeja kudaka ajöchükomo na Jaadanwawonno jüwakaadö. Kudaka ajöichü sadö tujunnena yaawö towadöödö adöödü eneedü wetö yaawö ö´düjai yeichü e´se´totoojo (ö´wasa´kö ajöichü, ö´wasa´koto yaawö jaji) mödööje yaawö ataame´da yeichü wetö kudaka ajöichü.
- **Tüdüüdü wetadawa´kajootojo medewaadi ene´matoojo yotonno yaawö yaimaja tüdüüdü tuna weichü e´nmenkatoojo dedewatö kanö.** Iyö eijaiña yaawö científicokomo wütooto´komooje ajichaajene owanomaiyeto yaawö medewaadi jökö yeichü yotonno tüdüyeto yaawö yötunnöi jo´wadö edö chuutakano tuna weichü. Edööje yeijökö yaawö weichojooje nadü tükoone´ma´dadeña wökaaje´da ne´a choone´madü se´kö na yaawö tükoone´majüünü medewaadi nünhato, mödööje yeijökö yaawö eda´chödü tujunnena yaawö.
- **Tüdüüdü chuuta kunnö´ajö eda´chotoojo.** Medewaadi na ötöda´chöjaicho yöötö nadü yöötödö yeicho´me.
- **Tü´tajö´nö tüdüüdü chu´nakaadü jökö yöötö naatodü eda´choto´meñe.** Jaada de´wonno ü´jonaadödö yootonno iññawonno jüwakaadödö na yaawö chuuta yootonno kudaka maja öönünhe´da naato. Tooni amoojato na na´kwakankomo weicho´komo (soodü, yantadükomo, tuna onkwe´danaadü, ojemma yaawö) mödöökomo tujunnena yaawö 82%je na´kwakankomo weichükoomo eda´choto´meñe, tönööamoje naatodü jadönñe´jüdüüdö töjeemajaamommaja yaawö. Chü´tammeküüdü tujunnena ö´wasa´kö eijaiñai yaawö töda´chöömü 80% je e´joye chuuta yootonno yaawö chuutankomo na´kwakankomo maja weicho´komoje naadü medewaadinña.
- **Woowanomatoojo tüdüüdü tujunne na yaawö na´kwaankomo weichojo eda´chödüjökö.** Edö ö´düjaiña yaawö ye´kwanakomo nonoodü´naköi. Ye´kwanakomo nonoodü jonno jü´wakadödö choone´madü mödö yaawö tukunnö´ato chuuta jataadü. Jü´wayenkomo sotto naato medewaadi´kwawono kedeña, mödöjejene owanomake´jünüche naato yaawö yootonkomo medewaadi eda´chödüüjökö, ataame´da yöötödö eijaicho yeicho´komooje.
- **Amo´chödü yaawö kudaka aninñajankomo amonno´jodü.** Chööwa´kö naato kudaka yeicho´komommaja yaawö medewaadinña aninñajano ötööne´ja´cha yaawö. Na´kwakankomo medewaadinñankomo ñannomma kanno wenesuweda de´wö yeichü yööje yaawö töda´chömje naato. Tüweiye na yaawö ekammajo´tojo aninñajaano kudaka eneejüdü tünkone´ma yeichü ñanno yotonkomo kudaka yeicho´komo maja yaawö.
- **Jenaadö ye´jüdü nünhemmaja tüdüüdü chuuta tunajakokono muuda soodünwawonno jüwakaadödö.** Edö yeichojo tükoone´mana, yaatameeajö mödö yootonkomo (biodiversidad) yootonno yaawö yeichükomo maja (estructura comunitaria). Edö choone´madü otonejajanña yaawö ooje kudaka, suudu, wayakani maja ataajöcha yeichüke jenaadö yei´jüdü nünhe´da. Tüweiye na yaawö chuuta ñaatü´totoojo mödööje iñammaja chuuta weichü ö´düjai na yootonno ooje´kö kudaka eijai natoodenha yaawö.
- **Edöökomo iye ako´tojo tü´tajö´nö e´se´totoojo tüdüüdü ataame´da yeichü wetö:** Ocotea cymbarum, Vochysia venezuelana (kudiyada tüdüemü), Acosmium nitens, Geonoma deversa (ma amotojo) yootonno yaawö Heteropsis flexuosa (wüwa tüka´emü, ma amotojo maja). Mödöje choneka´jökö i´ya´töjaina wö´kaaje´da tadawaju weichü ataame´da medewadinña, mödööjemmaja yaawö jata we´wa´tödü eduwa yöötö nadü komo.

LITERATURA CITADA

Aymard, G., S. Elcoro, E. Marín y A. Chaviel. 1997. Caracterización estructural y florística en bosques de tierra firme de un sector del bajo Río Caura, Estado Bolívar,

Venezuela. *En*: Huber, O. y J. Rosales (eds). Ecología de la Cuenca del Río Caura, Venezuela II. Estudios especiales. Scientia Guaianae, Vol. 7:143–169.

Berry, P. E., O. Huber y B. K. Holst. 1995. Floristic Analysis and Phytogeography. *En*: Berry, P. E., B. K. Holst, K. Yatskievych (eds.). Flora of the Venezuelan Guayana, Vol. 1. Introduction. Missouri Botanical Garden, Saint Louis, USA. pp. 161–191.

Bevilacqua, M. y J. G. Ochoa (eds). 2000. Informe del componente Vegetación y Valor Biológico. Proyecto Conservación de Ecosistemas Boscosos en la Cuenca del Río Caura, Guayana Venezolana. PDVSA-BITOR, PDVSA-PALMAVEN, ACOANA, AUDUBON de Venezuela y Conservation International. Caracas, Venezuela. 81 pp.

Bevilacqua, M. y J. Ochoa G. 2001. Conservación de las últimas fronteras forestales de la Guayana Venezolana: propuesta de lineamientos para la Cuenca del Río Caura. Interciencia 26(10): 491–497.

CVG-TECMIN. 1994. Informes de avance del Proyecto Inventario de los Recursos Naturales de la Región Guayana. Hojas NB-20: 1, 5, 6, 9, 10, 13 y 14. Gerencia de Proyectos Especiales. Ciudad Bolívar, Venezuela.

Dezzeo, N. y E. Briceño. 1997. La vegetación en la cuenca del Río Chanaro: medio Río Caura. *En*: Huber O. y J.Rosales (eds). Ecología de la Cuenca del Río Caura, Venezuela II. Estudios especiales. Scientia Guaianae, Vol. 7: 365–385.

Huber, O. 1996. Formaciones vegetales no boscosas. *En*: Rosales J.y O. Huber (eds.). Ecología de la Cuenca del Río Caura, Venezuela: I. Caracterización general. Scientia Guaianae, Vol. 6: 70–75.

Huber, O. y J. Rosales (eds). 1997. Ecología de la Cuenca del Río Caura, Venezuela. II Estudios Especiales. Scientia Guaianae No. 7: 1–473.

Huber, O., J. Rosales y P. Berry. 1997. Estudios botánicos en las montañas altas de la cuenca del Río Caura. *En*: Huber, O. and Rosales, J. (eds.). Ecología de la Cuenca del Río Caura, Venezuela II. Estudios esp.eciales. Scientia Guaianae, Vol. 7: 441–468.

Knab-Vispo, C. 1998. A rain forest in the Caura Reserve (Venezuela) and its use by the indigenous ye´kwana people. Doctoral Thesis. Universidad de Wisconsin-Madison, USA. 202 pp.

Machado-Allison, A. 1994. Factors affecting fish communities in the flooded plains of Venezuela. Acta Biol. Venez., 14(3):1–20.

Machado-Allison, A. 1999. Cursos de agua, fronteras y conservación. *En*: G. Genatios (ed). Ciclo Fronteras: Desarrollo Sustentable y Fronteras. Com. Estudios Interdisciplinarios, UCV. Caracas: 61–84.

Machado-Allison, A., B. Chernoff, C. Silvera, A. Bonilla, H. Lopez-Rojas, C. A. Lasso, F. Provenzano, C. Marcano y D. Machado-Aranda. 1999. Inventario de los peces de la cuenca del Río Caura, Estado Bolivar, Venezuela. Acta Biol. Venez., Vol. 19 (4):61–72.

Machado-Allison, A., B. Chernoff, R. Royero-Leon, F. Mago-Leccia, J. Velazquez, C. Lasso, H. López-Rojas, A. Bonilla-Rivero, F. Provenzano y C. Silvera. 2000. Ictiofauna de la cuenca del Río Cuyuní en Venezuela. Interciencia, 25(1): 13–21.

Marin, E. y A. Chaviel. 1996. Bosques de Tierra Firme. *En:* Rosales, J. y O. Huber (eds). Ecología de la Cuenca del Río Caura. Scientia Guaianae, 6: 60–65.

Miranda, M., A. Blanco-Uribe, L. Hernández, J. Ochoa y E. Yerena. 1998. No todo lo que brilla es oro: hacia un nuevo equilibrio entre conservación y desarrollo en las últimas fronteras forestales de Venezuela. Washington, DC: Inst. Rec. Mundiales (WRI). 59pp.

Novoa, D. 1990. El río Orinoco y sus pesquerías; estado actual, perspectivas futuras y las investigaciones necesarias. *En*: Weibezahn, F., H. Alvarez y W. Lewis (eds). El Río Orinoco como Ecosistema. Edelca, Fondo Ed. Acta Científica Venezolana, CAVN, USB: 387–406.

Peña, O. y O. Huber. 1996. Características Geográficas Generales. *En*: Rosales, J. y O. Huber (eds). Ecología de la Cuenca del Río Caura. Scientia Guaianae, 6: 4–10.

Rosales, J. 1996. Vegetación: los bosques ribereños. *En*: Rosales, J. y O. Huber (eds). Ecología de la Cuenca del Río Caura, Venezuela: I. Caracterización general. Scientia Guaianae, Vol. 6: 66–69.

Rosales, J. y O. Huber (eds). 1996. Ecología de la Cuenca del Río Caura, Venezuela. I. Caracterización General. Scientia Guaianae No. 6: 1–131.

Silva-Monterrey, N. 1997. La Percepción Ye'kwana del Entorno Natural. *En*: Huber, O. y J. Rosales (eds). Ecología de la Cuenca del Río Caura. Estudios Especiales Scientia Guaianae, 7: 65–84.

Tabla de Contenidos

Una Evaluación Rápida de los Ecosistemas Acuáticos de la Cuenca del Río Caura, Estado Bolívar, Venezuela

SPANISH

APÉNDICES

Prefacio

El Río Caura tiene más de 700 km de longitud, originándose en las tierras altas del Escudo de Guayana por encima de los 2000 metros sobre el nivel del mar. El área de la cuenca es de aproximadamente 45.336 km^2, la cual representa el 20% de la superficie total del Estado Bolívar o 5% de la superficie de Venezuela. La cuenca del Río Caura contiene miles de especies de plantas y animales, muchas de las cuales son endémicas del Escudo de Guayana o del Río Orinoco. Al mismo tiempo, el Río Caura es el albergue de muchas comunidades humanas que dependen del río para su subsistencia; comida, transporte, recreación, útiles domésticos y para su cultura. La cuenca forma parte de la Tierra Ye'kwana, una población muy cuidadosa del manejo y apreciación de sus recursos naturales. La alta calidad ambiental de la cuenca, especialmente desde el Raudal Cinco Mil hasta las cabeceras, es debida principalmente a su sabia dirección. En resumen, el Río Caura y su cuenca representa un hábitat crítico para todos los organismos de la región incluyendo las poblaciones humanas.

La cuenca del Río Caura representa una de las áreas más grandes y pristinas remanentes del bosque tropical representativo del Escudo de Guayana en el mundo. Los ambientes acuáticos del Río Caura, especialmente aquellos por arriba del Raudal Cinco Mil, están también en excelentes condiciones ambientales. La Cuenca del Río Caura provee una excelente oportunidad para el desarrollo de un plan integrado que provea conservación ambiental y uso sustentable de los ecosistemas y recursos.

A pesar de esto, la región se encuentra bajo amenaza inmediata que puede facilmente cambiar la naturaleza pristina del ambiente. Deforestación debido a extracción maderera está ocurriendo en las regiones al sur de la cuenca. Incrementando en la densidad poblacional, colonización y degradación de hábitats esta ocurriendo en la región norte. Existe un plan para el transvase de enormes cantidades de agua del Caura para el incremento del potencial hidroeléctrico del Caroní. La caceria ilegal y sobreexplotación pesquera debajo del Salto Pará, representan serias presiones sobre esta cuenca única. Con el propósito de investigar los efectos potenciales de estas amenazas sobre la biodiversidad y los ambientes acuáticos del Río Caura, un equipo de 17 personas incluyendo 13 científicos, representantes de Kuyujani, dos coordinadores logísticos, dos técnicos, un estudiante y un responsable de la información vía electrónica realizó una expedición AquaRAP en éste río, del 25 de Noviembre al 12 de Diciembre del año 2000. El equipo integró personas provenientes de Venezuela, Brasil y los Estados Unidos. Los científicos son especialistas en botánica terrestre, crustáceos decápodos, macroinvertebrados, ictiólogía, limnólogía, incluyendo química de agua y plancton. La expedición tuvo como objetivo principal estudiar del valor de la biologíco y de conservación de la región y como soluciones integrales pueden trabajar para preservar la máxima cantidad de biodiversidad en la presencia de las actuales y futuras amenazas.

La organización del informe comienza con un resumen ejecutivo, el cual incluye una breve descripción de las características físicas y terrestres de la región, así como también resúmenes de los reportes técnicos de las disciplinas científicas estudiadas y concluye con recomendaciones para una estrategia conservacionista. Después se presentan los resultados biológicos de botánica, limnología, invertebrados bénticos, crustáceos decápodos y peces. Los últimos capítulos resaltan las comonalidades entre los elementos florísticos y faunísticos en relación con la geología y ecología. Los reportes disciplinarios están escritos como trabajos científicos, cada uno incluyendo sus propias citas bibliograficas. Después de los trabajos científicos, nosotros incluimos un glosario de términos y apendices con los datos obtenidos.

Queremos explicar el uso de los términos de *diversidad* y *riqueza de especies* en este libro. Comúnmente *diversidad*, cuando es usado en el contexto de los organismos biológicos, se refiere al número y variedad de tipos de especies, organismos, taxa, etc. En la literatura ecológica *diversidad*, toma un significado más específico y ligeramente diferente, refiriéndose al número de entidades en combinación con sus abundancias relativas. En éste volumen *diversidad* es usado en dos formas: i) en la interpretación común o vernácula nosotros nos referimos ocasionalmente a la "*diversidad* de organismos," indicando el número y variedad de organismos; pero ii) en el sentido ecológico, nos referimos a "baja o alta *diversidad*," indicando "*diversidad*" como un resultado de la aplicación de una fórmula específica. Nótese que el uso vernacular, *diversidad* nunca es modificado por ser alta o baja o por comparación. De acuerdo a la literatura ecológica, el término *riqueza de especies* significa el número de especies. Nosotros usamos *riqueza de especies* en varios capítulos (p.e., Capítulo 4) cuando nos referimos al número de especies presentes en un hábitat o cuenca.

Este informe está dirigido a personas que toman decisiones, gerentes ambientales, agencias del gobierno y no gubernamentales, estudiantes y científicos. La información novedosa y los análisis presentados aqui tienen dos propósitos: (i) presentar un caso y una estrategia para los esfuerzos de conservación de la región; y (ii) proveer los datos científicos y análisis que estimulará investigaciones en ésta importante región. En éste volumen hemos intentado no solamente presentar un inventario de los organismos que hemos encontrado durante la expedición, sino utilizar ésta información, para evaluar estrategias de conservación basada en diferentes escenarios de amenaza ambiental. Damos la bienvenida a comentarios críticos en la medida de continuar la evolución de AquaRAP y los métodos para la evaluación de estrategias de conservación a partir de datos biológicos.

Barry Chernoff
Antonio Machado-Allison
Karen Riseng
Jensen Montambault

Participantes y Autores

Mariapia Bevilacqua (coordinador logístico, botánica)
Asociacion Venezolana para la Conservación de Áreas Naturales (ACOANA)
Av. Humboldt con calle Coromoto, Edif. Breto oficina 5, planta baja, Bello Monte Norte
Caracas, 1063-A
VENEZUELA
Email: mariapia@cantv.net

Barry Chernoff (ictiología, editor)
Dept. Zoology
Field Museum
1400 S. Lakeshore Dr.
Chicago, IL 60605
USA
Email: chernoff@fmnh.org

Wilmer Diaz (botánica)
Jardín Botánico del Orinoco
Calle Bolívar, Ciudad Bolívar
Edo. Bolívar
VENEZUELA
Email: jbov@telcel.net.ve

José V. García D. (invertebrados bénticos)
Instituto de Zoología Tropical, Facultad de Ciencias
Universidad Central de Venezuela
Apto. Correos 47058
Caracas, 1041-A
VENEZUELA
Email: jvgarcia@strix.ciens.ucv.ve.

Antonio Machado-Allison (ictiología, editor)
Instituto de Zoología Tropical
Universidad Central de Venezuela
Apto. Correos 47058
Caracas, 1041-A
VENEZUELA
Email: amachado@strix.ciens.ucv.ve

Célio Magalhães (crustáceos decápodos)
Instituto Nacional de Pesquisas Amazonicas
INPA/CPBA, Cx. Postal 478,
69011-970 Manaus
BRASIL
Email: celiomag@inpa.gov.br

Nigel Maxted (autor)
School of Biosciences
The Birmingham University
Edgbaston B15 2TT
Birmingham
UNITED KINGDOM

Alberto Marcano (ictiología)
Instituto de Zoología Tropical
Universidad Central de Venezuela
Apto. Correos 47058
Caracas, 1041-A
VENEZUELA

Jensen R. Montambault (coordinador, editor)
Conservation International
AQUARAP Program
1919 M Street NW, Suite 600
Washington, DC 20036
USA

Guido A. Pereira S. (invertebrados bénticos, crustáceos decápodos)
Instituto de Zoología Tropical, Facultad de Ciencias
Universidad Central de Venezuela
Apto. Correos 47058
Caracas, 1041-A
VENEZUELA
Email: gpereira@strix.ciens.ucv.ve

Geoffrey Petts (autor)
School of Geography and Environmental Sciences
The Birmingham University
Edgbaston B15 2TT
Birmingham
UNITED KINGDOM

Francisco Provenzano-Rizzi (ictiología)
Instituto de Zoologia Tropical
Universidad Central de Venezuela
Apto. Correos 47058
Caracas, 1041-A
VENEZUELA
Email: fprovenz@strix.ciens.ucv.ve

Lourdes Rico-Arce (autor)
Herbarium
Royal Botanic Gardens
Kew, Richmond, Surrey, TW9 3AB
UNITED KINGDOM

Karen J. Riseng (coordinador del reporte, limnología, editor)
Department of Ecology and Evolutionary Biology
University of Michigan
830 N. University
Ann Arbor, MI 48109
USA
Email: kjriseng@umich.edu

Angel Rojas (ictiología)
Instituto de Zoologia Tropical
Universidad Central de Venezuela
Apto. Correos 47058
Caracas, 1041-A
VENEZUELA

Judith Rosales (botánica)
Universidad Nacional Experimental de Guayana (UNEG)
Urbanización Chilemex, Calle Chile, Sede Uneg-Investigacion y Postgrado
Puerto Ordáz, Edo. Bolívar
VENEZUELA
Email: jrosales@uneg.edu.ve

Luzmila Sanchez (limnología)
Fundación La Salle
Estación de Investigaciones Hidrobiológicas de Guayana
UD-104 El Roble, Apto. 51
San Félix Edo. Bolívar
VENEZUELA
Email: luzsanchez@cantv.net

Brian Sidlauskas (ictiología)
Dept. Zoology
Field Museum
1400 S. Lakeshore Dr.
Chicago, IL 60605
USA
Email: bls@midway.uchicago.edu

John S. Sparks (limnología)
Museum of Zoology
Division of Fishes
University of Michigan
Ann Arbor, MI 48109-1079
USA
Email: jsparks@umich.edu.

Philip W. Willink (ictiología)
Dept. Zoology
Field Museum
1400 S. Lakeshore Dr.
Chicago, IL 60605
USA
Email: pwillink@fmnh.org

Grupo Kuyujani:
Alto Caura
Motoristas: Luís Flores, Wilfredo Flores y Carmelo Castro
Marineros: Rogelio Pérez, José Sarmiento, Justino Castro Bonifacio, Lucas González y Juan Núñez (también Coodinador de Kuyujani para expediciones)

Bajo Caura
Motoristas: Miguel Estaba, Wilfredo Flores, José Sosa Silva y Simón Caura
Marineros: Juan Núñez, Nelson Espinoza, Ernesto Sarmiento, Delfín Rivas, Eugenio García y Alberto Sarmiento (también ayudante de cocina)

Perfiles Organizacionales

CONSERVATION INTERNATIONAL

Conservation International (CI) es una organización internacional sin fines de lucro ubicada en Wahington, DC. CI cree que la herencia de las áreas naturales de la tierra deben ser mantenidas para el enriquecimiento espiritual, cultural y económico de las futuras generaciones. Nuestra mision es conservar esta herencia natural terrestre viva, la diversidad global y demostrar que las sociedades humanas son capaces de vivir armoniosamente con ellas.

Conservation International
1919 M Street NW, Suite 600
Washington, DC 20036
USA
Tel. 800-406-2306
Fax. 202-912-0772
Web. www.conservation.org
www.biodiversityscience.org

CONSERVATION INTERNATIONAL-VENEZUELA

Conservación Internacional (CI) inicia sus actividades en Venezuela en el 2000, con la finalidad de mantener la herencia natural de la tierra para que las generaciones futuras prosperen espiritual, cultural y económicamente. Nuestra experiencia nos señala que la conservación sólo tendrá éxito dentro del marco del desarrollo sostenible, lo cual entre otras cosas incluye la participación de las comunidades locales en actividades alternativas y creativas, como el ecoturismo, la capacitación para la conservación de los recursos naturales, el uso adecuado de los mismos y la educación ambiental, cuyos objetivos son evitar el uso destructivo de la tierra, la contaminación del agua, de la depredación de árboles y la pérdida de biodiversidad.

CI reconoce que no es una tarea individual y por eso construye alianzas estratégicas con actores sociales e institucionales para desarrollar actividades de conservación en base a criterios técnicos-científicos que respeten la diversidad sociocultural, desarrollen la creatividad local, evalúen hábitats clave, identifiquen amenazas y creen alternativas de ingresos a partir de la conservación de la biodiversidad.

Conservation International-Venezuela
Avenida Las Acacias
Torre La Previsora, Piso 15
Ofina Noreste, Noroeste
Urbanización Los Caobas
Caracas, 1050
VENEZUELA

THE FIELD MUSEUM

El Field Museum (FMNH) es una institución educativa dedicada a la diversidad y relaciones en la naturaleza y entre culturas. El Museo usa un alcance interdisciplinario combinando las áreas de Antropología, Botánica, Geología, Paleontología y Zoología, para lograr el incremento del conocimiento acerca del pasado, presente y futuro, del mundo físico, sus plantas, animales, personas, y sus culturas. Haciendo esto, le permite descubrir la extensión y características de la diversidad biológica y cultural; similaridades e interdependencies así que podemos entender mejor, respetar y celebrar la naturaleza y otras culturas. Sus colecciones que poseen más de 20 millones de ejemplares de todo el mundo, sus programas de aprendizaje público e investigación están intimamente relacionados y son inseparables para servir a un publico diverso en edades, experiencias y bases de conocimiento. El Museo Field publica una revista científica arbitrada llamada *Fieldiana*.

The Field Museum
1400 South Lake Shore Dr.
Chicago, IL 60657
USA
Tel. 312-922-9410
Fax. 312-665-7932
Web. www.fieldmuseum.org

INSTITUTO DE ZOOLOGÍA TROPICAL, UNIVERSIDAD CENTRAL DE VENEZUELA

El Instituto de Zoología Tropical (IZT) es un instituto de investigación de la Facultad de Ciencias, Universidad Central de Venezuela (UCV). Dentro de las ámplias disciplinas de la Zoología y Ecología, el IZT enfatiza educación e investigación en sistemática zoológica, parasitología, ecología teórica y aplicada, estudios ambientales y conservación. El IZT, es responsable de la administración del Museo de Biología de la UCV, el cual alberga algunas de las colecciones zoológicas más valiosas del mundo. Entre sus colecciones se encuentra la colección de peces de agua dulce, la cual es una de las más grandes en América Latina y la colección de mamíferos la cual es una de las más completas de Venezuela. El IZT también es responsable del Aquarium "Agustín Codazzi," el cual permanentemente disemina el conocimiento de los peces y la conservación de los ambientes acuáticos de Venezuela a través de sus exhibiciones y programas educativos gratuitos. El IZT publica además una revista arbitrada científica, *Acta Biologica Venezuelica*, fundada en 1951.

Instituto de Zoología Tropical
Universidad Central de Venezuela
Apto 47058
Caracas, 1041-A
VENEZUELA
Web. http://strix.ciens.ucv.ve/~instzool

ORGANIZATION KUYUJANI

Kuyujani es una organización que representa a dos pueblos indígenas de la región del Caura, los Ye´kuana y los Sanema, los cuales conforman 53 comunidades. Fue fundada el 8 de agosto de 1996. Tiene 6 coordinadores de áreas: educación, salud, ambiente y territorio, derechos humanos, desarrollo económico y cultura. Kuyujani trabaja en: 1) la defensa de los derechos y las tierras ancestrales de las comunidades indígenas ye´kuana y sanema en todos sus aspectos; 2) apoyo y fomento del diálogo y encuentros entre organizaciones sectoriales en la búsqueda de un desarrollo integral y autogestionario; 3) fomento de cursos, seminarios o talleres para la capacitación técnica y científica de los miembros de las comunidades; 4) establecimiento y mantenimiento de relaciones fraternales con las organizaciones indígenas regionales, nacionales e internacionales con el propósito de unir criterios y esfuerzos, en beneficio de los derechos e intereses de nuestras comunidades; 5) de igual manera establece relaciones con las organizaciones no indígenas cuyos objetivos sean afines.

Comunidades Santa Maria de Erebato and Boca de Nichare
Barrio Hueco Lindo, Callejón Los Teques
Ciudad Bolívar
VENEZUELA

Agradecimientos

Esta expedición de AquaRAP requirió de la dedicación de numerosas personas para que la misma fuera exitosa. Extendemos nuestras gracias a muchos individuos que proporcionaron el apoyo logístico y científico, servicios de comida, alojamiento, facilidades de reuniones y apoyo económico.

El equipo de AquaRAP desea dar la más profunda gratitud a los pueblos Ye'kuana y Sanema y la organización indígena Kuyujani por su invitación para la conducción de este estudio de biodiversidad acuática en sus territorios. El continuo apoyo y entendimiento del Capitán Perez, Alberto y Freddy Rodríguez, desde la planificación del proyecto, fue crítica para tener una operación sin mayores obstáculos, así como también, para asegurar que ésta información sea aplicada en planes de conservación con participación comunitaria.

Queremos dar gracias al Instituto de Zoología Tropical, Universidad Central de Venezuela, por su maravillosa hospitalidad, colaboraciones y apoyo científico. Estamos altamente agradecidos a Mariapia Bevilacqua y ACOANA (Asociación Venezolana para la Conservación de Areas Naturales) por su excelente planificación, apoyo logístico y experiencia científica. Agradecemos al Instituto Limnología del Orinoco de la Fundación La Salle por proveernos de dormitorios y facilidades de reunión al final de la expedición. También queremos extender nuestro agradecimiento a Jorge Luís de Akanan Tours por su apoyo logístico y quien con su cocinera, Tenilda Cranes, su ayudante Alberto Sarmiento y Jonas Cranes, proporcionaron la mejor fuente alimenticia para el equipo de campo.

Le damos gracias especiales a Dedemai y Simón Caura por recibirnos en Boca de Nichare y por su excelente guía. También queremos agradecer a Kuyujani por el uso de su infraestructura en El Playón. Extendemos tambien gracias a la Corporacíon Venezolana de Guyana (CVG), por el uso de su Estación de Campo en Entreríos y su personal local Jesús y Freddy.

Queremos darle las gracias al Capitán Raúl Arias de Raul Helicopters por su gran cortesía en los vuelos desde Entreríos a El Playón. Agradecimiento también para la compañia de aviación Comeravia, pilotos y personal de tierra, especialmente al Capitán Andrés Franco y María Isabel de Rivas, quienes realizaron un trabajo fantástico llevándonos con todo nuestro equipo desde Ciudad Bolívar a Entreríos.

Por toda nuestra travesía acuática damos gracias a los maravillosos motoristas y marineros. Motoristas en el Caura Superior fueron: Luis Flores, Wilfredo Flores Carmelo Castro y en el Bajo Caura: Miguel Estaba, Wilfredo Flores, José Sosa Silva y Simón Caura. Marineros en el Caura Superior fueron: Rogelio Pérez, José Sarmiento, Justino Castro Bonifacio, Lucas González y Juan Nuñez (Coordinador de Kuyujani para la expedición) y en el Bajo Caura fueron: Juan Nuñez, Nelson Espinoza, Ernesto Sarmiento, Delfín Rivas, Eugenio García y Alberto Sarmiento.

Damos gracias a Alberto Rodríguez y Freddy Rodríguez por la coordinación con las comunidades indígenas para la ayuda en el otorgamiento de su permiso para que nuestra expedición dentro de su territorio fuera un éxito. También, debemos agradecer a las siguientes agencias y oficinas del gobierno de Venezuela por el otorgamiento de los permisos correspondientes para la investigación de éstas maravillosas áreas acuáticas: Consejo Nacional de Investigaciones Científicas y Tecnológicas (CONICIT); Instituto Nacional de Parques (INPARQUES);

Oficina de Asuntos Indígenas del Ministerio de Educación, Cultura y Deportes (DAI); Oficina Nacional de Diversidad Biológica del Ministerio del Ambiente y los Recursos Naturales (MARN); Servicio Autónomo de Pesca del Ministerio de Industria y Comercio (SARPA); Consultoría Jurídica del CONICIT. Finalmente, queremos mostrar nuestra gratitud a los Dres. Carlos Genatios, Ministro de Ciencia y Tecnología y Héctor Navarro, Ministro de Educación por su gran ayuda.

Los autores del Capítulo 4 quieren agradecer la colaboración del Prof. Rafael Martínez por su ayuda en la identificación de los caracoles, las almejas y las sanguijuelas. Los autores del Capítulo 6, agradecen la ayuda de Mary Ann Rogers y Kevin Swagel por su desinteresada ayuda en el procesamiento de los ejemplares de peces, también la colaboración de John Friel y Richard Vari por la identificación de algunos ejemplares. Los autores del Capítulo 7, quieren dar gracias al CONICIT, British Council, la Universidad Experimental de Guayana, Universidad de Birmingham, Carlos Verbin, Claudia Knab-Vispo, Herbario Ovalles, Herbario Nacional de Venezuela, Herbario Regional de Guayana, Herbario de Guanare y el Kew Botanical Gardens.

El Programa AquaRAP en América del Sur y la expedición al Caura, han sido posible gracias al generoso apoyo de la Rufford Foundation. Fondos económicos adicionales para la expedición del Caura fueron otorgados por la Smart Family Foundation, la cual apoyó con tecnología y transmisión electrónica desde el campo y por la Fundación Científica y Educativa Comer. Finalmente, queremos dar gracias a John McCarter (Field Museum), Russ Mittermeier, Peter Seligmann, Anthony Rylands, Jorgen Thomsen y Leeanne Alonso (Conservation International) por su continuo interés, entusiasmo y apoyo para la conservación de ambientes acuáticos. Queremos extender nuestras especiales gracias a Ana-Liz Flores y Alfonso Alonso por revisar nuestro documento y a Alberto Rodriguez por las traducciones al Ye'kwana. Como siempre, deseamos agradecer la excelente revisión editorial y científica de Leeanne Alonso, Directora del Programa de Evaluación Rápida (RAP) de Conservation International, así como también sus guías y consejos para lograr el éxito de nuestro proyecto. Este reporte fue financiado por unas donaciónes generosas de la Gordon and Betty Moore Foundation y la Rufford Foundation.

Reporte a Primera Vista

UNA EVALUACIÓN RÁPIDA DE LOS ECOSISTEMAS ACUÁTICOS DE LA CUENCA DEL RÍO CAURA, ESTADO BOLÍVAR, VENEZUELA

Fecha del Estudio

25 Noviembre–12 Diciembre 2000

Descripción del sitio

La Cuenca del Río Caura, Estado Bolívar, Venezuela comprende un área extensa de bosques y ríos localizados sobre el Escudo de Guayana, una formación geológica muy antigua. El Río Caura constituye un tributario principal del Río Orinoco. Los bosques internos cubren aproximadamente 90% de la cuenca, mientras que el 10% remanente consiste de bosques ribereños inundables y otras formaciones vegetales no boscosas. La Cuenca del Río Caura alberga 30% de las especies de plantas registradas para Venezuela y 51.3% de las especies de Guayana, incluyendo 88% de las formas endémicas de plantas para la Guayana y 28% del total de peces dulceacuícolas de Venezuela. La extraordinaria diversidad biológica ha sido atribuida a la combinación de paisajes muy antiguos erosionados, convergencia de cuatro Provincias Geológicas y el marcado gradiente altitudinal (40–2350 msnm). Además de la gran diversidad y endemismo, el bosque en esta cuenca se encuentra en un 85% intacto y con muy poca intervención humana.

Razones para el estudio de AquaRAP

El objetivo principal de este estudio de AquaRAP fue el explorar y documentar la biodiversidad acuática de esta remota región en cooperación con el pueblo Ye'kuana. El manejo de esta cuenca pristina y del alto valor representa una tarea difícil. Esta área corresponde a la tierra de la Población Ye'kuana, la principal etnia en la zona. Los Ye'kuana son muy cuidadosos en el manejo y aprecio por sus recursos naturales. La alta calidad ambiental de esta región, especialmente aquella desde el Raudal Cinco Mil a las cabeceras, es debida en gran medida al excelente cuido y liderazgo de los Ye'kuana. Aunque un proyecto de transvase de aguas es menos inminente por ahora, la región todavía encara amenazas inmediatas debido al proyecto de construcción de una represa hidroeléctrica en el Salto Pará (cataratas que dividen la región entre zonas superiores e inferiores el área también es considerada sagrada para los Ye'kuana), el incremento de la presión pesquera comercial, y el aumento de turismo y agricultura en el Bajo Caura.

Resultados principales

El Río Caura posee una alta riqueza de especies de peces y diversidad de vegetación ribereña en comparación con otros tributarios similares de las cuencas del Amazonas y del Orinoco. Mientras que los invertebrados bénticos muestran una riqueza y abundancia típica de sistemas de ríos de las Guayana y Amazonas, el equipo de crustáceos decápodos encontró dos especies no descritas de camarones palaemónidos, uno de los cuales puede ser endémico para el área. La calidad del agua es típica de un sistema regido por las lluvias y que drenan este tipo de geología con bajo contenido de iones y bajo pH. Hay ríos de aguas claras y negras en este sistema. La mayor diversidad y abundacia de peces fue encontrada en el área de Cejiato en la región superior del Caura, mientras que el Río Takoto fue el más diverso en el Bajo Caura. Los datos indican que cada grupo principal de organismos poseen patrones específicos de distribución a lo largo del sistema del Caura. Los invertebrados se encontraron homogeneamente distribuidos, mientras que los peces y las plantas no lo fueron. Las plantas fueron más diversas en la región superior, mientras que los peces lo fueron en la inferior por debajo del Salto Pará. A pesar de esto, la región superior del Caura es más importante para la conservación de peces y plantas, debido a que las especies en los sectores superiores poseen distribuciones más restringidas.

Número de especies

Plantas: 399 especies
Peces: 278 especies
Insectos acuáticos: >87 especies
Moluscos:
gastrópodos 2 especies
bivalvos 1 especie
Crustáceos:
isopodos 1 especie
branchiopodos 1 especie
decápodos 10 especies (4 cangrejos, 6 camarones)
Anélidos: >3 especies

Nuevos registros para la Cuenca del Río Caura:

Peces: 110 especies
Camarones: 4 especies
Cangrejos: 2 especies

Nuevas especies descubiertas:

Camarones: *Pseudopalaemon* n. sp.
Peces: 10 especies

RECOMENDACIONES PARA LA CONSERVACIÓN

El Río Caura representa una región altamente pristina típica de las Guayanas y entonces provee de una excelente oportunidad para ser conservada. Aunque, la porción del río en su región inferior (debajo de Raudal Cinco Mil) posee una nivel actual o potencial de amenazas. Las recomendaciones más específicas incluyen:

- Parar los planes que pueden causar cambios en el sistema hidrológico natural de la Cuenca del Río Caura incluyendo represas, transvase de aguas y cambios mayores en el bosque ribereño
- Realizar un estudio de los peces migratorios
- Establecer un programa a 3-años sobre la pesca sustentable en la región inferior del Caura
- Establecer un programa de registro y revisión para la cuenca
- Desarrollar un corredor para proteger los bosques inundables (hasta el nivel de la máxima inundación en los últimos 50 años)
- Desarrollar programas comunitarios y educacionales públicos acerca de la importancia de los ecosistemas acuáticos
- Prohibir la introducción de especies exóticas
- Reestablecer el bosque ribereño el las áreas por debajo del Raudal Cinco Mil
- Desarrollar el manejo sostenible de explotación de plantas de uso doméstico y comercial tales como las especies: *Ocotea cymbarum, Vochysia venezuelana* (usadas para la construcción de curiaras), *Acosmium nitens, Geonoma deversa* (usadas en construcción de viviendas), y *Heteropsis flexuosa* (usada en la construcción de cestería y casas)
- Como parte de un plan general de conservación, las áreas siguientes necesitan protección immediata: Raudal Cejiato, Río Kakada, Remansos y bosques inundables cercanos a Entreríos, El Raudal Suajiditu y la región justamente por arriba del Salto Pará, El Playón y áreas cercanas, Los ríos Nichare y Tawadu, Las lagunas de rebalse cercanas a Boca de Nichare y El RíoTakoto cerca del Raudal Cinco Mil

Resumen Ejecutivo y Perspectiva

INTRODUCCIÓN

La cuenca del Río Caura esta localizada en la región este del Estado Bolívar, Venezuela. Es un área amplia y vasta de bosques y ríos relativamente pristina y es uno de los mayores tributarios del Río Orinoco (Rosales y Huber 1996). Bosques siempreverdes, de galería, inundables y sabanas cubren aproximadamente 90% de la cuenca (Marín y Chaviel 1996). La cuenca forma parte de las tierras de la Población Ye'kwana, quienes son muy celosos del manejo y apreciación de sus recursos naturales. La alta calidad ambiental de la cuenca del Río Caura, especialmente desde el Raudal Cinco Mil a las cabeceras, es debida al excelente manejo y dirección de esta comunidad étnica.

La Cuenca del Río Caura está localizada en el medio de la región que comprende el Escudo de Guayana en Venezuela (3°37'–7°47'N y 63°23'–65°35'W). Los ríos principales contenidos en la cuenca son los ríos Sipao, Nichare, Erebato y Merewari sobre el margen oeste y los ríos Tigrera, Pablo, Yuruani, Chanaro y Waña localizados en el margen este. El curso del Río Caura se extiende por más de 700 km, originándose en las tierras altas del Escudo de Guayana sobre los 2000 metros por encima del nivel del mar. El Río Caura atraviesa numerosas provincias fisiográficas desde las planicies aluviales cercanas a la boca en el Orinoco hasta los complejos rocosos del Escudo en el Sur. El área de la cuenca es de 45.336 km^2, la cual representa el 20% del área total del Estado Bolívar o 5% del área de Venezuela. La Cuenca del Río Caura es entonces la cuarta más grande en Venezuela, precedida en tamaño por las cuencas de los ríos Apure, Caroní y Orinoco (Peña y Huber 1996).

DIVERSIDAD BIOLÓGICA

La biodiversidad de la Cuenca del Río Caura no es conocida uniformemente en todos los grupos de plantas y animales. Mientras que mucha de la flora terrestre y fauna esta aceptablemente documentada (Rosales y Huber 1996; Huber y Rosales 1997), esto no es cierto para la fauna acuática (Machado-Allison et al. 1999). La vegetación del Caura es muy diversa; más de 1180 especies de plantas han sido identificadas en el área. Los datos existentes revelan que los bosques cubren aproximadamente 90% de la cuenca mientras que el remanente 10% consiste en bosques inundables y otras formaciones vegetales no boscosas (CVG-TECMIN 1994; Huber 1996; Marín y Chaviel 1996; Rosales 1996; Aymard et al. 1997; Dezzeo y Briceño 1997; Bevilacqua y Ochoa 2001).

La fauna terrestre es conocida e incluye aproximadamente 475 especies de aves, 168 especies de mamíferos, 13 especies de anfibios y 23 especies de reptiles (Bevilacqua y Ochoa 2001). Estos valores representan 30% de la especies registradas para el resto de Venezuela y 51% de las especies de la Guayana. Del total de vertebrados terrestres reportados para la cuenca, 5% son considerados bajo amenaza tanto nacional como internacionalmente (Bevilacqua y Ochoa 2000). Con la adición de 110 nuevos registros y 10 nuevas especies colectadas en este estudio, el número de peces continentales documentado con certeza se eleva a 278. Las 92 especies de invertebrados bénticos y 12 especies de crustáceos comprenden la primera información registrada para estos taxa.

Las relaciones biogeográficas de la flora y fauna indican el efecto del Salto Pará sobre la mezcla de especies. Arriba del Salto Pará, existe un más alto grado de endemismo que por debajo del salto. Por ejemplo en plantas, aproximadamente 88% de todas las especies endémicas de Guayana están presentes en la Cuenca del Río Caura y existe un nivel alto de endemismos en las tierras altas y comunidades tepuyanas (Berry et al. 1995; Huber et al. 1997; Bevilacqua y Ochoa 2001). Por encima del Salto Pará, la flora, invertebrados acuáticos y peces exhiben afinidades guayanesas o amazónicas. Mientras que por debajo del Salto Pará estos elementos incluyen formas presentes en el canal principal del Orinoco o los Llanos.

Diversidad Económica y Social

La estructura económica y social de las poblaciones humanas en la Cuenca del Río Caura es compleja y diversa. En las áreas superiores y medias (por encima del Salto Pará) existen culturas indígenas tales como los Ye'kuana y los Sanema (Yanomami), grupos étnicos los cuales mantienen sus tradiciones históricas; siembras de yuca y plátanos en "conucos," caza, pesca y recolección son actividades fundamentales para su vida doméstica y económica. Así que cada miembro de la población es responsable de acuerdo a su edad y sexo del desarrollo de cada una de ellas (Silva-Monterrey 1997). En las regiones bajas, donde la densidad poblacional es alta, los elementos tradicionales indígenas se mezclan con aquellos provenientes de culturas occidentales traídas por los "criollos." En estas zonas inferiores de la cuenca, la economía incorpora desarrollo forestal, agricultura, ganadería, y pesquerías así como también turismo y actividades artesanales. Las relaciones económicas en las regiones bajas de la Cuenca del Río Caura son más complejas. Los grupos indígenas usan una gran variedad de peces y otros recursos acuáticos para su subsistencia.

Pero más allá de las prácticas indígenas, las pesquerías comerciales para consumo humano usan una variedad menor de especies, principalmente: "cachamas" (*Colossoma macropomum*), "cajaros" (*Phractocephalus hemiliopterus*), "coporos" (*Prochilodus mariae*), "curbinatas" (*Plagioscion squamossissimus*), "laulaos" y "valentones" (*Brachyplatystoma* spp.), "morocotos" (*Piaractus brachypomus*), "palometas" (*Mylossoma* spp.), "rayaos" (*Pseudoplatystoma* spp.), "sapoaras" (*Semaprochilodus laticeps*) y "sardinatas" (*Pellona castelneana*) (Machado-Allison et al. 1999; Novoa 1990). Numerosas especies poseen un valor como peces ornamentales, una alternativa económica aún no empleada pero con gran potencial de ser desarrollada dentro de una industria sustentable para el desarrollo de las poblaciones indígenas del área. Las especies siguientes se han encontrado en el Río Caura y son reconocidas en el mundo de la acuariofilia: "tetras" (*Astyanax, Hemigrammus, Hyphessobrycon, Jupiaba, Moenkhausia*), "palometas" o "silver dollars" (*Metynnis, Myleus, Mylossoma*), "cíclidos" (*Aequidens, Apistogramma, Bujurquina, Mesonauta*), "caribes o piranhas" (*Pygocentrus* y *Serrasalmus*) y "cabeza pá bajo o headstanders" (*Anostomus* y *Leporinus*).

Mucha de la subsistencia de las comunidades humanas en la Cuenca del Río Caura, así como también las pesquerías comerciales del Río Orinoco, dependen directamente de la salud integral de la cuenca. El uso de plantas por las poblaciones indígenas es alta. Un total de 358 especies de los bosques de tierras bajas son conocidas de ser usadas en la cultura tradicional (Knap-Vispo 1998). Esto indica el papel fundamental que estos ecosistemas juegan en el mantenimiento de las culturas tradicionales (Bevilacqua y Ochoa 2001). Como un resultado de nuestra expedición, encontramos que los bosques inundables y hábitats protegidos en las orillas sirven de albergue (protección y crecimiento) de muchas especies acuáticas tales como la palometa, *Myleus rubripinnis*. El éxito de reclutamiento poblacional de peces en el Río Caura, contribuye a aproximadamente las 75.000 toneladas métricas de peces desembarcados en Maripa y Caicara del Orinoco.

Amenazas para la Región

La región del Escudo de Guayana de Venezuela ha sido afectada por grandes alteraciones ambientales provenientes del desarrollo de proyectos, industria y pesquerías y cacerías no reguladas. La explotación de minerales estratégicos tales como oro, diamantes y bauxita aunado al desarrollo y construcción de uno de los complejos de represas con fines hidroeléctricos más grandes del mundo ha producido: 1) biodegradación y destrucción de extensas áreas verdes en las cuencas de los ríos Caroní y Cuyuní (Machado-Allison et al. 2000); 2) contaminación mercurial de los ríos, bosques, fauna y humanos; 3) disminución de la calidad de agua debido al incremento de la sedimentación; y 4) pérdida de grandes cantidades de agua potable para uso doméstico (Machado-Allison 1994, 1999; Miranda et al. 1998).

La Cuenca del Río Caura está amenazada por actividades mineras, pesca ilegal y una propuesta represa. En estos momentos son explotados cientos de kilómetros de bosques con fines madereros, en áreas compartidas con la Cuenca del Paragua. Algunas actividades mineras también han sido introducidas en las cabeceras del Río Caura. En este momento, las actividades mineras son menores en comparación con cuencas vecinas tales como el Caroní y Cuyuní (Machado-Allison et al. 2000: fig. 1). Una merma dramática de los recursos pesqueros en las zonas bajas del Caura se piensa que es debida a la pesca comercial ilegal en territorios indígenas. La amenaza más importante a la Cuenca del Río Caura es la potencial construcción de una represa hidroeléctrica en el Salto Pará y el plan de transvase de aguas que acarrearía la pérdida de cerca del 75% del agua del Caura hacia el sistema Paragua-Caroní. Esto no solamente deprimiría drásticamente la cantidad de agua en el Caura, sino también resultaría en una alteración severa del ciclo hidrológico natural. La salud y mantenimiento de las poblaciones humanas, acuáticas y bosques ribereños completamente dependen de éste ciclo hidrológico natural.

En la región baja del Río Caura, la expansión de actividades turísticas, pesca deportiva y agricultura han impactado negativamente. Estas actividades deben ser estudiadas y reguladas con la finalidad de permitir un uso sostenible del ecosistema.

Oportunidades de Conservación

La Cuenca del Río Caura es un área silvestre y pristina extensa. Debido a la destrucción y desarrollo en regiones adyacentes en el norte del Escudo de Guayana (Ríos Caroní y Cuyuní), el Río Caura representa una oportunidad importante para preservar esta región particular. La composición de las comunidades de animales y plantas en la Cuenca del Río Caura no ocurre en otra parte en Venezuela o en las Guayanas. Es un área altamente diversa, con numerosas especies endémicas y comunidades únicas, tales como las comunidades de plantas acuáticas (Podostemonaceae) en rápidos y la flora sobre las islas rocosas en medio de los ríos. Protección a través programas de educación ambiental, actividades en las comunidades humanas y observaciones y registros a largo plazo son altamente recomendables. La regulación y revisión de las actividades pesqueras en la región es necesaria para poder alcanzar un nivel sostenible para los habitantes de la cuenca.

La Expedición AquaRAP

En respuesta a los planes de desarrollo en la región de la Cuenca del Río Caura, los cuales incluyen el transvase de aguas, deforestación, incremento de las poblaciones humanas y otras amenazas, un equipo científico multidisciplinario y multinacional realizó una expedición en la Cuenca del Río Caura enmarcado en el Programa de Evaluación Rápida del medio Acuático (AquaRAP) y con los propósitos siguientes: (i) describir los aspectos biológicos y ambientales del ecosistema acuático antes de una modificación a larga escala; y (ii) determinar, de ser posible, los cambios potenciales en los recursos acuáticos en respuesta a las amenazas percibidas. La expedición se realizó entre el 25 Noviembre al 12 Diciembre del año 2000 y estudió la sección del Río Caura entre el Raudal Cejiato en el Sur y el Río Mato en el Norte (Mapa). Para facilitar las comparaciones, el área estudiada fue dividida en ocho regiones: Río Kakada; Río Erebato; Río Caura desde Entrerios-Raudal Cejiato; Río Caura desde Entrerios-Salto Pará; Río Caura cerca de El Playón; Río Nichare; Río Caura en la vecindad del Raudal Cinco Mil; y el Río Mato (Mapa). La vegetación terrestre, acuática, características físico-químicas del agua, plancton, macroinvertebrados acuáticos y peces fueron evaluados. Estos inventarios, a la luz de las amenazas propuestas, serán utilizadas como la base de recomendaciones para la conservación e investigaciones futuras en el área.

El AquaRAP Sudamericano es un programa multinacional y multidisciplinario dedicado a la identificación de prioridades de conservación y manejo sustentable de oportunidades para los ecosistemas acuáticos en América Latina. La misión del AquaRAP es el determinar los valores biológicos y de conservación para ecosistemas dulceacuícolas a través de inventarios rápidos y transmitir la información a los gerentes ambientales, políticos, conservacionistas, científicos y agencias promotoras internacionales. El AquaRAP Sudamericano es un programa de colaboración coordinado por Conservación Internacional y el Field Museum.

El programa AquaRAP tiene un Comité Directivo internacional compuesto de científicos provenientes de siete países: Bolivia, Brasil, Ecuador, Paraguay, Perú, Venezuela y los Estados Unidos. El Comité Directivo revisa los protocolos para la evaluación rápida y asigna sitios prioritarios para estas investigaciones. Las expediciones de AquaRAP, las cuales involucra una gran colaboración con los investigadores de los países huéspedes, también promueve el intercambio internacional de información y oportunidades de entrenamiento. La información obtenida en estas expediciones es divulgada a través del *Boletín de Evaluación Biológica del RAP* editado por Conservación Internacional (*RAP Bulletin of Biological Assessment*), diseñado para los gerentes locales, políticos, líderes comunitarios y conservacionistas, los cuales pueden establecer prioridades de conservación y guías de acción a través de financiamientos en la región.

Una evaluación actual de la biodiversidad del Caura es necesaria. Esta parte de Venezuela no ha sido completamente estudiada, aunque la flora, aves y mamíferos son generalmente bien conocidos. Información adicional es necesaria antes de reconocer oportunidades de manejo. La sección siguiente resume los resultados de los capítulos técnicos basados en los análisis de los datos y colecta de ejemplares durante la expedición AquaRAP. Por favor vea los capítulos respectivos para una mayor información.

RESUMEN DE LOS CAPÍTULOS

Vegetación Terrestre y Acuática

Un total de 443 muestras fueron tomadas durante la expedición de una alta variedad de hábitats de bosque ribereño y comunidades acuáticas. Las muestras contienen 399 especies de plantas (291 por encima del Salto Pará y 185 en el Bajo Caura), todas las cuales están incluidas en la lista de 1180 especies conocidas para la cuenca. La investigación revela que existe una gran diversidad en la composición de comunidades vegetales características de climas húmedos con suelos de bajos nutrientes. La variación en la estructura de las comunidades comprende un gradiente de paisajes ribereños estructurados por la intensidad y duración de procesos erosivos. Estas características unidas con la convergencia de cuatro provincias geológicas y la presencia de un gradiente climático condicionado a la gran variación latitudinal (40 a 2350 m), determinan la existencia de una biodiversidad excepcional en la Cuenca del Río Caura.

La proporción de elementos endémicos es baja en las tierras bajas y bosques inundables. La región se encuentra dominada por las palmas: *Euterpe precatoria, Attalea maripa, Socratea exhorriza, Genoma baculifera* y *Bactris brongniartii.*

Sin embargo, la diversidad es relativamente alta y comparable con los valores registrados para ambientes similares en la Amazonia y Guyana. Las observaciones obtenidas en éste estudio sobre los ecosistemas inundables del corredor del Erebato, dominados por *Oenocarpus* y en sectores medios del Caura, dominados por *Mauritiella* son valores similares a otros corredores ribereños en el Escudo de Guayana. Una serie única de ensamblajes florales está asociada a las islas y encotrada principalmente en el Erebato y región superior y media del Caura. Estas islas posee tanto elementos de tierra firme como comunidades acuáticas densas de podostemonaceas asociadas a las aguas rapidas y rocosas.

Calidad de Agua

Cuarenta y tres sitios dentro de las regiones superior e inferior de la Cuenca del Río Caura fueron estudiadas para la obtención de varios parámetros limnológicos. La región superior del Río Caura es pristina, comprende una base más canalizada, sujeta a una variación hidrológica estacional mayor y como resultado una menor inundación del bosque. En este río tropical con bajos nutrientes, el objetivo principal fue colocado en analizar los siguientes parámetros: temperatura, conductividad, pH, profundidad de Secchi, carbono orgánico disuelto (COD), alcalinidad y sedimentos.

En general las aguas de la región son levemente ácidas y diluidas, con muy baja conductividad y alcalinidad; típicas de sistemas dominados por lluvias. Estos resultados son congruentes con aquellos obtenidos en estudios previos y pueden ser atribuidos a la muy antigua geología de la región. Las aguas fueron encontradas ser similares en estos parámetros tanto en la región superior como inferior de la cuenca; sin embargo, en tributarios ocasionales se encontró variación a esta generalidad. COD (descarga terrestre) varía a lo largo de la cuenca en ambos ríos de aguas transparentes y negras muestreados.

En general, la calidad de agua determinada fue "buena" para todos los lugares estudiados. La región inferior presentó una mayor perturbación que la región superior. Altas perturbaciones en los márgenes ribereños pueden resultar en un incremento de las descargas de sedimentos en el período de lluvias y decrecimiento de la descarga de formas biológicas importantes de carbono orgánico (COD) y causar un cambio en el hábitat de los organismos acuáticos. Estos cambios del hábitat pueden incluir cambios en la transparencia del agua, sustrato, canalización del río y el número y diversidad de hábitats que mantienen los actuales organismos acuáticos.

Macroinvertebrados Bénticos

La diversidad de los macroinvertebrados bénticos en el Río Caura fue evaluada en 25 localidades distribuidas tanto en la región superior como inferior del río. La comunidad de macroinvertebrados acuáticos está compuesta por insectos, caracoles, almejas, gusanos oligoquetos, sanguijuelas, turbelarios y crustáceos. Los insectos acuáticos están representados por varios ordenes los cuales con mayor diversidad fueron: Odonata, Diptera, Ephemeroptera, Hemiptera y Trichoptera. La composición específica observada es típica de ambientes pristinos, ya que se encuentran una alta diversidad de Odonata y Trichoptera, los cuales son habitantes de ambientes libre de contaminación o perturbación. La mayoría de los géneros de macroinvertebrados bénticos fueron encontrados homogéneamente distribuidos en la cuenca. La variación en diversidad de la fauna béntica parece estar relacionada con cambios en el oxígeno disuelto y turbidez del agua. Estos hallazgos son importantes ya que cambios drásticos en el ciclo hidrológico del agua, deforestación o minería podrían introducir material particulado suspendido e incrementar la conductividad del agua con el subsecuente cambio dramático en la comunidades bénticas.

Crustáceos Decápodos

Un total de diez especies de crustáceos decápodos fueron encontrados en las áreas muestreadas de la región media de la Cuenca del Río Caura, seis especies de Palaemonidae (camarones), una especie de Pseudothelphusidae (cangrejo) y tres especies de Trichodactylidae (cangrejos). La región río arriba de Salto Pará (Caura Superior) posee una baja riqueza, con cinco especies, comparada con la región inferior (Bajo Caura) la cual posee ocho especies. La fauna de decápodos de la Cuenca del Río Caura posee una posible especie endémica, la cual es un camarón no descrito del género *Pseudopalaemon* sp., mientras que otros son conocidos tanto de la región de los Llanos como de la Cuenca Amazónica. El número de especies y géneros de decápodos obtenidos en el Río Caura representa una muestra típica de diversidad y abundancia de los sistemas pertenecientes a las regiones Guayanesas y Amazónicas. El camarón *Macrobrachium brasiliense* fue la especie más frecuente y abundante encontrada en este estudio y puede ser considerada la especie más típica de éste sistema. Dos especies no descritas de camarones palaemonidos fueron colectadas. La abundancia de decápodos fue baja a moderada, probablemente reflejando las limitaciones del muestreo y las condiciones oligotróficas generales de los hábitats. De la estructura comunitaria de los decápodos así como también del hábitat compartido y su distribución, podemos concluir que las áreas estudiadas poseen una comunidad de decápodos saludable y permanecerá así, en la medida de que las condiciones prístinas sean preservadas en el tiempo.

Peces

Sesenta y cinco estaciones fueron muestreadas entre el Raudal Cejiato y Río Kakada en el sur hasta el Raudal Cinco Mil en el Norte. Un total de 278 especies fueron identificadas para la cuenca. El Orden Characiformes con 158 especies (56.8%) fue el más diverso, seguido de los Siluriformes (74, 26.6%), Perciformes (27, 9.7%), Gymnotiformes (9, 3.2%), Clupeiformes (3, 1.1%), Rajiformes y Cyprinodontiformes (2, 0.7%), Beloniformes, Pleuronectiformes y Synbranchiformes (1, 0.3%). La Familia Characidae con 113 especies (40% del total) fue la que mostró más especies. Los resultados añaden 54 especies para el Orden Characiformes y 39 especies de Characidae conocidas para la cuenca. Con respecto a otros grupos los incrementos fueron: Siluriformes, 74 vs 49 previamente reportados; Perciformes 27 vs 12. En total 110 nuevos registros son reportados para la cuenca. En las regiones superiores, Salto Pará (arriba), Raudal Cejiato y Río Kakada (playas y caño Suajiditu) fueron las más diversas y abundantes en especies con formas típicamente guayanesas. Las áreas por debajo del Salto Pará (El Playón), Ríos Tawadu-Nichare y Raudal Cinco Mil (incluyendo el Río Takoto) poseen alta diversidad y abundancia, con numerosas especies típicamente encontradas en las Llanos y Orinoco. Sin embargo, ambas áreas poseen sus propias características en términos de taxonomía, biogeografía e importancia en conservación. Un gran número de especies posee importancia económica como peces de consumo y ornamentales.

Basados en la riqueza de especies, diversidad y abundancia relativa de los peces y la dependencia de poblaciones humanas sobre las especies de importancia doméstica, ciertas áreas en las regiones superior e inferior del Caura deben ser protegidas y preservadas como parte de un plan general de conservación. Estas incluyen el Raudal Cejiato, Río Kakada, los remansos y lagunas de inundación cercanos a Entreríos, el Raudal Suajiditu y la región justo por encima del Salto Pará, El Playón y áreas cercanas, los ríos Nichare y Tawadu, las lagunas de rebalse cercanas a Boca de Nichare y el Río Takoto cerca del Raudal Cinco Mil.

Ecohidrología y Vegetación Ribereña

El trabajo prueba la posibilidad de una metodología que integra un alcance ecogeográfico, diseñado para conservación, con un alcance hidro-sistémico, diseñado para el estudio jerárquico de sistemas de ríos. En mismo realiza un análisis espacial utilizando un Sistema de Información Geográfica (SIG) con datos georeferenciados determinados en taxa blancos para el estudio de tendencias en diversidad de especies dentro de la cuenca (a lo largo del corredor ribereño del Caura usando la Familia Leguminosae) y entre cuencas (entre diferentes subcuencas del Amazonas y Orinoco usando géneros seleccionados de Ingeae).

Los análisis dentro y entre cuencas del número de especies, rareza, comonalidad y distinctividad geográfica indican la importancia de conservar los paisajes ribereños bajos en la Cuenca del Río Caura. Nuestros resultados demuestran la naturaleza crítica de conservación de la cuenca del Río Caura, debido no solo a la posesión de un gran número de especies, sino también porque dichas especies forman comunidades únicas las cuales se originaron a partir de en un número de diferentes cuencas y regiones fitogreográficas. El Río Caroní, el cual posee una flora similar al Caura, se encuentra altamente perturbado. La región más crítica de la Cuenca del Río Caura para su conservación son los sectores por debajo del Salto Pará. Los análisis dentro de cuencas muestran diferentes patrones regionales entre géneros seleccionados de Ingeae los cuales se relacionan con regiones fitogeográficas y tipos de ríos. Además a Río Caura, el Río Negro en Brasil y Venezuela es también importante para ser incorporado en un plan multinacional de conservación. El alcance demostró ser una buena herramienta de conservación para el análisis del valor de la biodiversidad ribereña. Es también apropiado para la selección de diversas áreas para conservación *in situ*.

Distribución de Peces y Patrones de Biodiversidad

Hipótesis nula concerniente a la distribución al azar de especies con respecto a subregiones y macrohábitats dentro de la Cuenca del Río Caura fueron examinadas con datos provenientes de 97 especies de invertebrados bénticos, 303 especies de plantas acuáticas y 278 especies de peces de agua dulce. El análisis de ocho subregiones divididas igualmente arriba y debajo de las catraratas indican que los invertebrados estaban distribuidos al azar con respecto a las subregiones. Los peces y plantas no lo estaban, aunque el efecto subregional de los patrones sobre las plantas fue mucho mayor que el mostrado por los peces. Más aún, los peces fueron menos ricos en especies arriba del Salto Pará que bajo. La conversión fue observada para plantas mientras que invertebrados fueron igualmente ricos. Efectos no azarosos del macrohábitat fueron encontrados en cada uno de los grupos con ciertas generalidades. Por ejemplo, especies de Odonata y Ephemeroptera se encuentra en hábitats de aguas rápidas con altos contenidos de oxígenos y están asociados que un ensamblaje mayor de peces que son encontrados en asociación con bancos densos de macrofitas, usualmente pertenecientes a la Familia Podostemonaceae. Los ensamblajes de peces demuestran una transición suave a lo largo de gradientes en varios macrohábitats (p.e., fondos de arena a fango y orillas). Al menos seis macrohábitats son necesarios para preservar 82% de las especies de peces.

ACTIVIDADES DE CONSERVACIÓN E INVESTIGACIÓN RECOMENDADAS

La base para recomendaciones de iniciativas en conservación e investigación pueden ser encontradas en la sección "Reporte a primera vista" y en los capítulos del éste informe. Las recomendaciones no están listadas en orden de prioridad.

- **Parar todos los planes de desarrollo que puedan causar cambios en el ciclo hidrológico natural en la Cuenca del Río Caura incluyendo represas, transvase de aguas y mayores cambios en el bosque ribereño.** Represas o transvase de aguas dañara el ciclo hidrológico el cual es importante en sistemas tropicales dominados por las lluvias. La salud de los bosques y el éxito en la recluta y mantenimiento de la biodiversidad dependen enteramente sobre el mantenimiento de un ciclo hidrológico natural.

- **Establecer un programa de registro y evaluación de peces migratorios.** Peces migratorios entran al Río Caura estacionalmente y usan las áreas de bosques inundados para su reproducción y crecimiento. Hábitats críticos por encima y bajo el Salto Pará debe ser determinados para asegurar la sobrevivencia de especies y las pesquerías.

- **Establecer un programa a tres años sobre pesquería sostenible en el Bajo Caura.** La evidencia indica que existe mucha actividad ilegal dentro del territorio Ye'kuana y una sobre pesca potencial en la región inferior del Río Caura por debajo del Salto Pará. La pesquería en esta región necesita ser vigilada y evaluada de forma tal de establecer programas de manejo técnico (cuotas, tipo de arte) para asegurar capturas sostenibles a perpetuidad.

- **Establecer un programa de seguimiento de registro de la hidrología y el restablecimiento de la estación hidrológica y limnológica en Entrerios.** Esta estación proveerá una plataforma valiosa para el acceso de personal científico para la realización de estudios de la región en mayor detalle y construir una base de datos a largo plazo para el ambiente de bosque lluvioso tropical. Dado a que la pristinidad de los ambientes acuáticos continuamente se están degradando a una velocidad alarmante a nivel mundial, la conservación de pocos pero extensas regiones no perturbadas que quedan, incluyendo la Cuenca del Río Caura, deben ser prioritarios en los protocolos de manejo de conservación.

- **Desarrollo de un corredor para la protección de los bosques inundables (más allá del ciclo de inundación de 50 años).** La Cuenca del Río Caura presenta una oportunidad única para salvar un recurso amenazado en Venezuela—el bioma Guayanés. De acuerdo con los deseos y costumbres de la población Ye'kuana, un corredor de protección debe ser establecido, al menos al nivel del ciclo de inundación ocurrido en los últimos 50 años, con la finalidad de proteger el bosque, el río y sus especies.

- **Establecer un plan de protección zonal para especies.** Las regiones por encima y abajo del Salto Pará poseen diferencias significantes en plantas y peces. Además, seis macrohábitats (p.e. rápidos, islas y remansos) son necesarios para proteger la existencia de al menos 82% de las especies de peces, incluyendo aquellas especies usadas pora consumo de las poblacione humanas y su comercio. Es necesario determinar la cantidad de área para cada uno de los hábitats que son necesarios para proteger mas del 80% de los animales y plantas de la cuenca. Estas áreas críticas deben ser consideradas como zonas de protección (ABRAE) con la mínima influencia o impacto humano. Debido a las diferencias en flora y fauna, zonas principales deben ser establecidas en áreas por encima y debajo del Salto Pará.

- **Establecer programas educativos y públicos acerca de la importancia de los ecosistemas acuáticos.** Esto necesita ser realizado principalmente en comunidades fuera del territorio Ye'kuana. Por debajo del territorio Ye'kuana existe un alto grado de degradación de hábitats en ambos ambientes de bosque y acuáticos. La mayoría de esta gente todavía depende del río para su existencia y, entonces, educación acerca de la protección de la cuenca es crítica para poder sostener o enriquecer su existencia.

- **Prohibir la introducción de especies exóticas acuáticas.** Existen suficiente explotación de recursos acuáticos nativos disponible en la Cuenca del Río Caura que abogan por prohibir la introducción de especies foráneas o exóticas. La fauna acuática del Río Caura es única dentro de Venezuela y debe ser protegida. Hay suficiente evidencia que muestra que la introducción de especies exóticas aceleran la pérdida de especies nativas y aceleran la degradación de hábitats naturales.

- **Restablecer los bosques ribereños por debajo de los rápidos de La Mura.** Regular y no permitir la expansión de la frontera agrícola y forestal dentro de la Cuenca del Río Caura. Esto es particularmente importante en las regiones del Bajo Caura. Estas áreas han sido degradadas con perdida sustancial de biodiversidad y estructura comunitaria. Esta degradación puede estar acoplada con la pérdida de la capacidad de recluta de especies de peces, camarones y cangrejos. Las técnicas de reforestación están disponibles y podrían ayudar a restaurar las comunidades nativas así como también potencialmente incrementar las pesquerías.

- **Desarrollar planes de manejo para la cosecha sostenible de especies de plantas.** Las especies de plantas *Ocotea cymbarum, Vochysia venezuelana* (usada para la fabricación de canoas), *Acosmium nitens, Geonoma deversa* (usadas en la construcción de casas), y *Heteropsis flexuosa* (usada en la fabricación de cestas y construcción) son ampliamente explotadas en el área. Así que un plan podría incrementar el crecimiento económico en una forma sostenible en la cuenca, así como también ayudar al apoyo de las comunidades humanas que actualmente aprovechan y dependen de estos recursos.

LITERATURA CITADA

Aymard, G., S. Elcoro, E. Marín y A. Chaviel. 1997. Caracterización estructural y florística en bosques de tierra firme de un sector del bajo Río Caura, Estado Bolívar,

Venezuela. *En*: Huber, O. y J. Rosales (eds). Ecología de la Cuenca del Río Caura, Venezuela II. Estudios especiales. Scientia Guaianae, Vol. 7:143–169.

Berry, P. E., O. Huber y B. K. Holst. 1995. Floristic Analysis and Phytogeography. *En*: Berry, P. E., B. K. Holst, K. Yatskievych (eds.). Flora of the Venezuelan Guayana, Vol. 1. Introduction. Missouri Botanical Garden, Saint Louis, USA. pp. 161–191.

Bevilacqua, M. y J. Ochoa G. (eds). 2000. Informe del componente Vegetación y Valor Biológico. Proyecto Conservación de Ecosistemas Boscosos en la Cuenca del Río Caura, Guayana Venezolana. PDVSA-BITOR, PDVSA-PALMAVEN, ACOANA, AUDUBON de Venezuela y Conservation International. Caracas, Venezuela. 81 pp.

Bevilacqua, M. y J. Ochoa G. 2001. Conservación de las últimas fronteras forestales de la Guayana Venezolana: propuesta de lineamientos para la Cuenca del Río Caura. Interciencia 26(10): 491–497.

CVG-TECMIN. 1994. Informes de avance del Proyecto Inventario de los Recursos Naturales de la Región Guayana. Hojas NB-20: 1, 5, 6, 9, 10, 13 y 14. Gerencia de Proyectos Especiales. Ciudad Bolívar, Venezuela.

Dezzeo, N. y E. Briceño. 1997. La vegetación en la cuenca del Río Chanaro: medio Río Caura. *En*: Huber O. y J.Rosales (eds). Ecología de la Cuenca del Río Caura, Venezuela II. Estudios especiales. Scientia Guaianae, Vol. 7: 365–385.

Huber, O. 1996. Formaciones vegetales no boscosas. *En*: Rosales J.y O. Huber (eds.). Ecología de la Cuenca del Río Caura, Venezuela: I. Caracterización general. Scientia Guaianae, Vol. 6: 70–75.

Huber, O. y J. Rosales (eds). 1997. Ecología de la Cuenca del Río Caura, Venezuela. II Estudios Especiales. Scientia Guaianae No. 7: 1–473.

Huber, O., J. Rosales y P. Berry. 1997. Estudios botánicos en las montañas altas de la cuenca del Río Caura. *En*: Huber, O. and Rosales, J. (eds.). Ecología de la Cuenca del Río Caura, Venezuela II. Estudios esp.eciales. Scientia Guaianae, Vol. 7: 441–468.

Knab-Vispo, C. 1998. A rain forest in the Caura Reserve (Venezuela) and its use by the indigenous ye´kwana people. Doctoral Thesis. Universidad de Wisconsin-Madison, USA. 202 pp.

Machado-Allison, A. 1994. Factors affecting fish communities in the flooded plains of Venezuela. Acta Biol. Venez., 14(3):1–20.

Machado-Allison, A. 1999. Cursos de agua, fronteras y conservación. *En*: G. Genatios (ed). Ciclo Fronteras: Desarrollo Sustentable y Fronteras. Com. Estudios Interdisciplinarios, UCV. Caracas: 61–84.

Machado-Allison, A., B. Chernoff, C. Silvera, A. Bonilla, H. Lopez-Rojas, C. A. Lasso, F. Provenzano, C. Marcano y D. Machado-Aranda. 1999. Inventario de los peces de la cuenca del Río Caura, Estado Bolivar, Venezuela. Acta Biol. Venez., Vol. 19 (4):61–72.

Machado-Allison, A., B. Chernoff, R. Royero-Leon, F. Mago-Leccia, J. Velazquez, C. Lasso, H. López-Rojas, A. Bonilla-Rivero, F. Provenzano y C. Silvera. 2000. Ictiofauna de la cuenca del Río Cuyuní en Venezuela. Interciencia, 25(1): 13–21.

Marin, E. y A. Chaviel. 1996. Bosques de Tierra Firme. *En:* Rosales, J. y O. Huber (eds). Ecología de la Cuenca del Río Caura. Scientia Guaianae, 6: 60–65.

Miranda, M., A. Blanco-Uribe, L. Hernández, J. Ochoa y E. Yerena. 1998. No todo lo que brilla es oro: hacia un nuevo equilibrio entre conservación y desarrollo en las últimas fronteras forestales de Venezuela. Washington, DC: Inst. Rec. Mundiales (WRI). 59pp.

Novoa, D. 1990. El río Orinoco y sus pesquerías; estado actual, perspectivas futuras y las investigaciones necesarias. *En*: Weibezahn, F., H. Alvarez y W. Lewis (eds). El Río Orinoco como Ecosistema. Edelca, Fondo Ed. Acta Científica Venezolana, CAVN, USB: 387–406.

Peña, O. y O. Huber. 1996. Características Geográficas Generales. *En*: Rosales, J. y O. Huber (eds). Ecología de la Cuenca del Río Caura. Scientia Guaianae, 6: 4–10.

Rosales, J. 1996. Vegetación: los bosques ribereños. *En*: Rosales, J. y O. Huber (eds). Ecología de la Cuenca del Río Caura, Venezuela: I. Caracterización general. Scientia Guaianae, Vol. 6: 66–69.

Rosales, J. y O. Huber (eds). 1996. Ecología de la Cuenca del Río Caura, Venezuela. I. Caracterización General. Scientia Guaianae No. 6: 1–131.

Silva-Monterrey, N. 1997. La Percepción Ye'kwana del Entorno Natural. *En*: Huber, O. y J. Rosales (eds). Ecología de la Cuenca del Río Caura. Estudios Especiales Scientia Guaianae, 7: 65–84.

Capítulo 1

Introducción a la Cuenca del Río Caura, Estado Bolívar, Venezuela

Antonio Machado-Allison, Barry Chernoff y Mariapia Bevilacqua

INTRODUCCIÓN

La región neotropical aún contiene cientos de miles de kilómetros cuadrados de áreas prístinas de bosque. Gran cantidad de esta región se encuentra en Brasil, Bolivia, Colombia, Guyana, Perú y Venezuela. Dichas regiones albergan la mayor diversidad de especies y biomasa de plantas, ecosistemas de vida salvaje y aguas dulces del planeta. Sin embargo, hoy día, el aumento de la demanda mundial de consumo y el incremento demográfico de poblaciones humanas, están acelerando la explotación de lo que, en un momento, fue la mayor reserva de alimento, minerales, belleza escénica, energía y biogenética del mundo (SISGRIL 1990; Bucher et al. 1993; Chernoff et al. 1996; Chernoff y Willink 1999; Machado-Allison 1999; Machado-Allison et al. 1999, 2000). Estudios de biodiversidad en Sur América, especialmente en las cuencas del Amazonas y del Orinoco son ahora, más que nunca, de gran importancia para poder integrar los intereses del desarrollo potencial económico con aquellos necesarios para promover la sostentibilidad biológica, como medio para reducir las amenazas actuales y cambios ambientales adversos (IUCN 1993; Aguilera y Silva 1997). Por otro lado y más importante aún es que estudios recientes han demostrado que Venezuela se encuentra entre los países con mayores niveles de diversidad biológica (Mittermeier et al. 1997, 1998).

Venezuela, debido al mantenimiento de una política de estado conservacionista, posee numerosas áreas protegidas o bajo régimen especial. Parques Nacionales, Reservas Forestales, Monumentos Naturales, Refugios etc., se encuentran distribuidos a lo largo del país. Estos incluyen: tierras andinas altas (paramos), sabanas (llanos), lagunas litorales, bosques nublados y bosques húmedos tropicales, todas ellas creadas siguiendo criterios o visiones dirigidas a preservar ecosistemas terrestres o humedales. Pocas áreas existen en donde se tomen en cuenta criterios de conservación de ecosistemas acuáticos y las cuencas a los cuales pertenecen.

A pesar de esto, Venezuela contiene una fracción importante de ecosistemas amenazados, la mayoría de los cuales corresponde a sistemas boscosos considerados como las últimas fronteras forestales del mundo tropical y en los cuales se presentan conflictos de tipo social y ambiental (Bevilacqua y Ochoa 2001). La Cuenca del Río Caura y los ecosistemas acuáticos distribuidos principalmente al sur del Orinoco se encuentran entre tales ecosistemas.

La Cuenca del Río Caura esta localizada en el sector occidental del Estado Bolívar. Es una vasta extensión de bosques y ríos, que constituye uno de los mayores afluentes del Río Orinoco (Rosales y Huber 1996). La primera expedición organizada al Río Caura fue dirigida por Chaffanjon (1889), siguiendo 20 años después por una expedición conjunta Venezolana-Inglesa dirigida por André. Sin embargo, los resultados de la última expedición nunca fueron formalmente publicados. En el siglo pasado, esfuerzos para el estudio del Caura han sido dirigidos por científicos de las universidades venezolanas e institutos de investigación. Esfuerzos paralelos para el desarrollo de plpanes de conservación para la Cuenca del Río Caura han sido realizados por Organizaciones No-Gubernamentales (ONG's) venezolanas como ACOANA y Econatura o instituciones como la Fundación La Salle, Jardín Botánico del Orinoco y la Corporación Venezolana de Guayana (CVG). Resultados de esos y otros proyectos en el Caura han sido publicados en dos volúmenes de *Scientia Guayanae* (Rosales y Huber 1996; Huber y

Rosales 1997). Publicaciones recientes acerca de la ictiofauna de la cuenca del Río Caura discuten diversidad de especies, biogeografía y conservación (Chernoff et al. 1991; Vari 1995; Balbas y Taphorn 1996; Lasso y Provenzano 1997; Bonilla-Rivero et al. 1999; Machado-Allison et al. 1999).

La Cuenca del Río Caura representa un área extensa pristina la cual corresponde parcialmente tierras del pueblo de Ye'kuana. Cambios recientes en la Constitución Venezolana da a los Ye'kuana y otros pueblos indígenas derechos sobre el futuro de sus tierras. La extremada y alta calidad de esta región, especialmente desde El Raudal Cinco Mil hasta las cabeceras, en gran parte existe, gracias a la excelente administración de recursos y la apreciación por los recursos naturales por parte de ellos.

HIDROLOGÍA

La cuenca del Río Caura está ubicada en la región media del Escudo Guayanés Venezolano (3°37'–7°47'N y 63°23'–65°35'W). Los principales ríos contenidos en la cuenca son el Sipao, el Nichare, el Erebato, y el Merewari, en su margen occidental y los Ríos Tigrera, Pablo, Yuruani, Chanaro, y Waña, en su margen oriental. El área de la cuenca cubre, aproximadamente, unos 45,336 km^2 (20% del total de la superficie del Estado Bolívar y 5% de la superficie territorial de Venezuela). Esta data ubica la cuenca del Río Caura como la cuarta cuenca más grande de Venezuela, precedida en tamaño por las cuencas de los Ríos Apure, Caroní, y Orinoco (Peña y Huber 1996).

El curso del río cuenta con más de 700km de longitud, originándose en las regiones montañosas del Escudo Guayanés a unos 2.000m por encima del nivel del mar. El Río Caura atraviesa varios tipos de ambientes fisiográficos, desde llanos aluviales cerca de la boca del Orinoco, hasta los complejos rocosos del Escudo Guayanés, al sur. Aproximadamente 90% de la cuenca está cubierta por diferentes tipos de bosques: siempre verdes, anegadizos, galería y sabanas (Marín y Chaviel 1996).

El promedio anual de lluvia varía entre 1.200 mm hacia la boca del Orinoco al norte, hasta 3.000–4.000 mm hacia las cabeceras al sur. Las diferencias en rango de las lluvias corresponden a las diferencias dramáticas en los cambios de estación climática entre las porciones norte y sur de la cuenca. En el norte, la estación de sequía se extiende desde enero hasta marzo, con una estación lluviosa extendiéndose desde abril hasta octubre. En el sur, hay una corta estación de sequía (enero-febrero) y una larga estación lluviosa durante el resto del año. El Río Caura aporta cerca de 3.500 m^3/s de agua al Orinoco. Esto ubica al Río Caura como el segundo afluente más importante de la margen del Orinoco que corresponde al Escudo Guayanés (Vargas y Rangel 1996).

La calidad de agua del Río Caura se considera buena (García 1996). El río ha sido tradicionalmente clasificado como un río de aguas negras según Sioli (1965), debido a su "aparente" coloración marrón, así como también sus bajos nutrientes, bajo pH, y alta transparencia. Sin embargo, el río no clasifica dentro de lo que realmente son aguas negras, si consideramos otros factores tales como las concentraciones de carbón y oxígeno (García 1996) y el examen del color del agua después de la filtración de sólidos en suspensión.

Fisiográficamente, la cuenca del Río Caura puede estar dividida en 3 secciones; a) El Bajo Río Caura, desde la confluencia con el Río Orinoco hasta el Salto Pará; b) El Medio Río Caura, desde Salto Pará hasta la confluencia con el Río Merewari y Río Waña (Guaña); y c) El Alto Río Caura, desde la confluencia de los Ríos Merewari y Waña hasta las fuentes en la Serranía del Vasade. El criterio aplicado en este informe para definir las subregiones fue principalmente la presencia de un accidente geográfico conocido como el Salto Pará. Nos referiremos a toda la zona arriba del Salto Pará como el "Alto Caura" y la zona abajo del Salto como "Bajo Caura," debido a que la logística no nos permitía subir hasta el verdadero Alto Caura.

GEOLOGÍA Y GEOMORFOLOGÍA

La geología de la cuenca del río Caura es parcialmente conocida. Mientras que las áreas medias y bajas (>1.000m) son bien conocidas, las regiones correspondientes al Alto Caura aún no han sido completamente exploradas desde el punto de vista tectónico y petrográfico (Colvee et al. 1990; Rincón y Estanga 1996). En el área del Caura se encuentra representadas 4 provincias geológicas (Imataca, Pastora, Cuchivero y Roraima) pertenecientes al Escudo de Guayana. Esta provincias geológicas, las cuales están han sido modificadas por una serie de procesos tectónicos, poseen rocas arqueozoicas y proterozoicas. La primera Provincia (Imataca) está caracterizada por abundancia de gneis, anfibolitas, intrusiones graníticas y cuarcitas ferruginosas. La segunda Provincia (Pastora) de edad Proterozoica, localizada en la región centro-occidental principalmente en los Río Icutú, Tudi y Erebato donde se observan mejor los afloramientos con abundantes rocas graníticas asociadas a gneis. La tercera Provincia (Cuchivero) formada entre 2000–1700 millones de años, es la de mayor extensión en la Cuenca. Se encuentra representada por un complejo de formaciones con mezcla de rocas de origen volcánico (ácidas) y rocas graníticas. Esta localizada principalmente en la zona media de Caura (Entreríos), Río Mato y la tierras altas del Caura (Meseta de Jaua) y Río Merewari. La cuarta Provincia (Roraima) perteneciente al Proterozoico Inferior y está caracterizada por rocas volcánicas y rocas sedimentarias volcaniclásticas incluyendo conglomerados y areniscas. Esta Provincia localizada cerca del Río Waña, Meseta de Jaua, Cerro Sarizariñana entre otros. Por otro lado es importante destacar sedimentos de origen fluviodeltaíco de la Formación Mesa, de finales del Terciario, los cuales se encuentran localizados hacia el norte de la cuenca, alrededores de Maripa y Aripao y la boca del Caura. Finalmente, debemos indicar la presencia de aluviones sedimentarios residuales de edades recientes (Cuaternario) los cuales son muy importantes en la conformación de los bosques inundables o de galería presentes en las áreas bajas de los Ríos Caura, Nichare y Tawadu (Colvee et al. 1990; Rincón y Estanga 1996).

Desde el punto de vista geomorfológico, en la cuenca del Caura destacan paisajes típicos del Escudo Guayanés y que reflejan los múltiples procesos geológicos y geomorfológicos ocurridos en el área desde el Precámbrico hasta el presente. Las rocas ígneo-metamórficas de las Provincias de Imataca, Pastora y Cuchivero se relacionan con el desarrollo de un paisaje caracterizado por montañas, plateaux, lomeríos y peniplanicies, mientras que las rocas sedimentarias del Roraima, han favorecido el modelado de altiplanicie o tepuyes. Aproximadamente, el 70% de la cuenca consiste en paisajes elevados y con altas pendientes principalmente ubicadas hacia el sur y zonas planas y más recientes hacia el norte.

BIODIVERSIDAD Y VALOR BIOLÓGICO

La biodiversidad de la cuenca del Río Caura no es conocida uniformemente. Mientras que mucha de la flora y fauna terrestre ha sido bien estudiada (Rosales y Huber 1996; Huber y Rosales 1997), esto no es cierto para la flora y fauna acuáticas (Machado-Allison et al. 1999). La fauna silvestre (principalmente vertebrados) de la región se encuentra moderadamente estudiada e incluye al menos 475 especies de aves, 168 de mamíferos, 13 de anfibios y 23 reptiles pertenecientes a 30 órdenes (Bevilacqua y Ochoa 2001). Esto corresponde a 30% de las especies registradas para Venezuela y 51.3% de las especies de la Guayana. Los órdenes con las mayores riquezas taxonómicas son: Aves (Passeriformes, Apodiformes, Falconiformes y Psittaciformes); Mamíferos (Chiroptera, Rodentia y Carnívora); Reptiles y Anfibios (Squamata y Anura). Del total de especies de vertebrados terrestres en el Caura, 5.2% son consideradas como amenazadas en el contexto nacional e internacional (Bevilacqua y Ochoa 2001). La sobrevivencia de estas especies amenazadas debe ser considerada en estufios futuros y planes de conservación.

La vegetación de la Cuenca del Río Caura es altamente diversa. Aproximadamente 88% de todos los géneros endémicos de Guayana se encuentran presentes y existe un alto grado de endemismo en la tierras altas o comunidades tepuyanas (Berry et al. 1995; Huber et al. 1997; Bevilacqua y Ochoa 2001). La flora es cáracterística de climas húmedos, sobre suelos pobres y un gradiente de paisajes determinado por los efectos de procesos erosivos prolongados (Rosales y Huber 1996). Esta particularidad, unida a la convergencia de cuatro provincias geológicas y la presencia de un gradiente bioclimático condicionado por una amplia variación altitudinal (40–2350 msnm), determinan la existencia en esta región de una diversidad biológica excepcional (Bevilacqua y Ochoa 2001). Los datos existentes revelan que aproximadamente un 90% de la superficie de la cuenca está cubierta por bosques de tierra firme y montanos, mientras que el resto del área (10%) corresponde a bosques ribereños y vegetación no boscosa (CVG-TECMIN 1994; Huber 1996; Marín y Chaviel 1996; Rosales 1996; Aymard et al. 1997; Dezzeo y Briceño 1997; Bevilacqua y Ochoa 2001)

Aunque, la información previamente expuesta nos da una buena idea general acerca del conocimiento actual de la biodiversidad en la Cuenca del Río Caura y los bosques circundantes, estamos lejos de tener una información completa e integrada que nos permita promover planes de conservación para ciertas áreas o ecosistemas. Tomando los peces como ejemplo, Mago-Leccia (1970), reportó la presencia de aproximadamente, unas 500 especies de peces dulceacuícolas principalmente habitantes de la cuenca del Río Orinoco, en tanto que Taphorn et al. (1997) aumentó ese número a casi 1.000. Con respecto a la información íctica del río Caura, Balbas y Taphorn (1996) reportan 135 especies, mientras Machado-Allison et al. (1999), incrementa este número a 191 y este reporte se eleva a 278. Esto indica que con tales incrementos de conocimiento, no hay indicio de que la tasa de aumento esté realmente nivelándose como resultado de los esfuerzos acometidos en colectas, especialmente en estas regiones (Chernoff y Machado-Allison 1990; Royero et al. 1992; Machado-Allison 1993). Además de esto, Mago-Leccia (1978), Chernoff et al. (1991) y Machado-Allison (1993), han sugerido que sólo un 30% de la fauna y la flora acuática existente (ver adelante) en aguas dulces pueden ser identificadas con precisión.

Nuestro conocimiento del Río Caura es un reflejo del mismo del Río Orinoco. La mayoría de la información reciente ha sido conocida debido a las grandes y numerosas colecciones, expediciones y esfuerzos de campo realizados por grupos nacionales e internacionales dirigidos hacia conservación y biodiversidad. Por ejemplo, en el Río Atabapo, Río Orinoco, en la Serranía La Neblina y Río Caura (Brewer-Carías 1988; Rosales y Huber 1996; Royero et al. 1992; Huber y Rosales 1997; Machado-Allison et al. 2000).

ECONOMÍA Y ESTRUCTURA SOCIAL

La base económica de los pobladores del área de la Cuenca del Caura es compleja y diversa. En las áreas superiores y medias (arriba del Salto Pará) se mantienen las culturas tradicionales indígenas de las Etnias Ye´kuana y Sanema (Yanomami) donde la recolección, siembra en conucos (Yuca y Plátano) y cacería-pesca forma parte fundamental de las actividades domésticas y económicas. El trueque o intercambio, forma parte fundamental de la relación económica y donde cada miembro posee sus responsabilidades según su edad y sexo (Silva-Monterrey 1997). Las zonas bajas y mayormente pobladas, poseen una mezcla de elementos tradicionales indígenas con una alta influencia de culturas occidentales traídas por los criollos. En esta región se incorporan actividades forestales, agrícolas, pesqueras, turísticas y artesanales, donde además de las posibilidades de intercambio de productos y servicios, pueden establecerse relaciones económicas más complejas.

Desde el punto de vista de éste informe y su objetivo conservacionista, nos interesan principalmente aquellas actividades que pueden ser potenciales riesgos al ecosistema y su biodiversidad (p.e. explotación forestal, contaminación agrícola y pesquerías). La mayoría de estas actividades están concentradas en las tierras bajas en la región norte de la cuenca, cercanas al río Orinoco. El uso agrícola (tubérculos, maíz y plantaciones frutales) y ganadería son las principales actividades en el área. Estas han causado pérdidas de cobertura de bosques e incrementado la fragmentacón del mismo, contaminación de suelos y aguas por insecticidas y haber incrementado la erosión cerca de Maripa.

Es crítico documentar actividades sostenibles y empresas que protejan la flora y fauna en toda la región (Chernoff et al. 1996). Como ejemplo podemos indicar que en la tradición Ye´kuana existen prohibiciones alimentarias históricas y que están dirigidas por un lado a impedir la sobreexplotación de un recurso y por el otro al acceso diverso de recursos alimentarios en forma diferencial (Silva-Monterrey 1997). La alimentación es percibida como una forma de mantener la vida, pero al mismo tiempo como una manera de reproducirse y de reproducir al medio ambiente; es decir, se come para vivir, pero por otra parte, se deja de comer para permitir que otros vivan o para evitar enfermedades. Su práctica incluye igualmente períodos de ayuno con el propósito de permitir que los recursos naturales puedan ser compartidos por los miembros de toda la comunidad (Silva-Monterrey 1997).

Desde el punto de vista pesquero existen muchas especies utilizadas por las etnias en los procesos de subsistencia. Sin embargo, los mayormente importantes desde el punto de vista comercial (consumo humano y ornamental) son: "cachamas" (*Colossoma macropomum*), "cajaros" (*Phractocephalus hemiliopterus*), "coporos" (*Prochilodus mariae*), "curbinatas" (*Plagioscion squamossissimus*), "laulaos" y "valentones" (*Brachyplatystoma* spp.), "morocotos" (*Piaractus brachypomus*), "Palometas" (*Mylossoma* spp.), "rayaos" (*Pseudoplatystoma* spp.), "sapoaras" (*Semaprochilodus laticeps*) y "sardinatas" (*Pellona castelneana*), forman parte de las pesquerías comerciales del Bajo Caura (Novoa 1990; Machado-Allison et al. 1999). Por otro lado, numerosas especies poseen alto valor económico, actividad aún no escogida como alternativa de desarrollo económico de las poblaciones indígenas en el área. Especies como "tetras" (*Astyanax, Hemigrammus, Hyphessobrycon, Jupiaba, Moenkhausia*), "palometas" o "silver dollars" (*Metynnis, Myleus, Mylossoma*), "cíclidos" (*Aequidens, Apistogramma, Bujurquina, Mesonauta*), "caribes" (*Pygocentrus* y *Serrasalmus*) y "mijes" (*Anostomus* y *Leporinus*) son comunmente conocidas en el mundo de la acuariofilia.

La explotación futura de la cuenca incluye deforestación adicional del bosque, alteración de la hidrología del Caura con fines de aprovechamiento hidroeléctrico, minería y turismo. Aún cuando estas propuestas de desarrollo tienen como objetivo legítimo el promover la activación y crecimiento de la economía regional y aumento de la calidad de vida de las poblaciones locales, la mayoría de los datos obtenidos indican que estos se basan en esquemas operativos y paquetes tecnológicos instrumentados con muy poco éxito en otros países y regiones de Venezuela (Miranda et al. 1998; Machado-Allison 1999; Machado-Allison et al. 1999, 2000; Bevilacqua y Ochoa 2001).

LAS AMENAZAS

La región venezolana del Escudo Guayanés en el Estado Bolívar, ha sufrido gran alteración ambiental debido a proyectos de desarrollo, la industria y actividades de pesca y caza no reguladas. La explotación de minerales estratégicos como el oro, el diamante, y la bauxita, aunado al desarrollo y construcción de la represa de Gurí, uno de los complejos hidroeléctricos más grandes del mundo, ha producido: 1) la biodegradación y destrucción de las extensas áreas verdes en las cuencas de los ríos Caroní y Cuyuní; 2) la contaminación de los ríos, vida salvaje y seres humanos por mercurio; 3) el aumento de cargas de sedimentación, afectando la calidad de agua de manera adversa; y 4) la pérdida de cantidades inmensas de agua destinadas al uso doméstico (Machado-Allison 1994; Miranda et al. 1998; Machado-Allison 1999).

La cuenca del Río Caura está mayormente intacta, sin embargo está cercana a áreas de mayor destrucción como la sección oriental del Escudo Guayanés. En la cuenca del Río Caura existe un área (cercanas a la cuenca del Río Paragua) de aproximadamente 100 kilómetros que están siendo objeto de aprovechamiento forestal y donde está ocurriendo una gran deforestación. Algunas actividades mineras están ocurriendo en las cabeceras del mismo. En cierto sentido, la actividad minera resulta menor aquí si la comparamos con las cuencas vecinas de los Ríos Caroní y Cuyuní, donde un alto nivel de destrucción está sucediendo (Machado-Allison et al. 2000: fig.1). Una dramática disminución de los recusos pesqueros en el Bajo Caura es debida aparentemente a pesquería comercial ilegal en territorios indígenas. Finalmente, La mayor amenaza que presenta la cuenca del Río Caura es la existencia de un plan de construir una nueva represa hidroeléctrica y desviar tanto como un 75% del caudal de agua del Río Caura al sistema fluvial Paragua-Caroní. Esto, no solamente reducirá drásticamente el caudal del Río Caura, sino que también alterará severamente su ciclo hidrológico. La salud y mantenimiento de la flora y fauna acuáticas, y ribereña depende completamente del ciclo hidrológico natural.

CONCLUSIONES GENERALES

La Cuenca del Río Caura es una vasta área de condiciones ambientales intactas y muy poco explorado por científicos. Debido a la destrucción y el desarrollo en regiones adyacentes al norte y sureste del Escudo Guayanés como son las del Río Caroní y Río Cuyuní, el Río Caura representa una importante oportunidad de preservar una región única. La composición de la comunidad de animales y plantas de la cuenca del Río Caura, no ocurre en ninguna otra parte de Venezuela o del Escudo Guayanés. Es un área de alta diversidad biológica, con muchas formas endémicas y comunidades únicas, ejemplo del cual es la flora acuática (Podostemonaceae) en rápidos y los bosques en islas rocosas. La protección através de programas permanentes de educación, de extensión comunitaria y monitoreo a largo plazo, son altamente recomendables. Resulta esencial el monitoreo y establecimiento de regulaciones de la actividad pesquera en la cuenca, a objeto de poder alcanzar un marco de sostenibilidad para los habitantes de la cuenca.

LITERATURA CITADA

Aguilera, M., y J. Silva. 1997. Especies y diversidad. Interciencia 22 (6): 289–298.

Aymard, G., S. Elcoro, E. Marín y A. Chaviel. 1997. Caracterización estructural y florística en bosques de tierra firme de un sector del bajo Río Caura, Estado Bolívar, Venezuela. *En:* Huber, O. y J. Rosales (eds.). Ecología de la Cuenca del Río Caura. Scientia Guaianae 7: 143–169.

Balbas, L., y D. Taphorn. 1996. La Fauna: Peces. *En:* Rosales, J., y O. Huber (eds.). Ecología de la Cuenca del Río Caura. Scientia Guaianae, 6: 76–79.

Berry, P. E, O. Huber y B. K. Holst. 1995. Floristic Analysis and Phytogeography. *En:* Berry, P. E., B. K. Holst, K. Yatskievych (eds.). Flora of the Venezuelan Guayana, Vol. 1. Introduction. Saint Louis, USA: Missouri Botanical Garden and Timber Press. Pp. 161–191.

Bevilacqua, M., y J. Ochoa G. 2001. Conservación de las últimas fronteras forestales de la Guayana Venezolana: propuesta de lineamientos para la Cuenca del Río Caura. Interciencia 26(10): 491–497.

Bonilla-Rivero, A., A. Machado-Allison, B. Chernoff, C. Silvera, H. López-Rojas y C. Lasso. 1999. *Apareiodon orinocensis*, una nueva especie de pez de agua dulce (Pisces: Characiformes: Parodontidae) proveniente de los ríos Caura y Orinoco, Venezuela. Acta Biol. Venez., 19(1): 1–10.

Brewer Carías, C. (ed.) 1988. Cerro de la Neblina. Resultados de la Expedición 1983–1987. Academia de Ciencias Físicas, Matemáticas y Naturales. Caracas

Bucher, E., A. Bonetto, T. Boyle, P. Canevari, G. Castro, P. Huszar y T. Stone. 1993. Hidrovia: un examen ambiental inicial de la via fluvial Paraguay-Paraná. Humedales para las Americas, publ. 10: 1–74.

Chaffanjon, J. 1889. L'Orinoque et le Caura. *En:* Castellana,m.A. 1986. Relation de voyages exécutés en 1886 et 1887. Hachette et Cie. Paris: Fund. Cult. Orinoco. 311pp.

Chernoff, B., y A. Machado-Allison. 1990. Characid fishes of the genus *Ceratobranchia*, with descriptions of new species from Venezuela and Peru. Proc. Acad. Nat. Sci. Philad., 142: 261–290.

Chernoff, B., A. Machado-Allison y W. Saul. 1991. Morphology variation and biogeography of *Leporinus brunneus* (Pisces: Characiformes: Anostomidae). Ichth. Explor. Freshwaters, 1(4): 295–306.

Chernoff, B., A. Machado-Allison y N. Menezes. 1996. La conservación de los ambientes acuáticos: una necesidad impostergable. Acta Biol. Venez., 16 (2): i–iii.

Chernoff, B., y P. Willink (eds.). 1999. A Biological Assesement of the Aquatic Ecosystems of the Upper Rio Orthon Basin, Pando, Bolivia. Bull. Biol. Asses., 15. 145 pp.

Colveé, P., E. Szczerban y S. Talukdar. Estudios y Consideraciones Geológicas sobre la Cuenca del Río Caura. *En*: Weibezahn, F., H. Alvarez y W. Lewis (eds.). *El Río Orinoco como Ecosistema.* Venezuela: Edelca, Fondo Ed. Acta Científica Venezolana, CAVN, USB pp. 11–44.

CVG-TECMIN. 1994. Informes de avance del Proyecto Inventario de los Recursos Naturales de la Región Guayana. Hojas NB-20: 1, 5, 6, 9, 10, 13 and 14. Ciudad Bolívar, Venezuela: Gerencia de Proyectos Especiales.

Dezzeo, N. and E. Briceño. 1997. La vegetación en la cuenca del Río Chanaro: medio Río Caura. *En:* Huber, O. and J. Rosales (eds.). Ecología de la Cuenca del Río Caura. Scientia Guaianae 7: 365–385.

García, S. 1996. Limnología. *En*: Rosales, J., and O. Huber, (eds.). Ecología de la Cuenca del Río Caura. Scientia Guaianae, 6: 54–59.

Goulding, M. 1980. The Fishes and the Forest: Explorations in Amazonian Natural History. Univ. Cal. Press. 280 pp.

Huber, O. 1996. Formaciones vegetales no boscosas. *En:* Rosales, J. and O. Huber (eds.). Ecología de la Cuenca del Río Caura. Scientia Guaianae 6: 70–75.

Huber, O., y J. Rosales (eds.). 1997. Ecología de la Cuenca del Río Caura. Estudios especiales. Scientia Guaianae, 7.

Huber, O., J. Rosales y P. Berry. 1997. Estudios botánicos en las montañas altas de la cuenca del Río Caura. *En*: Huber, O., y J. Rosales (eds.). Ecología de la Cuenca del Río Caura. Scientia Guaianae 7: 441–468.

IUCN. 1993. The Convention on Biological Diversity: An explanatory guide (Draft). IUCN Environmental Law Centre, Bonn. 143 pp. (mimeo).

Lasso, C., y F. Provenzano. 1997. *Chaetostoma vazquezi*, una nueva especie de corroncho del Escudo de Guayana, Estado Bolívar, Venezuela (Siluroidei-Loricariidae) descripción y consideraciones biogeográficas. Mem. Soc. Cienc. Nat. La Salle, 57(147): 53–65.

Machado-Allison, A. 1993. Los Peces de los Llanos de Venezuela: un ensayo sobre su Historia Natural. 2ª. Edition. Caracas: Consejo de Desarrollo Científico y Humanístico (UCV), Imprenta Universitaria. 121pp.

Machado-Allison, A. 1994. Factors affecting fish communities in the flooded plains of Venezuela. Acta Biol. Venez., 14(3):1–20.

Machado-Allison, A. 1999. Cursos de agua, fronteras y conservación. *En*: G. Genatios (ed.). Ciclo Fronteras: Desarrollo Sustentable y Fronteras. Caracas, Venezeula: Com. Estudios Interdisciplinarios, UCV. Pp. 61–84.

Machado-Allison, A., B. Chernoff, C. Silvera, A. Bonilla, H. Lopez-Rojas, C. A. Lasso, F. Provenzano, C. Marcano y D. Machado-Aranda. 1999. Inventario de los peces de la cuenca del Río Caura, Estado Bolívar, Venezuela. *Acta Biol. Venez.*, Vol. 19 (4):61–72.

Machado-Allison, A., B. Chernoff, R. Royero-Leon, F. Mago-Leccia, J. Velazquez, C. Lasso, H. López-Rojas, A. Bonilla-Rivero, F. Provenzano y C. Silvera. 2000. Ictiofauna de la cuenca del Río Cuyuní en Venezuela. Interciencia, 25(1): 13–21.

Mago-Leccia, F. 1970. Lista de los Peces de Venezuela: incluyendo un estudio preliminar sobre la ictiogeografía del país. Caracas: MAC-ONP. 283 pp.

Mago-Leccia, F. 1978. Los Peces de Agua Dulce del País. Cuadernos Lagoven, Caracas, 35 pp.

Marín, E. y A. Chaviel. 1996. Bosques de Tierra Firme. *En*: Rosales, J., y O. Huber (eds.) Ecología de la Cuenca del Río Caura. Scientia Guaianae, 6: 60–65.

Miranda, M., A. Blanco-Uribe, L. Hernández, J. Ochoa y E. Yerena. 1998. No todo lo que brilla es oro: hacia un nuevo equilibrio entre conservación y desarrollo en las últimas fronteras forestales de Venezuela. Washington, DC: Inst. Rec. Mundiales (WRI). 59pp.

Mittermeier, R.A., P. Robles y C. Goettsch. 1997. Megadiversidad: los países biológicamente más ricos del mundo. Cemex y Agrupación Sierra Madre, SC. México, 501 pp.

Mittermeier, R.A., N. Myers, P. Robles y C. Goettsch. 1998. Hotspots. Cemex y Agrupación Sierra Madre, SC. México, 430 pp.

Novoa, D. 1990. El río Orinoco y sus pesquerías; estado actual, perspectivas futuras y las investigaciones necesarias. *En*: Weibezahn, F., H. Alvarez y W. Lewis (eds). El Río Orinoco como Ecosistema. Caracas, Edelca, Fondo Ed. Acta Cienctífica Venezolana, CAVN, USB: 387–406.

Peña, O., y O. Huber. 1996. Características Geográficas Generales. En: Rosales, J., y O. Huber (eds). Ecología de la Cuenca del Río Caura. Scientia Guaianae, 6: 4–10.

Rincón, H., y Y. Estanga. 1996. Geología. *En*: Rosales, J., and O. Huber (eds.) Ecología de la Cuenca del Río Caura. Scientia Guaianae, 6: 20–28.

Rosales, J., y O. Huber (eds). 1996. Ecología de la Cuenca del Río Caura, Venezuela. I. Caracterización General. Scientia Guaianae No. 6: 1–131.

Royero, R., A. Machado-Allison, B. Chernoff y D. Machado. 1992. Los peces del Río Atabapo. Acta Biol. Venez., 14(1): 41–56.

Silva-Monterrey, N. 1997. La Percepción Ye´kwana del Entorno Natural. *En*: Huber, O., y J. Rosales (eds.). Ecología de la Cuenca del Río Caura. Estudios Especiales Scientia Guaianae, 7: 65–84.

Sioli, H. 1965. A Limnología e a sua importancia en pesquisas da Amazonia. Amazoniana, I: 11–35.

SISGRIL. 1990. Simposio Internacional sobre los Grandes Ríos Latinoamericanos. Interciencia, 15(6): 320–544.

Taphorn, D., R. Royero, A. Machado-Allison y F. Mago-Leccia. 1997. Lista actualizada de los peces de Agua Dulce de Venezuela. *En*: La Marca, E. (ed.) Vertebrados actuales y fósiles de Venezuela. Mérida, Venezuela. Vol. 1: 55–100.

Vargas, H., y J. Rangel. 1996. Hidrología y Sedimentos. *En*: Rosales, J., and O. Huber (eds.) Ecología de la Cuenca del Río Caura. Scientia Guaianae, 6: 48–54.

Vari, R. 1995. The Neotropical Fish Family Ctenoluciidae (Teleostei: Ostaripphysi: Characiformes): Supra and Infrafamilial Phylogenetic Relationships, with a Revisionary Study. Smith Contr. Zoology, 654:1–95.

Capítulo 2

Comunidades de Vegetación Ribereña de la Cuenca del Río Caura, Estado Bolívar, Venezuela

Judith Rosales, Mariapia Bevilacqua, Wilmer Díaz, Rogelio Pérez, Delfín Rivas y Simón Caura

RESUMEN

Un total de 443 muestras fueron tomadas durante la expedición de una variedad de habitats ribereños y acuáticos en la Cuenca del Río Caura. Las muestras contienen 399 especies de plantas (291 por encima del Salto Pará y 185 en el Bajo Caura), todas las cuales estan incluidas en la lista de 1180 especies conocidas para la cuenca. La investigación reveló que existe una gran diversidad en la composición de las comunidades florísticas típica de ambientes húmedos con suelos de bajos nutrientes. La variación en la estructura de la comunidad comprende un gradiente de ambientes ribereños estructurados por la intensidad y duración de los procesos erosionales. La diversidad de la geología basal así como también el gradiente climático asociado con la altitud (40–2350 m) también contribuye a la riqueza excepcional de la Cuenca del Río Caura.

La proporción de especies endémicas es baja en las tierras bajas y en los ecosistemas de bosques inundables en la cuenca. Esta región es dominada por las palmas: *Euterpe precatoria, Attalea maripa, Socratea exhorriza, Genoma baculifera* y *Bactris brongniartii*. La riqueza de especies es relativamente grande en comparación con valores similares provenientes de ambientes en el Amazonas y Guyana. La diversidad y riqueza de especies en los sistemas inundables del Erebato, es dominado por *Oenocarpus* y en el Caura Superior esta dominado por *Mauritiella*. Existen también valores similares comparado con otros sistemas ribereños del Esudo de Guayana. Una variable que da un valor único a los ensablajes de flora está asociado a islas en el Río Erebato y Río Caura Superior. Estas islas poseen tanto bosques de tierra firme y numerosas parches de plantas acuáticas de la familia Podostemonaceae ancladas a rocas en los rápidos.

INTRODUCCIÓN

El conocimiento de la diversidad florística y de los tipos de vegetación de la Cuenca del Río Caura se puede considerar avanzado, especialmente en los ecosistemas boscosos ribereños. Las contribuciones al estudio de la vegetación y la flora de la cuenca pueden correlacionarse con dos grandes períodos de exploración geográfica y científica en la región del Caura (Huber 1996). El primero corresponde a los siglos XVI y XIX, quienes en sus informes y publicaciones aportaron observaciones importantes a la caracterización de los ecosistemas boscosos para la época. El segundo período se desarrolla a partir de la década de los treinta, con las exploraciones geográficas recientes de la Región Guayana; los estudios forestales de la década de los cuarenta (Williams 1942; Veillon 1948); y, a partir de la década de los ochenta cuando la región del Caura toma relevancia para los objetivos de planificación del desarrollo regional y la conservación de la diversidad biológica (Steyermark y Brewer-Carias 1976; Lal 1990; CVG-TECMIN 1994; Berry et al. 1995; Briceño 1995; Huber 1995; Bevilacqua y Ochoa 1996; Huber 1996; Marín y Chaviel 1996; Rosales 1996; Aymard et al. 1997; Briceño et al. 1997; Dezzeo y Briceño 1997; Huber et al. 1997; Knab-Vispo et al. 1997; Rosales et al. 1997; Salas et al. 1997; Knab-Vispo 1998; Rosales 2000; Vispo 2000; Bevilacqua y Ochoa 2001; Rosales et al. 2002a, b; Knab-Vispo et al. en prensa; Vispo et al. en prensa).

Aproximadamente el 90% de la superficie de la Cuenca del Río Caura está cubierta por bosques de tierra firme y montanos, mientras que el resto del área corresponde a bosques ribereños y vegetación no boscosa. La región presenta una alta diversificación en sus formaciones vegetales, las cuales se desarrollan en climas húmedos, sobre suelos con bajo nivel nutricional y en un gradiente de paisajes determinado por los efectos de procesos erosivos prolongados y de una alta intensidad. Esta particularidad, unida a la convergencia de cuatro provincias geológicas y a la presencia de un gradiente bioclimático condicionado por una amplia variación altitudinal (40–2350 m), determinan la existencia en esta región de una diversidad biológica excepcional.

Una panorámica de análisis a escala de gran visión (1:2.000.000), permite caracterizar las formaciones boscosas de tierra firme en cuatro categorías: 1) bosques tropófilos macrotérmicos, formaciones leñosas asociadas con un clima marcadamente estacional, propio de las tierras bajas del sector norte de la cuenca, de carácter caducifolio a subsiempreverde; 2) bosques ombrófilos macrotérmicos, distribuidos principalmente en las grandes extensiones de tierras bajas de los sectores medio y alto de la cuenca, donde predominan temperaturas y precipitaciones altas; 3) bosques ombrófilos submesotérmicos asociados al dominio fisiográfico de los paisajes de tierras altas características de la Región Guayana; y 4) bosques ombrófilos mesotérmicos (similar al anterior).

En contraste a los bosques de tierra firme, los bosques ribereños inundables cubren una extensión relativamente pequeña de la cuenca, con una distribución asociada a las planicies de inundación del Río Caura, Río Erebato y sus canales de drenaje. No obstante, estos bosques cumplen un papel fundamental en la transferencia de energía y nutrientes entre los ecosistemas adyacentes, además de su importancia como reservorios de diversidad biológica, valor paisajístico y como fuentes de recursos para la población indígena local (Rosales 2000). La proporción de elementos endémicos es baja en las tierras bajas y en los bosques ribereños. Sin embargo, la diversidad es relativamente alta y comparable con valores registrados para bosques similares en la Amazonia y Escudo de Guayana (Klinge et al. 1995; Knap-Vispo 1998; Rosales et al. 1999; Rosales 2000; ter Steege 2000).

La vegetación no boscosa de la cuenca está integrada por formaciones dominadas por arbustos y hierbas, así como comunidades acuáticas asociadas con ríos y lagunas. Este tipo de vegetación incluye a las sabanas y morichales, cuyas distribuciones abarcan principalmente las tierras bajas colindantes con el Río Orinoco, en suelos aluviales y arenosos. Asimismo, resaltan los arbustales y herbazales asociados con las tierras altas tepuyanas, los cuales representan formaciones de alto valor biológico típicas del Escudo Guayanés (Berry et al. 1995; Huber 1995; Huber et al. 1997).

Las observaciones realizadas en este trabajo sobre los ecosistemas inundables, asociados a los corredores ribereños del Río Erebato y del Río Caura, confirman la diversidad relativamente alta de unidades ecológicas asociadas con formas de terreno específicas, que determinan la unicidad de esta cuenca, su diversidad en tipos de vegetación y la riqueza florística es similar con otros corredores ribereños tropicales (Rosales et al. 2001). Los valores biológicos y ambientales de la Cuenca del Río Caura, unidos a la presencia de grandes extensiones de bosques primarios, su diversidad cultural y su potencial para el desarrollo, catalogan a esta bioregión como una de las áreas de mayor relevancia en el contexto hemisférico (Bevilacqua y Ochoa 2001).

RESULTADOS

Descripción general de las subregiones y estaciones de trabajo

I. Subregión Caura Superior

El área de trabajo se caracteriza por presentar abundante pluviosidad durante todos los meses del año con un promedio de precipitación de 2.600 mm en los alrededores del Raudal Cejiato, hasta un promedio de 3.758 mm en los alrededores del sector Entreríos. Los valores más bajos de precipitación se observan en los meses de enero, febrero y marzo, de manera que el trabajo se realizó en la época de salida de lluvias. El bioclima es macrotérmico, es decir se caracteriza por un promedio anual de temperatura en el orden de los 25°C, con altitudes promedio alrededor de los 250 m sobre el nivel del mar. Los tipos de paisaje ribereño son característicos de ríos estructuralmente controlados con variaciones relacionadas a los gradientes hidráulicos (pendientes relativas), la carga de sedimentos y el grado de juventud o evolución del río en diferentes sectores. Las Figuras 2.1, 2.2 y 2.3 muestran ejemplos de relieves ribereños típicos y de los perfiles de vegetación asociada.

1. Río Kákada (AC03, AC04). Tributario del Río Erebato, que drena el sistema de areniscas de la Serranía Jaua-Sarisariñama (Formación Roraima). Sus aguas son muy ácidas, negras y poco turbias. Los paisajes de tierra firme predominantes que acompañan el curso de este tributario en los sectores trabajados son la Planicie antigua desarrollada sobre aluviones cuaternarios y el Lomerío asociado a la Provincia Cuchivero. Por otra parte, la Planicie aluvial reciente es el tipo de paisaje ribereño dominante que acompaña el curso del río en su confluencia con el Río Caura. Esta es de origen deposicional, desarrollada a partir de aportes aluviales, presentando una superficie plana de pendientes suaves. Los tipos de relieve observados en los sitios de trabajo incluyen la llanura de desborde y la vega. La llanura de desborde está limitada a las áreas afectadas por los desbordes provenientes de inundaciones periódicas (incluyendo períodos de retorno históricos de hasta 50 años) a ambos lados del río y de los pequeños tributarios. Se trata de áreas de deposición diferencial de aluviones arenosos, limosos y arcillosos, provenientes de aportes laterales del río que modelan un gradiente topográfico. Las posiciones mas elevadas son las formas de terreno de diques o bancos, de texturas franco arenosas con desniveles de 3–4 m. Estas preceden áreas depresionales con planos convexos, caracterizados por un microrelieve irregular distintivo de la combinación de procesos de disección (surcos de erosión) y deposición (cubetas marginales). El tipo de relieve de vega, incluye las deposiciones aluviales dentro del canal del río. Estas áreas están conformadas por barras laterales estrechas, sinuosas y alargadas con sustratos de texturas eminentemente arenosas y abundantes aportes orgánicos provenientes del mantillo de raíces y hojarasca de la vegetación ribereña. Por otra parte, en la confluencia del Caño Suajiditu la vega se presenta en forma de una barra lateral de textura arcillosa y después como una secuencia de barras de meandro. El desnivel del talud de la orilla es alrededor de 2–3 m de altura. El lecho del caño presenta mantos de grava y arenas, así como aportes de materia orgánica en diferentes estadios de descomposición.

2. Río Erebato (AC05, AC06, AC07). Tributario principal del Río Caura, que drena en parte de las provincias Roraima (Serranía Jaua-Sarisariñama), Pastora y Cuchivero (Sierra de Maigualida y Uasadi). Los paisajes de tierra firme predominantes que acompañan el curso del Río Erebato son la planicie antigua y el lomerío. Un aspecto a resaltar de este tributario es su carácter de río confinado y controlado estructuralmente, con alto gradiente y frecuencia de rápidos. El tener los tipos de relieve de llanura de desborde y vega poco expresivos, sugiere la juventud del sistema de drenaje. Ambas márgenes del río están caracterizadas por ser entalladas, con una topografía elevada predominando los hábitats de erosión sobre los deposicionales. Las vegas también tienen una baja expresión, así como también en los presentes en caños de confluencia con el Erebato conformando franjas angostas y sinuosas con muy pocos desniveles en sus orillas (2–3 m de altura). El fondo de los caños presenta mantos de grava y arenas, con abundante materia orgánica en descomposición.

El Río Erebato se caracteriza por la alta frecuencia de rápidos e islas rocosas de gran diversidad de tamaño y con formas predominantemente alargadas en el sentido de la corriente. El terreno de las islas incluyen playas rocosas además de barras laterales y frontales las cuales forman playas arenosas de texturas gruesas. Hacia el centro de la isla hay lomas, de forma aplanada o convexa y de microrelieve irregular.

3. Río Caura, Entreríos-Raudal Cejiato (AC01, AC02, AC08, AC09, AC10). Origen en sistemas de areniscas y rocas volcánicas de la Provincia Roraima (Formación Ichún y Grupo Roraima). Sus aguas son ácidas similares a las del Río Erebato, sin embargo presentan una mayor cantidad de sedimentos suspendidos, así como una mayor concentración de nutrientes. La red de drenaje es comparativamente más densa en el Río Caura con respecto al Río Erebato. El paisaje de tierra firme dominante es el lomerío alto en la margen occidental del río y lomerío bajo en la margen oriental. El Caura presenta gran heterogeneidad de hábitat ribereños. Esto caracteriza a sectores con desarrollo de planicie aluvial lo que indicaría madurez fluvial. Hay márgenes de erosión con taludes activos. También se encuentran márgenes deposicionales de meandros con barras de texturas arenosas, arcillosas y abundante materia orgánica. Las formas de terreno más importantes en la llanura de desborde incluyen: i) los diques o bancos elevados por sucesivos aportes laterales de carácter aluvial, ii) áreas depresionales con mayor escorrentía superficial (cubetas de desborde), iii) cubetas de decantación y iv) vegas, mayormente representadas por barras laterales deposicionales. En los caños y tributarios, hay un sustrato rocoso puesto al descubierto por la erosión, así como texturas gruesas provenientes de las deposiciones aluviales y con importantes aportes orgánicos de la vegetación ribereña. También se observaron complejos de islas y rápidos cuyas formas y relieves son similares a los descritos anteriormente.

4. Río Caura, Entreríos-Salto Pará (AC11, AC12, AC13, AC14). La característica fundamental es la confluencia de los sistemas hidrológicos Caura y Erebato. En general las aguas son ácidas, levemente de color marrón, bien oxigenadas, similares a las del Río Caura Superior. El área presenta una serie de rápidos hasta llegar al Salto Pará, con frecuencia de islas, afloramientos rocosos y áreas de remanso. El area del Salto Pará eta caracterizado por una mezcla de rápidos y saltos con un desnivel total de 50 m en sólo 2 km de recorrido. Las formas de terreno asociadas a la planicie aluvial reciente y las islas son similares a las descritos anteriormente.

II. Subregión Bajo Caura (Figuras 2.4 y 2.5)

El área del Bajo Caura presenta un gradiente bioclimático asociado principalmente a la distribución de la precipitación: 1.500–1.900 mm desde el Río Mato a la boca del Río Caura; 2.500–3.000 mm en el Río Nichare y 2.970–3.400 mm cerca del Salto Pará . La temperatura mantiene un régimen macrotérmico similar al descrito para el área de Entreríos, pero con fluctuaciones de 1 a 2 grados más calientes. Los tipos de paisaje ribereños dominantes desde el Salto Pará hasta el Raudal Cinco Mil, son característicos de ríos estructuralmente controlados con variaciones relacionadas a los gradientes altitudinales, la carga de sedimentos y la edad o evolución. Sin embargo, también se presenta un paisaje ribereño meándrico en los sectores de la cuenca media del Nichare hasta el Icutú y aguas abajo del Raudal Cinco Mil hasta la boca del Río Mato.

1. Río Caura, sector El Playón—Raudal Cinco Mil (BC01, BC02, BC03, BC15, BC16, BC17). El área se caracteriza por tener hábitats de alta dinámica hidráulica (turbulentos) y áreas de remansos. Geológicamente, domina la formación Provincia Imataca con rocas gneises, anfibolitas y cuarcitas ferruginosas en lomerío y la Provincia Cuchivero con rocas volcánicas y granitos en penillanura. Al igual que lo observado aguas arriba, el paisaje ribereño dominante es una planicie aluvial reciente. Similar a lo observado aguas arriba, se detectan vegas con barras laterales deposicionales.

2. Ríos Nichare, Tawadu e Icutú (BC04, BC05, BC06, BC07, BC08, BC09, BC10, BC11). En general las aguas son muy ácidas, bien oxigenadas, poco turbias, conductividad baja, más frias que el resto del Río Caura. Se observan dos paisajes ribereños. El primero es la planicie aluvial de tipo meándrica dentro de un valle desarrollado sobre aluviones cuaternarios y recientes, localizado en el Río Nichare hasta la desembocadura del Río Icutú y la boca del Río Tawadu. La diversidad de formas de terreno incluye barras laterales, barras de meandro y complejos de orillar, diques, cubetas de decantación y lagos de herradura (madre vieja). El sustrato es generalmente arcillo-limoso, con abundante materia orgánica tanto descompuesta (fangos orgánicos), como formando capas de hojarasca. El segundo paisaje está caracterizado por riberas disectadas del Río Tawadu, el cual se observa entallado dentro de un lomerío que se desarrolla sobre granitos de la Provincia Cuchivero. En dicho sector, el sustrato del lecho del río presenta grava, arena y cantos rodados de diferentes tamaños.

3. Río Caura desde Raudal Cinco Mil hasta Boca Río Mato (BC12, BC13, BC14). El paisaje ribereño representa el mayor desarrollo de la planicie aluvial en el Río Caura estudiado. El mismo corresponde al límite sur de la zona de confluencia del Bajo Caura, afectado por el efecto de remanso del Río Orinoco. El desarrollo de islas aluviales es común, así como las barras de canal y barras de meandro en el relieve de vega. En la llanura de desborde se observan lagunas originadas a partir de canales y lagos de herradura (en el Mato), igualmente diques o bancos y cubetas de decantación. La planicie se desarrolla sobre aluviones cuaternarios y recientes.

Descripción de los tipos de vegetación

Las formaciones vegetales inundables, se agrupan básicamente en cuatro tipos: 1) vegetación de tipo herbácea-arbustiva sucesional temprana en barras arenosas y remansos; 2) el bosque ribereño en las márgenes de caños y ríos; 3) los matorrales-herbazales en áreas de remanso; y 4) la vegetación acuática reófila. Todos ellos varían su fisionomía y estructura, en atención a la influencia de factores físico ambientales, tales como: inundaciones temporales, presencia de afloramientos y bloques rocosos en superficie, perturbaciones naturales, grado de material orgánico en suelos aluviales, litologías y topografía. A continuación se presta una descripción detallada de cada tipo de vegetación.

El primer tipo esta conformado por las comunidades más próximas al canal principal, en relieves de barras laterales deposicionales. Este tipo de vegetación incluye plantas sucesionales que crecen en suelos aluviales recientes. Esta vegetación sucesional temprana está conformada por especies herbáceas, elementos arbustivos, así como subfrútices dispersos y ocasionales, de altura comprendidas entre 1 a 3 m tales como: *Psidium* sp., *Mabea* sp., *Miconia* sp., *Vismia* sp., *Croton cuneatus, Calycolpus goetheanus, Myrcia splendens, Maytenus guyanensis* y una especie de Rubiacea (de flores blancas) de porte arbustivo bajo. Todas ellas incluidas en las familias Cyperaceae, Graminaeae, Rubiaceae, Labiatae y Gesneriaceae.

En condiciones de isla, esta vegetación está en contacto con un ecotono arbustivo transicional hacia el bosque bajo de poca extensión y desarrollo estructural, que generalmente ocupa el centro de la isla (Figuras 2.2 y 2.4). Debido a la inundación períodica de estas áreas, el ecotono arbustivo presenta regeneración de elementos leñosos, florísticamente relacionados con los bosques inundables de los márgenes.

El segundo tipo de formación vegetal, es el bosque ribereño inundable asociado a los relieves más elevados (bancos o diques) y terrenos depresionales (cubetas) (Figuras 2.1 y 2.5). En general, son bosques siempreverdes con alturas medias (18 y 25 m) en dos estratos. El superior formando un dosel continuo, que en algunos casos se presenta irregular, con especies como: *Pithecelobium cauliflorum, Macrolobium acaciiefolium, M. angustifolium, Eperua jehnmanii* spp. *sandwichii, Homalium guianense, Caraipa densifolia, Jacaranda copaia, Andira surinamensis, Eschweilera subglandulosa, Catostemma comune, Dialium guianense, Parkia pendula, Micranda minor, Virola surinamensis, Scheflera morototoni, Gustavia coriacea, G. poeppigiana, Tabebuia capitata, Pterocarpus* sp., *Chrysophyllum* sp., *Cupania* sp., *Sterculia* sp., *Lecythis* sp., *Abarema* sp., *Alexa confusa* y *Protium* sp. El estrato mas bajo de los dos, es comúnmente denso y se ubica entre 10 y 18 m. Este está constituido por especies tales como: *Phenakospermum guyanense, Rinorea flavescens, Swartzia schomburgkii, Cassipourea guianensis, Amphirrox latifolia, Psychotria* spp. *Erythroxylum* sp., *Diplasia* sp., *Alexa* sp., *Mabea* sp., *Chrysophyllum* sp., *Inga* sp., *Sterculia* sp., *Lecythis* sp., *Abarema* sp., *Alexa* sp., *Protium* sp., *Calycolpus* sp., *Myrcia* sp., *Ficus* sp. y *Zigia* sp. En ocasiones es posible observar un tercer estrato de escasos individuos con alturas entre 6 y 10 m dominados por algunos representantes de pocas especies. La presencia de emergentes es característica, con alturas entre 26 a 30 m entre los que destacan por ejemplo: *Parkia pendula, Micranda minor, Catostenma commune, Ceiba pentandra* y *Tabebuia* sp.

El sotobosque varía de medio a denso y en ocasiones se puede presentar ralo, conformado por individuos juveniles de la regeneración arbórea, así como hierbas, arbustos y subfrútices tales como: *Piper* sp., *Amphirrox* sp., *Costus* sp., *Heliconia* sp., *Renealmia* sp., *Tabernamontana* sp., *Eugenia* sp., *Miconia* sp., *Psychotria* sp., *Erythroxylum* sp. y *Diplasia* sp. Localmente ocurren colonias de cyperáceas en áreas donde existen claros, así como es posible observar también pequeños parches de musaceas, principalmente *Phenakospermum guyanense,* con alturas hasta de 8 m.

Las palmas son frecuentes y características de bosques ribereños en etapas juvenil y adulta. Las especies más frecuentes y notorias son: *Euterpe precatoria, Attalea maripa, Astrocaryum gynacanthum, Socratea exorrhiza* y *Desmoncus* sp. En ocasiones las palmas se presentan formando colonias, especialmente de *Geonoma deversa* y *Bactris brongniarti.* Las especies *Oenocarpus bacaba* y *O. bataua,* se presentan con mayor frecuencia en los bosques ribereños del Río Erebato.

En respuesta a un incremento de la humedad relativa, es posible observar abundancia de musgos, líquenes, aráceas y epífitas como bromelias y orquídeas. Los helechos también son frecuentes, con formas de vidas terrestres y epífitas. Las lianas y bejucos están especialmente asociadas a perturbaciones naturales como la caída de árboles "gaps". Igualmente, en áreas de mal drenaje. En estas últimas zonas se presentan enmarañadas conformando verdaderas "selvas de bejuco".

El bosque ribereño varía en estructura, disminuye su altura y simplifica la composición florística en condiciones de drenaje deficiente y zonas sometidas a inundaciones pro-

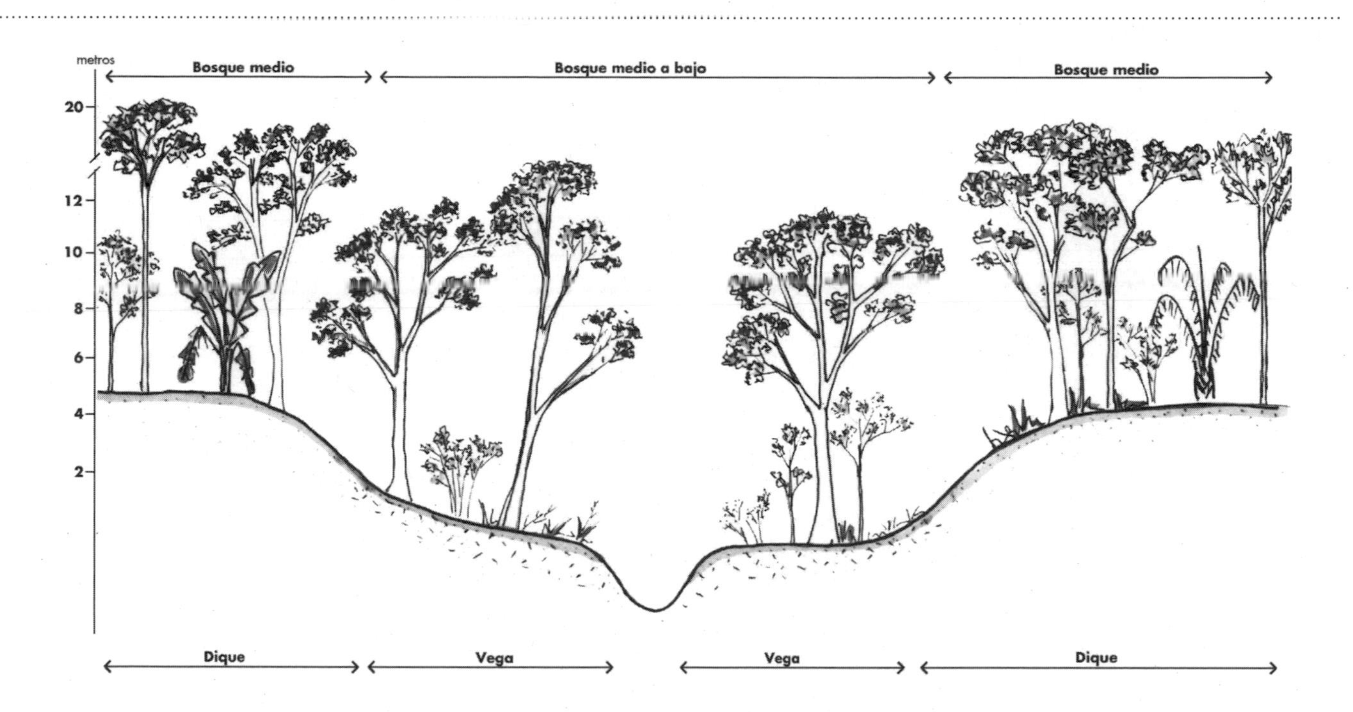

Figura 2.1. Diagrama del bosque inundable medio-alto en la confluencia del Caño Wididi y el Río Erebato (Punto de Georeferencia AC06). El suelo en la vega contiene material orgánico fino y el dique esta formado por material grueso.

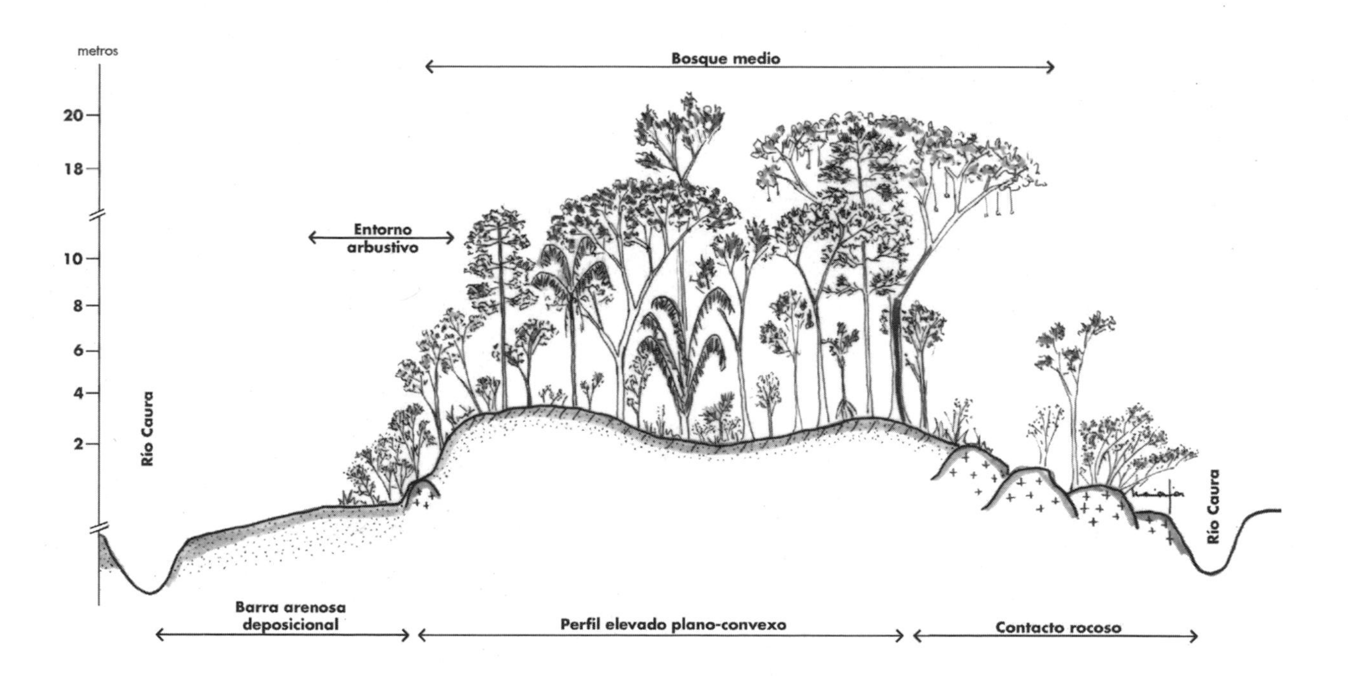

Figura 2.2. Diagrama del transecto de bosque medio en relieve de isla en rápidos del Río Caura (Punto de Georeferencia AC09).

Símbolos: cruces–afloramiento rocoso; líneas diagonales–aporte orgánico; y puntos–arena.

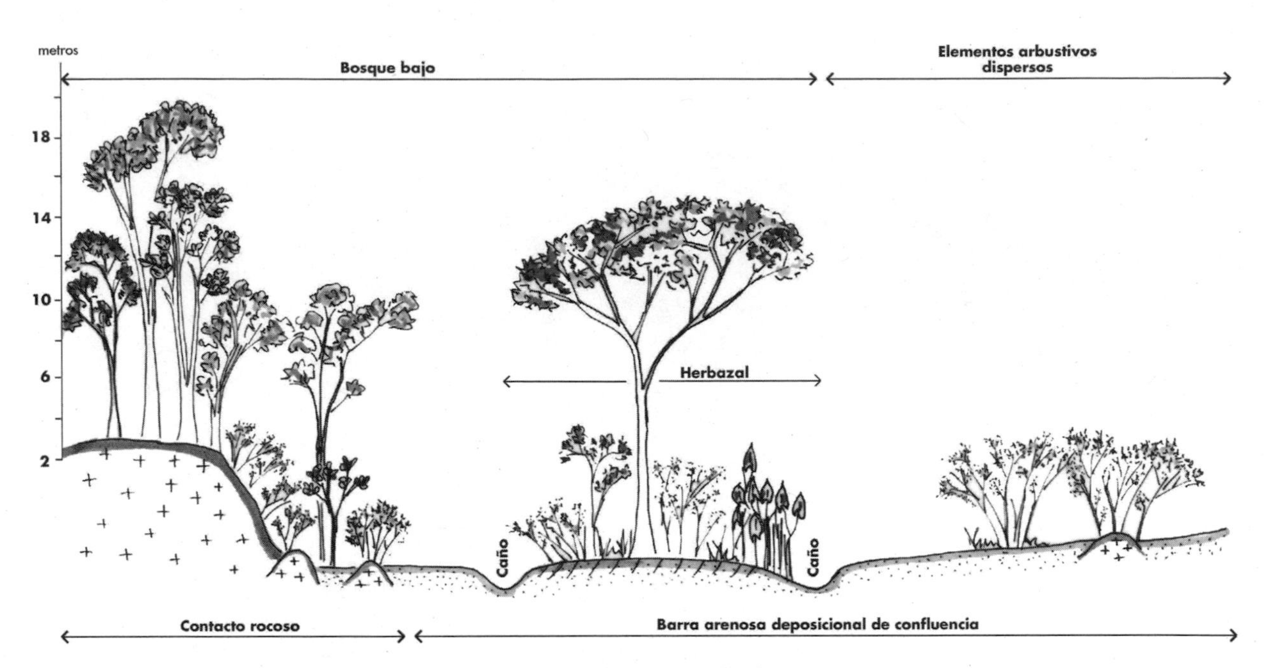

Figura 2.3. Diagrama del transecto de vegetación en la confluencia de Caño Cejiato con el Río Caura.

Símbolos: cruces–afloramiento rocoso; líneas diagonales–aporte orgánico; y puntos–arena.

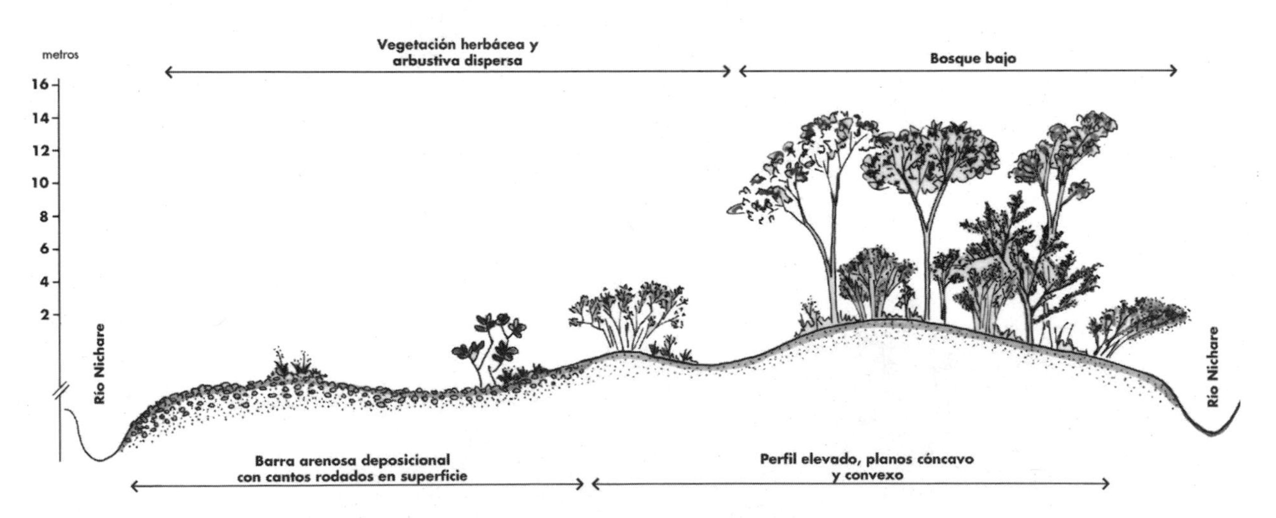

Figura 2.4. Diagrama del transecto de vegetación en relieve de isla en el Río Nichare próximo de la boca del Río Tawadu (Punto de georeferencia BC10). Los puntos indican arena.

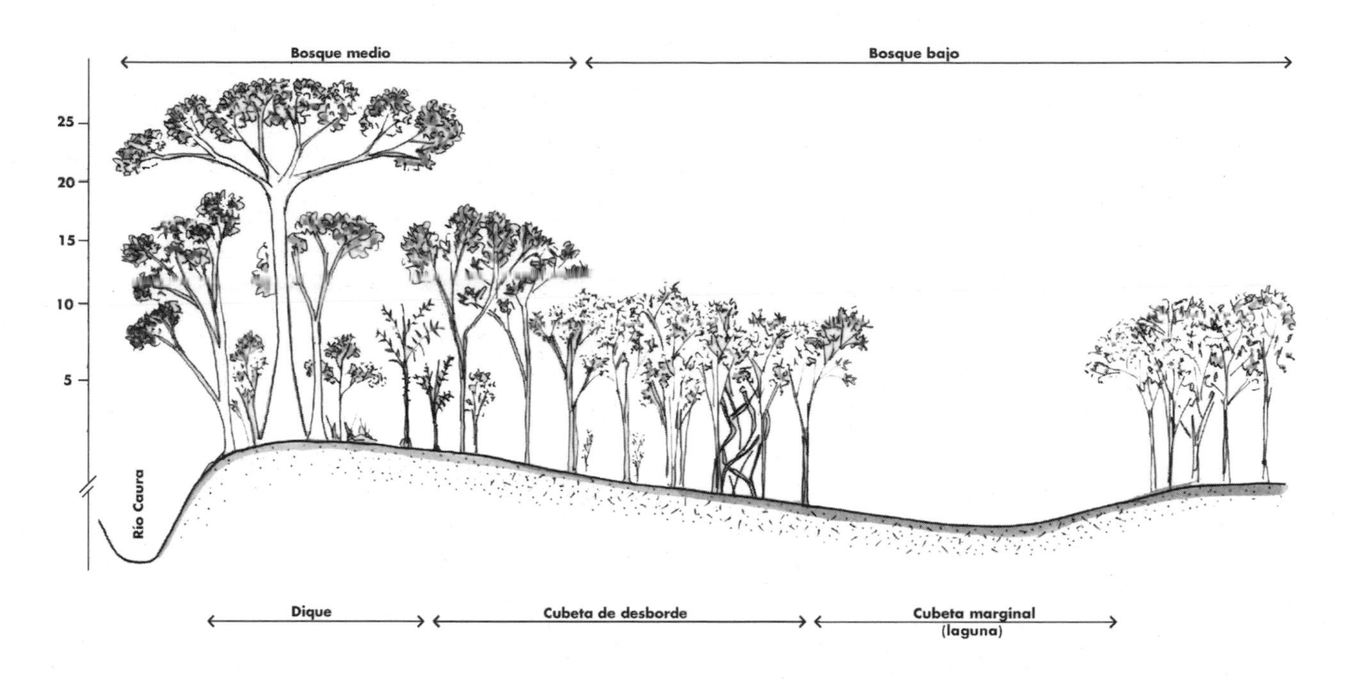

Figura 2.5. Diagrama del transecto de bosque ribereño medio y bajo asociado al relieve de dique y cubetas del Río Guaire, tributario al Río Caura (Punto de Georeferencia BC12).

Símbolos : líneas diagonales–finas y aporte orgánico; y puntos–arena.

longadas, en suelos de origen deposicional-residual (Figuras 2.1 y 2.5). En confluencias de caños y cubetas de decantación, se presentan parches de bosques bajos, dosel variable, con alturas comprendidas entre 8 y 15 m. Por otro lado, la cobertura en el sotobosque, puede variar de medio a ralo. El aporte de hojarasca al sustrato es importante fuente de nutrientes en todos estos bosques ribereños.

El tercer tipo de formación vegetal inundable, son los matorrales dominados ecológicamente por *Inga vera,* observado en márgenes deposicionales, playas de remanso y curvas de meandro. Abajo del Raudal Cinco Mil, los mismos hábitats son ocupados por otras especies de *Inga*, en conjunción frecuente con *Alibertia latifolia* y *Coccoloba obtusifolia*.

Esta formación puede alcanzar una altura de hasta 2 metros, presenta una cobertura densa formada a partir de una intrincada matriz de troncos fuertemente ramificados desde la base. Algunos bejucos, lianas y trepadoras, de las familias Apocynaceae, Vitaceae, Leguminosae, Convolvulaceae y Malvaceae, fueron observadas como elementos florísticos acompañantes, asociados principalmente al bosque ribereño en contacto con el matorral, formando mosaicos de "selvas de bejuco".

En el sector del Río Nichare y Río Tawadu se realizó una transecta en este tipo de hábitat, en una barra lateral deposicional expuesta, formada por una playa de aproximadamente 30 metros de ancho. El sustrato presentó una fracción importante de materia orgánica en combinación con aportes arcillosos que le conferían al suelo una contextura altamente fangosa. El matorral estaba expuesto totalmente y el elemento florístico dominante fue una especie de *Solanum* sp. conformando una comunidad ecológicamente equivalente al matorral de *Inga vera*. En esta área fue observada una comunidad herbácea sucesional temprana que antecede al matorral, conformada por especies herbáceas de las familias Cyperaceae, Graminaea, Rubiaceae, Labiatae, Gesneriaceae y como forma de crecimiento de mayor altura destaca *Montrichardia arborescens*, que tolera el ciclo de inundación. Estas comunidades que preceden al matorral, dependenden de la dinámica deposicional, la exposición de la barra lateral y el contacto con la lámina de agua. Finalmente, en transición hacia el bosque ribereño es posible observar un ecotono de elementos leñosos como *Genipa spruceana*, *Cecropia peltata, Myrcia* sp., *Corton* sp. y *Macrolobium acaciiefolium.*

El cuarto tipo de formación vegetal es la vegetación estrictamente acuática. La misma es escasa a lo largo del río Caura y Erebato. Básicamente, se limita a macrófitas reófilas de la familia Podostemonaceae, asociadas a los rápidos. Se colectaron un total de cinco especies; la mayor diversidad y abundancia fue observada en el sector desde Entreríos al Salto Pará. A pesar de lo anterior, estas comunidades son de gran importancia como indicadores ambientales por su fragilidad. Además estas comunidades proveen sustento y refugio para numerosos organismos acuáticos.

DISCUSIÓN

Se colectaron un total de 443 muestras en hábitats asociados a vegas, planicies de desborde, islas y rápidos. Se identificaron 399 especies (Apéndice 2). Se encontraron 291 especies en las áreas por encima del Salto Pará y 185 especies identificadas para el Bajo Caura. En general, la flora del Caura es similar al reportado en la literatura para los bosques neotropicales representado principalmente por las familias Leguminosae (*sensu lato*), Lauraceae, Anonaceae, Rubiaceae, Moraceae, Myristicaceae, Sapotaceae, Meliaceae, Euphorbiaceae, Chrysobalanaceae y Melastomataceae. Desde el punto de vista fitogeográfico, la flora del Caura Superior está mayormente relacionada con las provincias florísticas Amazónica y Guayanesa. En contraste, la del Bajo Caura, además de poseer elementos florísticos asociados a tales provincias, también incluye especies que la relaciona con la Provincia Llanera.

La diversidad gamma o regional de la vegetación ribereña del Caura es moderadamente alta. Los bosques inundables presentan elementos característicos de los igapó amazónicos (riachuelos asociados a aguas negras), aunque son también frecuentes especies de várzeas (areas inundables asociados a aguas negras) aumentando su frecuencia en aguas bajo del Salto Pará hasta la zona de confluencia Caura-Orinoco.

En general, las palmas *Euterpe precatoria, Attalea maripa, Socratea exhorriza, Geonoma baculifera* y *Bactris brongniartii*, son elementos florísticos característicos de los bosques ribereños del Caura. Por su parte, el género *Oenocarpus*, es frecuente en los bosques ribereños del Erebato y el género *Mauritiella*, es mayor en áreas inundables con suelos arenosos característicos del Caura Superior.

Eperua jehnmanii spp. *sandwichii*, es un elemento arbóreo muy frecuente de los bosques del Caura Superior, no se encuentra en los bosques ribereños del Bajo Caura. Mientras, *Ocotea cymbarum* y *Campsiandra laurifolia* solo aparece en el Bajo Caura.

En el Bajo Caura, abajo del Raudal Cinco Mil, se confirma la presencia de un nuevo conjunto florístico distinto al observado aguas arriba de este raudal. La comunidad está constituida principalmente por: *Campsiandra laurifolia, Piranhea trifoliata, Homalium racemosum, Mabea nítida, Gustavia augusta, Ruprechtia tenuiflora, Symmeria paniculata* y *Simira aristiguietae*. Mientras que la comunidad por arriba de los rápidos esta caracterizada por las especies: *Simira aristiguietae, Homalium racemosum, Gustavia augusta, Mabea nitida, Luelea candida, Ruprechtia tenuiflora, Macrolobium angustifolium, M. acaciaefolium, Montrichardia arborescens, Podostemonacea* sp.1, *Inga vera, Homalium guianense, Caraipa densifolia* y *Croton cuneatus*.

Las Podostemonaceae son frecuentes y diversas en los sistemas de rápidos arriba del Salto Pará como los raudales Culebra de Agua, Perro de Agua y Dimoshi. Se identificaron 5 especies: *Apinagia ruppioides, A. Staheliana, Mourera fluvilatilis, Rhyncoclacys* sp. y *Rhyncholacis* sp., además de un posible nuevo género, lo cual representa una alta variedad en comparación con otros ríos de la Guayana Venezolana.

Comunidades Ribereñas Rocosas: Ecosistemas de Gran Valor para la Conservación

Datos obtenidos durante la expedición AquaRAP, detectó una alta frecuencia de hábitats rocosos con unidades ecológicas propias en el Río Erebato y en menor proporción en el Río Caura. Un valor paisajístico excepcional a estos ríos es debida a la amplia variedad de comunidades vegetales asociadas con estos sistemas ribereños rocosos, en especial las comunidades de reófilas conformadas por plantas acuáticas de la Familia Podostemonaceae y la vegetación leñosa sobre las islas.

Una constante transformación de la cobertura vegetal en comunidades ribereñas es debida al intenso dinamismo asociado a perturbaciones fluviales periódicas en forma de inundación, sedimentación y erosión. Además, los suelos con bajo nivel nutricional y los afloramientos rocosos no sedimentarios, son factores determinantes en la génesis y variedad ecológica de las formaciones vegetales. El mosaico de cobertura vegetal, indicativo de este proceso, incluye: i) la presencia de comunidades herbáceas y arbustivas en diferentes grados de cobertura; y ii) bosques ribereños pioneros con diferentes etapas en la secuencia sucesional. Estos bosques van desde el matorral bajo hasta el bosque medio sucesional tardío con abundancia de especies trepadoras y lianas leñosas. La estructura y riqueza de especies, son en algunos casos, comparables con los bosques de ambientes ribereños colindantes de tierra firme.

Islas pequeñas situadas en rápidos, donde hay afloramientos rocosos, condiciones de alta humedad a lo largo de todo el año y recurrencia del impacto del ciclo hidrológico, tienen parches de bosques bajos y ralos, que adoptan fenologías particulares con formas retorcidas, hojas pequeñas coriáceas y un componente de epífitas y musgos muy importante. Esta fisionomía recuerda los bosques nublados transcicionales o una vegetación de tierras altas (> 2.000 msnm). Todas estas características le confieren a estas islas un gran valor biológico y paisajístico.

OPORTUNIDADES PARA LA CONSERVACIÓN

Además de los valores biológicos antes señalados, los bosques de la Cuenca del Río Caura aportan una serie de servicios ambientales que incluyen el almacenamiento y secuestro de carbono, el mantenimiento de reservorios para la biodiversidad, la conservación de suelos, la producción de agua, la regulación climática y la preservación de recursos alimenticios que sustentan poblaciones humanas locales en expansión así como también fauna silvestre (Rosales y Huber 1996; Silva 1996; Rosales et al. 1997; Silva 1997; Knab-Vispo 1998; Bevilacqua y Ochoa 2000; Centeno 2000; Vispo et al. 2002). La importancia actual y potencial de estos servicios ha sido resaltada a través de diversas iniciativas de conservación y desarrollo propuestas en los últimos años para esta región.

CONCLUSIONES Y RECOMENDACIONES

- Hábitats rocosos en islas en el Río Erebato y en la región superior del Río Caura poseen una comunidad vegetal y ecología especial y deben ser protegidos.

- Las áreas de rápidos en el Caura Superior deben ser protegidas por ser regiones cubiertas por macrofitas de la Familia Podostemonaceae y representan comunidades orgánicas especiales. Ellas cumplen un papel muy importante como refugios de una fauna zoológica única.

- Garantizar el mantenimiento del ciclo natural de las aguas ya que este esta intimamente asociado al intenso dinamismo observado en las vegetaciones ribereñas asociados. La construcción de represas o el transvase de aguas podría poner en peligro la comunidad vegetal de la cuenca.

- Regular y no permitir la expansión de la frontera agrícola y forestal dentro de la Cuenca del Río Caura. Esto es particularmente importante en las regiones del Bajo Caura.

LITERATURA CITADA

Aymard, G., S. Elcoro, E. Marín y A. Chaviel. 1997. Caracterización estructural y florística en bosques de tierra firme de un sector del bajo Río Caura, Estado Bolívar, Venezuela. *En:* Huber, O. y J. Rosales (eds.), Ecología de la Cuenca del Río Caura. Scientia Guaianae 7:143–169.

Bevilacqua, M., y J. Ochoa G. 1996. Areas Bajo Régimen de Administración Especial (ABRAE). *En:* Rosales, J. y O. Huber (eds.), Ecología de la Cuenca del Río Caura. Scientia Guaianae 6: 106–112.

Bevilacqua, M., y J. Ochoa G. (eds.). 2000. Informe del componente Vegetación y Valor Biológico. Proyecto Conservación de Ecosistemas Boscosos en la Cuenca del Río Caura, Guayana Venezolana. Caracas, Venezuela: PDVSA-BITOR, PDVSA-PALMAVEN, ACOANA, AUDUBON de Venezuela y Conservation International.

Bevilacqua, M., y J. Ochoa G. 2001. Conservación de las últimas fronteras forestales de la Guayana Venezolana: propuesta de lineamientos para la Cuenca del Río Caura. Interciencia 26: 491–497.

Berry, P. E, O. Huber y B. K. Holst. 1995. Floristic Analysis and Phytogeography. *En:* Berry P. E., B. K. Holst, K. Yatskievych (eds.), Flora of the Venezuelan Guayana, Vol. 1. Introduction. Saint Louis, USA: Missouri Botanical Garden and Timber Press. Pp. 161–191.

Briceño, J. A. 1995. Análisis Fitosociológico de los bosques ribereños del río Caura en el Sector Ceiato – Entrerios, Distrito Aripao del Estado Bolívar. Informe de Pasantía. Universidad de los Andes, Facultad de Ciencias Forestales, Mérida. Mimeografiado.

Briceño, E., L. Valvas y J. A. Blanco. 1997. Bosques ribereños del Bajo Río Caura: vegetación, suelo y fauna. *En:* Huber, O. y J. Rosales (eds.). Ecología de la cuenca del Río Caura. Scientia Guaianae 7: 259–289.

Centeno, J. C. 2000. Compensación de las emisiones de carbono provenientes del consumo de orimulsión: Viabilidad económica y política. Informe del Proyecto Conservación de Ecosistemas Boscosos en la Cuenca del Río Caura, Guayana Venezolana. Caracas, Venezuela: PDVSA-BITOR, CI, ACOANA, AUDUBON de Venezuela y PDVSA-PALMAVEN.

CVG-TECMIN. 1994. Informes de avance del Proyecto Inventario de los Recursos Naturales de la Región Guayana. Hojas NB-20: 1, 5, 6, 9, 10, 13 y14. Ciudad Bolívar, Venezuela: Gerencia de Proyectos Especiales.

Dezzeo, N., y E. Briceño. 1997. La vegetación en la cuenca del Río Chanaro: medio Río Caura. *En:* Huber, O. y J. Rosales (eds.). Ecología de la Cuenca del Río Caura. Scientia Guaianae 7: 365–385.

Huber, O. 1995. Vegetation. *En:* Berry P. E., B. K. Holst, K. Yatskievych (eds.). Flora of the Venezuelan Guayana, Vol. 1. Introduction. Saint Louis, USA: Missouri Botanical Garden and Timber Press. 97–160.

Huber, O. 1996. Formaciones vegetales no boscosas. *En:* Rosales, J. y O. Huber (eds.). Ecología de la Cuenca del Río Caura. Scientia Guaianae 6: 70–75.

Huber, O., J. Rosales y P. Berry. 1997. Estudios botánicos en las montañas altas de la cuenca del Río Caura. *En:* Huber, O. y J. Rosales (eds.). Ecología de la Cuenca del Río Caura. Scientia Guaianae 7: 441–468.

Klinge, H., J. Adis y M. Worbes. 1995. The vegetation of seasonal varzea forest in the lower Solimoes River, Brazilian Amazonia. Acta Amazonica 25:201–220.

Knab-Vispo, C. 1998. A rain forest in the Caura Reserve (Venezuela) its use by the indigenous Ye´kwana people. PhD Thesis. University of Wisconsin-Madison USA.

Knab-Vispo, C., J. Rosales y G. Rodríguez. 1997. Observaciones sobre el uso de plantas por los ye´kwana en el bajo río Caura. *En:* Huber, O. y J. Rosales (eds.). Ecología de la Cuenca del Río Caura. Scientia Guaianae 7: 211–257.

Knab-Vispo, C., J. Rosales, P. Berry, G. Rodríguez, Salas, I. Goldstein, W. Díaz y G. Aymard. Annotated floristic checklist of the riparian corridor of the lower and middle Río Caura with comments on animal use. Scientia Guaianae 13: (en prensa).

Lal, J. R. 1990. Estudios Fitosociológicos de varios tipos de bosque en la Reserva Forestal El Caura. Estado Bolívar. Informe de Pasantía, Facultad de Ciencias Forestales. Universidad de los Andes. Mérida. Mimeografiado.

Marín, E., y A. Chaviel. 1996. La vegetación: bosques de tierra firme. *En:* Rosales, J. y O. Huber (eds.). Ecología de la Cuenca del Río Caura. Scientia Guaianae 6: 60–65.

Rosales, J. 1996. Vegetación: los bosques ribereños. *En:* Rosales, J. y O. Huber (eds.). Ecología de la Cuenca del Río Caura. Scientia Guaianae 6: 66–69.

Rosales, J. 2000. An ecohydrological approach for riparian forest biodiversity conservation in large tropical river. PhD Thesis. School of Geography and Environmental Sciences, The University of Birmingham, Inglaterra.

Rosales, J., G. Petts y J. Salo. 1999. Riparian flooded forests of the Orinoco and Amazon river basins: a comparative review. Biodiversity and Conservation 8: 551–586.

Rosales, J., C. Knab-Vispo y G. Rodríguez. 1997. Bosques ribereños del bajo Río Caura entre el Salto Pará y los Raudales de Cinco Mil: su clasificación e importancia en la cultura ye´kwana. *En:* Huber, O. y J. Rosales (eds.). Ecología de la Cuenca del Río Caura. Scientia Guaianae 7:171–214.

Rosales, J., G. Petts y C. Knab-Vispo. 2001. Ecological gradients in riparian forests of the lower Caura River, Venezuela. Plant Ecology 152: 101–118.

Rosales, J., G. Petts, C. Knab-Vispo, J. Blanco, J. A. Briceño, E. Briceño, R. Chacón, B. Duarte, U. Idrogo, L. Rada, B. Ramos, H. Rangel y H. Vargas. 2002. Ecohydrological assessment of the riparian corridor of the Caura River in the Venezuelan Guayana Shield. Scientia Guaianae 13. (en prensa a).

Rosales, J., C. Vispo, N. Dezzeo, L. Blanco, C. Knab-Vispo, N. Gonzalez, C. Bradley, D. Gilvear, G. Escalante, N. Chacon y G. Petts. 2002. Riparian forests ecohydrology in the Orinoco River Basin. *En:* McClain, M. (ed.). The Ecohydrology of South American Rivers and Wetlands. UNESCO IHP Ecohydrology. Ecohydrology Programme. (en prensa b).

Salas, L., P. E. Berry y I. Goldstein. 1997. Composición y estructura de una comunidad de árboles grandes en el valle del Río Tabaro, Venezuela: una muestra de 18.75 ha. *En:* Rosales, J. y O. Huber (eds.). Ecología de la Cuenca del Río Caura. Scientia Guaianae 7: 291–308.

Silva, M. N. 1996. Etnografía de la Cuenca del Caura. *En:* Rosales, J. y O. Huber (eds.). Ecología de la Cuenca del Río Caura. Scientia Guaianae 6: 98-105

Silva, M. N. 1997. La percepción Ye'kwana del entorno natural. *En:* Rosales, J. y O. Huber (eds.). Ecología de la Cuenca del Río Caura. Scientia Guaianae 7: 65–84.

Steyermark, J., y C. Brewer-Carías.1976. La vegetación de la cima del macizo de Jaua. Boletín Sociedad Venezolana de Ciencias Naturales 22: 179–405.

ter Steege, H. 2000. Plant diversity in Guayana, with recommendations for a National Protected Area Strategy. Tropenbos Series 18:1–220.

Veillon, J. P. 1948. Cuenca del bajo y medio Caura. Estado Bolívar. Mapa Forestal. Caracas, Venezuela: Departamento de Divulgación Agropecuaria, Ministerio de Agricultura y Cría.

Vispo, C. 2000. Uso criollo actual de la fauna y su contexto histórico en el bajo Caura. Memorias Sociedad de Ciencias Naturales La Salle 149:115–144.

Vispo, C., J. Rosales y C. Knab-Vispo. Ideas on a conservation strategy for the Caura's riparian ecosystem. Scientia Guaianae 13: (en prensa).

Williams, L. 1942. Exploraciones Botánicas en la Guayana Venezolana. I. El medio y bajo Caura. Caracas Venezuela: Servicio Botánico, Ministerio de Agricultura y Cría.

Capítulo 3

Análisis Limnológico de la Cuenca del Río Caura, Estado Bolívar, Venezuela

Karen J. Riseng y John S. Sparks

RESUMEN

Parámetros limnológicos en cuarenta y tres sitios dentro de las regiones superiores e inferiores de la Cuenca del Río Caura, Estado Bolívar, Venezuela, fueron realizados. El Caura Superior es prístino, comprimida en una cuenca canalizada y sometida a una gran variación hidrológica estacional más que en Bajo Caura y como resultado una menor inundación de la floresta. Acá reportamos sobre los aspectos químicos de las aguas muestreadas, mientras que los componentes bióticos del sistema serán cubiertos por otros grupos (éste volumen). Generalmente, en estos ríos tropicales con bajos nutrientes, el énfasis fue colocado sobre el análisis de los parámetros siguientes: temperatura, conductividad, pH, transparencia secchi, carbón orgánico (COD), alcalinidad y sedimentos.

En general las aguas en esta región son levemente ácidas y diluidas, con muy baja conductividad y alcalinidad; típicas de un sistema dominado por lluvias. Estos resultados son congruentes con aquellos presentados en estudios previos y pueden ser atribuidos a la edad geológica antigua de la región. Los parámetros estudiados en las aguas fueron similares en toda la región superior e inferior de la cuenca; sin embargo, en tributarios ocasionalmente estudiados fueron encontradas variaciones a los valores generales obtenidos. COD (input terrestre) varía a lo largo de la cuenca, y ambas aguas claras y negras fueron estudiadas.

En general, la calidad del agua fue determinada como "buena" en todos los sitios estudiados en el momento de éste estudio. En las secciones bajas de la Cuenca del Río Caura existe una mayor perturbación que la superior. Perturbación alta en los márgenes de las aguas puede ocasionar un incremento en la incorporación de sedimento en el período lluvioso, una baja en la incorporación de importantes formas biológicas de carbón terrestre (COD) y causar un cambio del hábitat de los organismos acuáticos. Estos cambios en el hábitat pueden incluir cambios en la transparencia, sustrato, canalización del río y el número y diversidad de hábitats que pueden mantener la comunidad de organismos acuáticos presentes.

Nuestras recomendaciones sobre conservación son: Represas o transvase producirá cambios en el ciclo hidrológico altamente importante en sistemas tropicales dominados por las lluvias. También recomendamos la creación de una estación de monitoreo biológico a lo largo del canal principal del río Caura, posiblemente en Salto Pará. Esta estación proveerá una plataforma valiosa para científicos que puedan acceder y estudiar esta región con mayor detalle y poder construir una base de datos (a largo plazo) para estos ambientes de bosque lluvioso tropical. Dado a que lo prístino de los hábitats continentales acuáticos se están degradando continuamente y a tasas alarmantemente aceleradas en el mundo, preservar pocas, pero grandes regiones no perturbadas remanentes, incluyendo la Cuenca del Río Caura, debe ser una prioridad en los protocolos de manejo conservacionista.

INTRODUCCIÓN

El equipo de AQUARAP fue enviado al Rió Caura para obtener el valor del área para implementar esfuerzos de conservación futura. La Cuenca del Río Caura representa una región relativamente prístina en Venezuela y posee un gran potencial para explotación futura (pesquerías, minería y potencial hidroeléctrico) de éste importante recurso natural. Los objetivos del muestreo limnológico son: 1) obtener información sobre las condiciones generales y calidad del agua en toda la cuenca; 2) examinar las tendencias limnológicas a lo largo del río y sus tributarios; y 3) proveer una descripción cuantitativa de los hábitats para los análisis organísmicos.

La cuenca del Río Caura posee 45.3 km^2 y es parte de la gran Cuenca del Río Orinoco (1.000.000 km^2), la tercera cuenca en importancia en América del Sur (AquaRAP map). El Orinoco posee un caudal o descarga que lo ubica en tercer lugar en todo el mundo cercano a 1.100 km3/año (Paolini et al. 1987), con máximos en mediados de Agosto y comienzos de Septiembre (Depetris y Paolini 1991). La mayoría de los tributarios del Río Caura corren sobre el Escudo de Guayana. Ríos que drenan el Escudo de Guayana pueden ser caracterizados como mayormente ácidos, bajos en conductividad y bajos en sólidos totales suspendidos, comparados con ríos que drenan los Andes o Llanos (Depetris y Paolini 1991). Las cabeceras del Río Caura se originan en la porción sur del Estado Bolívar, cercano a la frontera con Brasil. La cuenca está bordeada por la Sierra de Pakaraima en el Sur y la Sierra de Maigualida en el Oeste. Mientras que el origen de tributarios por encima de los 2000msnm está presente, la gran mayoría de la cuenca se encuentra en niveles por debajo de los 500msnm. Muchos rápidos están presentes a lo largo de la cuenca y sus tributarios. Grandes cataratas, p.e Salto Pará, separan las regiones superiores e inferiores de la cuenca. Por encima de las cataratas, los mayores tributarios son el Río Erebato (~10.5 km^2 de cuenca). El Río Yuruaní (2.8 km^2) también se une al Caura arriba de las cataratas. Por debajo del Salto Pará, los principales tributarios son el Río Nichare (4.0 km2) y el Río Mato (2.6 km^2).

Estacionalidad

El Río Caura es una sistema tropical dominado por las lluvias y con fuerte ciclicidad estacional. El clima en la región es determinado ser un sistema húmedo tropical, cálido (media anual de temperatura del aire varía entre 25.7 a 27.7°C) con una media anual de humedad de 75% en Maripa (Bajo Caura) y 83–85% en Entreríos y Salto Pará (Vargas y Rangel 1996a). La estación lluviosa se extiende aproximadamente de Mayo a Octubre con una precipitación anual total de 1980 mm (Maripa), 2970 mm (Salto Pará) y 3260 mm (Entreríos) (Vargas y Rangel 1996a). La máxima descarga ocurre entre Julio y Agosto. El Río Caura puede experimentar una diferencia de 10 variación en la descarga anual, cerca de 7.5 m^3 máximo en su boca (Lewis et al. 1987). Estas grandes fluctuaciones en descarga se trasladan a dramáticos cambios anuales en el nivel del agua, por ejemplo 7 metros calculados en 20 años de registros en Maripa (Vargas y Rangel 1997b) y grandes inundaciones forestales en áreas donde el paisaje es favorablemente inclinado. El Bajo Caura, especialmente por debajo del Raudal 5000, posee grandes extensiones de bosques ribereños inundables (Rosales et al. 2001) comparados con sitios en la región superior y entonces el río posee una fuerte interacción con los sistemas terrestres.

Relación con el sistema terrestre

El sistema del Caura contiene tributarios con aguas "claras" y "negras" y fluye sobre una base extremadamente antigua formación geológica (Escudo de Guayana). El color en los ríos de aguas negras está generalmente asociado con ácidos húmicos y flúvicos que se acumulan proveniente de la descomposición de material orgánico. Estos compuestos pueden filtrarse de material terrestre que se acumula en el río y dentro del ambiente acuático y son evidenciados por el alto contenido en carbono orgánico disuelto (COD). Una vez en el sistema, el COD es una fuente importante de carbono para la producción microbiana. COD puede ser producido por la degradación de material orgánico dentro del río, tales como las macrofitas, o por material fuera del río, como la hojarasca. Para una revisión de la transformación del COD es ríos ver Allan (1995). Con respecto a las especies de carbono, el carbono orgánico disuelto (COD) es el forma dominante en los ríos que drenan el Escudo de Guayana, mientras que el carbono inorgánico disuelto la forma dominante de carbono en ríos y tributarios que drenan los Andes y Llanos en la Cuenca del Orinoco (Depetris y Paolini 1991). Los lagos de inundación contribuyen al COD del sistema riverino (p.e. medio Orinoco, Castillo 2000), cuando ellos se encuentran llenos y drenan al río en el período lluvioso. La cantidad de inundación del bosque, la morfología del río y bosque ribereño contribuyen significantemente en la interacción entre los sistemas acuáticos y terrestres.

Una vista general de la ecología del Río Caura puede ser encontrada en *Scientia Guaianae*, No. 6 y 7, incluyendo limnología (García 1996; Ibañez y Lara 1997) y resultados de estudios de química de aguas realizado en el Bajo Caura fueron publicados por Lewis (1986), Lewis et al. (1987) y Lewis et al. (1986). Acá reportamos sobre los aspectos químicos de las aguas muestredas, mientras que los components bióticos del sistema estás cubiertos por otros equipos en este volúmen. En este río de bajos nutrientes (Lewis 1986; García 1996) el objetivo fue colocado en el análisis de los siguientes parámetros: temperatura, conductividad, pH, profundidad de Secchi, carbono orgánico disuelto, oxígeno disuelto, alcalinidad y sedimentos.

MÉTODOS

Un analizador de calidad de agua Horiba fue utilizado para obtener conductividad (mS/cm), oxígeno disuelto (mg/l), y temperatura del agua (deg C). El pH fue medido usando un analizador portátil Orion 250. La transparencia fue medida utilizando un disco de secchi (8"). La turbidez fue registrada en forma cualitativa como baja, media, o alta. Adicionalmente, el porcentaje de sombreado sobre el agua fue registrado cuando el sol se encontraba directamente perpendicular.

Todas las muestras de agua fueron colectadas de cerca de 10 cm por debajo de la superficie. Después de la colecta, el agua fue filtrada utilizando un papel Whatman GF/F con porosidad nominal 0.7mm. El color del agua filtrada fue registrado como claro, o muy levemente marrón (vsb), levemente marrón (sb), marrón claro (lb), o marrón (b). Filtros fueron analizados para obtener partículas y el filtrado fue analizado para alcalinidad y COD. Material sedimentario fue medido usando filtros previamente pesados GF/F 47mm los cuales fueron secados a 40°C por 48 h después de usados. Los filtros secos fueron repesados en una Balanza Metler. La diferencia de pesos fue dividida por el volumen filtrado para el material particulado más grande que el tamaño de poro nominal del filtro (0.7mM) por litro de muestra.

Filtrados para los análisis de COD fueron preservados con 1ml HCl por ml de muestra y almacenados en un vial de vidrio de borosilicato de centelleo. El filtrado para análisis de alcalinidad fue almacenado en botellas de HDPE de 60mL. El agua filtrada fue analizada para COD y alcalinidad en el Laboratorio George Kling en la Universidad de Michigan. COD fue medido con un analizador Shimadzu 5000 TOC. Alcalinidad fue medida por titulación directa al punto de inflección usando un sistema de Radiómetro autotitulación con 0.25N o 0.1N HCl.

RESULTADOS Y DISCUSIÓN

El canal principal del Río Caura progresivamente incrementa en amplitud a medida que nos movemos aguas abajo. Existen numerosos hábitats únicos y raros dentro de la cuenca en adición al canal principal, incluyendo lagunas, lagos de inundación, y numerosos tributarios pequeños incluyendo ambos tipos de aguas claras y negras. El incremento de la perturbación externa es evidente río abajo, especialmente cercanos a la confluencia con el Orinoco. A lo largo de la cuenca, un total de 43 lugares fueron estudiados correspondiendo a 31 estaciones o puntos de georeferencia. La elevación de la región muestreada varío desde los 500m en el Alto Caura a menos de 100m en la porción inferior del Caura. Una lista completa de resultados de la química del agua es presentada en los Apéndices 3, 4 y 5 y resumida en la Tabla 3.1 a continuación.

Para todos los sitios muestreados, el agua estaba diluida, un resultado atribuible mayormente a la geología regional (antigua formación geológica, p.e. Escudo de Guayana) y la época del año cuando se realizo el estudio. En este estudio las muestras fueron colectadas entre Noviembre-Diciembre, inmediatamente posterior al final del período lluvioso. La temperatura incrementa en aproximadamente 2 grados centígrados desde la región más superior a la estación más baja muestreada. Conductividad fue generalmente muy baja, con pocas excepciones. El agua resultó levemente ácida, mayormente en áreas lagunares y generalmente bien oxigenada, cercana a saturación debajo de cataratas y rápidos. Areas lacustrinas fueron mucho menos oxigenadas, con lagunas completamente aisladas (p.e., ningún flujo hacia o entre al momento de la colecta) extremadamente pobres en oxígeno en este momento del año. Los pequeños tributarios y caños generalmente poseen baja turbidez y alta transparencia.

En la región superior de la Cuenca del Caura (p.e., encima de Salto Pará), 21 estaciones limnológicas fueron muestreadas dentro de 14 localidades georeferenciadas. El agua fue moderadamente ácida, con un pH variando desde 5.0 a 5.9. Conductividad fue extremadamente baja, variando desde 6 a 12 mS. El agua de la región fue generalmente bien oxigenada, sin embargo, niveles de saturación fueron solamente encontrada inmediatamente por debajo de los rápidos o cataratas. Debemos hacer notar que algunos sitos del Alto Caura en términos de unicidad de hábitats, los cuales incluyen el Río Cacara (aguas negras) y el Yurawani (aguas verdosas, con una comunidad de algas muy productiva).

En la región del Bajo Caura (p.e., por debajo de las cataratas de Salto Pará) 22 sitios limnológicos fueron muestreados dentro de 17 localidades georeferenciadas. El agua en esta región es similar a la observada en la región superior, excepto que la temperatura es mayor en aproximadamente 2°C sobre la mayoría de los sitios muestreados aguas arriba. Los sitios muestreados incluyen muchos sobre el canal principal, lagunas semi o completamente aisladas y remansos y un número de pequeños y grandes tributarios. Tributarios en ambas riberas fueron muestreados, también como sistemas de aguas claras y negras. El agua en la región resultó moderadamente ácida a ácidas, con un intervalo de pH desde 4.4 a 6.0. Conductividad fue generalmente baja, aunque moderada en algunos tributarios (p.e., Río Takoto) y varió desde 9 a 41 mS/cm. La temperatura de las aguas en la región varía desde 23 a 26°C, con el Río Tabaro (tributario de aguas negras) como la más fría del área (<23°C). El Bajo Caura incluye un mayor número de ambientes acuáticos que la región superior, incluyendo un gran número de lagunas y áreas inundables.

Encontramos valores similares a estudios previamente realizados en la Cuenca del Río Caura (Lewis 1986; Lewis et al. 1987; García 1996; Ibañez y Lara 1997). Existen interesantes tendencias limnológicas y observaciones en el COD, pH, conductividad, temperatura y alcalinidad. Lewis (1986) encontró valores de COD de 417–1417 mM en la boca del Río Caura. El máximo de COD en este estudio estuvo en 916 mM obtenido en un lago temporal dentro de una isla. El grupo botánico determinó que este es un lago completamente conectado con el río en el período de lluvias. Como en el Orinoco, lagos de inundación (Castillo 2000), el COD se concentrará previo a que el lago drene al río. Esto demuestra un importante recurso de carbono para el ecosistema acuático. La conductividad, pH y alcalinidad fueron similares a los datos publicados (García 1996; Ibañez

Tabla 3.1. Promedio de temperatura (Temp), pH, conductividad (Cond), oxígeno disuelto (D.O.), alcalinidad (Alc), carbono orgánico disuelto (DOC) y sedimentos por categoría. Note que en algunos casos los datos están ausentes para algún sitio en particular, para una lista completa ver los Anexos. El lago en el Bajo Caura está completamente inundado al subir las aguas (determinado por el grupo de botánica).

	Número de Sitios Promedio	Temp (°C)	pH	Cond (μS/cm)	D.O. (mg/l)	Alc (μeq/L)	DOC (μM)	Sediment (mg/L)
Promedio de todos los sitios	43	24.9	5.60	13.0	6.7	105	266	11.0
Intervalo		22.7–26.8	4.33–5.98	6–41	0.9–9.1	20–375	98–915	1.3–21.9
Desviación de estándar		1.1	0.34	7.4	1.4	74	149	6.4
Caura Superior (arriba Salto Pará)								
Todos los sitios	21	24.9	5.60	9.9	6.6	71	223	11.1
Ríos								
Caura arriba	8	26.0	5.78	10.7	7.0	78	275	12.5
Kakada	1	23.9	5.04	6.0	6.4	20	233	4.2
Erebato	3	25.0	5.78	9.3	6.7	68	191	7.4
Yuruani	1	25.4	5.68	12.0	6.4	94	278	21.8
Riachuelos	5	24.0	5.39	9.0	6.5	68	193	13.7
Caños	3	23.9	5.46	10.3	4.9	79	130	6.8
Bajo Caura (debajo Salto Pará)								
Todos los sitios	22	26.2	5.69	10.3	8.0	73	272	11.4
Ríos								
Bajo Caura	4	24.9	5.59	16.0	6.9	129	294	10.9
Nichare	2	24.6	5.61	12.0	6.9	98	187	17.3
Tawadu	5	23.2	5.76	9.6	8.1	61	352	1.3
Takoto	1	25.4	5.99	41.0	6.8	375	183	16.2
Icutú	1	24.8	5.87	14.0	6.8	126	184	13.8
Mato	1	25.9	5.82	20.0	6.2	165	528	21.9
Caños	4	24.8	5.80	21.5	7.3	174	199	9.5
Lagunas	2	26.0	4.90	12.0	4.1	87	272	5.1
Lago inundado	1	26.6	4.34	19.0	0.9	86	915	10.6
Riachuelo	1			35.0		280	200	

y Lara 1997). En este estudio el pH incrementa constantemente río abajo, empezando en 5.5 aguas arriba y terminando en 7.2 en el Bajo Caura. La conductividad en general es baja (10–15 mS/cm) pero con valores altos encontrados en un río cerca del Salto Pará (35 mS/cm) y cerca del Río Takoto (41 mS/cm). El río Takoto drena y corre sobre una formación geológica diferente, la Provincia de Imataca. Esta provincia está caracterizada por la abundancia de gneiss, anfibolitas, intrusiones graníticas y cuarcitas ferruginosas. La temperatura del Río Caura generalmente incrementa río abajo. La temperatura es menor en rápidos y pequeños tributarios de bosque o ríos con suficiente volúmen tales el Erebato y el Nichare, los cuales producen un efecto sobre la temperatura general del canal principal del Caura. Mientras que la alcalinidad en general varió desde 20–375 meq/L en la cuenca, ésta generalmente se encuentra entre 60–90 meq/L, lo cual representa un valor muy bajo. La mayor alcalinidad fue encontrada en caños y tributarios.

CONCLUSIONES Y RECOMENDACIONES PARA CONSERVACIÓN

Hemos encontrado toda el área muestreada en ésta expedición de AquaRAP relativamente intacta. La región superior del Caura corresponde a un área prístina, mientras que en la región inferior, aunque todavía pristine, existen ciertos grados de perturbación. Deforestación en tierras bajas de ríos de la Amazonia alteran el suelo y la química de agues, incluyendo nutrientes disueltos y particulados (Neill et al. 2001) Las regiones muestreadas en este estudio experimentan poco o mínimo impacto proveniente de deforestación, erosion u otra influencia externa que puede afectar negativamente la calidad del agua en la cuenca. Esfuerzos de conservación de este sistema riverino deben incluir:

- el mantenimiento de la zona riberena, ya que es un amortiguador natural para el sistema acuático;
- el mantenimiento del ciclo hidrológico natural; la inundación del bosque es claramente una fuente importante de carbono y otros nutrientes fundamentales para los organismos acuáticos;
- el mantenimiento de la cuenca superior y su alta calidad ambiental, es crucial para la calidad total de la cuenca;
- el establecimiento de una estación de registro científico permanente en el Caura cerca de Salto Pará que permita acceso al Alto Caura. Perturbaciones del ciclo hidrológico representa una amenaza enorme al ecosistema y estos incluyen los proyectos de represamiento o transvase de las aguas a lo largo del río.

LITERATURA CITADA

Allan, J. D. 1995. Stream Ecology: Structure and function of running waters. London England: Chapman and Hall.

Castillo, M. M. 2000. Influence of hydrological seasonality on bacterioplankton in two neotropical floodplain lakes. Hydrobiologia 437: 57–69.

Depetris, P. J., y J. E. Paolini. 1991. Biogeochemical Aspects of South American Rivers: The Paraná and the Orinoco. *En:* Degens, E. T., S. Kempe, J. E. Richey (eds.). Biogeochemistry of Major World Rivers. Scope 42: 105–122.

García, S. 1996. Limnología. *En:* Rosales, J. y O. Huber (eds.). Ecología de la Cuenca del Río Caura. Scientia Guaianae 6: 54–59.

Ibañez, A. M., y J. I. Lara. 1997. Algunos aspectos fisicoquimicos y biologicos de las aguas del Río Caura (Venezuela), en su parte media. *En:* Rosales, J. y O. Huber (eds.). Ecología de la Cuenca del Río Caura. Scientia Guaianae 6: 34–39.

Lewis, W. M. 1986. Nitrogen and phosphorus runoff losses from a nutrient-poor tropical moist forest. Ecology 67(5): 1275–1282.

Lewis, W. J., S. K. Hamilton, S. L. Jones y D. D. Runnels. 1987. Major element chemistry, weathering, and element yields for the Caura River drainage, Venezuela. Biogeochemistry 4: 159–181.

Lewis, W. M., J. F. Saunders, S. N. Levine y F. H. Weibezahn. 1986. Organic carbon in the Caura River, Venezuela. Limnol. Oceanogr. 31(3): 653–656.

Neill, C., L. A. Deegan, S. M. Thomas y C. C. Cerri. 2001. Deforestation for pasture alters nitrogen and phosphorus in small Amazonian streams. Ecological Applications 11(6): 1817–1828.

Paolini, J. E., R. Hevia y R. Herrera. 1987. Transport of carbon and minerals in the Orinoco and Caroní Rivers during the years 1983–84. *En:* Degens, E. T., S. Kempe, Gan Weibin (eds.). Biogeochemistry of Major World Rivers. Scope 64: 325–38.

Rosales, J., G. Petts y C. Knap-Vispo. 2001. Ecological gradients within the riparian forests of the lower Caura River, Venezuela. Plant Ecology 152: 101–118.

Vargas, H., y J. Rangel. 1996a. Clima: Comportamiento de las variables. *En:* Rosales, J. y O. Huber (eds.). Ecología de la Cuenca del Río Caura. Scientia Guaianae 6: 34–39.

Vargas, H. y J. Rangel. 1996b. Hidrología y Sedimentos. *En:* Rosales, J. y O. Huber (eds.). Ecología de la Cuenca del Río Caura. Scientia Guaianae 6: 48–53.

Capítulo 4

Diversidad de Macroinvertebrados Bentónicos de la Cuenca del Río Caura, Estado Bolívar, Venezuela

José Vicente García y Guido Pereira

RESUMEN

La diversidad de macroinvertebrados del Río Caura fue estudiada en 25 localidades distribuidas en las regiones superiores y bajas del río. La comunidad de macroinvertebrados acuáticos estuvo compuesta de insectos, caracoles, almejas, gusanos oligoquetos, sanguijuelas, turbelarios y crustáceos. Los insectos acuáticos están representados por varios ordenes; de los cuales los más diversos fueron: Odonata, Diptera, Ephemeroptera, Hemiptera y Trichoptera. La composición de especies observada es típica de ambientes pristinos, ya que ellas contienen una gran diversidad de Odonata y Trichoptera, los cuales son habitantes de ambientes libres de contaminantes o perturbación. La mayoría de los géneros de macroinvertebrados bénticos en la cuenca fueron encontrados homogéneamente distribuidos. La variación en diversidad de la fauna béntica parece estar relacionada a cambios en el nivel de oxígeno disuelto y turbidez del agua. Estos hallazgos son importantes debido a que cambios drásticos en el ciclo hidrológico, deforestación o minería podrían introducir material particulado suspendido e incrementar la conductividad causando un subsecuente cambio dramático en las comunidades bénticas.

INTRODUCCIÓN

La Cuenca del Río Caura conforma una gran región de bosques los cuales permanecen en condiciones prístinas y constituye uno de los principales tributarios del Río Orinoco en Venezuela. En esta región, la flora y fauna terrestre es bastante conocida; mientras que en el caso de la fauna acuática, solamente los peces se conocen escasamente (Machado-Allison et al. 1999, 2003). Otros grupos tales como los cangrejos, camarones e insectos acuáticos son totalmente desconocidos a pesar de su importancia en la estructura comunitaria de esos ecosistemas. Los organismos bentónicos son muy importantes en la transferencia de energía a través de los niveles tróficos en sistemas acuáticos, muchos de ellos incorporan material alóctono y transforman y transfieren material autóctono dentro del sistema (Wiggins 1927; Wallace and Merrit 1980). Algunos grupos tales como los insectos acuáticos, se alimentan de algas unicelulares, bacterias, hongos, plantas vasculares y detrito de hojas, zooplancton, otros invertebrados y peces pequeños (McCafferty 1981), mientras que otros grupos tales como las almejas, los caracoles y algunos insectos son filtradores de seston. Por otra parte, ellos constituyen los principales componentes alimenticios de los estadíos juveniles de peces (Lowe-McConnell 1975; Machado-Allison 1992), cangrejos, camarones y aves (Epler 1995). Adicionalmente, la composición de especies en estas comunidades es una evidencia de la calidad del agua de los ambientes en los cuales ellos habitan y en consecuencia, son importantes indicadores del nivel de perturbación o contaminación en cualquier cuerpo de agua (Wiggins 1927; Hynes 1970; Epler 1995).

Este trabajo representa un rápido diagnóstico de la biodiversidad de la fauna de macroinvertebrados bentónicos con la finalidad de evaluar el grado de degradación. De ser posible, estableceremos una comparación con otros grandes ríos cercanos tales como Orinoco y Caroní. Finalmente evaluaremos la importancia de esos grupos en el establecimiento de estrategias de conservación para esta región.

MATERIALES Y MÉTODOS

Un total de 25 localidades geo-referenciadas, 11 en el Medio Caura y 14 en el Bajo Caura, fueron muestradas durante el diagnóstico. Estas localidades y las variables fisicoquímicas son mostradas en los Apéndices 3, 4 y 5. Los métodos de muestreo incluyeron una red de Surber, una draga Eckman, filtración a través de tamices graduados, búsqueda por método visual y varias clases de redes incluyendo una red de mano, redes pequeñas de acuario y un chinchorro. Adicionalmente, se colectó hojarasca, rocas y troncos sumergidos y búsqueda por método visual. Todos los organismos fueron fijados en el campo en etanol al 70%. Los sedimentos fueron fijados en formol al 5% y preservados en etanol al 70%. Esta técnica no fue útil para preservar turbelarios, debido a la desintegración de los tejidos durante el transporte al laboratorio. Todas las muestras fueron preservadas posteriormente en etanol al 90%. Para la identificación, un microscopio estereoscópico, un microscopio óptico y claves convencionales hasta el nivel de género fueron utilizados para la mayoría de los grupos (Richardson 1905; Johannsen 1937a, b; Van Deer Kuyp 1950; Needham y Westfall 1955; Hilsenhoff 1970; Peters 1971; Bryce y Hobart 1972; Benedetto 1974; Flint 1974; Edmunds et al. 1976; Hulbert et al. 1981a, b; McCafferty 1981; Limongi 1983; Stehr 1987; Kensley 1989; Daigle 1991; Epler 1995, 1996; Milligan 1997). Para las larvas de quironómidos (Chironomidae), fueron seguidos los métodos descritos por Bryce (1972), hirviendo las cabezas de las larvas en una solución de KOH por 3 a 5 minutos y montándolas en láminas con polivinil-lactofenol. Este medio también fue utilizado para observar las setas de oligoquetos. Un Análisis de Componentes Principales basado en el número de géneros de insectos y clases para los otros grupos; así como algunas variables fisicoquímicas tales como oxígeno disuelto, turbidez, y conductividad para cada localidad geo-referenciada fue realizado utilizando el programa MVSP versión 3.1 (Kovach 1998), con la finalidad de determinar la existencia de algunos patrones de distribución y la afinidad de los grupos a estas variables fisicoquímicas. Adicionalmente, la riqueza total y el índice de diversidad de Sannon-Wiener fueron calculados para cada localidad.

RESULTADOS

La comunidad de macro-invertebrados bénticos de la Cuenca del Río Caura está compuesta de insectos acuáticos, caracoles, almejas, lombrices acuáticas, turbelarios y crustáceos. Los principales microhábitats son: vegetación acuática enraizada, vegetación marginal, lechos rocosos, troncos sumergidos, remansos con lechos de hojarasca, playas arenosas, playas fangosas, charcas marginales y charcas sobre lajas (grandes rocas). Esta gran heterogeneidad de hábitats se encontró tanto en el canal principal del río, así como en los tributarios.

Los insectos acuáticos están representados por los siguientes órdenes: Odonata, Ephemeroptera, Hemiptera, Trichoptera, Coleoptera, Diptera, Plecoptera, Neuroptera, Lepidoptera y Collembola. Los moluscos acuáticos muestran una baja diversidad y están representados por dos especies abundantes de *Pomacea* (*Limnopumus)* (Gastropoda, Ampullariidae), dos especies muy abundantes de *Doryssa* (Gastropoda, Melaniidae) y una especie de *Eupera* (Bivalvia, Sphaeriidae), con una baja abundancia. Los crustáceos están representados por una especie de isópodo (*Exocorallana berbicensis*) de la familia Corallanidae y una especie de conchostráceo (*Cyclestheria hislopi*) de la familia Cyclestheriidae, ambos encontrados sólo el Río Takoto en el bajo Caura. Las lombrices acuáticas están representadas por especies de las familias Lumbriculidae y Enchytraeidae, y una especie no determinada. Las sanguijuelas son de una especie (*Haementeria tuberculifera*) de la familia Glossiphoniidae, y una especie no determinada. Los turbelarios acuáticos (Plathyhelminthes) se observaron en tanto en el Caura Superior como en el Bajo Caura. Una lista completa de clases, familias y géneros es mostrada en el Apéndice 6.

En el caso de los insectos acuáticos, los Odonata están representados por siete familias con al menos 19 géneros (Fig. 4.1); de los cuales un género se encontró solamente en el Caura Superior y dos de ellos solamente en el Bajo Caura. Los Odonata se encontraron en la mayoría de los hábitats muestreados. Los Ephemeroptera están representados por siete familias y 12 géneros (Fig. 4.1), tres de ellos encontrados solamente en el Bajo Caura y uno en el Caura Superior. Los géneros *Thraulodes* y *Ulmeritus* son particularmente abundantes y distribuidos en toda la cuenca. Los Hemiptera también están distribuidos por toda la cuenca

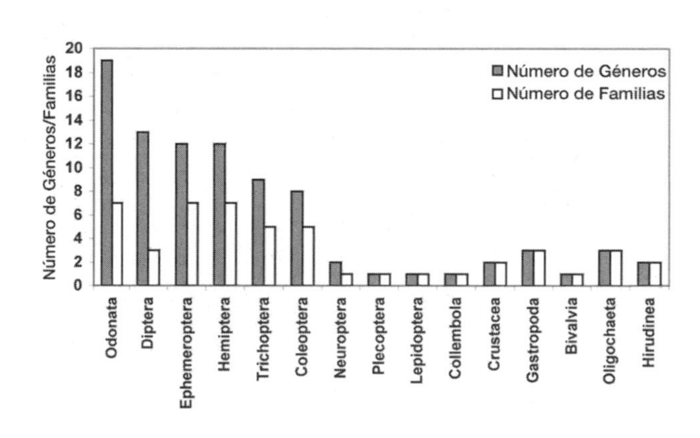

Figura 4.1. Número de familias y géneros en cada orden encontradas en la Cuenca del Río Caura.

y están representados por 12 géneros en siete familias (Fig. 4.1). El Orden Trichoptera está representado por nueve géneros en cinco familias con una abundancia moderada. Coleópteros de al menos ocho géneros y cinco familias se encontraron distribuidos por toda la cuenca (Fig. 4.1). El género *Megadytes* es particularmente abundante cuando está presente. Los dípteros son abundantes y están representados por especies de 13 géneros en tres familias; siendo la familia Chironomidae la más diversa. El Orden Plecóptera está representado por abundantes organismos de una especie del género *Anacroneuria*. Neurópteros de la familia Corydalidae fueron encontrados principalmente en el Caura Superior, mientras que lepidópteros de la familia Pyralidae se encontraron distribuidos por toda la cuenca. Una especie de *Willowsia* (Colembolla, Entomobryidae) fue encontrada solamente en la Laguna de La Ceiba (Bajo Caura). En orden de importancia, Odonata representa el 22% del total de géneros, seguido por Diptera con 16%, Ephemeroptera y Hemiptera con 13%, y Trichoptera y Coleoptera con 10 y 9% respectivamente (Fig. 4.2).

Se realizaron dos Análisis de Componentes Principales. En el primero se incluyó el número de géneros en cada una de las localidades con la finalidad de observar la existencia de algunos patrones de distribución. En el segundo análisis, además, se incluyen las variables fisicoquímicas por cada localidad para observar si algunos taxa muestran asociación con alguna de las variables fisicoquímicas oxígeno disuelto, turbidez, y conductividad. Debido a que los valores de pH son muy bajos y muy similares en todas las localidades los mismos no fueron considerados en el análisis. Los resultados de las contribuciones de cada variable en los dos análisis se muestran en las Tablas 4.1 y 4.2.

El primer ACP muestra que los primeros tres componentes principales retienen aproximadamente el 76% de la variabilidad total. Los resultados (Fig. 4.3) indican que la distribución es homogénea en la mayoría de los órdenes. Sin embargo, hay cinco localidades (AC06, AC09, AC14, BC05 y BC08) que se muestran altamente correlacionadas con Odonata y Ephemeróptera, dos localidades (AC05, AC11) altamente correlacionadas con Diptera y cuatro localidades (BC12, BC13, BC15 and BC17) con correlación negativa para esos grupos, pero positiva para Hemíptera.

En el segundo análisis, los primeros tres componentes retienen cerca del 84% de la variabilidad total. En este análisis (Fig. 4.4) muestra una correlación altamente positiva entre Odonata, Ephemeróptera y el oxígeno disuelto y una correlación altamente negativa entre esos grupos, la conductividad y la turbidez. Los otros grupos bentónicos de la fauna no se encuentran relacionados a ninguna variable fisicoquímica; sin embargo todos los grupos se alejan de la conductividad.

DISCUSÍON

El Río Caura presenta en general bajo pH y conductividad; por lo cual las variaciones en la diversidad de los grupos de la fauna parecen estar relacionados con cambios en la

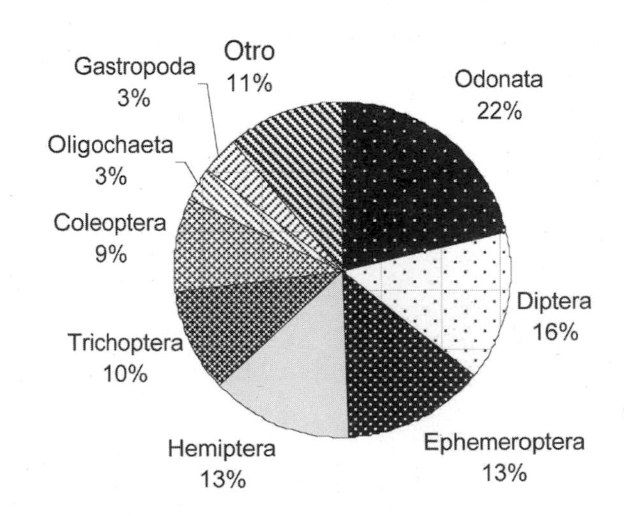

Figura 4.2. Porcentaje del total de géneros en cada orden o clase encontrados en la Cuenca del Río Caura.

Tabla 4.1. Autovalores y porcentaje de variación explicados por los tres primeros ejes de componentes principales usando el número de géneros encontrados en cada sitio geo-referenciado en el primer ACP.

	Ejes de componentes principales		
	Ejes 1	Ejes 2	Ejes 3
Autovalores	9.01	4.46	3.05
Porcentaje	41.39	20.48	14.00
Porcentaje Acumulativo	41.39	61.87	75.87
Cargas de las variables de ACP			
Odonata	0.84	-0.15	-0.25
Ephemeroptera	0.47	0.36	0.42
Hemiptera	-0.07	-0.59	0.64
Trichoptera	0.14	0.18	0.47
Coleoptera	0.08	-0.15	0.07
Diptera	-0.17	0.64	0.13
Plecoptera	0.10	0.09	0.05
Neuroptera	0.04	0.00	0.09
Lepidoptera	0.02	0.10	0.09
Crustacea	0.01	-0.06	0.01
Gastropoda	-0.02	0.06	0.23
Bivalvia	-0.04	-0.02	0.06
Oligochaeta	0.01	0.08	0.18
Hirudinea	0.03	0.01	0.04

Tabla 4.2. Autovalores y porcentaje de variación explicados por los tres primeros ejes de componentes principales usando el número de géneros y las variables fisicoquímicas para cada sitio geo-referenciado en el segundo ACP.

	Ejes de components principales		
	Ejes 1	**Ejes 2**	**Ejes 3**
Autovalores	52.39	10.45	6.71
Porcentaje	65.32	13.03	8.37
Porcentaje Acumulativo	65.30	78.33	86.70
Cargas de las variables de ACP			
Odonata	-0.01	0.71	-0.44
Ephemeroptera	-0.01	0.47	0.05
Hemiptera	-0.01	-0.17	-0.39
Trichoptera	0.05	0.10	-0.01
Coleoptera	-0.04	0.04	-0.15
Diptera	-0.05	-0.02	0.53
Plecoptera	0.00	0.10	0.02
Neuroptera	-0.02	0.03	-0.02
Lepidoptera	-0.01	0.02	0.06
Crustacea	0.00	0.00	-0.05
Gastropoda	-0.01	-0.04	-0.03
Bivalvia	0.03	-0.04	0.02
Oligochaeta	-0.01	0.03	0.00
Hirudinea	0.01	0.03	0.00
Conductividad	0.99	-0.01	0.02
Oxígeno disuelto	0.10	0.33	0.18
Turbidez	0.05	-0.33	-0.55

concentración de oxígeno disuelto y la turbidez del agua. Estos resultados son muy importantes debido a que un cambio drástico en el ciclo anual de descarga de las aguas, la deforestación y la minería podrían introducir material particulado en suspensión; así como un aumento en la conductividad del agua con consiguiente cambio dramático en las comunidades bentónicas.

La mayoría de los géneros de los macroinvertebrados bentónicos en el Río Caura se encuentran distribuidos tanto en la sección Superior como en la Baja de Río Caura. Estos grupos habitan en una gran diversidad de microhábitads con un alto grado de repetición a lo largo del río, tanto en el canal principal como en los rápidos, caños y tributarios. El ACP no detecta asociación entre la mayoría de las localidades geo-referenciadas y los componentes bentónicos de la fauna; por lo cual concluimos que la comunidad es bastante homogénea. Solamente cuatro ordenes de insectos acuáticos (Odonata, Ephemeroptera, Hemiptera and Diptera) muestran afinidad con algunas localidades.

La mayor diversidad de Odonata y Ephemeroptera se encuentra en el caño Wididikenü (Río Erebato), raudales Paují y Raudal Culebra de Agua (Río Caura), Raudal Dimoshi (Río Tawadu) y Río Icutú. Las dos zonas con alta diversidad de dípteros están presentes en el Raudal Perro (Río Erebato) y la Isla Fiaka (Río Caura). En el caso de Hemiptera, la mayor diversidad se encontró en la Laguna de La Ceiba, Río Mato y Takoto, todos en el Bajo Caura.

Los caracoles del género *Doryssa*, muy abundantes en toda la cuenca, han sido previamente reportados en el alto Orinoco y Alto Siapa (Estado Amazonas), habitando lechos rocosos con aguas ácidas (pH 5.7) (Martínez and Royero 1995). Mientras que caracoles del género *Pomacea*, los cuales se encontraron muy abundantes en el Bajo Caura, son componentes frecuentes de la fauna bentónica del Bajo Orinoco.

La composición de especies es típica de un ambiente pristino, debido a que encontramos una alta diversidad de Odonata y Trichoptera; los cuales son habitantes de ambientes libres de contaminación o perturbación (Wiggins 1927; Daigle 1991,1992). Algunos coleópteros de la familia Elmidae, los cuales no son capaces de tolerar contaminación por jabones o detergentes (Epler 1996), se encontraron habitando la cuenca. Lo mismo puede ser aplicado a algunos géneros de Ephemeroptera y Plecoptera (Hilsenhoff 1970; Edmunds et al. 1976). Los ordenes Odonata, Ephemeroptera y Trichoptera son los componentes principales de la fauna de macroinvertebrados bénticos en orden de importancia. Esto último sugiere que el Río Caura presenta una alta calidad del agua.

La mayoría de las localidades muestran índice similares de diversidad de Shannon-Wiener. Sin embargo, hay algunas localidades con altos valores de estos índices tales como: Raudal Perro (AC05), Raudal Ceijato (AC08), Raudal Paují (AC09), Zona arriba del Salto Pará (AC12), Raudal Culebra de Agua (AC14), Raudal Tajañano (BC06), Caño arriba del Raudal 5000 (BC15) y Río Takoto (BC17), indicando esto que esas localidades deben ser consideradas en las estrategias de conservación.

Los Odonata y Ephemeroptera, poseen especies relativamente frágiles y su presencia indican alta calidad ambiental de los ecosistemas (grado de pristinidad). Las áreas con la más alta diversidad de estos grupos son: caño Wididikenü (ACO6), Raudal Paují (AC09), Raudal Culebra de Agua (AC14), Raudal Dimoshi (BC05) y Río Icutú (BC08). Consideramos que éstas áreas deben ser consideradas prioritarias en el establecimiento de las estrategias de conservación de la Cuenca del Río Caura.

Finalmente, existen muy pocos trabajo sobre diversidad en los otros grandes ríos cercanos como el Orinoco y el Caroní, pero basándonos en los trabajos de Vásquez et al. (1990) y Marrero (2000), podemos concluir que la diversidad del Caura es mayor que la del Caroní y similar a la del Orinoco.

Diptera
Ephemeroptera
Trichoptera
Odonata
Hemiptera
AC05
AC11
AC08
BC06
BC11
BC07
AC03
AC12
BC02
AC10
AC04
AC14
BC08
AC06
BC05
AC09
AC01
BC13
BC12
BC15
BC17
Eje 2 (61,87%de la varianza total)
Eje 1 (41,39% de la varianza total)

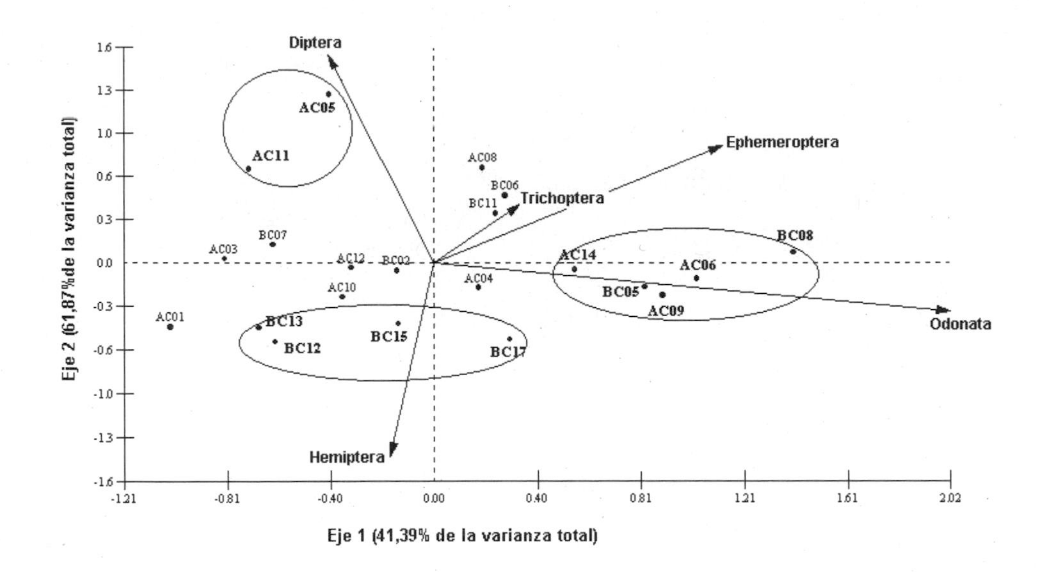

Figura 4.3. Biplot de las variables en los ejes 1 vs. 2 del primer ACP incluyendo el número de géneros por cada localidad geo-referenciada.

Odonata
Ephemeroptera
O2
Trichoptera
Plecoptera
Coleoptera
Diptera
Bivalvia
Conductabilidad
Hemiptera
Turbiedad
Eje 2 (78,33 % de la varianza total)
Eje 1 (65,30 % de la varianza total)

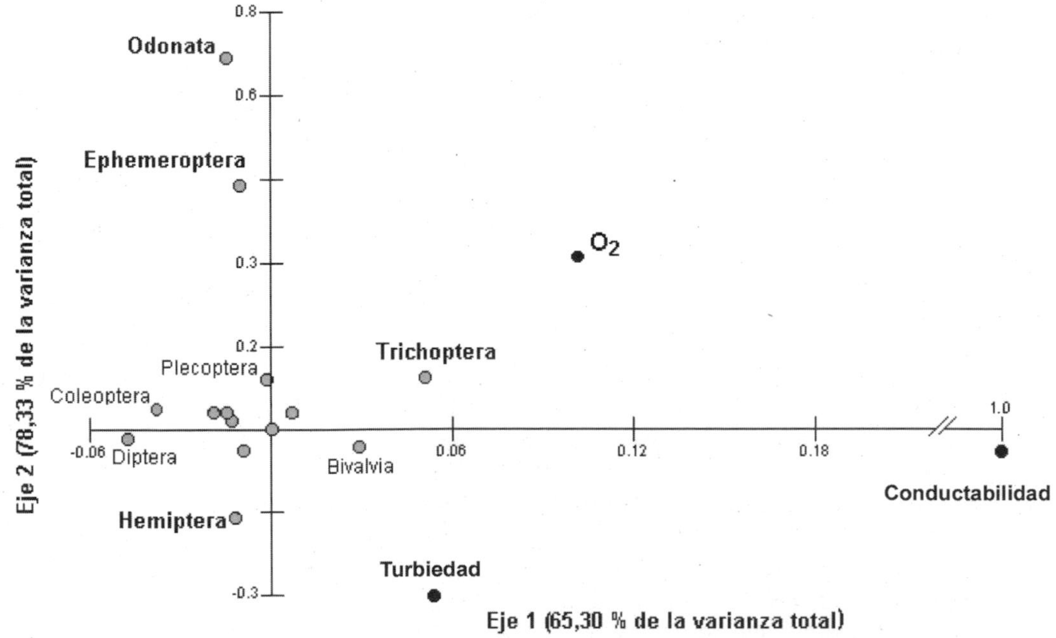

Figura 4.4. Gráfico de los ejes 1 vs. 2 del Segundo ACP utilizando el número de géneros y las variables fisicoquímicas en cada localidad geo-referenciada.

CONCLUSIONES Y RECOMENDACIONES

La composición de especies observada es típica de ambientes prístinos, ya que ellos contienen una alta diversidad de taxa normalmente asociados con ambientes libres de perturbación.

Los Ordenes Odonata, Ephemeroptera y Trichoptera son los principales componentes en orden de importancia, los cuales indican que la cuenca posee una gran calidad de agua.

Aunque ocho localidades mostraron un alto índice de diversidad de Shannon-Wiener, prioridades de conservación deben ser dadas a aquellas áreas que poseen gran diversidad de Odonata y Ephemeroptera: Caño Wididikenü, Raudal Paují, Raudal Culebra de Agua, Raudal Dimoshi y Río Icutú. Los Odonata y Ephemeroptera son indicadores ambientales importantes de ambientes pristinos.

La variación en la diversidad de la fauna béntica parece estar relacionada a cambios en el oxígeno disuelto y turbidez del agua. En consecuencia, cambios en el ciclo anual del agua, deforestación o minería generará cambios dramáticos en las comunidades de macroinvertebrados de la Cuenca.

LITERATURA CITADA

Benedetto, L. 1974. Clave para la determinación de los plecópteros suramericanos. Studies on the Neotropical Fauna 9: 141–170.

Bryce, D., y A. Hobart. 1972. The biology and identification of the larvae of the Chironomidae (Diptera). Entomologist's Gazette 23: 175–215.

Daigle, J. J. 1991. Florida Damselflies (Zygoptera): A species key to the aquatic larval stages. Department of Environmental Regulation. Florida State. Technical Series Vol. 11, No 1.

Daigle, J. J. 1992. Florida Dragonflies (Anisoptera): A species key to the aquatic larval stages. Department of Environmental Regulation. Florida State. Technical Series Vol. 12, No 1.

Edmonson, W. T. (ed.). 1959. Fresh-Water Biology. Second Edition. John Wiley & Sons, Inc., New York. 1242p.

Edmunds, G. F., Jensen, S. L., y L. Berner. 1976. The mayflies of North and Central America. University of Minnesota Press, Minneapolis. 330p.

Epler, J. H. 1995. Identification manual for the larval Chironomidae (Diptera) of Florida). Final Report DEP Contract Number WM579. Department of Environmental Protection. Florida State, Tallahassee. 320p.

Epler, J. H. 1996. Identification manual for the water beetles of Florida (Coleoptera: Cryopidae, Dytiscidae, Elmidae, Gyrinidae, Haliplidae, Hydraenidae, Hidrophilidae, Noteridae, Psephenidae, Ptilodactyllidae, Scirtidae). Final Report DEP Contract Number WM621. Department of Environmental Protection. Florida State, Tallahassee.

Flint, O. S. 1974. Studies of Neotropical Caddisflies, XVII: The Genus *Smicridea* from North and Central America (Trichoptera: Hydropsichidae). Smithsonian Contributions to Zoology 167. Smitsonian Institution Press, Washington.

Hilsenhoff, W. L. 1970. Key to genera of Wisconsin Plecoptera (stonefly) nymphs Ephemeroptera (mayfly) nymps and Trichoptera (caddisfly) larvae. Research Report 67. Departement of Natural Resources, Maddison, Wisconsin. 68p.

Hulbert, S. H., G. Rodríguez y N. W. Dos Santos (eds.) 1981. Aquatic Biota of Tropical South America. Parte 1. Arthropoda. San Diego University Press, San Diego. 323p.

Hulbert, S. H., G. Rodríguez y N. W. Dos Santos (eds.). 1981. Aquatic Biota of Tropical South America. Parte 2. Anarthropoda. San Diego University Press, San Diego. 298p.

Hynes, H.B. 1970. The ecology of running waters. University of Toronto Press, Toronto.

Johannsen, O. A. 1937. Aquatic Diptera part III. Chironomidae: Subfamilies Tanypodinae, Diamesinae, and Orthocladiinae. Cornell University Experiment Station Memoir 205. New York.

Johannsen, O. A. 1937. Aquatic Diptera part IV. Chironomidae: Subfamily Chironominae. Cornell University Experiment Station Memoir 210. New York.

Kensley, B., y M. Schotte. 1989. Guide to the marine isopod crustaceans of the Caribbean. Smithsonian Institution Press, Washington. 308p.

Kovach, W. L. 1998. MVSP, a multivariate statistical package for Windows. Version 3.0. Kovach Computing Services, Pentraeth, Wales.

Limongi, J. 1983. Estudio morfo-taxonomico de náyades en algunas especies de Odonata (Insecta) en Venezuela. Trabajo de Grado, Universidad Central de Venezuela. 100p.

Lowe-McConnell. 1975. Fish communities in tropical freshwaters. Cap. 3, Equatorial forest rivers: ecological conditions and fish communities. Logman, New York. Pp. 56–89.

Machado-Allison, A. 1992. Larval ecology of fishes of the Orinoco Basin. In: Hamlett, W. (ed.) Reproductive Biology of South American Vertebrates. Springer Verlag, 45–59.

Machado-Allison, A., B. Chernoff, C. Silvera, A. Bonilla, H. López-Rójas, C. A. Lasso, F. Provenzano, C. Marcano y D. Machado-Aranda. 1999. Inventario de los peces de la cuenca del Río Caura, Estado Bolívar, Venezuela. Acta Biol. Venez., Vol. 19 (4): 61–72.

Machado-Allison, A., B. Chernoff, F. Provenzano, P. Willink, A. Marcano, P. Petry y B. Sidlauskas. 2003. Inventario, Abundancia Relativa, Diversidad, e Importancia de los Peces de la Cuenca del Río Caura, Estado Bolívar, Venezuela. *En:* Chernoff, B., A. Machado-Allison, K. Riseng, y J. R. Montambault (eds.). A Biological Assessment of the Aquatic Ecosystems of the Caura River Basin, Bolívar State, Venezuela. RAP Bulletin of Biological Assessment,

No. 28. Conservation International, Washington, DC. Pp. 158–169.

Marrero, C. 2000. Biomonitoreo de poblaciones de insectos acuáticos bentónicos en ríos de la cuenca del Río Caroní, para detectar bioacumulación de subproductos mercuriales. Reporte Final. Project number 96001791 CONICIT.

Martínez, R., y R. Royero. 1995. Contribución al conocimiento de *Diplodon (Diplodon) granosus granosus* Brugiere (Bivalvia: Hyriidae) y *Doryssa hohenackeri happleri* Vernhout (Gastropoda: Melaniidae) en el alto Siapa (Departamento de Río Negro), Estado Amazonas, Venezuela. Acta Biol. Venez., 16: 79–84.

McCafferty, W. P. 1981. Aquatic Entomology. Jones and Bartlett Publishers, Boston. 448p.

Merrit, R. W., y K. W. Cummins. 1978. An introduction to the aquatic insects of North America. Kendall-Hunt Publishing Co., Dubuque, Iowa. 441p.

Milligan, M. R. 1997. Identification manual for the aquatic Oligochaeta of Florida. Volume 1. Freshwater Oligochaetes. Final Report DEP Contract Number WM550. Department of Environmental Protection. Florida State, Tallahassee.

Needham, J. G., y M. J. Westfall. 1955. A manual of the Dragonflies of North America, including the Grater Antilles and provinces of Mexican border. University of California Press, Berkeley.

Peters, W. L. 1971. A revision of the Leptophlebiidae of the West Indies (Ephemeroptera). Smithsonian Contributions to Zoology 62. Smithsonian Institution Press, Washington.

Richardson, H. 1905. Isopods of North America. Bulletin No. 54. United States National Museum, Washington. 727p.

Stehr, F. W. (ed.) 1987. Immature Insects. Vol. 1. Kendall-Hunt Publishing Co., Dubuque, Iowa. 754p.

Van Deer Kuyp. 1950. Mosquitoes of the Netherlands Antilles and their hygienic importance. Studies on the Fauna of Curacao and other Caribbean Islands, 23: 37–114.

Vásquez, E., L. Sánchez, L. E. Pérez y L. Blanco. 1990. Estudios hidrobiológicos y piscicultura en algunos cuerpos de agua (ríos, lagunas y embalses) en la cuenca baja del Río Orinoco. *En*: Weibezahn, F. H., H. Alvarez and W. Lewis (eds.). El Río Orinoco como ecosistema. Resultatos del Simposio: Ecosistema Orinoco: conocimiento actual y necesidades de estudios futuros. XXXVI Convención anual AsoVAC Caracas. 430p.

Wallace, J. B., y W. Merrit. 1980. Filter-feeding ecology of aquatic insects. Annual Review of Entomology 25: 103–132.

Waltz, R. D., y W. P. McCafferty. 1979. Freshwater springtails (Hexapoda, Collembola) of North America. Purdue University Agricultural Experiment Station Research Bulletin 960. West Lafayette, Indiana.

Wiggins, G. B. 1927. Larvae of the North American caddisfly genera (Trichoptera). University of Toronto Press, Toronto. 401p.

Capítulo 5

Inventario de los Crustáceos Decápodos de la Cuenca del Río Caura, Estado Bolívar, Venezuela: Riqueza de Especies, Hábitat, Aspectos Zoogeográficos e Implicaciones de Conservación

Célio Magalhães y Guido Pereira

RESUMEN

Un número total de diez especies de crustáceos decápodos se encontraron en el área inspeccionada del Río Caura; seis especies camarones palaemónidos, una especie de cangrejo Pseudothelphusidae y tres especies de cangrejos Trichodactylidae. La región río arriba del Salto Pará (Caura Superior) mostró baja riqueza, con cinco especies, mientras que se encontraron ocho especies en la región río abajo del Salto Pará. De la fauna de decápodos de la Cuenca del Río Caura hay posiblemente una encontrada en el estudio, especie endémica, el camarón no descrito *Pseudopalaemon* sp., mientras las otras especies encontrados se conocen para la región de los Llanos o la Cuenca del Amazonas. El número de especies de decápodos y géneros encontrados en el Río Caura representa una muestra típica de la diversidad y abundancia de sistemas de ríos interiores de las Guayanas y regiones de la Amazonía. El camarón *Macrobrachium brasiliense* es la especie más frecuente y abundante encontrada en este estudio y puede ser considerada la especie más típica del sistema. Se colectaron además dos especies no descritas de camarones palaemónidos. La abundancia de decápodos fue de baja a moderada, probablemente como un reflejo de los problemas de muestreo y la condición general oligotrófica del medio ambiente. La estructura de la comunidad de decápodos así como la repartición de hábitat y distribución, nos permiten concluir que representa una comunidad saludable y estable siempre y cuando las condiciones prístinas de la región sean preservadas.

INTRODUCCIÓN

Los decápodos de agua dulce venezolanos se conocen bastante bien debido a las numerosas contribuciones de Rodríguez (1980, 1982a, b, 1992, y otros) y Pereira (1985, 1986, 1991). Hay sin embargo, pocos estudios faunísticos acerca de este grupo para una región particular o cuenca del hidrográfica del país; de estos estudios se cuenta hasta ahora para la península de Paria (López y Pereira 1994) y el Delta del Río Orinoco (López y Pereira 1996, 1998).

La composición de la fauna de decápodos del Río Caura se conoce de colecciones esporádicas. Tales colecciones revelan que existen cinco especies cohabitando en esta cuenca: el camarón *Macrobrachium brasiliense* (Heller), *Palaemonetes carteri* Gordon, y una especie no descrita de *Macrobrachium* (ver Rodríguez 1982b), el cangrejo Trichodactylidae *Valdivia serrata* (Rodríguez 1992) y una especie de cangrejo Pseudothelphusidae, *Fredius stenolobus* que Rodríguez y Suárez (1994) describieron para la cuenca. Se espera que este número incremente en la medida que la región sea mejor explorada. La expedición de AquaRAP para realizar una evaluación biológica del medio Río Caura, efectuada entre noviembre-diciembre, 2000 ofreció la oportunidad para tal exploración. En este capítulo, informamos nuestros resultados sobre la riqueza de especies, hábitat y distribución longitudinal de la fauna de decápodos en el Río Caura. También se comentan implicaciones de conservación acerca de este grupo particular.

MÉTODOS

La fauna del decápodos del Río Caura se evaluó basados en el material colectado en 55 estaciones en 31 puntos georeferenciados entre las latitudes 05º29,563'N a 07º11,890'N (AquaRAP Map), siguiendo el protocolo de muestreo de AquaRAP (Chernoff y Willink 2000). Las colecciones se hicieron entre el 25 noviembre al 10 de diciembre, del 2000. El área explorada fue arbitrariamente dividida en dos grandes regiones separadas por la presencia de una gran catarata, el Salto Pará. La zona río arriba del Salto Pará (Caura Superior) incluye el Río Caura, Río Erebato, Río Kakada y Río Yuruani, y se extiende desde el Salto Pará hasta el Raudal Cejiato (Río Caura) y Río Kakada. La región río abajo del Salto Pará (Bajo Caura) comprende el Río Caura, Río Nichare, Río Tawadu, Río Takoto y Río Mato.

Intentamos tomar muestras de decápodos en todos los hábitats georeferenciados. Estos hábitats consistieron en ríos, quebradas de bosque, rápidos, lagunas y lagos marginales. En cada hábitat, se realizaron colectas en tantos microhábitats como fuese posible identificar, por ejemplo vegetación marginal, piedras, en Podostemonaceae (vegetación arraigada acuática), playa arenosa, hojarasca sumergida y restos de troncos sumergidos. Como herramientas de colección se emplearon redes de mano, redes de arrastre de 2 y 5 m de largo, trampas y las manos desnudas. Algunos impedimentos del muestreo son el tiempo de colección diurno, mientras los crustáceos tienen hábitats nocturnos, el tiempo de colección corto, dificultades empleando redes en sustratos rocosos y leñosos. El tiempo de muestreo fue de 1 a 3 horas. El estudio es estrictamente cualitativo y ningún esfuerzo fue hecho para estandarizar los esfuerzos. El análisis de composición y similaridad entre subregiones se realizó con el análisis de agrupación UPGMA empleando el Coeficiente de Similitud de Jaccard aplicado a los datos de presencia/ausencia (Ludwig y Reynolds 1988). Seis subregiones fueron reconocidas: Erebato (Ríos Erebatoy Kakada); Caura (Río Caura arriba del Salto Pará); Playón (Río Caura y riachuelos de bosque en el área del Salto Pará); Nichare (Río Nichare, Río Tawadu y Río Icutu); Cinco Mil (Río Caura y Río Takoto); y Caumato (Río Caura y Río Mato por debajo del Raudal Cinco Mil). También, un Análisis de Componentes Principales (ACP) fue realizado de manera de resumir la información en pocos componentes y para determinar las variables que contribuyen con la mayor varianza entre localidades. Los datos consisten en abundancia de la especie y datos fisicoquímicos (oxígeno disuelto, turbidez, conductibidad, tipo del fondo, PH y temperatura) como las variables, para cada localidad muestreada. Se empleó el programa MVSP. 3.1 para computador personal (Kovach 1998).

Los ejemplares se preservaron en Etanol 70%, las identificaciones se realizaron en el laboratorio empleando las descripciones de Rodríguez (1980, 1982a, b, 1992); Rodríguez y Suárez (1994), Pereira (1986) y Magalhães y Türkay (1996a, b). Los ejemplares se depositaron en las colecciones de Crustáceos del Museo de Biología de la Universidad Central de Venezuela (Caracas - Venezuela) y el Instituto Nacional de Pesquisas Amazônia (Manaus - Brasil).

RESULTADOS Y DISCUSIÓN

Riqueza de especies y distribución longitudinal

La fauna de decápodos colectada por la expedición de AquaRAP consistió en diez especies de camarones y cangrejos, incluidas en tres familias y siete géneros (Tabla 5.1, Apéndice 11). Todos los camarones pertenecen a la familia Palaemonidae, mientras que una especie de cangrejo pertenece a la familia Pseudothelphusidae y tres especies a Trichodactylidae. La composición específica de la fauna de decápodos varió entre las regiones del Alto y Bajo Caura. Por encima del Salto Pará se encontraron sólo cinco especies: los camarones, *Macrobrachium brasiliense* y *Macrobrachium.* sp. 1, y los cangrejos *Popiana dentata, Valdivia serrata* y *Fredius stenolobus*. En la región abajo del Salto Pará, se encontró un número mayor de especies de camarones (*Macrobrachium amazonicum, M. brasiliense, Palaemonetes carteri, Palaemonetes mercedae* y *Pseudopalaemon* n. sp.) mientras sólo una especie adicional de cangrejo Trichodactylidae (*Forsteria venezuelensis*). En la región del Bajo Caura, la riqueza de las especies incrementa hacia las tierras más bajas de la cuenca. Con la excepción de *Macrobrachium.* sp. 1, todas las especies que ocurren sobre el Alto Caura, también están presentes río abajo del salto, además de otras especies típicas de las tierras bajas. Éste es el caso de los camarones *Macrobrachium amazonicum* y *Palaemonetes carteri* y el cangrejo *Forsteria venezuelensis.*

El camarón *Macrobrachium brasiliense* es el taxon más frecuente en las colecciones. Un estimado preciso sobre su abundancia no se puede ofrecerse debido a que los esfuerzos de captura no son homogéneos. Sin embargo, parece ser que en general la abundancia de todos los taxa es normalmente baja, salvo algunas estaciones dónde el camarón *M. brasiliense* aparece con una abundancia moderada. La densidad de los cangrejos es normalmente baja indiferentemente de la región. Sin embargo, esta evaluación es un compromiso debido a las dificultades en colectar estos animales de hábitos nocturnos.

La diversidad de la fauna de decápodos en la cuenca del Caura puede ser considerada como moderada. Aunque no hay muchos estudios sistemáticos para los afluentes mayores del Orinoco y la cuenca del Amazonas, el número de especies de decápodos y géneros encontrados en el Río Caura representan una muestra típica representativa de la diversidad y abundancia de los sistemas de ríos interiores de las Guayanas y regiones de la Amazonía. Estudios similares llevados a cabo en los sistemas fluviales Tahuamanu/Manuripe (afluentes secundarios del Río Madeira) en Bolivia (Magalhães 1999), y en el Río Pastaza en Ecuador/Perú (AquaRAP 1999), también resultaron en diez especies para cada región, aunque con una composición taxonómica

Tabla 5.1. Lista de especies de crustáceos decápodos colectados por la expedición AquaRAP en la Cuenca del Río Caura (Estado Bolívar, Venezuela), Noviembre–Diciembre 2000, de acuerdo a las regiones examinadas.

Taxa	Caura Superior	Bajo Caura	Registradas previamente para la cuenca	Nuevos registros para la cuenca
Palaemonidae (camarones)				
Macrobrachium amazonicum		X		X
Macrobrachium brasiliense	X	X	X	
Macrobrachium sp. 1	X			X
Palaemonetes carteri		X	X	
Palaemonetes mercedae		X		X
Pseudopalaemon n. sp.		X		X
Pseudothelphusidae (cangrejos)				
Fredius stenolobus	X	X	X	
Trichodactylidae (cangrejos)				
Forsteria venezuelensis		X		X
Poppiana dentata	X	X		X
Valdivia serrata	X	X	X	

diferente. Sin embargo la curva de acumulación de especies (Figure 5.1) sugiere que el número total no debe ser mucho mayor a las diez especies encontradas, puesto que la curva alcanza la asíntota después del día 13 de muestreo. La curva también indica que la riqueza de especie es más alta en la parte baja de la cuenca: el gráfico despliega una asíntota entre el tercero y séptimo día de muestreo que corresponde con el período de estudio en la parte superior del Caura; después de ese período, la curva aumenta en la medida que la expedición procedía río abajo.

La región superior del Caura muestra una riqueza de especies más baja que la región aguas abajo del salto. La zona justo bajo el salto, (El Playón), también tiene un número bajo de especies, y posee las mismas especies que río arriba del salto; ambas subregiones son muy similares a la subregion Erebato (Tabla 5.2; Figura. 5.2). Estas subregiones tienen un patrón similar de disponibilidad del hábitat para los decápodos, esto es, río, rápidos y quebradas del bosque

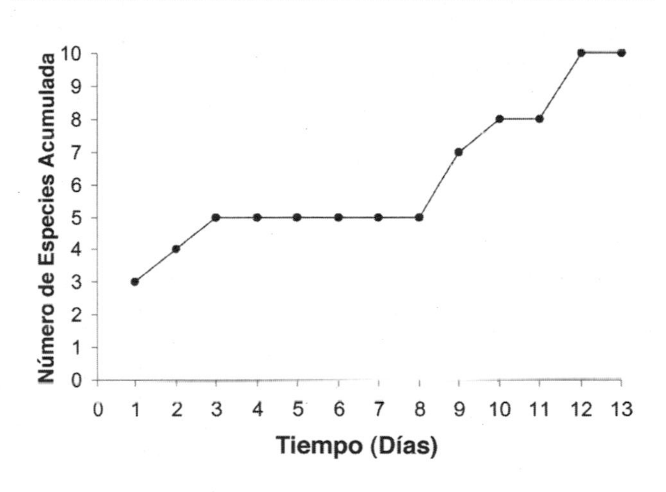

Figura 5.1. Curva de acumulación de especies de crustáceos decápodos colectados en la Cuenca del Río Caura (Estado Bolívar, Venezuela) durante la expedición AquaRAP, Noviembre–Diciembre 2000.

Tabla 5.2. Coeficientes de Similaridad Binarios de Jaccard de las regiones dentro de la Cuenca del Río Caura para la fauna de crustáceos decápodos colectados durante la expedición AquaRAP, Noviembre–Diciembre 2000.

	Caura	Erebato	Playón	Nichare	Cinco Mil	Caumato
Caura	1.000					
Erebato	0.800	1.000				
Playón	1.000	0.800	1.000			
Nichare	0.571	0.500	0.571	1.000		
Cinco Mil	0.667	0.571	0.667	0.625	1.000	
Caumato	0.286	0.250	0.286	0.333	0.571	1.000

que podrían explicar el patrón de similaridad de especies encontrado.

La subregión comprendida entre la subcuenca del Río Nichare y Río Caura justo después de la boca del río Nichare se caracteriza por la presencia de especies de las tierras bajas. Las similaridades en la composición de especies entre estas subregiones incrementa en la medida que aparecen más especies progresando río abajo. La subregión del Río Nichare presenta el número más alto de especies (siete) y fue el único lugar en el cual la nueva especie de camarón palaemónido, *Pseudopalaemon* n. sp., fue encontrada. Además, esta subregión muestra una composición faunal muy similar a la mayoría de las otras subregiones (Fig. 5.2) lo cual la hace un área potencial donde dirigir los esfuerzos de conservación a futuro.

El conocimiento de la composición de especies de decápodos de la Cuenca del Río Caura se ha incrementado grandemente con los resultados de la expedición de AquaRAP ahora totalizando once especies. Además de las cinco especies previamente conocidas (tres camarones palaemónidos, un cangrejo pseudotelfúsido y un cangrejo tricodáctilo) eran conocidos (Rodríguez 1982b; Rodríguez y Suárez 1994). Este número ha incrementado en más del doble con la aparición de cuatro especies adicionales de camarones y dos de cangrejos nuevos para la cuenca (ver Tabla 5.1) *Macrobrachium* sp. Una especie no descrita reportado por Rodríguez (1982) para el Río Tauca (Tributario del Bajo Caura) no fue colectada en esta expedición.

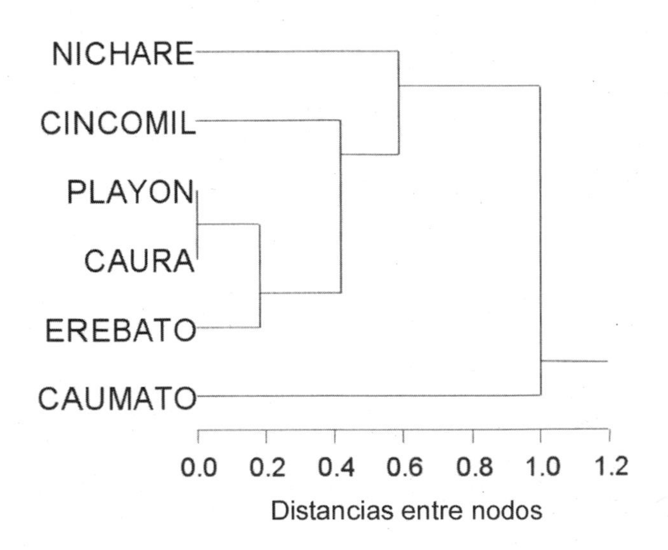

Figura 5.2. Análisis de Agrupamiento (UPGMA) de las regiones dentro de la Cuenca del Río Caura concerniente a la fauna de crustáceos decápodos colectados durante la expedición AquaRAP, Noviembre–Diciembre, 2000. (Distancias Euclideanas basadas en Coeficientes de Similaridad Binaria de Jaccard; Método de enlace: media; número de observaciones: 10).

Hábitat

Los hábitats principales explorados fueron los riachelos y caños en el bosque, el cauce del río principal, rápidos e islas. En la región del Caura Superior, los decápodos se encontraron principalmente en los pequeños cauces de bosque preferencialmente en fondos arenoso y rocoso, pero también se encontraron en los rápidos e islas rocosas a lo largo del cauce principal de los ríos. Los cangrejos pseudotelfúsidos normalmente se encontraron asociados al substrato rocoso en el cauce principal o en los arroyos del bosque; en este último, también pueden habitar en los agujeros presentes en los restos troncos de árboles sumergidos. Los juveniles de camarones y de cangrejos Trichodactylidae normalmente se encontraron entre la hojarasca sumergida y restos de madera mientras los adultos se encuentran principalmente en microhábitats crípticos tales como los agujeros en los troncos del árbol sumergidos. La Tabla 5.3 resume la presencia de las especies según el hábitat, microhábitat y tipo del fondo.

Los distintos hábitats ocupados por los crustáceos son en general los mismos en toda la región. La ausencia de especies en algunas estaciones podría ser bien debida a limitaciones del muestreo más que cualquier otra razón. *Fredius stenolobus* no se colectó en las estaciones del Río Kakada a pesar de la presencia de substratos rocoso en. Sin embargo, poco tiempo y las colectas diurnas pueden haber limitado la colecta de esta especie en esas estaciones.

Las preferencias del hábitat en la región del Bajo Caura son similares a las encontradas en el Caura Superior. Sin embargo, ambientes tales como las playas arenosas y arenoso-fangoso, lagunas rocosas con vegetación marginal, lagos internos y marginales, se encontraron más frecuentemente en esta región, particularmente en la subregion río abajo del Río Nichare. Estos hábitats estaban normalmente habitados por las especies de tierras bajas, por ejemplo: *Macrobrachium amazonicum* presente en las playas arenoso-fangoso, lagos y lagunas rocosas con vegetación marginal, mientras *Palaemonetes carteri* aparece en una laguna. En general, la abundancia fue baja, pero moderada para *Macrobrachium brasiliense* y posiblemente para *Poppiana dentata* que se encuentra en los arroyos del bosque con fondo arenoso y muchos restos vegetales.

Análisis

El Análisis de Componentes Principales (ACP) muestra que 92.10% de la variación total son explicados por los tres primeros componentes principales (Tabla 5.4) Debido a la alta varianza descrita por algunos de los componentes principales, el análisis provee una rasonable descripción de la correlación estructural entre las variables bióticas de las regiones del Caura Superior y Bajo Caura. El Componente 1 explica 69.9% de variación total; sin embargo, este componente se caracteriza por poseer una sola variable con un autovalor alto: *Macrobrachium brasiliense* con 0.99, mientras

Tabla 5.3. Distribución de los crustáceos decápodos de acuerdo con los hábitats acuáticos, microhábitats y tipos de fondo obtenidos en la Cuenca del Río Caura (Estado Bolívar, Venezuela) durante la expedición AquaRAP, Noviembre–Diciembre 2000.

	Hábitat					Microhábitat						Fondo				
	Arroyo bosque	Laguna	Pozo	Rápidos	Río	Vegetación sobrepuesta	Podeoste-monaceae	Rocas	Playa arenosa	Troncos y ramas sumergidos	Restos leñosos	Arcilla	Grava	fangoso	Rocoso	Arenoso
Palaemonidae (camarones)																
Macrobrachium amazonicum		x	x		x	x			x	x		x			x	x
Macrobrachium brasiliense	x	x	x		x		x	x	x	x	x	x	x	x	x	x
Macrobrachium sp. 1	x									x			x	x	x	x
Palaemonetes carteri		x								x		x				
Palaemonetes mercedae			x		x	x					x	x			x	
Pseudopalaemon n. sp.	x				x					x	x		x	x		x
Pseudothelphusidae (cangrejos)																
Fredius stenolobus	x			x	x			x		x	x		x		x	x
Trichodactylidae (cangrejos)																
Forsteria venezuelensis	x				x				x	x	x		x	x		x
Poppiana dentata	x	x			x					x	x	x		x		x
Valdivia serrata	x		x		x					x	x	x	x	x	x	x

el resto de las variables poseen valores de 0% o cercanos a 0%. Esto puede explicarse porque *M. brasiliense* es la especie más común y dispersa en el sistema, está presente en casi todas las estaciones y como tal puede ser considerada la más típica y representativa del sistema del Caura.

El Componente 2 explica 12.67% de la variación total; se caracteriza por el hecho que algunas especies tienen valores altos y positivos de varianza, tal y como *Macrobrachium amazonicum, M. brasiliense, Palaemonetes mercedae* y *Pseudopalaemon* n. sp., mientras otro grupo de especies muestra valores de variación que van de 0% o cerca de 0% positivos y negativos, tales como: *Macrobrachium* sp.1, *Palaemonetes carteri, Poppiana dentata, Valdivia serrata* y *Forsteria venezuelensis*. Finalmente, de todas las variables fisicoquímicas, sólo la temperatura, muestra una carga positiva alta. Este eje podría interpretarse como el comunitario o de componente biológico. Este toma en cuenta las especies ampliamente distribuidas con abundancia relativamente alta, especies restringidas con abundancia baja y probablemente las interacciones entre ellas. Finalmente, la temperatura parece ser un factor importante que podría explicar patrones de distribución.

Tabla 5.4. Autovalores y porcentage de variación explicados por los tres primeros components usando abundancia de especies y variables fisico-químicas en cada stación de georeferencia muestreada durante la expedición AquaRAP a la Cuenca del Río Caura, Venezuela, Noviembre–Diciembre 2000.

	Ejes de Componentes Principales		
	Ejes 1	Ejes 2	Ejes 3
Autovalores	460.196	83.288	62.082
Porcentaje	69.99	12.67	9.44
Porcentaje Acumulativo	69.99	82.66	92.10
Cargas de las variables PCA			
Macrobrachium brasiliense	0.99	0.41	0.04
Macrobrachium amazonicum	-0.04	0.86	-0.39
Macrobrachium sp.	-0.01	0.00	0.00
Palaemonete carteri	0.00	0.00	0.00
Palaemonetes mercedae	0.00	0.27	-0.12
Pseudopalaemon n. sp.	0.00	0.14	-0.06
Poppiana dentata	0.08	-0.06	-0.19
Valdivia serrata	0.04	-0.03	-0.04
Forsteria venezuelensis	0.04	0.00	0.00
Tipo de Fondo	-0.02	0.03	-0.11
pH	0.01	0.08	0.17
Oxígeno Disuelto	0.01	0.06	0.27
Temperatura	-0.01	0.39	0.82
Turbidez	0.00	0.08	0.12

El Componente 3 explica 9.44% de la varianza total, posee autovalores altos y positivos para las variables fisicoquímicas (pH, oxígeno disuelto, temperatura y turbidez) y negativo para el tipo del fondo. Siguiendo un razonamiento similar al empleado en el componente anterior, éste eje podría ser interpretado como el componente abiótico que refleja la heterogeneidad ambiental a lo largo del sistema superior y bajo del Caura. Cuando las localidades son graficadas sobre los ejes 2 y 3, se representan de una manera general los componentes bióticos y abióticos del sistema (Figura 5.3). Puede verse que las variables fisicoquímicas (excepto el tipo de fondo) descansan sobre el origen en el cuadrante derecho, con la temperatura mostrando el valor más alto. Tres especies localizadas en el cuadrante derecho inferior muestran progresivamente valores altos desde *Pseudopalaemon* n. sp., luego *Palaemonetes mercedae* y finalmente *Macrobrachium amazonicum* con el valor más alto. Entonces, junto con *Macrobrachium brasiliense* estas tres especies juegan un papel importante en este ecosistema. *Macrobrachium amazonicum* parece tener alguna relación con la temperatura y turbiedad, lo cual se ha observado intuitivamente en el campo.

En conclusión el análisis de PCA sugiere a *Macrobrachium brasiliense* como la especie más típica el sistema estudiado y que los componentes 2 y 3 pueden reflejar características del comunitarias del sistema del Río Caura.

Aspectos Zoogeográficos

La fauna del decápodos del Río Caura consiste de elementos amazónicos y del Orinoco. Salvo posiblemente *Pseudopalaemon* n. sp., ninguna especie de decápodo es endémica a la cuenca. *Pseudopalaemon* n. sp. podría ser endémico de la cuenca del Río Caura, pero esta declaración es prematura puesto que los camarones palaemónidos usualmente tienen una distribución más amplia. La presencia de esta especie en otros ríos de la Región de Guayana puede eventualemente ser registrada en el futuro con colectas más intensivas. El camarón *Macrobrachium brasiliense* tiene una distribución muy amplia a lo largo de las cuencas de ríos de América del Sur (Holthuis 1952; Rodríguez 1981; Coelho y Ramos-Porto 1985) y es una de las pocas especies de palaemónidos encontrado en altitudes sobre los 300 m. La otra especie, el *Macrobrachium* sp. 1 que no está descrita (G. Pereira, en preparación), también parece ser una especie de ríos de tierras altas, puesto que se ha colectado principalmente en los cauces superiores de la cuenca del Río Caroní. Su presencia en la cuenca del Río Caura sugiere que su distribución es más amplia en los ríos del Escudo de Guyana. Los camarones *Macrobrachium amazonicum* y *Palaemonetes carteri* son especies de distribución amplia. El primero, es una especie muy común en las tierras bajas del Orinoco, Amazonas, Paraguay y Paraná (Holthuis 1952; Rodríguez 1980, 1981, 1982b; Coelho y Ramos-Porto 1985; López y Pereira 1996, 1998; Ramos-Porto y Coelho 1998), pero hasta ahora no se conocía para el Río Caura. *Palaemonetes carteri* también está presente en la Guyana, Surinam, Guayana francesa, otras áreas de la Amazonía y cuenca del Orinoco (Holthuis

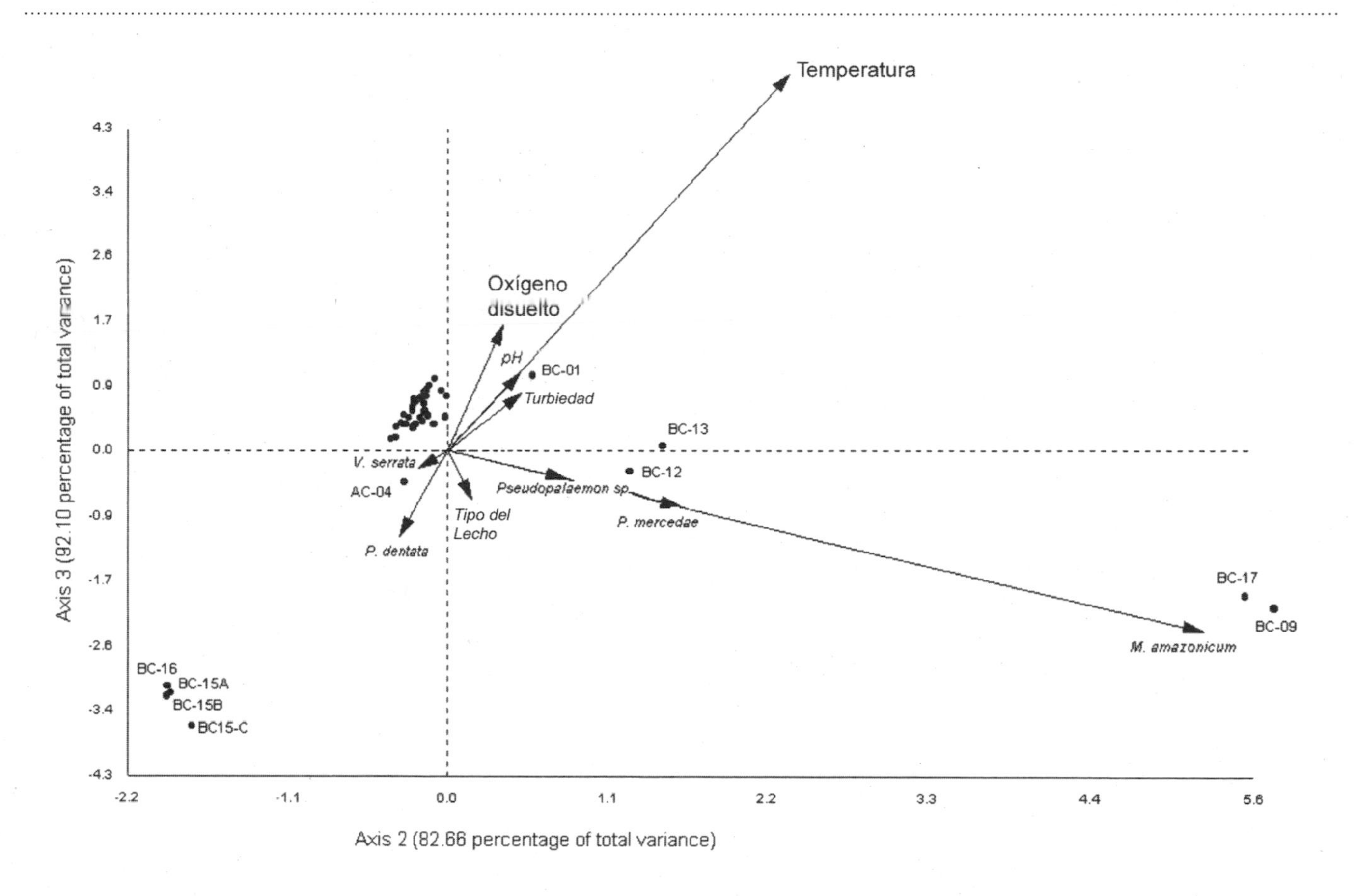

Figura 5.3. Biplot de variables en los ejes 2 vs. 3 del ACP usando abundancias de especies y variables fisico-químicas para cada estación georeferenciada muestreada durante la expedición AquaRAP a la Cuenca del Río Caura, Venezuela, Noviembre–Diciembre 2000.

1952; Rodríguez 1981; Coelho y Ramos-Porto 1985; López y Pereira 1996, 1998; Ramos-Porto y Coelho 1998), incluso ya había sido reportada para el cauce bajo del Río Caura (Rodríguez 1982b).

Tres especies de cangrejos encontradas en el Río Caura tienen distribución Orinoco-Amazónica. *Valdivia serrata* es un cangrejo tricodctílido con una distribución amplia en ambas cuencas y su presencia en el bajo Río Caura ha sido reportada previamente (Rodríguez 1992). *Poppiana dentata* era conocido en un área estrecha cerca de la costa norte de América del Sur, desde Venezuela a la Guayana francesa (Rodríguez 1992) y *Fredius stenolobus* restringido a la cuenca del Río Caura (Rodríguez y Suárez 1994). Sin embargo, colecciones recientes indican que las últimas dos especies también se encuentran en otros ríos de la guayana venezolana y la cuenca amazónica (Jennifer Meri y C. Magalhães, datos inéditos). La presencia de *Poppiana dentata* en el Río Caura, particularmente en las áreas superiores del Caura, sugiere que su distribución también abarca otros afluentes del sur del Río Orinoco.

El cangrejo tricodáctilido *Forsteria venezuelensis* es endémico de la cuenca del Río Orinoco (Rodríguez 1980, 1992), pero su presencia en el Río Caura se verificó primera vez por durante la expedición de AquaRAP. Esta parece ser una especie típica de las tierras bajas, lo cual explica su ausencia en el área superior del Caura.

IMPLICACIONES PARA CONSERVACIÓN

Los hábitats ocupados por estos crustáceos a lo largo del área inspeccionada estaban en condiciones sanas, la abundancia normalmente baja de especimenes puede ser debido a la situación oligotrófica global de estas áreas. Las posibles amenazas a la fauna de decápodos incluyen deforestación de la vegetación ribereña y colmatación por sedimentos de los cauces del bosque. Sin embargo, esto no se notó en el área evaluada y por lo tanto no constituye en la actualidad una amenaza a la comunidad de decápodos. Aunque las especies de crustáceos no se explotan comercialmente en la región, ellos juegan un papel importante en los procesos ecológicos del ambiente y son sensitivos a los impactos causados por deforestación

La subregión del Río Nichare es un área importante para la conservación de decápodos en el Sistema del Caura. Además del alto número de especies, hay también la posibilidad de albergar una especie endémica de camarón palaemónico.La composición faunística de la subregión del Río Nichare muy similar a la de otras subregiones (Fig. 5.2). La conservación de esta área podría eventualmente contribuir a la preservación de la mayoría de las especies de camarones y cangrejos de la Cuenca del Río Caura.

Recomendaciones de conservación e investigación

Las recomendaciones de conservación para el grupo sigue las recomendaciones generales para ésta región. Recomendaciones específicas acerca de los crustáceos son:

- Desarrollar un censo detallado de los afluentes pequeños ("quebradas" o "riachuelos") de manera de precisar aquellos que podrían actuar o bien como refugios para la reproducción o reservorios de las poblaciones de camarones y cangrejos; entonces se podrían proponer las medidas de conservación específicas.

- Desarrollar un inventario extenso y más detallado de todos los grupos de macro y micro crustáceos para evaluar la composición específica de la comunidad del crustáceo más precisamente a lo largo de la cuenca.

- Desarrollo de un plan piloto de censo, monitoreo e investigación básica de la comunidad de decápodos del río Nichare, como ambiente típico de la cuenca del Caura.

LITERATURE CITED

Chernoff, B., y P. W. Willink. 2000. AquaRAP sampling protocol/Protocolo de amaostragem do AquaRAP. *En*: Willink, P.W. et al (eds.). A biological assessment of the aquatic ecosystems of the Pantanal, Mato Grosso do Sul, Brasil. RAP Bulletin of Biological Assessment 18: 241–242. (Appendix 1)

Coelho, P. A., y M. Ramos-Porto. 1985. Camarões de água doce do Brasil: Distribuição geográfica. Revista brasileira de Zoologia 2(6): 405–410.

Holthuis, L. B. 1952. A general revision of the Palaemonidae (Crustacea Decapoda Natantia) of the Americas. II. The subfamily Palaemonidae. Occasional Papers, Allan Hancock Foundation 12:1–396.

Kovach, W. L. 1998. A Multi-Variate Statistical Package for Windows. (ver. 3.1). Anglesey, Wales: Kovach Computing Services.

López, B., y G. Pereira. 1994. Contribución al conocimiento de los crustaceos y moluscos de la Peninsula de Paria / Parte I: Crustacea: Decapoda. Memoria de la Sociedad de Ciencias Naturales La Salle 54(141): 51–75.

López, B., y G. Pereira. 1996. Inventario de los crustaceos decapodos de las zonas alta y media del Delta del Rio Orinoco, Venezuela. Acta Biologica Venezuelica 16(3): 45–64.

López, B., y G. Pereira. 1998. Actualización del inventario de crustáceos decápodos del Delta del Orinoco. Pp. 76–85. *En*: Sánchez, J. L. L., I. I. S. Cuadra y M. D. Martínez (eds.). El Rio Orinoco. Aprovechamiento Sustentable. Caracas, Venezuela: UCV. (Memorias de las Primeras Jornadas Venezolanas de Investigacion sobre el Río Orinoco).

Ludwig, J. A. y J. F. Reynolds. 1988. Statistical Ecology. A Primer on Methods and Computing. New York USA: John Wiley and Sons.

Magalhães, C. 1999. Diversity and abundance of decapods crustaceans in the rio Tahuamanu and rio Manuripi basins. *En*: Chernoff, B. y P. W. Willink (eds.). A biological assessment of the aquatic ecossystems of the Upper Río Orthon basin, Pando, Bolivia. Bulletin of Biological Assessment 15. Washington DC: Conservation International. Pp. 35–38, Appendix 5.

Magalhães, C., y M. Türkay. 1996a. Taxonomy of the Neotropical freshwater crab family Trichodactylidae I. The generic system with description of some new genera (Crustacea: Decapoda: Brachyura). Senckenbergiana biologica 75(1/2): 63–95.

Magalhães, C., y M. Türkay. 1996b. Taxonomy of the Neotropical freshwater crab family Trichodactylidae II. The genera *Forsteria*, *Melocarcinus*, *Sylviocarcinus*, and *Zilchiopsis* (Crustacea: Decapoda: Brachyura). Senckenbergiana biologica 75(1/2): 97–130.

Pereira, G. 1985. Freshwater shrimps from Venezuela III: *Macrobrachium quelchi* De Man and *Euryrhynchus pemoni* n. sp. (Crustacea: Decapoda: Palaemonidae). Proceedings of the Biological Society of Washington 3: 615–621.

Pereira, G. 1986. Freshwater shrimps from Venezuela I: Seven new species of Palaemoninae (Crustacea: Decapoda: Palaemonidae). Proceedings of the Biological Society of Washington 99(2): 198–213.

Pereira, G. 1991. Camarones de agua dulce de Venezuela II: Nuevas adiciones en las familias Atyidae y Palaemonidae (Crustacea, Decapoda, Caridea). Acta Biologica Venezuelica 13(1-2): 75–88.

Ramos-Porto, M., y P. A. Coelho, 1998. Malacostraca – Eucarida. Caridea (Alpheoidea excluded). *En*: Young, P. S. (ed.). Catalogue of Crustacea of Brazil. Rio de Janeiro, Museu Nacional. p. 325–350. (Série Livros n. 6)

Rodríguez, G. 1980. Crustaceos Decapodos de Venezuela. Caracas,Venezuela: IVIC.

Rodríguez, G. 1981. Decapoda. *En*: Aquatic Biota of Tropical South America, Part 1: Arthropoda. Hurlbert, S. H., G. Rodríguez y N. D. Santos (eds.). San Diego State USA: San Diego University. Pp. 41–51.

Rodríguez, G. 1982a. Les crabes d'eau douce d'Amerique. Famille des Pseudothelphusidae. Collection Faune Tropicale, 22. Paris, France: Editions Office de la Recherche Scientifique et Technique Outre-mer (ORSTOM).

Rodríguez, G. 1982b. Fresh-water shrimps (Crustacea, Decapoda, Natantia) of the Orinoco Basin and the Venezuelan Guayana. Journal of Crustacan Biology 2(3): 378–391.

Rodríguez, G. 1992. The Freshwater Crabs of America. Family Trichodactylidae and Supplement to the Family Pseudothelphusidae. Collection Faune Tropicale, 31. Paris, France: Editions Office de la Recherche Scientifique et Technique Outre-mer (ORSTOM).

Rodríguez, G., y H. Suarez. 1994. *Fredius stenolobus*, a new species of freshwater crab (Crustacea: Decapoda: Pseudothelphusidae) from the Venezuelan Guiana. Proceedings of the Biological Society of Washington 107: 132–136.

Capítulo 6

Inventario, Abundancia Relativa, Diversidad e Importancia de los Peces de la Cuenca del Río Caura, Estado Bolívar, Venezuela

Antonio Machado-Allison, Barry Chernoff, Francisco Provenzano, Philip W. Willink, Alberto Marcano, Paulo Petry, Brian Sidlauskas y Tracy Jones

RESUMEN

Durante 21 (Noviembre–Diciembre 2000) días fue investigada una región del Río Caura. 65 estaciones de colecta comprendidas entre el Raudal Cejiato (Caura) y Río Kakada (Erebato) en el Sur hasta el Raudal Cinco Mil (La Mura) en el Norte fueron realizadas. Un total de 278 especies de peces fueron identificadas para los diferentes ríos de la Cuenca del Río Caura. El Orden Characiformes, con 158 especies (56.8%) fue el más diverso. Le siguen en importancia los órdenes Siluriformes (74, 26.6%), Perciformes (27, 9.7%), Gymnotiformes (9, 3.2%), Clupeiformes (3, 1.1%), Cyprinodontiformes (2, 0.7%), Rajiformes (2, 0.7%), Beloniformes (1, 0.4%), Pleuronectiformes (1, 0.4%) y Synbranchiformes (1, 0.4%). La Familia Characidae resultó la mejor representada en este estudio, con 113 especies que representa aproximadamente 40.7% del total identificado. En este sentido nuestros resultados incrementan en 54 especies las conocidas para el Orden Characiformes y 39 especies de Characidae para la Cuenca. Para los bagres y peces eléctricos (Siluriformes y Gymnotiformes) hemos incrementado el número de especies conocidas de 49 a 72. Mientras que los Perciformes 27 son ahora confirmados en lugar de 12 previamente reportados. En total, 110 nuevos registros son dados para la cuenca. En la región superior, el Salto Pará, Raudal Cejiato y Río Kakada (Playa y caño Suajiditu) resultaron las más ricas, con algunas especies típicas de las guayanas. Las áreas por debajo del Salto Pará (El Playón), los ríos Tawadu-Nichare y Raudal 5.000 (incluyendo el río Takoto) mostraron una diversidad alta, con numerosas especies típicas de los llanos y Orinoco. Sin embargo, ambas áreas poseen sus propias características y tienen importancia taxonómica, biogeográfica y de conservación. Un gran número de especies poseen importancia económica tanto como comida u ornamental. Pocas especies son utilizadas por las comunidades indígenas.

Basadas en la riqueza, diversidad y abundancia relativa de los peces y la dependencia de las poblaciones humanas locales sobre los recursos pesqueros en ciertas áreas del Río Caura, estas deben ser preservadas como parte de un programa general de conservación. Estas áreas incluyen en la región superior al Raudal Cejiato, Río Kakada, los remansos y lagunas de inundación cercanas a Entreríos, el Raudal Suajiditu y la región superior inmediata al Salto Pará. En la región baja del Caura debemos proteger las áreas el Playón y zonas cercanas, los ríos Nichare y Tawadu, las lagunas de inundación cerca de Boca de Nichare y el Río Takoto cerca del Raudal Cinco Mil.

INTRODUCCIÓN

Venezuela, al igual que otros países americanos situados en la región tropical, todavía posee áreas prístinas extensas que incluyen geográficamente varios centenares de miles de kilómetros cuadrados, algunos de ellos compartidos por varios países (Brasil, Bolivia, Colombia, Ecuador, Guyanas, Perú y Venezuela) y que alberga la mayor diversidad y volumen de biomasa verde, fauna silvestre y agua dulce del planeta. Sin embargo, hoy día numerosas razones se han esgrimido a favor de un desarrollo o explotación acelerada de éste inmenso reservorio de materia prima alimentaria, mineral, escénica, cinegética, energética y biogenética (SISGRIL 1990; Bucher et al. 1993; Machado-Allison 1994; Chernoff et al. 1996, 1999; Machado-Allison 1999; Machado-Allison et al. 1999a, b).

La región de la Guayana Venezolana (Edo. Bolívar), no escapa a estos procesos y riesgos. La explotación de sus minerales tales como: oro, diamantes, bauxita y además de la construcción de uno de los mayores sistemas hidroeléctricos del mundo (Guri), han producido: 1) la biodegradación y destrucción de inmensas áreas vegetales en la Cuenca de los ríos Caroní y Cuyuní; 2) la contaminación mercurial de nuestros ríos, animales silvestres y el hombre; 3) el incremento de la sedimentación; y 4) la pérdida de la calidad y volumen de agua de numerosas fuentes hídricas importantes en el país (Machado-Allison 1994). En este caso, la cuenca del Río Caura ha sido considerada recientemente para el desarrollo de un proyecto ingenieril que transvasaría sus aguas hacia la cuenca del Caroní, debido a la pérdida de caudal de este último causada por actividades antrópicas en sus cabeceras.

El conocimiento de la ictiofauna de agua dulce de Venezuela y en particular de los peces de la Guayana Venezolana se ha incrementado debido a la creciente atención que se le ha prestado al tema de la biodiversidad y al resultado de las numerosas colecciones (expediciones) realizadas en la cuenca del Orinoco. Por ejemplo, Mago-Leccia (1970), reportó para Venezuela hace menos de tres décadas alrededor de 500 especies de agua dulce. Recientemente, Taphorn et al. (1997), incrementaron el número a 1.065 especies exclusivas de agua dulce y 119 especies marinas que ocasionalmente penetran aguas dulces. Sin embargo, se han dado proyecciones que indican que estos números podrían alcanzar las mil doscientas especies (Chernoff y Machado-Allison 1990; Machado-Allison et al. 1993), tomando en consideración que es ahora cuando se están estudiando áreas de alta diversidad y con muchas implicaciones biogeográficas. Además, Mago-Leccia (1978), Chernoff et al. (1991) y Machado-Allison (1993), han sugerido que sólo el 30% de nuestras especies de peces dulceacuícolas pueden ser identificadas con precisión, lo que pudiera indicar que muchas más especies podrían ser endémicas, especialmente en áreas como la Guayana y Amazonas venezolanos. Recientemente, se han publicado algunos estudios que incorporan discusiones acerca de la diversidad de especies o discuten aspectos biogeográficos de la ictiofauna de la cuenca del Río Caura (Chernoff et al. 1991; Buckup 1993; Balbas y Taphorn 1996; Lasso y Provenzano 1997; Bonilla-Rivero et al. 1999; Machado-Allison et al. 1999b).

El objetivo principal de éste trabajo es presentar un inventario de las especies de peces del río Caura, colectadas durante la expedición, dar observaciones acerca de la diversidad, abundancia relativas, relaciones biogeográficas, importancia económica y social y conservación con la finalidad de incrementar el conocimiento de la zona, requisito indispensable para la toma de decisiones que afectan los ecosistemas acuáticos.

MATERIAL Y MÉTODOS

Area de estudio

La cuenca del Río Caura abarca la región central del Escudo de la Guayana Venezolana (3°37'–7°47'N y 63°23' y 65°35'W). Sus límites son: al noreste la Cuenca del Río Aro; al este y sureste limita con la Cuenca del Río Paragua; al sur y suroeste con las cuencas de Río Uraricoera y Avarís (Brasil), al suroeste y oeste con el Río Ventuari y finalmente al norte y noroeste con el Río Orinoco y Cuchivero respectivamente. Sus principales afluentes son los ríos Sipao, Nichare, Erebato y Merewari, por su margen oeste y los ríos Tigrera, Pablo, Yuruani, Chanaro y Waña por su margen este. La superficie aproximada es de 45.336 km² (20% de la superficie total del Estado Bolívar y 5% del Territorio Nacional), lo que lo sitúa por su extensión como la tercera cuenca del país, después de las cuencas del Río Apure y Caroní (Peña y Huber 1996). Características fisiográficas, hidrológicas y climáticas de la cuenca están descritas en el Capíto 1 de éste volúmen (Machado-Allison et al. 2003). (ver también Peña y Huber 1996; Rosales y Huber 1996; Huber y Rosales 1997).

Estaciones Ictiológicas

65 colectas de peces fueron realizadas en diferentes macrohábitats desde el Río Kakada (Erebato) y Raudal Cejiato (Caura) en el sur, hasta el Raudal Cinco Mil o La Mura en el norte incluyendo las desembocaduras de los ríos Nichare y Tawadu (Tabaro) (Fig 6.1; Apéndice 7). Los criterios para escoger las estaciones en general siguen el protocolo propuesto por el programa AquaRAP (Chernoff y Willink 2000). La región fue dividida arbitrariamente en dos macroáreas identificadas como Alto Caura (AC) o Caura Superior y Bajo Caura (BC). El límite entre las dos fue un accidente geográfico importante, las cataratas que forman el Salto Pará. El Caura Superior fue a su vez subdividida en 14 ecoregiones o georeferencias (AC01-AC14) y el Bajo Caura en 17 ecoregiones o georeferencias (BC01-BC17), las cuales se encuentran descritas en Apéndice 1.

Las Estaciones correspondiente a peces están identificadas como Ictiología (ICT-xxx). Se realizaron 31 colecciones en la región superior (ICT-01 a ICT-31) y 34 colecciones en la región inferior (ICT-32 a ICT-65). Las coordenadas fueron determinadas gracias al uso de GPS calibrados. En cada estación se obtuvieron variables ecológicas y una descripción del hábitat. Estas incluyeron: descripción fisionómica y florística, márgenes del río, tipo de fondo, tipo de hábitat, clasificación de aguas (claras, negras, turbias, ácidas), velocidad, pH, transparencia, conductividad y temperatura (Ver Apéndices 1, 3, 4, 5).

Los peces fueron colectados mediante el uso de artes de pesca convencionales como chinchorros de playa y redes experimentales de ahorque, atarrayas, trampas, red de arrastre y redes de mano. Los ejemplares fueron preservado en solución de formol amortiguada al 10% y posteriormente colocados en alcohol etílico (70%). Los ejemplares fueron enviados

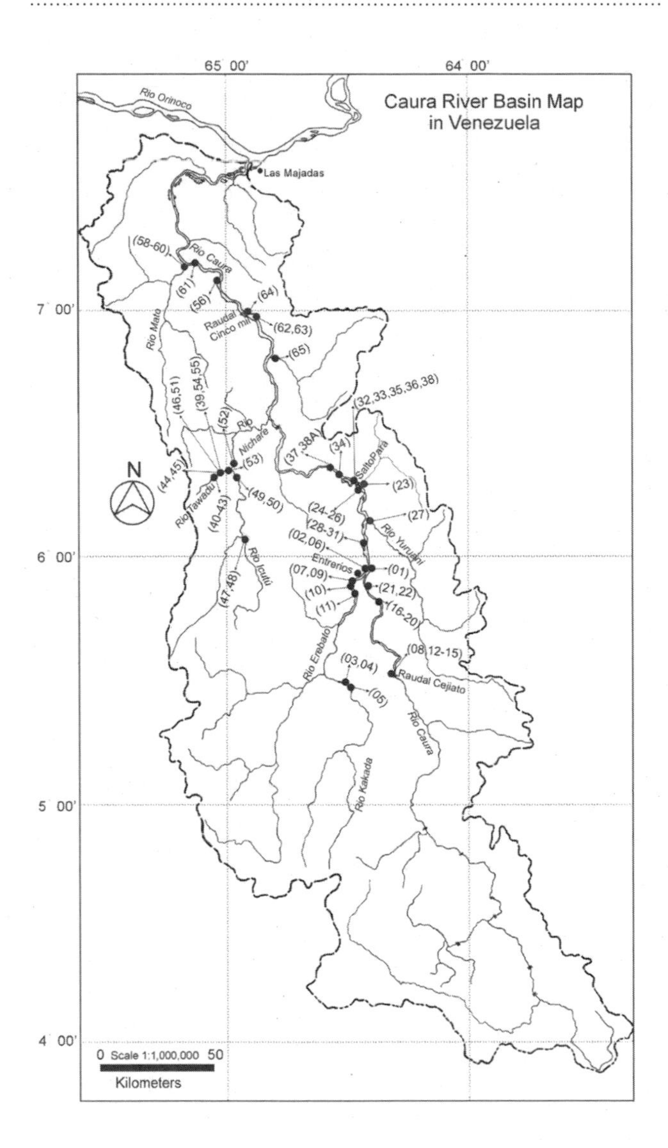

Figura 6.1. Mapa de la Cuenca del Río Caura, Estado Bolívar, Venezuela. Los círculos oscuros muestran las localidades muestreadas para peces. Los números representan las colecciones de peces (ICT) dadas en el Apéndice 7, en conjunto con los números de georeferencia y las coordenadas exactas.

al Field Museum (Chicago, USA), donde fueron separados e identificados. Posteriormente las muestras fueron divididas y distribuidas a los museos: Museo de Biología (UCV) en Caracas, Field Museum (Chicago, USA), Museu de Zoologia, Universidade de São Paulo (São Paulo, Brasil) y Museo de Historia Natural, Universidad Mayor de San Marcos (Lima, Perú). Algunos ejemplares de gran tamaño fueron preparados para la obtención de esqueletos y estudiuos anatómicos y depositados en el Museo de Biología de la UCV.

La identificación de los ejemplares fue realizada con sumo cuidado pero en una forma rápida. Trabajos generales como: Eigenmann (1918–1929), Eigenmann y Myers (1929), Gery (1977), fueron utilizados, pero preferencia se le dió a trabajos de revisión tales como: Buckup (1993); Machado-Allison y Fink (1996); Lasso y Machado-Allison (2000); Machado-Allison et al. (1993a); Mago-Leccia (1994); Nijsen y Isbruker (1980); Vari (1992,1995) y descripciones originales tales como Chernoff y Machado-Allison (1999); Bonilla-Rivero et al. (1999); Lasso y Provenzano (1997) y muchos otros. Las especies fueron identificadas en lo posible hasta la categoría más inferior (usualmente género y especie). Sin embargo, en algunos casos, la identificación fue imposible. Escogimos un alcance conservativo y no incluimos todos los taxa que colectamos, eliminando de nuestros análisis, aquellos taxa cuyas identificaciones fueran ambiguas o desconocidas para nuestro inventario. Algunas especies tal como *Ancistrus* sp. A, poseen suficientes caracteres que los separan de sus congéneres previamente descritos y reportados en la literatura. Estas formas son potenciales nuevas especies aún no descritas.

RESULTADOS Y DISCUSIÓN

Diversidad y Distribución: general

Un total de 19.266 ejemplares pertenecientes a 278 especies fueron capturados e identificadas en la expedición. Trabajos previos Balbás y Taphorn (1996) y Machado-Allison et al. (1999a,) reportaron 130 y 191 especies respectivamente, incluyendo áreas cercanas al Orinoco (Bajo Caura), no colectadas en éste estudio. Nuestros resultados, incrementan en más de 30% las especies previamente conocidas. Esto demuestra la importancia de nuestra expedición debido a que el área de estudio sólo representa una subsección modesta de la totalidad de la cuenca. El número de especies obtenidas puede ser colocado en un contexto más general coparandolo con inventarios publicados como resultado de estudios en cuencas adyancentes (Tabla 6.1).

Como ha sido establecido en otros sistemas estudiados (Río Tahuamanu, Chernoff et al. 1999), es difícil establecer con cierta certeza el grado de endemismo del sistema del Caura. Nuestra carencia de conocimientos sobre la distribución actual de las especies en la mayoría de los sistema acuáticos en América del Sur y en particular aquellos del Escudo de Guayana, hace esta tarea muy difícil. Sin embargo, podemos indicar que estamos en la presencia de un sistema con una alta riqueza y variedad de especies, con un aporte excepcional de nuevos registros y con altas posibilidades de describir nuevas especies. Usando un alcance conservador, documentamos un total de 110 especies no previamente reportadas para el Río Caura (Apéndice 8). Además, tenemos certeza de la existencia de al menos 10 especies nuevas incluidas en los géneros: *Apareiodon, Aphyocharax, Astyanax, Bryconops, Harttia, Imparfinis, Moenkhausia* y *Paravandellia* las cuales se encuentran en proceso de descripción (p.e Chernoff et al. 2003).

No tenemos duda en indicar que es muy posible que un estudio más detallado podrá arrojar nuevos conocimientos de la ictiofauna del Caura. Basamos este estimado sobre las

Tabla 6.1. Número de especies obtenidas para varios ríos y lagunas en la región del Escudo de Guayana y Cuenca del Orinoco en Venezuela.

Rio/Sistema	Cuenca	# Especies	Referencias
Atabapo	Orinoco	169	Royero et al. 1986
Caroní	Orinoco	120	Lasso 1989; Lasso et al. 1991
Caroní (Río Claro)	Orinoco	81	Taphorn y García 1991
Caroní (Lagunas)	Orinoco	54	Rodriguez y Lewis 1990
Caura	Orinoco	130(450?)	Balbas y Taphorn 1996
Caura	Orinoco	191	Machado-Allison et al. 1999b
Cuyuní	Essequibo	136	Machado-Allison et al. 2000
Suapure	Orinoco	140	Lasso 1992
Llanos	Apure	226	Machado-Allison et al. 1993b
Caura	Orinoco	278	Este estudio

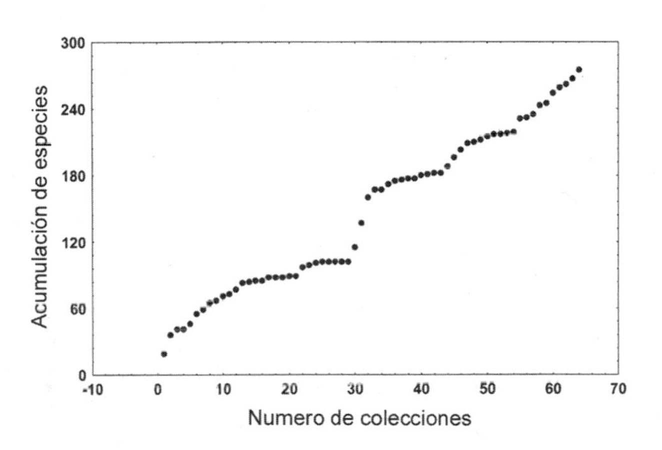

Figura 6.2. Curva de acumulación de especies colectados en el Río Caura, Estado Bolívar, Venezuela, durante la Expedición AquaRAP, Noviembre–Diciembre 2000.

curvas de acumulación obtenidas (Fig. 6.2). Después de 21 días de muestreo, la tasa de acumulación de especies nuevas para la expedición no disminuye o llega al plateu. La curva muestra un incremento en la región inferior del río donde la complejidad y diversidad de hábitats es mayor. Como por ejemplo capturamos 60 especies en la última colección en el río Takoto, de las cuales, al menos 10 especies adicionales, fueron añadidas a la lista de la expedición.

Importancia Biológica y económica

Los peces capturados durante nuestra expedición (Apéndice 8) muestra una gran variedad de grupos tróficos y ecológicos funcionales, por ejemplo predadores—*Acestrorhynchus, Hoplias, Serrasalmus, Hydrolicus, Crenicichla*; herbívoros—*Piaractus, Leporinus*; detritívoros—*Bunocephalus, Cyphocharax, Curimata, Prochilodus, Semaprochilodus*; planctonívoros—*Anchoviella*; parásitos—*Exodon, Paravandellia* y *Vandellia*; e insectívoros como la gran mayoría de las especies de los géneros *Astyanax, Jupiaba, Moenkhausia, Hemigrammus, Hyphessobrycon, Knodus,* etc. Estos últimos se encuentran intimamente asociados al bosque ribereño del cual obtienen gran parte de su alimento en forma alóctona. También, existe un gran número de especies asociadas a los parches de vegetación acuática presentes en los rápidos principalmente representadas por especies de hábitos riacofílicos como son las pertenecientes a los géneros *Anostomus, Ancistrus, Characidium, Leporinus, Melanocharacidium* y *Parodon* aunque hay algunas excepciones importantes que se encuentran mencionadas más adelante.

Debemos anotar igualmente, que en el Caura se encuentra representadas especies migratorias como por ejemplo: *Piaractus, Curimata, Prochilodus* y *Semaprochilodus.* Mientras que especies de este grupo se encuentran ampliamente distribuidas en la Cuenca del Río Orinoco, ellas penetran al Caura y sus tributarios.

También fueron colectadas especies de importancia comercial tales como: *Piaractus, Prochilodus, Ageneiosus, Hoplias, Hydrolicus, Myleus, Plagioscion* y ornamentales como: *Apistogramma, Geophagus, Guianacara, Hemigrammus, Jupiaba, Moenkhausia, Pyrrhulina, Ramirezella, Rivulus Satanoperca* y *Xenagoniates.*

Finalmente, otros poseen importancia taxonómica particular tales como miembros de los géneros: *Ancistrus, Astyanax, Apareiodon, Aphyocharax, Bryconops, Doras, Harttia, Hypostomus, Moenkhausia, Pimelodella* y *Satanoperca*, que poseen representantes que pueden ser nuevas especies para la ciencia y ayudarán a documentar relaciones biogeográficas entre las cuencas adyancentes.

Riqueza de especies

Un análisis de la ictiofauna muestra que las 278 especies estan ubicadas en 10 órdenes y 36 familias. La mayor riqueza se encontró en el Orden Characiformes con 158 especies (56.8%) del total. Le siguen los Siluriformes (74, 26.6%), Perciformes (27, 9.7%), Gymnotiformes (9, 3.2%), Clupeiformes (3, 1.1%), Ciprinodontiformes (2, 0.7%), Rajiformes (2, 0.7%), Beloniformes (1, 0.4%), Pleuronectiformes (1, 0.4%) y Synbranchiformes (1, 0.4%). La Familia Characidae resultó la mejor representada en este estudio, con 113 especies que representa aproximadamente 41% del total. En este sentido nuestros resultados incrementan en 54 especies las conocidas para el Orden Characiformes y 39 especies de Characidae para la Cuenca. En cuanto a los otros grupos; 74 especies de Siluriformes fueron identificadas comparado con 49 previamente conocidos y 27 especies de Perciformes contra 12 previamente reportados.

Comparación entre las dos regiones (Superior y Inferior)

Como se ha establecido anteriormente, el estudio fue dividido arbitrariamente en dos regiones tomando al Salto Pará, cómo el límite entre ambas. En la región superior se capturaron un total de 4.659 (24.1%) peces, pertenecientes a 103 (37%) especies, mientras que en la región inferior fueron colectados 14.607 (75.9%) ejemplares, pertenecientes a 226 (81.3%) especies. Tomando en cuenta las especies únicas para cada región, la región superior posee 52 especies (18.3%), mientras que se identificaron 175 (62.6%) especies únicas para la región inferior. Un total de 51 (18.7%) especies, son comunes para ambas regiones.

Existen varios factores que podrían explicar la diferencia en riqueza y abundancia relativa entre estas dos áreas. Posiblemente la más importante es el accidente geográfico constituido por el Salto Pará cuya estructura física forma una barrera natural para la dispersión de especies de peces. Esto es cierto para especies que penetran al sistema del Caura provenientes del Orinoco y no pueden subir más alla de este accidente geográfico. Otros factores como por ejemplo: la heterogeneidad de hábitats, disponibilidad de alimentos, gradiente del río y grado general de oligotrofía de los ambientes, también pueden haber actuado promoviendo estas diferencias. Las aguas superiores por arriba del Salto Pará son mayormente cristalinas, ácidas y con muy baja conductividad (García 1996). Sólo, en aquellos ambientes en los cuales había evidencia de acumulación de detritus proveniente de la descomposición de materia orgánica (p.e. Playas en Raudal Cejiato), mostraron una abundancia moderada y alta diversidad de especies. Condiciones similares de ambientes en el sector inferior dio resultados parecidos (García 1996), permitiendo concluir que las condiciones ecológicas seguramente juegan un papel muy importante en la distribución y abundancia de especies actuales (ver adelante).

La Tabla 6.2 (Apéndice 7), muestran un resumen del número total de especies capturadas por estación y sus abundancias relativas. Las estaciones que mayormente aportaron riqueza y abundancia relativa en la región superior fueron ICT-23, ICT-25, ICT-10, ICT-12, ICT-02, ICT-13, ICT-05 e ICT-07. Las primeras dos corresponden al área inmediatamente por arriba del Salto Pará, caracterizadas por áreas de lajas, abundantes plantas acuáticas y fondos con arena y piedras. Las estaciones ICT-12 e ICT-13 corresponden a las playas y boca del riachuelo en Raudal Cejiato. ICT-02, es un remanso de aguas negras cerca de Entrerios e ICT-10 es un remanso de aguas negras en el río Erebato. Finalmente, las estaciones ICT-03 (playa) en el Río Kakada y la Estación ICT-05, caño Suajiditu en el mismo río, poseen cierta importancia. Debemos también hacer notar que en los remansos y las playas correspondientes a Cejiato y Suajiditu, se encontraron numerosos juveniles de varias especies, lo que sugiere que estas zonas pueden estar actuando como áreas de desarrollo (*nursery*) en períodos reproductivos. En resumen para la sección superior tanto la riqueza de especies como el número de ejemplares capturados señala a tres áreas principalmente: i) Areas inmediatamente cercanas al Salto Pará; ii) El área correspondiente al Raudal Cejiato; y iii) El área del Río Kakada (Playa y caño Suajiditu).

Las estaciones que mayormente aportaron riqueza y abundancia relativa en la región inferior fueron: ICT-65, ICT-51, ICT-59, ICT-34, ICT-52, ICT-62, ICT-48, ICT-32-33, ICT-63, ICT-47 e ICT-35. La más rica en términos del número de especies corresponde al Río Takoto (ICT-65), un área excepcionalmente heterogénea (rápidos, pozos, playas arenosas y remansos) y a la cual se le debe prestar atención para futruros planes de conservación. Las áreas cercanas al Raudal Cinco Mil (ICT-62 y 63), también presentan alta riqueza y una moderada abundancia. En estas áreas se colectaron especies únicas para la cuenca. Las estaciones ICT-32, 33, 34 y 35 corresponden a la región inferior del Salto Pará que incluye El Playón y un caño tributario cercano. Finalmente, las Estaciones ICT-47, 48, 51 y 52 corresponden a diversas zonas en el Río Tawadu (Tabaro) y Nichare incluyendo playas y lagunas. Todas estas áreas son puntos clave para el desarrollo de planes de conservación.

Especial atención se le debe prestar a la Estación ICT-51 (Laguna) en el Río Nichare ya que posee importancia local. La laguna es comunmente utilizada por pescadores y la presencia de peces de gran talla, nos permite recomendar su protección como reserva pesquera.

Aspectos biogeográficos

Pocos estudios han presentado información sobre distribución y relaciones geográficas de los peces de aguas continentales de Venezuela. Mago (1970 y 1978) fue el primero en construir varias zonas biogeográficas, entre las cuales ubica a la Cuenca del Río Caura dentro de la Provincia Guayanesa. Posteriormente Chernoff et al. (1991), Lasso et al. (1991) y Lasso y Provenzano (1997), entre otros, han construido diverss hipótesis entre elementos faunísticos del Orinoco-Amazonas y Guayanas. Sin embargo, poco podemos indicar acerca de relaciones biogeográficas a menos que tengamos información confiable acerca de las relaciones filogenéticas de los diversos grupos de peces presentes.

Los datos obtenidos en este estudio indican que la mayoría de las formas presentes en el área superior: arriba del Salto Pará están más asociados a especies o formas presentes en cuencas vecinas del Caroní y Cuyuní en las regiones guayanesas de Venezuela y Guyana. Especies como *Bryconops* cf. *colaroja*, *Harttia* sp., *Chaetostoma vasquezi*, *Lebiasina uruyensis*, *Aequidens chimantanus* y *Guianacara geayi* permiten documentar esta hipótesis.

Con respecto a las especies que se colectaron en la región inferior del Caura, nos permiten hipotetizar una mayor relación con formas de la Orinoquía y Llanos y formas presentes en la región superior del Orinoco y Casiquiare-Río

Tabla 6.2. Resumen de las estaciones, especies, abundancias relativas y tipo de hábitat (h) colectados durante la Expedición AquaRAP, Noviembre Diciembre 2000. Abrev. Bw–Remanso; C–Caño; R–Rapidos; Pa–Playa Arenosa; Pr–Playa Rocosa; Pm–Playa Fangosa; L–Laguna; Ia–Isla Arenosa; Ir–Isla Rocosa; T–Tributario (riachuelo).

ALTO CAURA				BAJO CAURA			
Estaciones	**Especies**	**Ejemplares**	**h**	**Estaciones**	**Especies**	**Ejemplares**	**h**
ICT-01	18	220	C	ICT-32/33	37	519	Pa
ICT-02	25	139	R	ICT-34	46	720	C
ICT-03	12	151	Pa	ICT-35	31	344	Ir
ICT-04	9	31	R	ICT-36	13	238	Pr
ICT-05	22	447	C	ICT-37	24	504	C
ICT-06	13	41	C	ICT-38	10	107	Ir
ICT-07	16	493	R	ICT-38a	2	10	C
ICT-08	20	220	Pa	ICT-39	14	164	Pr
ICT-09	15	38	C	ICT-40	4	90	R
ICT-10	29	392	C	ICT-41	15	246	R
ICT-11	18	113	C	ICT-42	5	112	C
ICT-12	28	236	Pa	ICT-43	2	2	R
ICT-13	25	159	T	ICT-44	8	284	Ir,R
ICT-14	9	20	T	ICT-45	15	58	C,R
ICT-15	7	28	R,Pr	ICT-46	12	84	L
ICT-16	9	178	Pa	ICT-47	35	563	Pa
ICT-17	12	50	R	ICT-48	41	693	C
ICT-18	11	14	Pr	ICT-49	33	848	Pa
ICT-19	9	52	Pa	ICT-50	22	326	Bw
ICT-20	16	201	Bw	ICT-51	50	1977	L
ICT-21	--	--		ICT-52	44	740	Ir
ICT-22	5	15	C	ICT-53	29	1100	Pm
ICT-23	32	429	R	ICT-54	22	304	Pm
ICT-24	--	--		ICT-55	9	39	C
ICT-25	30	359	Pr	ICT-56	36	811	L
ICT-26	13	222	Bw	ICT-58	6	15	Pm
ICT-27	14	62	Bw	ICT-59	46	590	Pm
ICT-28	16	174	P	ICT-60	21	234	Pm
ICT-29	16	61	R	ICT-61	20	201	Pa
ICT-30	9	20	T	ICT-62	43	1288	Ia
ICT-31	6	13	R	ICT-63	36	433	C
				ICT-64	25	281	R,Pr
				ICT-65	62	780	R,Pa

Negro (Amazonas). Entre las primeras están: *Acestrorhynchus falcatus, A. microlepis, Anchoviella jamesi, Aphyocharax alburnus, Boulengerella lucia, Brycon bicolor, Bryconops alburnoides, Bujurquina mariae, Eingenmannia virescens, Gymnotus carapo, Hypostomus plecostomus, Microphilipnus ternetzi, Orinocodoras eigenmanni, Ochmacanthus alternus, O. orinoco, Pellona castelneana, Potamorhina altamazonica, Potamotrygon dorbigny, Prochilodus mariae, Pygocentrus cariba, Semaprochilodus laticeps, Raphyodon vulpinnus, Sorubim lima, sternopygus macrurus* y *Triportheus albus.* Las especies con relación al Alto Orinoco y Amazonas están: *Micoschemobrycon casiquiare, Serrasalmus* sp., *Anostomus anostomus, A. ternetzi, Bryconops alburnoides, Cynodon gibbus, Hydrolicus armatus, Crenicichla lenticulata, Exodon paradoxus* e *Hyphessobrycon serpae,* entre otros.

Importancia Económica y Social

Peres y Terbourgh (1995), Goulding (1980, 1981) y Goulding et al. (1988), documentan no solamente la importancia de los ríos en la estructuración y asentamiento de comunidades humanas en la Amazonia, sino también el incremento de su dependencia sobre los recursos acuáticos para su mantenimiento. En la Cuenca del Río Caura, un aspecto de suma importancia es lo referente a las condiciones sociales y tradiciones de las poblaciones autóctonas. Los Ye'kuana y los Sanema, son habitantes ancestrales de la Cuenca. Todas sus actividades domésticas, comerciales y creencias mítico-religiosas, están íntimamente relacionadas con el río, su fauna y flora, por lo que su conservación o uso sostenible es sumamente importante para el mantenimiento de éstas culturas.

Muchas de las especies de peces encontradas en el Río Caura tienen alto valor, tanto como peces para el consumo humano, o como parte de las pesquerías ornamentales. Poca atención se ha prestado a ellos en la región superior del Caura, debido principalmente a lo dificultoso de su acceso gracias al accidente geográfico del Salto Pará, pero también a los deseos expresados por los pueblos Ye´kuana y Sanema.

En la región inferior de Caura, su confluencia con el Orinoco y lagunas de rebalse, la actividad de pesca comercial se ha desarrollado intensivamente especialmente sobre especies tales como: "cachamas" (*Colossoma macropomum*), "morocotos" (*Piaractus brachypomus*), "sapoaras" (*Semaprochilodus laticeps*), "coporos" (*Prochilodus mariae*), "palometas" (*Mylossoma duriventre*), "caribe colorado" (*Pygocentrus cariba*), "valentones" , "dorados" y "laolaos" (*Brachyplatystoma filamentosum, B. rosseauxi* y *B. vaillanti*), "rayaos" (*Pseudoplatystoma fasciatum* y *P. tigrinum*), "cajaro" (*Phractocephalus hemiliopterus*) y "curbinata" (*Plagioscion squamossissimus*) (Novoa y Rámos 1978; Novoa 1982). A pesar de su importancia de esta pesquería, no existen datos confiables sobre volúmenes de captura, precios, vías o rutas de comercialización, etc. de esta actividad. Más aún, no existe información acerca de aspectos biológicos de importancia para el desarrollo de una actividad pesquera sostenible, como son: crecimiento, migraciones, períodos o edad mínima reproductiva, fecundidad, hábitos, etc.

La presión pesquera en áreas superiores y medias del Caura se realiza sobre toda la ictiofauna con el propósito de obtener proteína de sustento para las poblaciones indígenas locales. Aunque en menor grado que en la región inferior, tenemos conocimiento de capturas masivas obtenidas por el uso del "barbasco" en caños causando un daño considerable a los recursos. Esta actividad tradicional, promotora de potenciales daños ambientales, ha sido denunciada y materia de discusión en recientes talleres realizados con las diferentes etnias que habitan la cuenca (ACOANA, com. per.). Por otro lado, se realizan faenas de pesca con anzuelo dirigida a especies de porte mayor como por ejemplo: "aimaras" y "guavinas" (*Hoplias macrophthalmus* y *H. malabaricus*), "pacus" (*Myleus rubripinnis, M. asterias* y *M. torquatus*), "bagres" (*Ageneiousus* sp.), "guitarrillos" (*Pseudodoras* sp. y *Doras* sp.) "caribes" (*Serrasalmus rhombeus*) y "payaras" (*Hydrolicus armatus* y *H. tatauaia*). No existen datos existen acerca de la utilización de redes en las regiones media y superior del Caura.

Por otro lado, hemos obtenido información (*internet*) que especies como "pavones" (*Cichla orinocensis, C. temensis* y *C. monoculus*), "sardinatas" (*Pellona castelneana*) y "payaras" (*Hydrolicus armatus* y *H. tatauaia*), son anunciadas en paquetes turísticos promoviendo la pesca deportiva y el turismo en la Baja Caura. Esto requiere estudios cuidadosos debido a que no hay datos disponibles de la cantidad de peces deportivos que son capturados. Sin embargo, muchos de los anuncios describen valores de capturas diarias, que de ser ciertos, pueden causar daños a las poblaciones de peces.

Finalmente, es importante destacar que podría haber un potencial desarrollo de una pesquería ornamental como mascotas o cultivo de peces (alimento) en el río Caura. Numerosas especies presentes tanto en las regiones superiores como inferiores poseen valor internacional como peces ornamentales. Especies como: *Ammocryptocharax elegans, Ancistrus* sp., *Anostomus anostomus, A. ternetzi, Aphyocharax alburnus, A. erythrurus, Apistogramma iniridae, Brachychalcinus opercularis, Bryconops giacopini, Bujurquina mariae, Carnegiella strigatta, Caenotropus labyrinthicus, Chaetostoma vasquezi, Chalceus microlepidotus, Corydoras blochii, C. boehlkei, C. bondi, C. osteocarus, Eigenmannia virescens, Exodon paradoxus, Farlowella vittata, Guianacara geayi, Hyphessobrycon bentosi, H. serpae, Jupiaba zonata, Leporinus arcus, L. brunneus, L. grandti, L. maculatus, Melanocharacidium dispiloma, M. nigrum, Moenkhausia collettii, M. copei, M. lepidura, M. oligolepis, Myleus rubripinnis, M. asterias, Nannostomus erythrurus, Ramirezella newboldi, Poptella longipinnis, Potamotrygon schoederi, Pygocentrus cariba, Pyrrhulina brevis, Rineloricaria fallax, Tatia galaxias, T. romani* y *Xenagoniates bondi,* pueden ser encontradas en los negocios dedicados a la venta mascotas o de peces ornamentales. Esta actividad económica, social y científicamente regulada, puede ser una alternativa económica válida para incrementar la calidad de vida de las poblaciones indígenas locales.

Piscicultura extensiva, podría igualmente ser una alterativa con la incorporación de numerosas lagunas de inundación que posee la cuenca en sus áreas medias e inferiores. Zonas estudiadas durante nuestra expedición muestran algunas de estas áreas importantes y que son actualmente utilizadas con fines de capturas de peces comestibles.

En resumen, podemos indicar que los datos sobre las pesquerías en estas zonas (altas y bajas) no existen o son poco confiables. La carencia de los mismos hace difícil el establecimiento de programas que permitan un mejor y adecuado manejo del recurso. La discusión que presentamos permitirá el estímulo de investigaciones dedicadas a la obtención de información actualizada y confiable acerca de los recursos explotados y así poder garantizar un manejo sostenible de la pesquería presente en la Cuenca del Río Caura.

Hábitats críticos

Fueron identificados varios hábitats críticos los cuales necesitan de una continua revisión (monitoreo) para garantizar la sobreviviencia de las comunidades de peces de agua dulce y el mantenimiento de la espectacular biodiversidad. Estas son las mismas áreas que apoyan la reproducción y crecimiento de especies de importancia económica tanto de consumo como ornamental. Los hábitats en general están descritos en detalle en varios de los capítulos del presente volumen en cada uno de los tópicos tratados. Sin embargo, en lo que respecta a los peces en particular podemos agruparlos en tres grandes grupos: i) Las áreas inundables incluyendo los bosques, sabanas y lagunas; ii) los rápidos; y iii) Los caños y playas asociadas a remansos.

Las áreas de bosque inundable incluyendo lagunas y sabanas inundables asociadas, son posiblemente las zonas más críticas y altamente en peligro tanto en la región media como baja del Caura, como ha sido demostrado en otros trabajos (Welcomme 1979; Goulding 1980; Lowe-McConnell 1987; Chernoff et al. 1991; Machado-Allison 1993, 1994). Estas áreas proveen refugio (permanente o temporal) a más del 60% de las especies colectadas en estos sistemas tropicales. Más aún, muchas especies incluyendo: "cachamas," "morocotos" y "palometas" (*Colossoma, Mylossoma, Myleus* y *Piaractus*) y "palambras" (*Brycon* spp.), ingieren frutas y semillas como fuente de alimento cuando éstas caen al agua, o cuando el bosque se inunda y los peces tienen acceso al suelo de tierra firme (Goulding 1980; Machado-Allison 1982, 1993). Tanto el bosque, como la fauna que vive en ellos (incluyendo los peces), están íntimamente asociados a las inundaciones periódicas. Mantenimiento de esta asociación coevolutiva entre los diferentes elementos que habitan los bosques inundables, al menos temporalmente, es crítico para la sobrevivencia del ecosistema como un todo, incluyendo las comunidades de peces.

Como ha sido detectado en otros sistemas como el Río Madeira (Brasil) (Goulding 1979) y el Tahuamanu-Manuripi (Chernoff et al. 1999), el Río Caura y Erebato poseen un bosque inundable relativamente angosto. Esto es mayormente cierto en sus áreas superiores. Dada esta característica, existe muy poco nivel de amortiguación entre las zonas de explotación forestal o agrícolas y estas zonas inundables críticas. Las zonas bajas y cercanas a la desembocadura en el Orinoco incluyen zonas más extensas. Sin embargo, la actividad extractiva y otros factores antrópicos son más intensos promoviendo contaminación doméstica, industrial y agrícola.

Otros hábitats importantes dentro de la cuenca son los rápidos en las áreas arriba del Salto Pará y en los río Nichare y Tawadu. Estos rápidos poseen una cobertura extensa por plantas acuáticas de la Familia Podostemonaceae. La comunidad de peces que explota esas áreas de rocas y macrofitas son únicas para la cuenca del Río Caura tanto como podemos determinar. Los peces están íntimamente asociados con estos ecosistemas y usan las macrofitas para protección o alimento. El grado de cobertura vegetal también produce refugio de las fuertes corrientes debido a que algunos peces, como por ejemplo especies de los géneros *Crenicichla* o *Jupiaba*, los cuales son característicos de remansos o caños y no son frecuentemente encontradas en aguas rápidas. Cambios sutiles en los niveles naturales del agua en esas regiones producirían enormes daños a estas comunidades que dependen exclusivamente del sistema natural periódico de lluvias y sequía en la zona.

Finalmente, es necesario indicar unos pocos hábitats especializados y de importancia íctica. La región superior del Caura y Erebato está caracterizada por tener bajos niveles de nutrientes (baja conductividad) disueltos o suspendidos en la columna de agua, así que la acumulación de mayor diversidad y biomasa de peces se encuentra en zonas donde las aguas depositas gran cantidad de restos orgánicos y detritus durante los períodos lluviosos; usualmente áreas protegidas tales como caños, lagunas y playas asociadas a remansos. Esta situación fue observada para el Raudal Cejiato, Kakada y Suajiditu, donde las aguas depositan grandes cantidades de material orgánico y se forman zonas de gran acumulación de detritus sobre playas arenosas. Estas áreas son particularmente importantes en la región superior ya que son utilizadas como áreas reproductivas o refugio (*nursery*) para estadios jóvenes de peces. Al igual que los rápidos, estas zonas dependen grandemente en el sistema climático periódico.

En las regiones bajas del Caura, este fenómeno sucede en zonas cercanas a El Playón y los ríos Nichare y Tawadu, donde en las áreas de playas asociadas a remansos y de lagunas o madreviejas existe una enorme diversidad y acumulación de biomasa íctica apreciable incluyendo muchas especies de importancia comercial y donde las comunidades indígenas obtienen sustento.

CONSERVACIÓN

Los datos aportados y el análisis de la información lograda en otros sistemas, permiten sugerir que la conservación de la biodiversidad en los ecosistemas acuáticos es uno de los retos más importante y difícil al cual se enfrenta el mundo hoy día (Chernoff et al. 1996; Machado-Allison et al. 1999a). Uno de los principales obstáculos se basa en el nivel del conocimiento actual referente al uso de estos ecosistemas como parte del desarrollo humano, explotación, manejo y conservación de recursos naturales incluyendo el agua (IUCN 1993; Gleick 1998).

De acuerdo a la información disponible la biodiversidad de nuestros ecosistemas acuáticos es pobremente conocida comparado con ecosistemas similares terrestres tropicales (p.e. bosque tropical). La carencia de conocimientos incluye sistemática y taxonomía de las esppecies, sus relaciones filogenéticas y biogeográficas (Böhlke et al. 1978; Mago-Leccia 1978; Chernoff et al. 1991; Mago-Leccia 1994) y sus ecologías o historias de vida (Goulding 1979; Lundberg et al. 1987; Winemiller 1989; Machado-Allison 1992, 1993; Menezes y Vazzoler 1992). Poco se conoce acerca de las interacciones de los organismos entre ellos y con el ambiente físico y esta está limitada a un minúsculo grupo de hábitats o estaciones climáticas temporales (Lowe-McConnell 1967; Goulding 1980; Lowe-McConnell 1984, 1987; Machado-Allison 1993).

Nuestro estudio basado en diversidad y abundancia relativa de peces y su relación con las poblaciones humanas, mostró que ciertas áreas tanto en la región superior como inferior deberían ser protegidas y formar parte de planes especiales de conservación como por ejemplo: el Raudal Cejiato, Río Kakada, los remansos y lagunas de rebalse cercanos a Entreríos, Raudal Suajiditu y la región inmediata superior del Salto Pará, el área del El Playón, los ríos Nichare y Tawadu, las lagunas de reblase en la Boca del Nichare y el Río Takoto cercano al Raudal Cinco Mil.

Por otro lado es importante señalar que cualquier programa que permita el cambio en la periodicidad natural climática e hidraúlica afectará los ambientes naturales. El mayor impacto será sobre áreas llanas y de aguas rápidas (raudales) donde existe un microhábitat especial estructurado por la presencia de plantas acuáticas de la Familia Podostemonaceae.

Para poder tomar decisiones importantes que puedan afectar irreversiblemente los ecosistemas acuáticos (p.e. transvase, canalización, represamiento de aguas) como los que han sido propuestos para el Sistema del Caura y cuencas adyacentes, el desarrollo de planes como el de Hidrovia (Bucher et al. 1993) que pretende interconectar las tres principales cuencas suramericanas, o regionalmente el desarrollo del "Eje Orinoco-Apure," es necesario tener información ecológica, económica y social confiable (SISGRIL 1990; Bucher et al. 1993; Machado-Allison 1994; Aguilera y Silva 1997; Machado-Allison 1999). Es imperativo tener una apreciación de las complejidades ecológicas que poseen nuestros ecosistemas acuáticos y las historias o ciclos de vida de nuestros organismos, para poder estar en conocimiento del daño a causar y poder de alguna manera desarrollar planes alternativos.

El conocimiento aportado por este trabajo sobre la diversidad íctica y los ecosistemas acuáticos dentro de la Cuenca del Río Caura, su potencial económico e importancia para las poblaciones humanas discute fuertemente para el desarrollo de planes de conservación inmediatos. Los esfuerzos de conservación en la Cuenca del Río Caura pueden proveer reglas principales hacia la conservación de ecosistemas acuáticos en Venezuela. Esfuerzos de conservación en gran escala son necesarios inmediatamente desde las perspectivas de la sociedad y de la ciencia. Los ecosistemas acuáticos de Venezuela son recursos naturales críticos que mantienen tanto poblaciones humanas como fauna silvestre. Ellos ayudan a proveer seguridad alimentaria y aguas limpias así como también la conservación de la vida silvestre. La alta calidad de vida humana en Venezuela está íntegramente asociada al mantenimiento y conservación de nuestros ecosistemas tropicales

CONCLUSIONES Y RECOMENDACIONES

La región estudiada en la Cuenca del Río Caura, Estado Bolívar, Venezuela es posiblemente un punto clave (*hotspot*) de biodiversidad. Hemos identificado 278 especies de peces en la región lo cual representa aproximadamente el 28% de las especies conocidas de agua dulce para Venezuela.

La diversidad en la Cuenca debe incrementarse con un mayor muestreo íctico. Numerosas áreas no fueron colectadas como por ejemplo el Alto Caura y Erebato y Bajo Caura cerca del Orinoco. Los análisis muestran que a pesar del esfuerzo pesquero, las últimas colecciones continuaban incrementando el número de especies conocidas para el área a una tasa considerable.

La Ictiofauna posee un ensamblaje interesante de especies con asociaciones biogeográficas guayanesas en las áreas superiores al Salto Pará y formas relacionadas con el Orinoco y Amazonas en las regiones inferiores.

Muchas especies de peces poseen importancia económica, comercial de consumo u ornamentales. Los peces son actualmente utilizados para la subsistencia de las poblaciones indígenas, principalmente en las áreas superiores. El desarrollo de una pesquería ornamental puede ser una alternativa económica y social para las poblaciones indígenas locales. También recomendamos el desarrollo de una pesquería para incrementar el consumo local y regulaciones estrictas de las capturas para la exportación a otras áreas del país.

El desarrollo de estudios sobre las pesquerías en las áreas del Bajo Caura es necesario y urgente. Datos bioecológicos que aporten información para el manejo sostenible del recurso son imprescindibles.

Zonas con hábitats críticos tales como rápidos, bosques inundados o lagunas y caños deben ser protegidos. En lugar de poroponer la creación de un Parque Nacional, debe considerarse el uso multiple de las zonas con cierta restricción en la modificación de los mismos. Se debe proteger las pocas planicies de inundación especialmente en aguas superiores del Caura y Erebato, así como también el Raudal Cejiato, Río Kakada, los remansos y lagunas de inundación cercanos a Entreríos, el Raudal Suajiditu y las región inmediata superior al Salto Pará, el Playón y areas cercanas, los ríos Nichare y Tawadu, las lagunas cercanas a Boca de Nichare y el río Takoto cerca del Raudal Cinco Mil.

Diversos programas deben ser desarrollados conjuntamente con las etnias y poblaciones locales con el propósito de intercambiar conocimientos sobre las relaciones existentes entre el mantenimiento de los hábitats y la diversidad de peces. Programas educativos deben ser desarrollados e involucrar residentes locales y pescadores con la finalidad de monitorear las poblaciones de peces y los hábitats donde son capturados. Estos programas deberán promover el aprendizaje de las diferentes especies de peces y como reconocer especies nuevas o raras para llamar la atención de los científicos.

LITERATURA CITADA

Aguilera, M., y J. Silva 1997. Especies y Biodiversidad. Interciencia, 22(6):289–298.

Balbas, L., y D. Taphorn. 1996. La Fauna: Peces. *En*: Rosales, J. y O. Huber (eds.). Ecología de la Cuenca del Río Caura. Scientia Guaianae, 6: 76–79

Böhlke, J., S. Weitzman y N. Menezes. 1978. The status of systematic studies of South American fresh water fishes. Acta Amazonica, 8: 657–677.

Bonilla-Rivero, A., A. Machado-Allison, B. Chernoff, C. Silvera, H. López y C. Lasso. 1999. *Apareiodon orinocensis* una nueva especie de pez de agua dulce (Pisces, Characiformes, Parodontidae) proveniente de los ríos Caura y Orinoco, Venezuela. Acta Biol. Venez., 19(1): 1–10.

Bucher, E., A. Bonetto, T. Boyle, P. Canevari, G. Castro, P. Huszar y T. Stone. 1993. Hidrovía. Un examen ambiental inicial de la vía fluvial Paraguay-Paraná. Humedales para las Américas, Publ. 10:1–10.

Buckup, P. 1993. Review of the characidiin fishes (Teleostei: Characiformes), with descriptions of four new genera and ten new species. Ichth. Explor. Freshwaters, 4(2): 97–154.

Chernoff, B., y A. Machado-Allison. 1990. Characid fish of the genus *Ceratobranchia* with description of new species from Venezuela and Peru. Proc. Acad. Nat. Sci. Phil., 142:261–290.

Chernoff, B., A. Machado-Allison y W. Saul. 1991. Morphology variation and biogeography of *Leporinus brunneus* (Pisces: Characiformes: Anostomidae). Ichth. Explor. Freshwaters, 1(4): 295–306.

Chernoff, B., y A. Machado-Allison. 1999. *Bryconops colaroja* y *B. colanegra,* two new species from the Cuyuni and Caroní drainages of South America. Ichth. Explor. Freshwaters., 10(4):355–370.

Chernoff, B., A. Machado-Allison y N. Menezes. 1996. La conservación de los ambientes acuáticos: una necesidad impostergable. Acta Biol. Venez., 16 (2): i–iii.

Chernoff, B., A. Machado-Allison, F. Provenzano, P. Willink y P. Petry. 2003. *Bryconops* n. sp. a new species from the Caura River Basin of Venezuela (Characiformes, Teleostei). Ich. Explor. Freshwaters 13:4.

Chernoff, B., P. Willink, J. Sarmiento, S. Barrera, A. Machado-Allison, N. Menezes y H. Ortega. 1999. Fishes of the rios Tahuamanu, Manuripi and Nareuda, Dpto. Pando, Bolivia: Diversity, Distribution, Critical Habitats and Economic Value. *En*: Chernoff, B. y P. Willink (eds.). A Biological Assessment of the aquatic Ecosystems of the Upper Río Orthon Basin, Pando, Bolivia. RAP Bull. Biological Assessment 15. Pp. 39–46.

Eigenmann, C. 1918-1928. The American Characidae (I–IV). Mem Mus. Comp. Zool., 43.

Eigenmann, C., y G. Myers. 1929. The American Characidae (V). Mem Mus. Comp. Zool., 43: 429–574

García, S. 1996. Limnología. *En*: Rosales, J. y O. Huber (eds). Ecología de la Cuenca del Río Caura. Scientia Guaianae, 6: 54–59.

Gery, J. 1977. The Characoids of the World. TFH Publications, Neptue City, NY. 672 pp.

Gleick, P. 1998. The World's Water: The Biennial Report on Freshwater Resources. Island Press. Washington, DC.

Goulding, M. 1979. Ecologia da Pesca do Rio Madeira. Cons. Nac. Des. Cient. e Tec., INPA.

Goulding, M. 1980. The Fishes and the Forest: Explorations in Amazonian Natural History. Univ. Cal. Press.

Goulding, M., M. L. Carvalho y E. G. Ferrerira. 1988. Río Negro: rich life in poor water: Amazonian diversity and foodchain ecology as seen through fish communities. SPB Academic Publ. The Hague, Netherlands.

Huber, O., y J. Rosales (eds). 1997. Ecología de la Cuenca del Río Caura, Venezuela. II. Estudios Especiales. Scientia Guianae, 7: 1–473.

IUCN. 1993. The Convention on Biological Diversity: An explanatory guide (Draft). IUCN Environmental Law Centre, Bonn, Germany. 143 pp. (mimeo).

Lasso, C. 1989. Los Peces de la Gran Sabana, Alto Caroní, Venezuela. Mem. Soc. Cienc. Nat. La Salle, 49–50 (131–134): 208–285.

Lasso, C. 1992. Composición y aspectos ecológicos de la ictiofauna del Bajo Suapure, serranía Los Pijiguaos (Escudo de Guayana), Venezuela. Mem. Soc. Cienc. Nat. La Salle, 52(138): 5–54.

Lasso, C., y A. Machado-Allison. 2000. Sinopsis sobre las especies de la Familia Cichlidae en la Cuenca del Orinoco. Conicit, Caracas.

Lasso, C., A. Machado-Allison y R. Pérez. 1991. Consideraciones zoogeográficas de los peces de la Gran Sabana (Alto Caroní) Venezuela, y sus relaciones con las cuencas vecinas. Memoria, Soc. Cien. Nat. La Salle IL y L: 1-21.

Lasso, C., y F. Provenzano. 1997. *Chaetostoma vasquezi*, una nueva especie de corroncho del Escudo de Guayana, Estado Bolívar, Venezuela (Siluroidei-Loricariidae) descripción y consideraciones biogeográficas. Mem. Soc. Cienc. Nat. La Salle, 57(147):53–65.

Lowe-McConnell, R. 1964. The fishes of the Rupununi Savanna District of British Guiana. Pt.1. Grouping of fish species and effects of the seasonal cycles on the fish. Journ. Linn. Soc. (Zool.), 45:103–144.

Lowe-McConnell, R. 1969. Some factors affecting fish populations in amazonian waters. Atas do Simposio sobre a Biota Amazonica, 7:177–186.

Lowe-McConnell, R. 1987. Ecological Studies in Tropical Fish Communities. Cambridge Univ. Press.

Lundberg, J., J. Baskin y F. Mago-Leccia. 1979. A preliminary report on the first cooperative U.S. - Venezuelan ichthyological expedition to the Orinoco River. 14 p. (Mimeo).

Machado-Allison, A. 1982. Estudios sobre la Subfamilia Serrasalminae (Teleostei-Characidae). Parte I. Estudio comparado de los juveniles de las "cachamas" de Venezuela (Géneros *Colossoma* y *Piaractus*). Acta Biol. Venez., 11(3): 1–102.

Machado-Allison, A. 1992. Larval ecology of fishes of the Orinoco Basin. *En*: Hamlett, W. (ed.). Reproductive Biology of South American Vertebrates Springer Verlag, 45–59.

Machado-Allison, A. 1993. Los Peces del Llano de Venezuela: un ensayo sobre su Historia Natural. (2nda. Edición) Consejo de Desarrollo Científico y Humanístico (UCV), Imprenta Universitaria, Caracas.

Machado-Allison, A. 1994. Factors affecting fish communities in the flooded plains of Venezuela. Acta Biol.Venez., 15(2):59–75.

Machado-Allison, A. 1999. Cursos de agua, fronteras y conservación. *En*: Genatios, G. (ed.). Ciclo Fronteras: Desarrollo Sustentable y Fronteras. Com. Estudios Interdisciplinarios, UCV. Caracas: 61–84.

Machado-Allison, A., y W. Fink. 1996. Los Peces Caribes de Venezuela. Univ. Central de Venezuela-Conicit.

Machado-Allison, A., B. Chernoff, P. Buckup y R. Royero. 1993a. Las especies del género *Bryconops* Kner, 1859 en Venezuela (Teleostei-Characiformes). Acta Biol. Venez., 14(3):1–20

Machado-Allison, A., F. Mago-Leccia, O. Castillo, R. Royero, C. Marrero, C. Lasso y F. Provenzano. 1993b. Lista de especies de peces reportadas en diferentes cuerpos de agua de los bajos llanos de Venezuela. *En*: Machado-Allison, A. (ed.) Los Peces del Llano de Venezuela: un ensayo sobre su Historia Natural. (2da. Edición) Consejo de Desarrollo Científico y Humanístico (UCV), Imprenta Universitaria, Caracas. pp. 129–143.

Machado-Allison, A., J. Sarmiento, P.W. Willink, B. Chernoff, N. Menezes, H. Ortega, S. Barrera y T. Bert. 1999a. Diversity and Abundance of Fishes and Habitats in the Rio Tahuamanu and Rio Manuripi Basins (Bolivia). Acta Biol. Venez., 19(1): 17–50.

Machado-Allison, A., B. Chernoff, C. Silvera, A. Bonilla, H. Lopez-Rojas, C. A. Lasso, F. Provenzano, C. Marcano y D. Machado-Aranda. 1999b. Inventario de los peces de la cuenca del Río Caura, Estado Bolívar, Venezuela. Acta Biol. Venez., Vol. 19 (4):61–72.

Machado-Allison, A., B. Chernoff, R. Royero-León, F. Mago-Leccia, J. Velazquez, C. Lasso, H. López-Rójas, A. Bonilla-Rivero, F. Provenzano y C. Silvera. 2000. Ictiofauna de la cuenca del Río Cuyuní en Venezuela. Interciencia, Vol. 25(1):13–21.

Machado-Allison, A., B. Chernoff y M. Bevilacqua. 2003. Introducción a la Cuenca del Río Caura, Estado Bolívar, Venezuela. *En:* Chernoff, B., A. Machado-Allison, K. Riseng y J. R. Montambault (eds.). A Biological Assessment of the Aquatic Ecosystems of the Caura River Basin, Bolívar State, Venezuela. RAP Bulletin of Biological Assessment 28. Washington, DC, USA: Conservation International. Pp. 123–128.

Mago-Leccia, F. 1970. Lista de los Peces de Venezuela. Minist. Agric. y Cría, Ofic. Nac. Pesca. Caracas.

Mago-Leccia, F. 1978. Los Peces de agua dulce de Venezuela. Cuadernos Lagoven, Caracas.

Mago-Leccia, F. 1994. Electric Fishes of the Continental Waters of America. Biblioteca Acad. Ciec. Fis. Mat. y Nat., Vol. XXIX, 206 pp.

Meneses, N., y P. Vanzoler. 1992. Reproductive characteristics of Characiformes. *En*: Hamlett, W. (ed). Reproductive Biology of South American Vertebrates. Springer-Verlag, Chap.4:60–70.

Nijsen, H., y J. Isbruker. 1980. A review of the genus *Corydoras* Lacepede, 1803 (Pisces, Siluriformes, Callichthyidae) Bijdragen tot de Dierkunde, 50(1):190–220.

Novoa, D. 1982. Los Recursos Pesqueros del Río Orinoco y su Explotación. Corp. Venez. Guayana (CVG).

Novoa, D., y F. Rámos. 1978. Las Pesquerías Comerciales del Río Orinoco. Corp. Venez. Guayana.

Peña, O., y O. Huber. 1996. Características Geográficas generales. *En*: Rosales, J. y O. Huber (eds). Ecología de la Cuenca del Río Caura. Scientia Guaianae, 6: 4–10.

Peres, C. A., y J. W. Terborgh. 1995. Amazonian nature reserves: an analysis of the defensibility status of existing conservation units and design criteria for the future. Conservation Biology 9: 34–45.

Rodríguez, A., y W. Lewis. 1990. Diversity and species composition of fish communities of Orinoco floodplain lakes. Nat. Geograp. Res., 6(3): 319–328.

Royero, R., A. Machado-Allison, B. Chernoff y D. Machado-Aranda. 1986. Peces del Río Atabapo. Territorio Federal Amazonas, Venezuela. Acta Biol. Venez., 14(1): 41–55.

Rosales, J., y O. Huber (eds.). 1996. Ecología de la Cuenca del Río Caura, Venezuela. I. Caracterización General. Scientia Guaianae, No. 6: 1–131.

SISGRIL. 1990. Simposio Internacional sobre Grandes Ríos Latinoamericanos. Interciencia, 15(6). 193 pp.

Taphorn, D., y J. García. 1991. El río Claro y sus peces, con consideraciones de los impactos ambientales de las presas sobre la ictiofauna del Bajo Caroní. Biollania, 8:1–15.

Taphorn, D., R. Royero, A. Machado-Allison y F. Mago-Leccia. 1997. Lista actualizada de los peces de agua dulce de Venezuela. *En*: La marca E (ed.) Vertebrados actuales y fósiles de Venezuela, Vol. 1, Mérida, Venezuela: 55–100.

Vari, R. 1992. Systematics of the Neotropical Characiform Genus *Cyphocharax* Fowler (Pisces:Ostariophysi). Smith.Contr. Zool. 529:1–137.

Vari, R. 1995. The Neotropical Fish Family Ctenoluciidae (Teleostei:Ostariophysi; Characiformes: supra and infrafamilial phylogenetic relationships with a revisionary study. Smith. Contr. Zoology, 654:1–97.

Welcomme, R. 1979. Fisheries Ecology of Floodplain Rivers. London Logman.

Winemiller, K. 1989. Pattern of variation in life story among South American fishes in seasonal environments. Oecologia, 81:225–241.

Capítulo 7

Metodologías Ecohidrológicas y Ecohidrográficas aplicadas a Conservación de la Vegetación Ribereña: el Río Caura como Ejemplo

Judith Rosales, Nigel Maxted, Lourdes Rico-Arce y Geoffrey Petts

RESUMEN

Este trabajo combina análisis de ecohidrográfica y estudios jerárquicos de sistemas riverinos para su utilización aplicable a conservación. Realizamos un análisis espacial utilizando Sistemas de Información Geográfico (SIG) con datos contentivos de georeferencias de *taxa-indicadores* para el estudio de tendencias en diversidad de especies a lo largo del corredor de bosque ripa-rino utilizando la familia Leguminosae. Entonces comparamos estos resultados entre diferentes subcuencas de los ríos Amazonas y Orinoco usando generos seleccionados de la Tribu Ingae. Los análisis entre cuencas y dentro de cuencas del número de especies, rareza, comonalidad y distinctividad dentro e intercuencas indican la importancia de conservación de los paisajes de bosque ribereño bajo en el Río Caura. Nuestros resultados demuestran que la conservación de la Cuenca del Río Caura es crítica, no solamente porque el Caura posee un gran número de especies, sino que también las especies conforman una ciomunidad única originarias de un número de cuencas y fitoregiones diferentes. El Río Caroní, el cual posee una flora similar al Caura se encuentra altamente perturbado. Las secciones más críticas para conservación de la cuenca del Río Caura son los sectores por debajo del Salto Pará. El análisis intercuencas muestra patrones regionales diferentes entre los géneros de Ingae seleccionados que los relacionan a regiones fitogeográficas y tipos de río. Además del Río Caura, el Río Negro en Brasil y Venezuela también es importante para realizar un esfuerzo de conservación multinacional. Demostramos que el método ecohidrológico representa una herramienta de conservación importante para analizar la diversidad en bosques ribereños. Es también útil para ayudar a seleccionar y conservar áreas ribereñas de reserva *in situ.*

INTRODUCCIÓN

Como ha sido definido por la Convención sobre la Diversidad Biológica (UNEP 2000) la conservación *in situ,* significa: i) la conservación de ecosistemas y hábitats naturales; y ii) el mantenimiento y recuperación de poblaciones viables de especies en sus sistemas naturales; y en el caso de especies domesticadas o cultivadas, iii) recuperación en los alrededores donde ellas han desarrollado sus propiedades distintivas. El establecimiento de una red de áreas protegidas ha sido propuesto en el Artículo 8 de la Convención y es visto como vital para la conservación de los recursos naturales y culturales del mundo. Los valores de las áreas protegidas van desde la protección de hábitats naturales y sus floras y faunas asociadas, hasta el mantenimiento de una estabilidad ambiental en las regiones cercanas.

Indicadores usados para la selección de reservorios para la conservación *in situ* incluyen: riqueza de especies, distinción biológica, estado de conservación actual y valor económico (Dinerstein et al.1995; Maxted et al. 1997b). Los datos requeridos para lograr estos valores son, sin embargo, difíciles de evaluar. Es por ello que se ha sugerido que las estrategias de conservación para plantas deben priorizar la preservación de tantas especies como sea posible (particularmente endémicas) (Prance 1997).

En este contexto, un alcance ecogeográfico como el propuesto por Maxted et al. (1997b) para la conservación genética *in situ*, pudiera ser útil para el establecimiento de prioridades de conservación de ecosistemas boscosos ribereños. Ecogeografía es definida por Maxted et al. (1995), como un proceso de obtención y síntesis de información ecológica, geográfica y taxonómica para un taxón particular. Esta incluye aspectos de su biología tales como: taxonomía, ecología, distribución, fenología, biología reproductiva, diversidad genética, almacenaje de semillas, comportamiento, entre otros). La selección de *taxa-indicadores* es importante en regiones tropicales debido, a que en muchas áreas, la composición florística es desconocida y no es práctico darle a todos los taxa igual prioridad. Sin embargo, una selección objetiva de *taxa-indicadores* esta basada en principios lógicos científicamente repetibles y económicos relacionados a los valores percibidos de las especies en el rango de las comunidades escogidas para conservación. Después de seleccionar *taxa-indicadores* para conservación, es necesario llevar a cabo un registro ecogeográfico y un análisis de los datos obtenidos (Maxted et al. 1997c). Los valores resultantes mencionados anteriormente pueden ser usados para asistir en la formulación de prioridades de conservación.

En ecosistemas ribereños, un enfoque ecogeográfico (Maxted et al. 1995, 1997a) puede ser unido a un estudio jerárquico de los ríos (Petts y Amoros 1996). En este enfoque, el corredor ribereño (canal del río y las tierras adyacentes afectadas por las características de hábitats ribereños) es considerado en escalas a través de un espacio anidado y progresivamente es escalas más limitado:

- Red de drenaje de la cuenca;
- Sectores funcionales o segmentos del río, los cuales son definidos por criterios hidrológicos y geomorfológicos;
- Unidades 'ecológicas' funcionales asociadas con ambientes caracterizados por comunidades de plantas típicas indicadoras de condiciones de hábitat localizados o particulares.

La terminología exacta es de importancia menor y generalmente varía en otras estrategias jerárquicas donde: 'tramo', 'sección' o 'sector' son mencionados usualmente como sinónimos (Curry and Slater 1986; Frissel et al. 1986; Gregory et al. 1991). La característica clave de asociación en de los enfoques hidrosistémicos y ecogeográficos es aquel en la cual la localización de cualquier especies de planta a los largo de un bosqe ribereño puede ser descrito de acuerdo con una serie de descriptores jerárquicos del ambiente ribereño. Cada comunidad debe entonces estar unido a factores de influencia probable en un arreglo de escalas.

Este trabajo pretende probar la utilidad de este enfoque ecohidrológico con datos provenientes de la cuenca del Río Caura con la finalidad de determianr prioridades de conservación. Un análisis de la posibilidad del uso de éste enfoque mixto para las bioregiones suramericanas fue realizado mediante el contraste con resultados previos obtenidos del uso de inventarios extensivos de especies de árboles y datos ambientales relacionados en áreas muestreadas en la Cuenca del Río Caura (Rosales 2000).

MÉTODOS

Registro Ecogeográfico de los Hidrosistemas Fluviales

La metodología propuesta sigue un enfoque espacial. Un Sistema de Información Geográfico (SIG) es usado para georeferenciar datos provenientes de diferentes fuentes. Un análisis espacial de los patrones de riqueza de especies y de las variaciones ambientales pertenecientes a bases de datos específicas fue obtenida usando versiones digitalizadas de redes de drenaje en las cuencas de los ríos seleccionados. El *software* usado es (i) ARCView 3.1, (ii) EXCEL y ACCESS para el manejo de las bases de datos y (iii) SPSS para los cálculos estadísticos.

El enfoque hidrosistémico funcional (Petts and Amoros 1996) es usado para el análisis espacial dentro de la cuenca del Río Caura (un tributario de aguas negras del Río Orinoco drenando el Escudo de Guayana, Rosales et al. 2002b). La comparación entre cuencas incluye tributarios de los ríos Amazonas y Orinoco dependiendo de criterios de clasificación de aguas como es revisado por Rosales et al. (1999).

El corredor ribereño en el Río Caura fue clasificado por Rosales et al. (2003) en unidades, sectores y paisajes funcionales. El corredor ribereño indica un área del río, sus bancos y las tierras cercanas incluyendo el río y su canal junto con su vida silvestre asociada y el ecosistema ribereño adyacente (siguiendo a Angold et al. 1995). Unidades funcionales representan diferentes unidades geomorfológicas p.e. remansos, barras de meandro, terrazas y cubetas o pantanos. Sectores funcionales son definidos como segmentos del río diferenciados por discontinuidades espaciales en dos variables hidrológicas – la descarga estimada, la cual incrementa después de la confluencia con un tributario y la pendiente del valle, la cual está relacionada con la presencia de extensos rápidos, caídas y cataratas. Un paisaje ribereño funcional representa un grupo de sectores ribereños funcionales espacialmente conectados y relacionados a procesos fluviales similares dominantes, tipos de vegetación y patrones de uso de la tierra. (Rosales et al. 2003). Este une sectores funcionales espacialmente contiguos y conectados que comparten sectores funcionales similares así como grupos de unidades funcionales.

Sectores y paisajes ribereños funcionales (para un detalle mayor ver Rosales et al. 2002) son considerados como elementos básicos para un modelo de Sistema de Información Geográfico (SIG) en un análisis dentro de una cuenca, mientras que los corredores ribereños completos son usados para un análisis entre cuencas. Una investigación de la variación taxonómica en *taxa-indicadores* en relación a los cambios en los sectores y paisajes funcionales (análisis dentro de la cuenca) y diferentes corredores ribereños (análisis entre cuencas) es conducido usando procedimientos generales indicados en la metodología propuesta por Maxted et al. (1997b) para la conservación genética *in situ*:

Fase 1 incluye: (i) delimitación de las áreas o regiones a trabajar , (ii) selección de *taxa-indicadores*, (iii) identificación de información del taxon (iv) diseño, y (v) construcción de bases de datos.

Fase 2: (i) colección de datos ecogeográficos, (ii) selección de ejemplares representativos, (iii) verificación de datos, (iv) anális de datos geográficos, ecológicos y taxonómicos, y (v) establecimiento del estado actual de conservación.

Fase 3: (i) Una síntesis ecogeográfica, y (ii) identificación de las prioridades de conservation.

I. Análisis dentro de la cuenca

a) **Fase 1. La selección del área de estudio y *taxa-indicadores***
El área de estudio escogida para el análisis dentro de cuencas fue el Río Caura. La Familia Leguminosae fue seleccionado como el *taxon-indicador* por las razones siguientes:

1. La Familia Leguminosae es la más comun y poseen la más alta riqueza entre las especies de árboles de los bosques ribereños de las cuencas del Amazonas y Orinoco (Rosales et al. 1999; Rosales et al. 2002).
2. Aunque las especies de Leguminosae no siempre son las más importantes en términos de abundancia de individuos, ellas ocupan todas las unidades funcionales presentes en los corredores ribereños. Algunas de estas especies son indicadoras de ciertas unidades funcionales (p.e. en sectores estructuralmente controlados con baja sinuosidad: la especie *Macrolobium angustifolium* es dominante en las cubetas y las confluencias de caños y la especie *Inga vera* es dominante en las barras arenosas y barras de meandros). Las especies mencionadas podrían también ser consideradas como especies clave de esos ambientes o unidades funcionales dada su gran biomasa y contribución a la productividad y cadenas alimenticias.
3. Muchas de las especies ribereñas dentro de las Leguminosae son usadas por las comunidades humanas locales y pueden ser consideradas como taxa de alta importancia económica a niveles locales y regionales en la Cuenca del Río Caura y áreas adyacentes ribereñas a lo largo del Río Orinoco (Rosales 1990; Knab-Vispo et al. 1997; Rosales et al. 1997).
4. Numerosas especies de leguminosas poseen una importancia funcional debida a su asociación con bacterias fijadoras de nitrógeno.

b) **Fase 2. Colección de datos ecogeográficos**
La información de todas las especies de leguminosas colectadas en la cuenca del Río Caura ha sido obtenida de los ejemplares depositados en los herbarios siguientes:

- Herbario Nacional de Venezuela (VEN)
- Herbario Ovalles – Facultad de Farmacia, Universidad Central de Venezuela (MYF)
- Herbario Universidad Nacional Experimental Ezequiel Zamora (PORT)
- Herbario Regional de Guayana, Jardin Botánico del Orinoco (GUYN)
- Royal Botanic Gardens, Kew (K)

También uamos información provenientes de herbarios en las siguientes localizaciones electrónicas:

- TROPICOS database, Missouri Botanical Garden (http://mobot.mobot.org/Pick/Search/pick.html)
- El Vascular Plants Type Catalogue, en el New York Botanical Garden, donde están depositados muchos de los tipos de las especise de Leguminosae colectadas en el Neotrópico (http://www.nybg.org/bsci/hcol/vasc/tflow.html)

Las variables siguientes fueron obtenidas y registradas en la base de datos: subfamilia, especies, colector, número de colección, hábitat, sitio, latitud y longitud. Esta base de datos fue relacionada como puntos de una capa temática en un proyecto de Sistema de Información Geográfica (SIG) desarrollado en ARCView. El sistema también usa capas vectoriales de mapas del Río Caura previamente digitalizados que son mostrados en la Figura 7.1 (basado en Rosales 2003). Las capas poseen: (i) el canal principal dividido en sectores entre la confluencia de los grandes o principales tributarios o la presencia de extensos rápidos y saltos y (ii) canales principales de tributarios. La capa correspondiente al canal principal esta subdividida en 15 sectores funcionales los cuales han sido clasificados por Rosales et al. (2002) como paisajes funcionales ribereños homogéneos, tomando en consideración una serie de variables ambientales asociadas. La tabla asociada maneja una base de datos con las siguientes variables: sector funcional, descarga estimada, pendiente, ancho, potencia unitaria del rio y sinuosidad.

Los siguientes paisajes funcionales a lo largo del Caura fueron descritos por Rosales et al. (2003):

- Sectores 1 a 3 se encuentran localizados en el Bajo Caura entre los rápidos Cinco Mil y la boca del Caura. Están influenciados por el efecto de represamiento del pulso de inundación del Orinoco el cual controla las variaciones hidrológicas en las características de las inundaciones y la bioquímica del plano de inundación del Caura. Estos sectores están caracterizados por valles no confinados hasta valles rocosos confinados en un patrón de canal

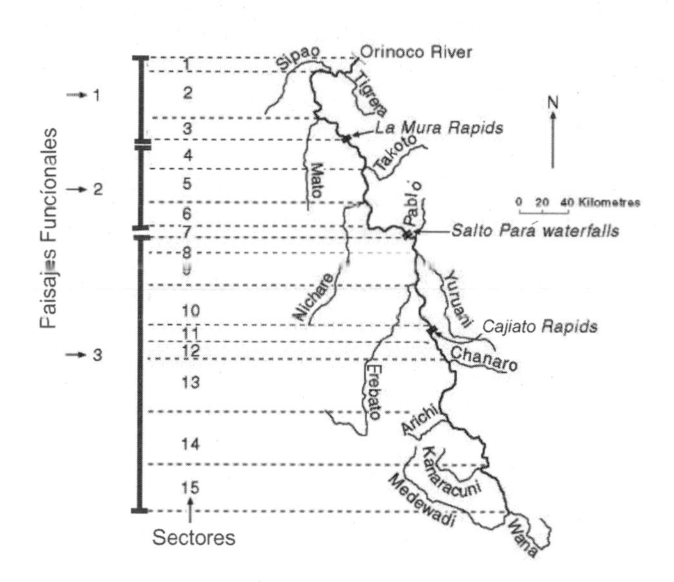

Figura 7.1. Versión digitalizada de la Cuenca del Río Caura mostrando la localización de sectores y los paisajes ribereños.

simple desde sinuoso hasta meándrico. El clima es una transición desde macrotérmico húmedo a seco, en altitudes de 20 a 50 msnm. La geología del valle está dominada por la Formación Mesa. Estos sectores están caracterizados por la presencia de las siguientes unidades geomorfológicas: pantanos de confluencia, madreviejas, canales abandonados, lagunas de inundación, bancos, diques, sistemas de barras de meandro, islas aluvionales y barras arenosas laterales.

- Sectores 4 a 6 están localizados en el Bajo Caura, entre los rápidos Cinco Mil y Salto Pará. Ellos presentan un valle rocoso confinado, patrones de canal simple hasta multiramificado. El clima varía desde macrotérmico húmedo a una transcición húmeda a seca. Las altitudes varian desde 60 a 100 msnm. La geología del valle está dominada por el Grupo Cuchivero. El area presenta las siguientes unidades geomorfológicas: i) unidades erosionales y de canal (márgenes rocosos, bancos y barras arenosas laterales) y ii) depósitos de la planicie aluvial producto de sedimentación vertical además de unidades dentro de la vega y el canal (diques, pantanos, pantanos de confluencia, áreas de remanso y barras de meandro).
- Sectores 7 a 15 están localizados en el Caura Superior, entre Salto Pará y la confluencia con el Río Kanaracuni. El río se presenta dentro de un valle rocoso confinado, con patrones de canal simple o multiramificados. El clima en macrotérmico húmedo. Altitudes variando desde 200 a 500 msnm, y la geología del valle está dominada por rocas de Cuchivero. El sector presenta las siguientes unidades geomorfológicas: i) unidades erosionales y de canal (márgenes rocosos, bancos y barras arenosas laterales); y ii) depósitos de la planicie aluvial producto de sedimentación vertical además de unidades dentro de la vega y el canal (diques, pantanos, pantanos de confluencia, áreas de remanso y barras de meandro).

c) **Fase 3. Análisis de datos ecogeográficos y establecimiento de prioridades de conservación.**
Análisis en el SIG permiten la busqueda espacial de las bases de datos de las leguminosas a lo largo de las tres principales áreas de variación ambiental que ocurren dentro de los paisajes ribereños. Los registros seleccionados de las bases de datos fueron resumidos como el número total de especies por sector funcional y paisaje ribereño considerando las subfamilias: Caesalpinioideae, Mimosoideae y Papilionoideae. También fueron resumidos la tribu Ingeae y el género *Inga* dentro de las Mimosoideae.

Valores de los sectores funcionales y paisajes funcionales ribereños para la conservación le fueron dados de acuerdo a los siguientes criterios: la contribución numérica a la biodiversidad, lo distintivo en términos de la composición de especies (rareza, unicidad) y por último la presencia y abundancia de *taxa-indicadores*. Algunos criterios dentro de los muchos reconocidos para la selección de los *taxa-indicadores* para conservación de la biodiversidad fueron usados para un exámen preliminar de la viabilidad de la metodología (Spellerberg 1996; Maxted 1997d):

Presencia en los sectores de las principales especies *Inga vera* y *Macrolobium angustifolium* consideradas como clave debido a su dominancia en hábitats funcionales de barras y pantanos respectivamente (Rosales et al. 2001; Rosales 2002).

Presencia y abundancia de las especies: *Acosmium nitens* y *Eperua jenmanii* ssp. *sandwithii*, las cuales son socio-economicamente utilizadas particularmente por poblaciones humanas locales (Knab-Vispo et al. 1997; Rosales et al. 1997).

Presencia y abundancia de especies amenazadas p.e. *Acosmium nitens*, la cual frecuentemente es utilizada para la construcción de viviendas en estas áreas.

II. El análisis entre cuencas

a) **Fase 1. La selección de las áreas de estudio y de taxa-indicadores**
Las áreas selecionadas fueron las cuencas del Amazonas y Orinoco (ver Rosales et al. 1999) y el *taxon-indicador* fue la Tribu Ingeae dentro de la familia Leguminosae, subfamilia Mimosoideae que incluye seis géneros:

Macrosamanea, Zygia, Hydrochorea, Albizia, Abarema y *Cedrelinga.*

La Tribu Ingeae fue seleccionada debido a que posee un alto porcentaje de especies asociadas a hábitats ribereños en el Neotrópico (Rico-Arce 1987; Guinet y Rico-Arce 1988; Barneby y Grimes 1996, 1997; Rico-Arce 2000).

b) **Fase 2. Obtención de datos ecogeográficos**

Para este propósito, la presencia de especies ribereñas para un número seleccionado de ríos fue registrada a partir de la revisión más reciente de la literatrura donde se encontraron datos ecogeográficos y mapas de distribución de los géneros de Ingeae en América excluyendo el género *Inga* (Barneby y Grimes 1996, 1997).

c) **Fase 3. Análisis de datos ecogeográficos y establecimiento de prioridades de conservación**

Los ríos seleccionados para este estudio pertenecen a los siguientes tipos:

1. Ríos de aguas turbias o blancas que transportan aguas de los Andes ricas en nutrientes y sedimentos
    - Amazonas Central y Bajo (entre el Solimões y el Tapajos)
    - Bajo Río Madeira
    - Río Japurá

2. Ríos de aguas negras o transparentes oligotróficas drenando el Macizo de Guayana y las tierras bajas del Amazonas.
    - Río Caura
    - Río Caroní
    - Alto Orinoco-Casiquiare
    - Río Negro

    Después de la muestra de especies por ríos, los datos fueron registrados en una tabla en términos de presencia-ausencia por cada río y los siguientes análisis fueron llevados a cabo:

1. Riqueza, como el número total de especies

2. Porcentajes de rareza y comonalidad como el porcentaje de especies por río clasificadas como especies muy raras o muy comunes respectivamente de acuerdo a cuatro intervalos relativos la rareza:
    - Muy raro—su presencia en un solo río
    - Raro—su presencia en dos a tres ríos
    - Común—su presencia en cuatro a cinco ríos
    - Muy común su presencia en más de seis ríos

3. Distinción Geográfica: como la separación obtenida a partir de diferentes cuencas utilizando un análisis TWINSPAN (Gauch 1979) y un análisis de Correspondencia Canonica usando CANOCO (ter Braak 1988) con la presencia-ausencia de especies por cuenca. Las técnicas de ordenamiento son útiles para explicar resultados del estudio de gradientes ambientales a diferentes escalas (Greig-Smith 1983) y ellos pueden revelar patrones de distribución de especies relacionado a ambientes y/o singularidad evolutiva.

Una evaluación final es conducida mediante un análisis de la riqueza de especies, rareza y comonalidad asi como también distinctividad usando distribución de especies como indicadores para la conservación.

RESULTADOS

I. Análisis dentro de la Cuenca

Riqueza y rareza de especies

Quinientas treinta y ocho muestras de leguminosas leñosas fueron registradas provenientes de las áreas bajas dentro de toda la Cuenca del Río Caura. Aunque este número refleja un bajo esfuerzo de colecta en comparación con el tamaño de la cuenca (45.336 km^2), es el mejor realizado hasta los momentos. Como fue señalado por Huber (1995), el mayor esfuerzo de colecta en la Guayana Venezolana ha sido concentrado a lo largo de ríos. Debido a esto, las 341 colecciones (Apendice 9) incluídas en una franja de 5 km de ancho a lo largo de los sectores en aproximadamente 500 km del corredor ribereño del Caura, son consideradas como altamente representativo de éste estudio preliminar. Una riqueza total de 110 especies de leguminosas leñosas es resumida en la Tabla 7.1. Esta tabla también presenta los resultados de la riqueza de especies, número de colecciones (relacionadas con el esfuerzo de colecta) y la fracción entre riqueza de especies/número de colecciones por sector y agrupadas por ambientes ribereños. El análisis por sector, indica que el número de especies es dependiente del esfuerzo de colecta, encontrándose una correlación positiva fuerte hacia el número de colecciones (R^2 = 0.96). La correlación es baja cuando usamos el nivel de paisaje ribereño (R^2 = 0.84) dado que la probabilidad de colecta de la misma especie en diferentes paisajes incrementa. Este nivel es considerado más apropiado para el análisis dado que el paisaje constituye un nivel functional del ecosistema ribereño. No existe una correlación significativa entre la longitud del canal del río para cada sector y el número de colecciones o el número de especies. Un resultado similar fue encontrado al nivel de paisaje funcional. En todas las relaciones, la fracción entre el número de colecciones/longitud del corredor y número de especies/longitud del corredor parece ser alto en el paisaje ribereño intermedio Cinco Mil-Salto Pará. Este paisaje ribereño, ha sido, el mas intensivamente estudiado en varios proyectos (Williams 1942; Knab-Vispo 1998; Rosales 2000).

La Figura 7.2 muestra algunos de los resultados preliminares encontrados sobre las relaciones de algunas medidas hidrológicas a nivel de sector (descarga estimada, ancho del canal, potencia unitaria del río e índice de sinuosidad) y dos medidas de diversidad (riqueza de especies y riqueza de especies/número de colecciones). Las tendencias indican que los sectores que poseen valores intermedios en las variables hidrológicas poseen alta riqueza en especies. Sin embargo, la fracción de riqueza de especies/número de colecciones nos da una tendencia negativa. Aunque debemos enfatizar el carácter preliminar de estas tendencias, los resultados parecen tener importancia para futuros estudios.

Cuando evaluamos por paisaje, el número de especies fue más alto en los paisajes intermedios (Cinco Mil-Salto Pará). La Figura 7.3 también muestra que este patrón se repite separadamente para las diferentes subfamilias, exceptuando la Papilionoideae. Dentro de las Mimosoideae, el mismo patrón es observado en el género *Inga*, mientras que toda la Tribu Ingeae, mostró baja riqueza en paisajes del bajo Caura (Boca del Caura-Cinco Mil). La fracción riqueza de especies/número de colecciones sin embargo, resultó en valores relativamente altos en sectores río arriba, pero cuando fue evaluado por paisaje, fueron encontrados valores mayores en los paisajes intermedios (Tabla 7.1).

El porcentaje de similaridad entre los grupos de sectores ribereños es mostrado en la Figura 7.4. Los tres paisajes ribereños poseen similar porcentaje de especies comunes, sin embargo, la comparación Cinco Mil-Boca del Caura con Cinco Mil-Salto Pará, resultó en los más bajos porcentajes. La comparación entre Cinco Mil-Salto Pará y Salto Pará-Kanaracuni resultó en los mayores. Las subfamilias Caesal-pinioideae y Mimosoideae muestran el mayor contraste si se comparan con la Papilionoideae. En terminos de la rareza

Tabla 7.1. Resumen de riqueza de especies a lo largo del corredor ribereño del Río Caura. L1, L2 y L3 representan los paisajes ribereños, los números debajo representan los sectores funcionales.

	L-1			L-2			L-3				
Especies	**1**	**2**	**3**	**4**	**5**	**6**	**7**	**10**	**11**	**12**	**13**
Número total de especies por sector (110)	14	24	9	26	48	19	34	15	14	14	15
Total de especies por paisaje ribereño		34			74				62		
Total de colecciones por sector	23	36	15	39	65	26	58	21	21	17	20
Total de colecciones por paisaje		74			130				137		
Relación del número de especies/total colecciones por sector	0.6	0.7	0.6	0.7	0.7	0.7	0.6	0.7	0.7	0.8	0.18
Tamaño del corredor (km)		134			127				189		
Relación del número de colecciones/Tamaño del corredor		0.55			1.02				0.72		
Relación del número de especies/total colecciones por paisaje		0.46			0.57				0.45		
Relación del número de especies/Tamaño del corredor		0.25			0.58				0.33		
Relación del número de especies/número de colecciones. Tamaño del corredor x 100		0.34			0.45				0.24		
Número de especies únicas al paisaje y porcentaje de especies únicas por número de especies presentes por tamaño de corredor		9 20%			28 32%				25 22%		

relativa, un número alto de especies únicas fue encontrada en los paisajes del sector Cinco Mil-Boca del Caura.

Distribución de especies indicadoras

La distribución dentro del Caura de especies clave consideradas como importantes para la conservación local (*Acosmium nitens* y *Eperua jenmanii* ssp. *sandwithii*) es mostrada en la Figura 7.5. Si *Acosmium nitens*, es seleccionada para conservación, ella sólo ocurre en sectores del Bajo Caura. Sectores uno a cuatro, muestran la presencia de especies típicas de Igapó (Prance 1979; Kubitski 1989), indicando una baja recurrencia de estos hábitats. *Eperua jehnmanii* ssp. *sandwithii*, una especie de particular importancia económica para las poblaciones Ye'kwana sólo está presente en sectores del Caura Superior. Esta especie está presente, pero no domina, en los diques de ésta región del Caura (Briceño 1995). También, se le encuentra en tierras altas boscosas dando un amplio margen de oportunidades si la misma es considerada para programas de conservación. *Inga vera*, una especie indicadora de barras laterales y barras de meandro y *Macrolo-*

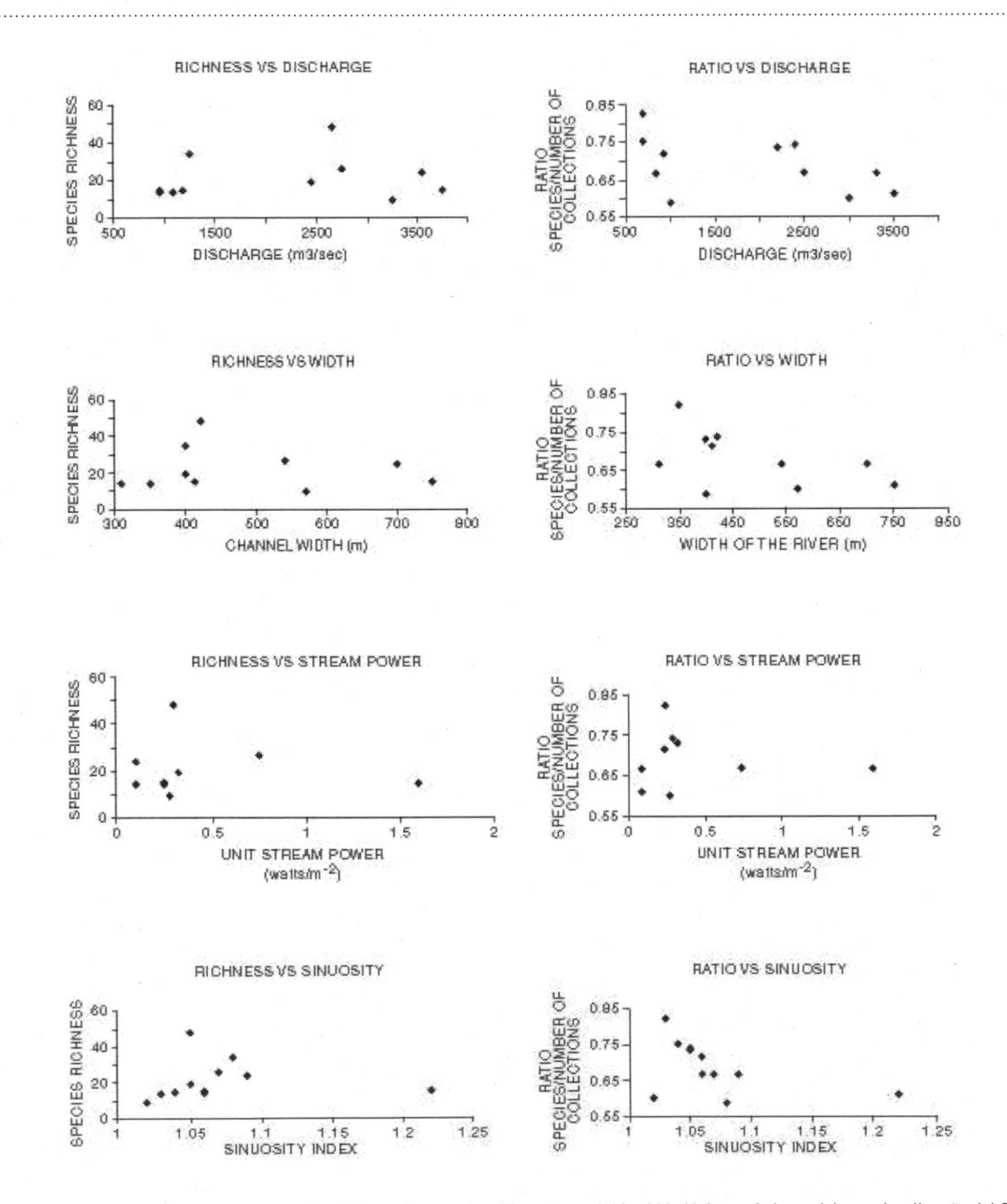

Figura 7.2. Relaciones de la diversidad de especies de leguminosas y las diferentes variables hidrológicas a lo largo del corredor ribereño del Caura.

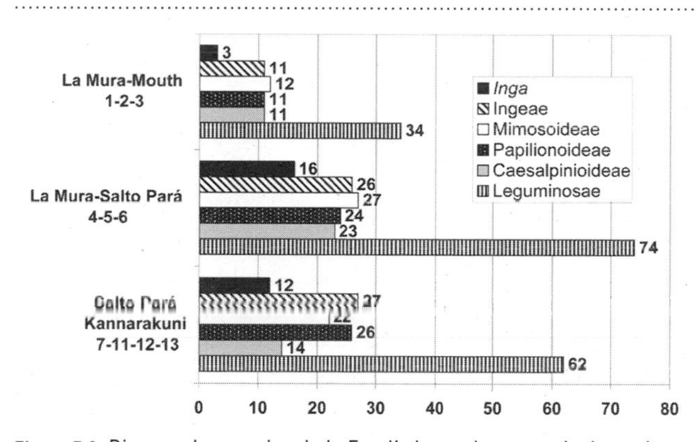

Figura 7.3. Riqueza de especies de la Family Leguminosae y el género *Inga* en paisajes ribereños diferentes en el corredor ribereño del Río Caura. La Mura = Raudal Cinco Mil.

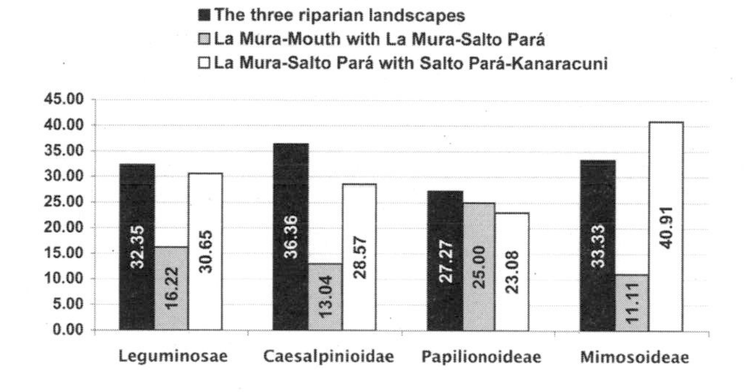

Figura 7.4. Porcentajes de similaridad en la composición de especies entre diferentes ambientes ribereños a lo largo del corredor ribereño del Caura. La Mura = Raudal Cinco Mil.

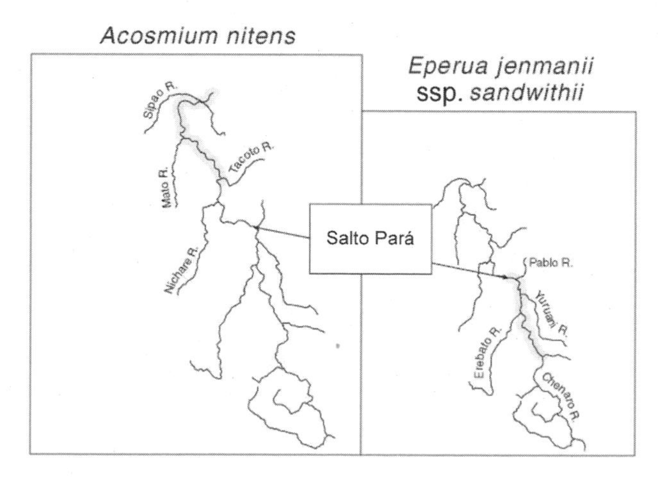

Figura 7.5. Distribución de dos especies de leguminosas a lo largo del corredor ribereño del Caura.

bium angustifolium, una especie indicadora de pantanos, son sin embargo ampliamente distribuida en sectores rubereños superiores. Sin embargo, estas especies poseen baja frecuencia en el Río Erebato, principal tributario del Caura, dominado por alta velocidad de corriente y bajo desarrollo de estructuras deposicionales.

Siguiendo este enfoque, de los datos mostrados en el Apéndice 9, es possible hacer una evaluación similar en la distribución observada para cada sector de otras especies indicadoras tales como aquellas utilizadas por las poblaciones humanas (Knab-Vispo *et al* 1997):

Directamente:

1. En la construcción de canoas: *Andira surinamensis, Sclerolobium guianense, Macrolobium multijugum*;
2. En construcción de viviendas: *Etaballia dubia, Cassia moschata, Dialium guianense, Centrolobium paraense, Swartzia arborescens, Acosmium nitens, Sclerolobium guianense*;
3. Como alimento: *Dialium guianense, Hymenaea courbaril, Dypterix punctata, Inga alba, Inga bourgoni, Inga coruscans, Inga crocephala, Inga densiflora, Inga dumosa, Inga edulis, Inga laurina, Inga leiocalycina, Inga nobilis, Inga oerstediana, Inga pilosula, Inga splendens, Inga thibaudiana, Inga umbellifera.*

Indirectamente:

1. Comestibles por especies de peces importantes en las pesquerías comerciales y deportivas: *Campsiandra laurifolia, Dialium guianense, Macrolobium angustifolium, Macrolobium acaciaefolium, Macrolobium multijugum, Sclerolobium chrysophyllum, Sclerolobium guianense, Senna silvestris* var. *silvestris, Dalbergia glauca, Dalbergia hygrophylla, Dypterix odorata, Dypterix punctata, Swartzia leptopetala, Abarema jupunba, Hydrochorea corymbosa, Inga vera* y otras referidas en 3 (sección anterior), *Newtonia suaveolens, Parkia pendula, Zygia cataractae, Zygia latifolia, Zygia unifoliolata.*

II. Análisis entre cuencas

Una lista de 51 especies of Ingeae (Apéndice 10) obtenidas de mapas de distribución publicados por Barneby y Grimes (1996, 1997) es ordenada de acuerdo a una clasicación usando TWINSPAN (Figura 7.6). Todas estas especies son reportadas como características de hábitats ribereños. Un resumen del número de especies, así como del porcentaje, rareza y comonalidad están dados en la Tabla 7.2. Se resalta el bajo número de especies en las cuencas de los ríos Caroní y Caura, valores altos de rareza para los ríos Negro, Madeira y Japurá y una alta comonalidad para el Caura, Caroní y Madeira.

En términos de singularidad geográfica, la Figura 7.6 muestra los resultados de la clasificación TWINSPAN. Una diferenciación entre las regiones fitogeográficas amazónicas y guayanesas puede ser observada en el gráfico. En la región amazónica se separa al Río Negro (aguas negras) como un

Tabla 7.2. Resúmen de los indicadores de conservación para los diferentes corredores ribereños en las cuencas del Orinoco y Amazon usando los resultados resumidos en la base de datos de Ingae.

	Negro	Alto Orinoco-Casiquiare	Amazon	Madeira	Japurá	Caroní	Caura
Número de especies	37	27	27	21	20	13	9
Porcentaje de rareza	11	7	4	10	10	0	0
Porcetaje de comonalidad	16	22	22	29	20	38	56

grupo distinto de los ríos de aguas blancas o turbias (Japurá, Madeira y Amazonas), mientras que los ríos de aguas negras de la región guayanesa aparentemente se diferencian en componentes Este y Oeste de la Región de Guayana siguiendo a Huber (1995).

El Análisis de Correspondencia Canónica refleja que los tres principales ejes de variación explican el 70% de la varianza de la distribución de las especies con un autovalor de 1.146, los primeros dos ejes explicando el 53% de la varianza. La Figura 7.7 muestra la distribución de los ríos a lo largo de éstos ejes mostrando una distinción geográfica entre los ríos. Esta distinción probablemente está representando la evolución de las cuencas del Amazonas-Orinoco. El Río Negro y Alto Orinoco-Casiquiare, serían las tierras bajas más recientes drenando arenas blancas podzólicas separadas de las regiones antiguas del Escudo donde el Caura, Caroní (Escudo de Guayana) y el Madeira (tierras bajas que drenan el Escudo Brasilero) están localizados. La cercana relación entre el Río Japurá con el Río Negro y Casiquiare puede probablemente ser consecuencia de vínculos fitogreográficos previos. En la actualidad, un tributario del Caquetá, tributario del Japurá, drena también suelos podzolicos de arenas blancas (Duivenvoorden 1995).

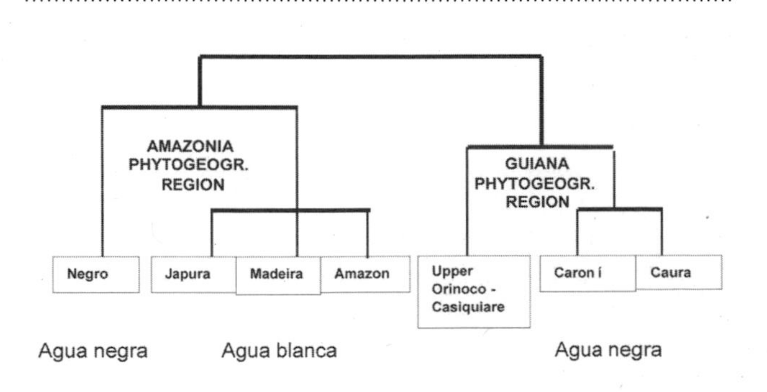

Figura 7.6. Clasificación de corredores ribereños dentro de las cuencas de los ríos Orinoco y Amazon River de acuerdo con un análisis TWINSPAN de presencia-ausencia de especies de géneros seleccionados de Ingeae.

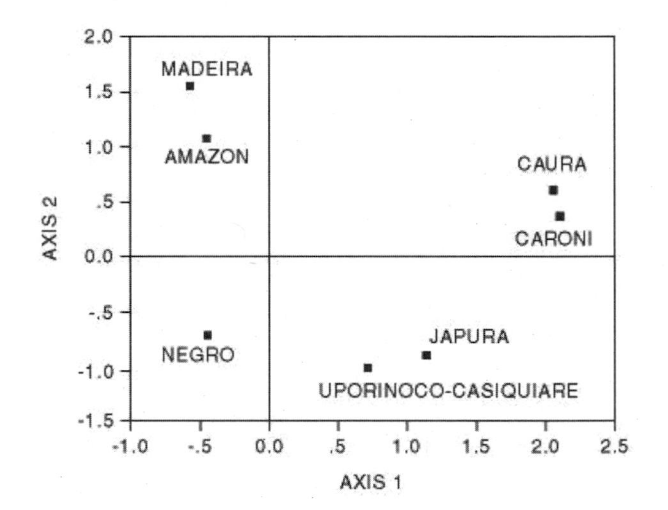

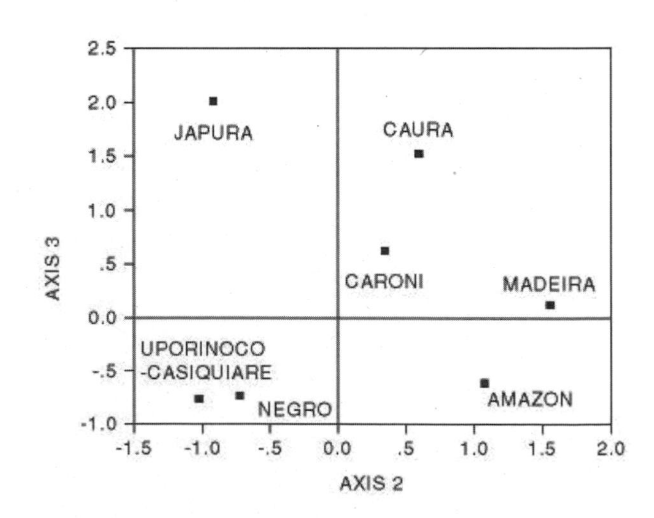

Figura 7.7. Resultados de un Análisis de Correspondencia Canónica de corredores ribereños en las cuencas de los ríos Orinoco y Amazonas usando presencia-ausencia de especies de géneros seleccionados de Ingeae.

DISCUSIÓN

Prioridades de Conservación para la Cuenca del Río Caura

Los resultados del análisis de datos ecogeográficos son comparables con aquellos obtenidos a partir de exámenes completos de árboles de los paisajes ribereños usando 74 lotes de 0.1 ha (Rosales 2000; Rosales et al. 2002b). Estos autores dan un total de número de especies de 56, 191 y 180 en 2.3, 2.8 y 2.3 has respectivamente, con densidades de especies de 24.4, 68.2 and 78.3 species/ha. Los resultados también muestran que en los paisajes ribereños del Bajo Caura, representado por la zona de confluencia bajo el efecto de agua represada por el Orinoco (Rosales et al. 2001), posee la menor diversidad, pero es más único a nivel de la cuenca, que aquellos asociados a suelos rocosos y canales estructuralmente controlados. Estos últimos, presentan sin embargo, los mayores valores de diversidad pero con taxa más ampliamente distribuidos.

En terminos de valores de conservación, los resultados indican que de los tres paisajes ribereños a lo largo del Corredor ribereño del Caura, los paisajes La Boca-Cinco Mil y Cinco Mil-Salto Pará, deben tener la más alta prioridad dado que juntos preservarían una mayor riqueza de especies y rareza relativa. Aunque los paisajes ribereños aguas arriba del Salto Pará posee un alto número de especies, también posee la mayor similaridad con el sector Cinco Mil-Salto Pará.

Si la recurrencia de sectores y paisajes funcionales es tan importante como la recurrencia de especies al evaluar una estrategia de conservación de la biodiversidad (Walker 1995), entonces el paisaje ribereño del bajo Caura cercano a la Boca de Caura, aunque poseen la menor diversidad, es único en términos de su baja recurrencia a nivel de la cuenca. Ciertas unidades funcionales dentro del Río Caura poseen el potencial para desaparecer, cuando se percibe la vulnerabilidad de los ecosistemas ribereños a cambios potenciales producidos por alteración ambiental, tal como por ejemplo la proyectada conexión Caura-Paragua (Rosales et al. 2001). Algunas de las especies indicadoras de éstas unidades poseen importancia económica para las poblaciones indígenas locales y las unidades funcionales a su vez dependen de la recurrencia de los sectores o paisajes ribereños donde ellas están presentes. Este es el caso de especies clave en las unidades de cubeta o barras de canal, *Macrolobium multijugum* e *Inga vera* respectivamente, las cuales están asociadas con sectores con baja corriente. Ríos cercanos como el Erebato, tributario del Caura en el Caura Medio no pueden asegurar el mantenimiento de tales unidades funcionales en esta región si se produce una diversión de aguas hacia el Paragua. Por otro lado, el potencial incremento del urbanismo a lo largo del Orinoco, incluyendo el Bajo Caura incrementará la amenaza sobre el estatus de conservación de una especie importante como *Acosmium nitens* usada para la construcción de viviendas. El uso experimental sostenible de especies económicamente valiosas como el ejemplo anterior y la restauración de sus hábitats debe ser considerada para asegurar la preservación de la unicidad del paisaje ribereño del Bajo Caura, pero el mantenimiento de niveles adecuados de riqueza de especies requerirían de monitoreo continuo.

Evaluación de las prioridades de conservación entre las cuencas del Amazonas y Orinoco

El alto porcentaje de comonalidad obtenida en el Río Caura is muy importante para su estado de conservación aunque el mismo no posea un gran rareza o riqueza de especies. Dado que el Río Caroní se encuentra altamente perturbado, el Caura posee el potencial para preservar un gran número de especies representativas del Escudo de Guayana y Amazonas. Por otro lado, la región del Río Negro ha sido considerada por muchos biogeógrafos caracterizada como un área de alto endemismo y diversidad, así como también la región del Alto Orinoco-Casiquiare (Prance 1982; Kubitzki 1989; Pires-O'Brien y O'Brien 1995). Aunque limitada por el bajo número de ríos analizados, los datos también reflejan una diversidad geológica, la cual esta relacionada no sólo con la fertilidad, sino también con la dinámica de los ríos y posiblemente con la evolución de las cuencas. Las cuencas del Caroní y Caura forman un grupo distinto que ocurre al Norte y Este de la Región Fitogeográfica de la Guayana (Huber 1995), mientras que la región del Alto Orinoco - Casiquiare y Río Negro pertenecen a la Provincia Oeste de la Guayana, siendo cercanos al Japurá debido a sus afinidades fitogeográficas. El Amazonas y Madeira forman otro grupo de ríos distintos. En término de la vegetación ribereña, los canales estructuralmente controlados del Caroní y Caura parecen ser un grupo intermedio entre los altamente fértiles bosques inundables dinámicos del Madeira, Amazonas y Japurá, y los altamente oligotróficos bosques de Igapó característicos del Río Negro y Alto Orinoco-Casiquiare. Los resultados de los análisis usando un enfoque ecohidrológico sugiere que el Río Negro en Brasil y Venezuela también tendría una alta prioridad en planes de conservación, debido a su alto número de especies, alta rareza y alta distinción. Dado las divisiones potenciales de las cuencas del Amazonas y Orinoco, compartidas por diferentes países, es necesario un razonamiento holístico, una estrategia regional multinacional para la conservación del corredor ribereño. La aplicación del enfoque ecohidrográfico para otros estudios presentado en este trabajo puede ser útil en el desarrollo de una estrategia para la conservación de la biodiversidad ribereña.

La importancia de la metodología en la selección de reservas para conservación in situ de biodiversidad ribereña

Nuestros resultados indican que el enfoque ecohidrológico puede tener utilidad para ayudar a una rápida evaluación en la selección de reservas *in situ* para la conservación de la biodiversidad ribereña. Los estudios dentro de la cuenca previamente realizados y los aportados por la expedición AquaRAP, han dado resultados similares al compararlos con aquellos obtenidos del análisis extensivo de parcelas en la cuenca del Río Caura, mientras que el análisis entre cuencas refleja patrones reportados por fitogeógrafos en términos de la importancia de las regiones donde los ríos fueron analizados. La metodología ecohidrográfica propuesta acá también ofrece una vía para integrar la diversidad genética. La dinámica e historia de conexión y fragmentación de los diferentes sectores ribereños puede ser asociada con niveles de diversidad genética. Por ejemplo, perturbaciones intermedias en las tasas de la dinámica del canal pueden ayudar a interpretar la localización de altos niveles de heterozygosis en taxa de las Leguminosae como fue encontrado a lo largo del Bajo Río Solimões en el Amazonas (Hill et al. 1978). Aunque limitado en términos de calidad y cantidad la información que puede ser encontrada en herbarios (los cuales frecuentemente carecen de localidades precisas) o en la literatura, el enfoque presentado en este trabajo podría ser útil en la planificación y conducción de evaluaciones rápidas de los valores de conservación tales como los "*Aquatic Gap*" (Haverland y Muir 1998) o AquaRAP.

LITERATURA CITADA

Angold, P., A. Gurnell y P. Edwards. 1995. Information from river corridor surveys. Water and Environmental Management 9, 489–498.

Barneby, R., y J. Grimes. 1996. Silk Tree, Guanacaste, Monkey's Earring: A generic system for the synandrous Mimosaceae of the Americas, Part I: *Abarema, Albizia*, and allies. Memories of the New York Botanical Garden 74, 1–292.

Barneby, R., y J. Grimes. 1997. Silk Tree, Guanacaste, Monkey's Earring: a generic system for the synandrous Mimosaceae of the Americas. Part II: *Pithecellobium, Cojoba*, and *Zygia.* Memories of the New York Botanical Garden 74, 1–292.

Briceño, A. 1995. Análisis fitosociológico de los bosques ribereños del Río Caura en el Sector Ceiato-Entreríos. Distrito Aripao del Edo. Bolívar. Forest Engineer Thesis, Universidad de los Andes, Mérida, Venezuela.

Curry, P., y F. Slater. 1986. A classification of river corridor vegetation from four catchments in Wales. Journal of Biogeography 13, 119–132.

Dinerstein, E., D. Olson, D. Graham, A. Webster, S. Primm, M. Bookbinder y G. Ledec. 1995. A conservation assessment of the terrestrial ecoregions of Latin America and the Caribbean. Published in association with The World Wildlife Fund. The World Bank, Washington, DC.

Frissel, C., W. Wiss, C. Warren y M. Huxley. 1986. A hierarchical framework for stream classification: viewing streams in a watershed context. Environmental Management 10, 199–214.

Gauch, H. 1979. TWINSPAN a FORTRAN program for two-way indicator species analysis and classification. Cornell University. New York.

Gregory, S., F. Swanson, W. Arthur McKee y K. Cummins. 1991. An ecosystem perspective of riparian zones: focus on links between land and water. BioScience 41, 540–551.

Greig-Smith, P. 1983. Quantitative Plant Ecology. Blackwell Scientific Publications, Oxford.

Guinet, Ph., y L. Rico-Arce. 1988. Pollen characters in the genera *Zygia, Marmaroxylon and Cojoba* (Leguminosae, Mimosoideae, Ingeae) a comparison with related genera. Pollen and Spores 30, 313–328.

Haverland, P., y T. Muir. 1998. Aquatic GAP—current status and next steps. GAP Annual Meetings - Available at http://www.gap.uidaho.edu/GAP.

Hill, R., G. Prance, S. Mori, W. Steward, D. Shimabukuru y J. Bernardi. 1978. Estudo eletroforetico da dinamica de variacao genetica em tres taxa ribeirinhos ao longo do rio Solimoes, America do Sul. Acta Amazonica 8, 183–199.

Huber, O. 1995. Vegetation. *En:* Berry, P. B. Holst y K. Yatskievych (eds.) Flora of the Venezuelan Guayana. Vol. 1. Introduction. Missouri Botanical Garden and Timber Press. Oregon. USA: pp. 97–160.

Knab-Vispo, C. 1998. A rain forest in the Caura Reserve (Venezuela) and its use by the indigenous Ye'kwana people. PhD Thesis, University of Wisconsin, Madison.

Knab-Vispo, C., J. Rosales y G. Rodríguez. 1997. Observaciones sobre el uso de las plantas por los Ye'kwana en el bajo Caura. *En*: Huber, O. y J. Rosales (eds.). Ecología de la cuenca del Río Caura. II. Estudios específicos. pp. 215–257. Scientia Guaianae, 7. Ediciones Tamandúa. Venezuela.

Kubitzki, K. 1989. The ecogeographical differentiation of Amazonian inundation forests. Plant Systematics and Evolution 162, 285–304.

Maxted, N., B. Ford-Lloyd y J. Hawkes. 1997a. Plant Genetic Conservation, the *in situ* approach. Chapman and Hall, London.

Maxted, N., V. Ford-Lloyd y J. Hawles. 1997b. Complementary conservation strategies. *En*: Maxted, N., B. Ford-Lloyd y J. Hawkes (eds.). Plant Genetic Conservation, the *in situ* approach. Chapman and Hall, London. Pp. 15–39.

Maxted, N., L. Guarino y M. Dullo. 1997c. Management and monitoring. *En:* Maxted, N., B. Ford-Lloyd y J. Hawkes (eds.). Plant Genetic Conservation, the *in situ* approach. Chapman and Hall, London. Pp. 144–159.

Maxted, N., J. Hawkes, L. Guarino y M. Sawkins. 1997d. The selection of plant conservation targets. Genetic Resources Crop Evolution 7,1–12.

Maxted, N., M. van Slageren y J. Riham. 1995. Ecogeographic surveys. *En:* Guarino, L., V. Ramanatha y R. Reid (eds.). Collecting Plant Genetics Diversity. CAB International, Wallingford. Pp. 255–285.

Petts, G., y C. Amoros. 1996a. Fluvial Hydrosystems: a management perspective. *En:* Petts, G. y C. Amoros (eds.). Fluvial Hydrosystems. London, UK: Chapman and Hall. Pp. 263–278.

Pires-O'Brien, M., y O. O'Brien. 1995. Ecologia e modelamento de florestas tropicais. Ministerio da Educacao e do Desporto, Facultade de Ciencias Agrarias do Para. Belem.

Prance, G. 1979. Notes on the vegetation of Amazonia III. The terminology of Amazonian forest types subject to inundation. Brittonia 31, 26–38.

Prance, G. 1997. The conservation of botanical diversity. *En*: Maxted, N., B. Ford-Lloyd y J. Hawkes (eds.) Plant Genetic Conservation, the *in situ* approach pp. 1–14. Chapman and Hall, London.

Prance, G. (ed) 1982. Biological diversification in the Tropics. Columbia University Press. New York, USA.

Rico-Arce, L. 1987. Generic patterns in the tribe Ingeae with emphasis in Zygia-Caulathon. Bulletin IGSM 15, 51–69.

Rico-Arce, L. 2000. New combinations in Mimosaceae. Novon 9, 554–556.

Rosales, J. 1990. Análisis florístico-estructural y algunas relaciones ecológicas en un bosque estacionalmente inundable en la boca del Río Mapire, Edo. Anzoategui. MSc. Thesis. Instituto Venezolano de Investigaciones Científicas, Venezuela.

Rosales, J. 2000. An ecohydrological approach for riparian forest biodiversity conservation in large tropical rivers. PhD Thesis, University of Birmingham, UK.

Rosales, J., M. Bevilacqua, W. Díaz, R. Pérez, D. Rivas y S. Caura. 2003. Comunidades de vegetación ribereña del Río Caura. *En*: Chernoff, B., A. Machado-Allison, K. Riseng, and J. R. Montambault (eds). A Biological Assessment of the Aquatic Ecosystems of the River Caura Basin, Bolívar State, Venezuela. RAP Bulletin of Biological Assessement 28. Conservation International, Washington, DC. Pp. 127–136.

Rosales, J., y O. Huber (ed). 1996. Ecología de la cuenca del Río Caura. I. Caracterización general. Scientia Guaianae 6. Ediciones Tamandua, Caracas.

Rosales, J., C. Knab-Vispo y G. Rodríguez. 1997. Los bosques ribereños del bajo Caura entre el Salto Para y Los Raudales de La Mura: su clasificación e importancia en la cultura Ye'kwana. *En*: Huber, O. y J. Rosales (eds.). Ecología de la Cuenca del Río Caura, II. Estudios Específicos pp. 171–213. Scientia Guaianae 7. Ediciones Tamandua, Caracas.

Rosales, J., G. Petts y C. Knab-Vispo. 2001. Ecological gradients in riparian forests of the lower Caura River, Venezuela. Plant Ecology 152 (1): 101–118.

Rosales, J., G. Petts, C. Knab-Vispo, J. A. Blanco, A. Briceño, E. Briceño, R. Chacón, B. Duarte, U. Idrogo, L. Rada, B. Ramos, J. Rangel y H. Vargas. 2002. Ecohydrological Assessment of the Riparian Corridor of the Caura River in the Venezuelan Guayana Shield. *En*: Vispo, C. y C. Knab-Vispo (eds). Plants and Vertebrates of the Caura's Riparian Corridor: Their Biology, Use and Conservation. Scientia Guianae 13.

Rosales, J., G. Petts y J. Salo. 1999. Riparian flooded forests of the Orinoco and Amazon river basins: a comparative review. Biodiversity and Conservation 8, 551–586.

Spellenberg, I. 1996. Conserving biological diversity. *En:* Spellenberg, I. (ed.) Conservation Biology. Longman. Edinburg. Pp. 25–35.

ter Braak, C. 1988. CANOCO - a FORTRAN Program for Canonical Community Ordination by [Partial] [Detrended] [Canonical] Correspondence Analysis, Principal Components and Redundancy Analysis. Agricultural Mathematics Group. Wageningen.

UNEP. 2000. Documents of the Convention of Biological Diversity and the Conferencies of the Parties COP (1992–2000). Secretariat of the Convention on Biological Diversity. http//www/biodiv.org/.

Walker, B. 1995. Conserving biological diversity through ecosystem resilience. Conservation Biology 9, 747–752.

Williams, L. 1942. Exploraciones Botánicas en la Guayana Venezolana. I. El medio y bajo Caura. Servicio Botánico-Ministerio de Agricultura y Cría.

Capítulo 8

La Distribución de Peces y los Patrones de Biodiversidad en la Cuenca del Río Caura, Estado Bolívar, Venezuela

Barry Chernoff, Antonio Machado-Allison, Philip W. Willink, Francisco Provenzano-Rizzi, Paulo Petry, José Vicente García, Guido Pereira, Judith Rosales, Mariapia Bevilacqua y Wilmer Díaz

RESUMEN

Probamos la hipótesis nula concerniente a la distribución al azar de especies con respecto a subregiones y macrohábitats dentro de la Cuenca del Río Caura mediante el examen de datos proveniente de 97 especies de invertebrados bénticos, 399 especies de plantas acuáticas y 278 especies de peces de agua dulce. El análisis de ocho subregiones divididas igualmente arriba y debajo de las cataratas indican que los invertebrados estaban distribuidos al azar con respecto a las subregiones. Los peces y plantas no coparten esta distribución azarosa, aunque el efecto subregional de los patrones sobre las plantas fue mucho mayor que el mostrado por los peces. Más aún, los peces mostraron una menor riqueza en especies arriba del Salto Pará que bajo. La conversión fue observada para plantas mientras que los invertebrados fueron igualmente ricos. Efectos no-azarosos del macrohábitat fueron encontrados en cada uno de los grupos con ciertas generalidades. Por ejemplo, especies de Odonata y Ephemeroptera se encuentran en hábitats de aguas rápidas con altos contenidos de oxígeno y están asociados que un ensamblaje mayor de peces y bancos densos de macrofitas, usualmente pertenecientes a la Familia Podostemonaceae. Los ensamblajes de peces demuestran una transición suave a lo largo de gradientes en varios macrohábitats (p.e., fondos de arena a fango y orillas). Al menos seis macrohábitats son necesarios para preservar 82% de las especies de peces.

INTRODUCCIÓN

La cuenca del Río Caura representa una prístina y relativamente extensa región, la cual alberga miles de especies de plantas y animales. Durante la expedición AquaRAP, más de 90 especies de invertebrados bénticos, 399 especies de plantas y 278 especies de peces provenientes de los ecosistemas acuáticos y zonas inundables fueron colectados. Además de la cantidad de nueva información acerca de la distribución y presencia de nuevas especies, es crítico desde el punto de vista de la conservación, el probar hipótesis acerca de la naturaleza de las distribuciones de animales y plantas dentro de la cuenca.

En un patrón de distribución heterogéneo de la flora y la fauna dentro de la cuenca del Río Caura, las subregiones y macrohábitats podría indicarnos importantes ramificaciones para la determinación de recomendaciones de conservación. Por ejemplo, si las especies estuvieran homogéneamente distribuidas, entonces un área particular conservada podría ser establecida y efectivamente proteger la mayoría de las especies. Sin embargo, como la distribución de las especies entre las subregiones o entre los macrohábitats se presentan diferentes y formando parches, entonces un área simple, aparte de la región entera, podría no proveer el deseado nivel de protección. Chernoff y Willink (2000) y Chernoff et al. (1999, 2001a) demostraron como podemos usar información sobre la relativa distribución heterogénea entre subregiones o macrohábitats para predecir posibles cambios faunísticos en respuesta a amenazas ambientales específicas y tales análisis pueden y deben ser llevados a cabo dentro del marco de un programa de evaluación rápida.

Este trabajo comienza con el examen de dos hipótesis nulas críticas para la conservación de peces de agua dulce de la cuenca del Río Caura: que los peces están distribuidos al azar entre (i) ocho subregiones; y (ii) 20 macrohábitats. Compararemos los resultados de los peces con aquellos provenientes de los invertebrados bénticos y las plantas.

MÉTODOS

Las colecciones provenientes del Río Caura fueron divididas en dos regiones principales – arriba (Caura Superior) y debajo del Salto Pará (Bajo Caura). Ocho subregiones geográficas fueron posteriormente designadas (Mapa 1). Las cuatro regiones arriba del Salto Pará, son: Kakada, Erebato, Entreríos-Cejiato y Entreríos-Salto Pará. Las cuatro regiones del Bajo Caura por debajo del Salto Pará son: El Playón, Nichare, Cinco Mil, y Mato.

En cada localidad de colecta un número de variables ecológicas, tales como tipo de fondo, tipo de orilla, tipo de hábitat, etc. fue registrado. Nosotros podemos categorizar cada estación de acuerdo con su tipo de macrohábitat. Veinte macrohábitats principales fueron identificados. No existen lagos verdaderos, o cuerpos de agua con cuencas de drenaje endorreicas; en lugar de lagos existen lagunas con una pequeña conexión o temporalmente aislados del río. En cada localidad evaluamos cuando hierbas acuáticas estaban presentes, cuando existía bosque de galería o si existía vegetación inundada. Vegetación inundada incluye parches vegetales adosados a rocas en rápidos (p.e., Podomostemacea) o flotantes (p.e., *Eichhornia*).

Con la finalidad de determinar si el número de colecciones por región o por macrohábitat estaba afectando los estimados de riqueza de especies, una regresión linear entre grupos combinados fue calculada. La línea de regresión es forzada lógicamente a través del origen (p.e., Chernoff and Willink 2000). Debido a que el análisis de varianza (ANOVA) de la regresión fue significante mediante el examen de la pendiente contra la hipótesis nula de cero, un análisis de covariancia (ANCOVA) fue realizado, para ver sí otros efectos hipotetizados (p.e. cabeceras *vs* tierras bajas) eran significantes. En ANCOVA, la variable cualitativa grupal (i.e., grupo elevacional) es incluida como la variable independiente, el número de especies es la variable dependiente y el número de colecciones sirven de covariados. Si el Estadístico-F del ANCOVA es significante, dos otras pruebas deben ser llevadas a cabo para determinar si la diferencia es atribuible a la diferencia de medias de la variable independiente. El primer examen es probar la hipótesis nula en la cual las varianzas dentro de grupos son iguales. La segunda prueba la hipótesis nula de homogeneidad de las pendientes dentro de grupos. Si una de ellas falla en rechazar ambas hipótesis nulas, entonces la significancia del Estadístico-F puede ser atribuida a las diferencias indicadas por la variable.

Chernoff et al. (1999, 2000), seleccionaron el Indice de Similaridad de Simpson, S_s, como el más consistente con los datos colectados durante los inventarios rápidos o con datos puntuales. El Indice de Simpson usa el siguiente formato de tabla para calcular la similaridad entre dos listas o muestras de especies:

		Muestra 1	
		1	0
Muestra 2	1	*a*	*b*
	0	*c*	*d*

donde ***a***, es el número de correspondencias positivas o especies presentes en ambas muestras; ***b***, es el número de especies presentes en la muestra 2 y ausentes de la muestra 1; ***c***, es el inverso de ***b***; y ***d***, es el número de correpondencia negativass o especies ausentes en ambas localidades. El Indice de Similaridad de Simpson, $S_s = a/(a+b)$, donde $b < c$, o $S_s = a/n_s$ donde n_s, es el número de especies presentes en la más pequeña de las dos listas. El denominador del índice elimina interpretación de los negativos – ausencia de especies. Los 0's en las matrices son realmente artefactos de codificación o reservas para la ausencia de datos.

Con la finalidad de interpretar la similaridad observada de dos muestras, ambas tomadas de un universo fijo más grande (p.e la lista de todas las especies capturadas en el Caura) nosotros realizamos un procedimiento en cuatro pasos. En el **Paso 1,** nosotros calculamos S_s reduciendo vía rarefacción el número de especies en la muestra más grande hasta igualar el número de especies en la muestra más pequeña, n_s. Esta rarefacción y cálculo de S_s es repetido 500 veces. A partir de 500 simulaciones una media de similaridad, S'_s, es calculada y reportadas en tablas de similaridad. Este procedimiento es repetido para calcular un S'_s para cada par de muestras en el análisis.

Interpretar la significancia de las medias de similaridades entre muestras requiere simulaciones a través de un intervalo del número de especie encontradas en las muestras de subregiones, tipos de agua y macrohábitats. En el **Paso 2,** simulamos 200 pares al azar mediante *bootstrapping* con reemplazo a partir del grupo de todas las especies capturadas durante la expedición, con la constricción que cada muestra al azar contiene un número fijo de especies para cada punto dado en éste intervalo. Para cada par al azar de los 200 calculamos su Similaridad de Simpson. Estas 200 similaridades al azar se aproximan a una distribución normal, de la cual calculamos una media y una desviación estándar debida a causas al azar; acá denominada media de silimaridad al azar, S^*_n, donde n representa el número de especies presentes en la muestra.

Distribuciones de similaridad al azar fueron generadas a intervalos de 10 especies con la finalidad de estimar S^*_n y sus desviaciones estándar para muestras que contienen entre 20 y 140 especies. Este intervalo de listas-tamaño al azar incluye el número actual de especies observadas en las subregiones y en macrohábitats. En el **Paso 3,** las medias y las desviaciones estándar son graficadas en contra del número de especies presentes en la muestra. A medida que el número de especies presentes en la muestra aumenta la similaridad observada debida a efectos al azar también incrementa pero las varianzas disminuyen

En el **Paso 4,** comparamos las media de similaridad observada S'_s, calculada de rarefacción (paso 1) con el valor predecido de S^*_n y su desviación estándar. Usando un alcance paramétrico de dos colas, calculamos la probabilidad de obtener la similaridad observada al azar, a partir del número de desviaciones estándar que la similaridad observada fuera tanto arriba como debajo de la media de la distribución del *bootstrap* al azar. Esta probabilidad fue obtenida por interpolación de los valores presentados en Rohlf y Sokal (1995: tabla A). La significancia de los valores de probabilidad fue ajustada con la técnica secuencial de Bonferoni (Rice, 1989) debido a que cada muestra está involucrada en multiples comparaciones. El procedimiento secuencial de Bonferoni es conservativo, haciéndo más difícil el rechazar la hipótesis nula. Seleccionamos el nivel P = 0.01 como nuestro criterio de rechazo de una hipótesis nula. El valor, 0.01, fue dividido por el número de comparaciones fuera de la diagonal presentes por arriba o debajo del triángulo de la matriz de similaridades observadas. Este nuevo resultado es usado como criterio para evaluar la hipótesis nula que $S'_s = S^*_n$. Por ejemplo, en el triángulo inferior de la Tabla 8.3, existen 28 similaridades. Con la finalidad de rechazar la hipótesis nula de ésta prueba de dos colas, S'_s debe ser mayor que 3.5 desviaciones estándar sobre o debajo S^*_n así que $P<0.0002$.

Si S'_s es encontrado ser significativamente diferente de S^*_n, entonces nosotros rechazamos la hipótesis nula y concluimos que la *similaridad* observada no es debida a efectos al azar. Sin embargo, si S'_s falla dentro de los efectos al azar o si S'_s es más grande que la media al azar fallamos en rechazar la hipótesis nula concerniente a las *muestras* – que dos de estas *muestras* son iguales. En el primer caso concluimos que las dos listas son sacadas homogéneamente de una distribución al azar mayor. En el último caso, concluimos que la similaridad es debida a dependencia biológica o correlación, tal como subgrupos similares. Esto es, una población forma la fuente poblacional de otra. Si S'_s, es significativamente menor que S^*_n, rechazamos la hipótesis nula para igualdad de similaridad y para igualdad de muestras. Nosotros entonces podemos buscar razones biológicas o ambientales para estas diferencias.

Si descubrimos que las similaridades no son al azar, podemos investigar cuando los patrones de distribución de especies en relación con variables ambientales, no es al azar. La medida de desorden de la matriz como ha sido prouesta por Atmar y Patterson (1993), calcula la entropía de una matriz como medida de la temperatura. Temperatura mide la desviación desde el orden completo (0°) al completo desorden (100°), en la cual las células de la matriz son análogas a la posición de moléculas de gas en un contenedor rectangular. Después que el contenedor ha sido llenado a su máximo hasta llenar la esquina superior izquierda (por convención), la distribución de células llenas y vacias determina el grado de desorden en la distribución de las especies y corresponde a temperatura que produciría el grado de desorden (Atmar y Patterson 1993). Para probar si la temperartura puede ser obtenida debido a efectos al azar, 500 simulaciones *Monte Carlo* de matrices geométricamente similares y tomadas al azar, fueron calculadas. La significancia de las temperaturas observadas es definida en relación a la variancia de las distribuciones simuladas. Debido a que nosotros no estamos usando este procedimiento para probar específicamente cuando un patrón no-azaroso puede ser adscrito a algún subgrupo anidado o cambio clinal, las modificaciones propuestas por Brualdi y Sanderson (1999), no son requeridas. *Software* para calcular matrices desordenadas está disponible de Atmar y Patterson en la siguiente dirección de internet: http://www.fieldmuseum.org.

Relaciones entre regiones y macrohábitats son resumidas en diagramas ramificados. Dos tipos de procedimientos fueron usados ambos involucrando criterios de parsimonia. El primero es una red de mínima "evolución" calculada de una matriz de distancia como 1 menos la media del Indice de Simpson rarificado (=1- S'_s). El segundo método usa la presencia de especies como caracteres compartidos en un *Análisis de Parsimonia de Camin-Sokal* (Sneath y Sokal 1973) debido a que éste no permite reversiones, entonces, ausencias compartidas no son tomados como caracteres. Paup* 4.0b fue usado para este análisis.

RESULTADOS

Regiones

Un total de 67 estaciones de colecta fueron realizadas durante la expedición resultando en la captura de 278 especies de peces. Un total de 31 colecciones fueron muestreadas incluidas dentro de 14 áreas georeferenciadas río arriba de las cataratas y 35 colecciones dentro de 17 áreas georeferenciadas fueron obtenidas río abajo de las cataratas.

La aparente riqueza de especies de peces fue mucho mayor por debajo del Salto Pará que arriba con 226 y 103 especies capturadas, respectivamente. La riqueza de especies no fue igual entre las regiones (Fig. 8.1). El número de especies por región varía desde 26 a 120. Sin embargo, encontramos casi doble de especies por región en tierras bajas (media = 104.8) que las encontradas por arriba de las cataratas (media = 54.5). Hay, sin embargo, un efecto muy fuerte de esfuerzo de colecta sobre el número de especies capturadas (Fig. 8.1). El análisis de regresión combinado demuestra que la riqueza de especies es una función linear significante del número de colecciones realizadas para cada área georeferenciada, con una pendiente diferente a cero ($F_{1,6}$= 26.2, P<.002). Los resultados del ANCOVA corrige por los efectos de esfuerzo de colección y rechaza la hipótesis nula de igualdad de medias de riqueza de especies arriba y debajo de las cataratas ($F_{1,5}$ = 45.4, P<.002). Los resultados reflejan las diferencias en medias debido a que no podemos rechazar la hipótesis nula de homogeneidad de varianzas o de pendientes (P>.24). Entonces, para ocho colecciones por región, el predecido y valor de confidencia de 95% en la región superior del Caura

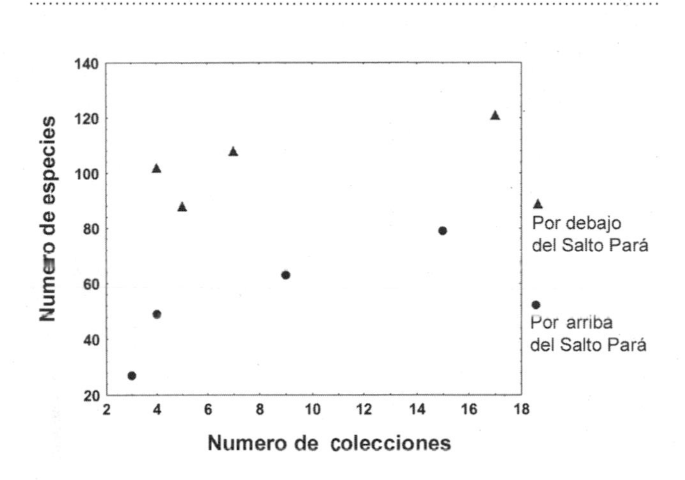

Figura 8.1. Riqueza de especies como una función del número de colecciones para subregiones del Río Caura River realizadas arriba (círculos) y abajo (triángulos) del Salto Pará.

es 55.4 ± 21 especies comparadas con 104.3 ± 20 especies en el Bajo Caura.

La disparidad en riqueza de especies de peces entre las cabeceras y las tierras bajas (n=157) es aparente de la lista total de especies (Apéndice 7). En la medida que procedemos hacia las cabeceras de los ríos, por arriba de las grandes cataratas y dentro del piedemonte, esperamos encontrar pocas y mucho más restringidas especies, que especies en las tierras bajas sin barreras (Lowe McConnell 1987). Entonces, arriba del Salto Pará (Caura Superior) podemos predecir que nuestras muestras deberían contener: 1. especies con amplias distribuciones altitudinales, compartidas por ambas, cabeceras y tierras bajas; y 2. especies con limitadas preferencias altitudinales, encontradas en las tierras bajas, especialmente aquellas con fondos fangosos. El Bajo Caura comparte 52 especies de peces, de las cuales podemos observar su similaridad, S'_s, ser 50.5%. La sección del Río Caura arriba del Salto Para (Caura Superior) posee un gradiente ambiental alto, con muchas rocas, rápidos y fondos arenosos y márgenes inundables angostos (Machado et al. 2003). De las

Tabla 8.1. Especies de peces encontrados sólo en el Caura Superior por arriba del Salto Pará.

Una o dos Regiones (n=39)	
Acestrorhynchus	cf. *apurensis*
Aequidens	sp.
Ageneiosus	sp.
Ancistrus	sp. A
Ancistrus	sp. B
Anostomus	*anostomus*
Apareiodon	sp.
Brachychalcinus	*orbicularis*
Cetopsorhamdia	cf. *picklei*
Chaetostoma	*vasquezi*
Characidae	sp. A
Characinae	sp. A
Corydoras	cf. *osteocarus*
Creagrutus	sp.
Ctenobrycon	*spilurus?*
Cyphocharax	cf. *festivus*
Doras?	sp.
Farlowella	*oxyrryncha*
Geophagus	sp.
Guianacara	cf. *geayi*
Gymnotus	*carapo*
Harttia	sp.
Hemigrammus	sp. B

Una o dos Regiones (n=39) (continúa)	
Hemiodus	cf. *unimaculatus*
Hemiodus	*goeldii*
Hypostomus	cf. *ventromaculatus*
Imparfinis	sp. B
Jupiaba	cf. *zonata*
Jupiaba	sp. B
Knodus	sp. C
Melanocharacidium	*melanopteron*
Moenkhausia	cf. *grandisquamis*
Moenkhausia	cf. *miangi*
Moenkhausia	sp. B
Myleus	*asterias*
Myleus	*torquatus*
Phenacogaster	sp. B
Pimelodus	cf. *ornatus*
Plagioscion	cf. *auratus*
Tres o más Regiones (n=7)	
Aphyocharax	sp.
Bryconops	sp. A
Corydoras	*boehlkei*
Crenicichla	*saxatilis*
Guianacara	*geayi*
Knodus	cf. *victoriae*
Rineloricaria	*fallax*

51 especies capturadas sólo arriba de las cataratas, muchas son representantes de taxa ampliamente distribuidos (p.e., *Myleus asterias, Cyphocharax* cf. *festivus, Gymnotus carapo, Crenicichla alta, Plagioscion* cf. *auratus*). Sin embargo, un gran número de especies se encontraron solamente en la región superior y características de los ambientes descritos. Por ejemplo, aquellas caracterizadas por el ambiente de piedemonte, en lugar de hábitats de las tierras bajas. Estas incluyen: *Ancistrus* spp., *Cetopsorhamdia* cf. *picklei, Chaetostoma vasquezi, Hartia* sp., *Hypostomus* cf. *ventromaculatus, Rineloricaria fallax, Apareiodon* sp., *Leporinus arcus, L.* cf. *granti, Melanocharacidium melanopteron* y *Guianacara* sp. Además, hay un número de especies usualmente asociados con fondos arenosos que tienden a no ocurrir en las amplias tierras bajas de fondos fangosos y planicies inundables del Río Orinoco tales como: *Corydoras* spp., *Imparfinis* sp. B., *Characidium* spp., *Knodus* spp. y *Geophagus* sp.

También deberíamos esperar que especies verdaderas de cabeceras tienen una distribución restringida debido a que las regiones de cabeceras se encuentran relativamente aisladas una de otra, actuando similar a las islas (Lowe-McConnell 1987). De las 39 especies colectadas solo en las cabeceras, sólo siete fueron encontradas que habitan en tres o más de las cuatro regiones (Tabla 8.1). Aunque pocas especies aparentemente muestran un patrón de distribución limitado o por parches producto de artefactos (p.e., *Myleus* spp., *Pimelodus* cf. *ornatus*), la mayoría no lo son. El resultado es que más del 80% de las especies encontradas sólo en la región superior fue capturada en una sola región. Aunque la continua colecta podría incrementar indudablemente la distribución de los taxa colectados (Alroy 1992; Chernoff y Willink 2000), así como también añadir más especies a las previamente conocidas, dudamos seriamente que la mayoría de las especies estarán ubicuosamente distribuidas. De ahí que concluyamos que las regiones arriba de las cataratas contienen una combinación de especies con amplias tolerancia altitudinales así como también especies que prefieren hábitats típicos de alturas o cabeceras.

La ictiofauna de las tierras bajas es muy rica, incluyendo 226 species. Los peces no estuvieron distribuidos tan ampliamente como hubieramos esperado en el Bajo Caura (p.e., en el Pantanal; Chernoff y Willink 2000). Sólo 26% de ellas fueron colectadas en tres o cuatro de las regiones bajas. Sólo 28 species (12.5%) fueron colectadas en cada uno de las cuatro regions bajas. Hay 174 especies de peces solo colectadas en las tierras bajas. La fauna de las tierras bajas contienen especies que están distribuidas en el Río Caura, otros ríos del Escudo de Guayana y el Río Negro (p.e., *Microschemobrycon* spp., *Leporinus brunneus, Ammocryptocharax elegans, Anostomus ternetzi, Serrasalmus* sp., *Crenicichla* cf. *lenticulata, C.* cf. *wallacei*) y aquellos que son típicos de las principales áreas de inundación del Río Orinoco (p.e., *Anchoviella* spp., *Pellona castelneana, Pygocentrus cariba, Aphyocharax erythrurus, Triportheus albus, Sorubim lima, Pimelodus blochii, Bujurquina mariae, Achirus* sp.). Treinta y siete de las especies sólo encontradas en tierras bajas (=16.8%) estaban distribuidas en tres o cuatro regiones (Tabla 8.2) costituyendo un ensamblaje de especies ornamentales o minúsculas (p.e., *Anostomus ternetzi, Microschemobrycon callops, Ramirezella newboldi, Parvandellia* sp., *Apistogramma* cf. *indirae, Paravandellia* sp.). Pero también encontramos especies de gran porte (> 200 mm SL, e.g., *Hydrolycus tatauaia* o *Hypostomus* cf. *plecostomus*) y una gran diversidad de especialistas tróficos, desde comedores de fango (iliófagos) y herbivoros (p.e., *Curimata incompta*) a piscivoros (e.g., *Serrasalmus* sp.).

El patrón de similaridades, S'_s (Tabla 8.3) demuestra que el efecto del Salto Pará sobre la estructura de las comunidades de peces en el Río Caura. Entre las regiones arriba de las cataratas, los coeficientes son significativamente diferente del azar y son positivos. Esto significa que las regiones del Caura Superior están positivamente correlacionados o biológicamente dependientes una de la otra (Chernoff et al. 1999; Chernoff y Willink 2000). Buena evidencia de esto puede encontrarse en el número de especies que compartidas relativo a aquellas que son compartidas unicamente. Aunque las cuatro regiones arriba de las cataratas comparten altos porcentajes de la totalidad de especies, el número de especies compartidas unicamente entre esas regiones es excepcionalmente bajo, menos que cuatro y modalmente cero. Este resultado indica que las especies no están segregándose o fragmentando diferenciálmente áreas arriba de las cataratas: no existe evidencia de cambios de especies o fronteras de transición. Este resultado es consistente con regiones que poseen grupos de relaciones anidadas entre ellas (ver abajo).

Nuestros análisis de rarefacción y simulación de similaridades, demuestran que de 16 coeficientes de similaridad para las regiones separadas por el Salto Pará, solo un simple coeficiente es significantemente diferente de la media de similaridad al azar (Tabla 8.3). Las especies encontradas arriba y debajo de las cataratas provienen enteramente del grupo de especies que son encontradas en cinco o más ocho regiones (Tabla 8.4). Esto es, esencialmente aquellas especies que son comunes. En un caso, la similaridad de Entreríos-Cejiato a la del Río Mato es significante ($P<.01$) y es negativa – mucho menor que la esperada debida a un proceso al azar. Relaciones negativas indican desplazamiento o fuerte arreglo regional de taxa (Chernoff et al. 1999). Así es que, la muestra del Río Mato, la cual está localizada río abajo, incluye muchas especies provenientes del Río Orinoco que penetran sólo parcialmente hacia arriba del Río Caura (p.e., el caribe, *Pygocentrus cariba*, el bagre, *Xyliphius* cf. *melanopterus*).

En las regiones del Bajo Caura, cuatro de los seis coeficientes de similaridad son significativamente diferentes del azar ($P<.01$) y son positivos. Los dos coeficientes al azar comparan el Río Mato con las subregiones de El Playón y Nichare. Estas son dos subregiones del Bajo Caura las cuales son más distantes de la subregión del Río Mato; la subregión del Raudal Cinco Mil es positiva y significantemente corelacionada con el Río Mato. La significante similaridad entre Río Mato y Cinco Mil es debida a que elementos del Río Orinoco que caracterizan grandemente a la fauna del Río Mato se extienden río arriba hasta los rápidos del Cinco Mil.

Tabla 8.2. Peces capturados solamente en tres o cuatro regions del Bajo Caura por debajo del Salto Pará (n=37).

Ancistrus	sp. C
Anostomus	*ternetzi*
Aphyocharax	*alburnus*
Apistogramma	sp. A
Astyanax	sp.
Brycon	*pesu*
Bryconamericus	cf. *cismontanus*
Characidae	sp. B
Corydoras	cf. *bondi*
Creagrutus	cf. *maxillaris*
Curimata	*incompta*
Cyphocharax	*oenas*
Farlowella	*vittata*
Hemigrammus	cf. *tridens*
Hemiodus	*unimaculatus*
Hydrolycus	*tatauaia*
Hyphessobrycon	*minimus*
Hypostomus	cf. *plecostomus*
Ancistrus	sp. C
Anostomus	*ternetzi*

Jupiaba	*polylepis*
Knodus	sp. B
Leporinus	cf. *maculatus*
Microschemobrycon	*callops*
Microschemobrycon	*casiquiare*
Microschemobrycon	*melanotus*
Moenkhausia	cf. *lepidura* D
Moenkhausia	*copei*
Ochmacanthus	*alternus*
Paravandellia	sp.
Pimelodella	cf. *cruxenti*
Pimelodella	cf. *megalops*
Ramirezella	*newboldi*
Rineloricaria	sp.
Serrasalmus	sp. A
Steindachnerina	*pupula*
Tetragonopterus	*chalceus*
Triportheus	*albus*
Vandellia	*sanguinea*

Tabla 8.3. Número de especies compartidas (triángulo superior) y media del Coeficiente de Similaridad de Simpson, S's (triángulo inferior) entre regiones del Río Caura. Los Coeficientes de Similaridad que son mostrados en negritas son significativamente diferentes de similaridad al azar (P<0.001). Las celdas sombreadas representan comparaciones entre regiones debajo de las cataratas. Abreviaciones: Ent-Cejiato–Río Caura entre Entreríos y Raudal Cejiato; Ent-SP–Río Caura entre Entreríos y Salto Pará; n–número de especies; u–número de especies únicas; %u–porcentaje de especies únicas.

	Kakada	Erebato	Ent-Cejiato	Ent-SP	Playón	Nichare	Cinco Mil	Mato
Kakada	1	25	25	18	11	12	7	4
Erebato	**50.67**	1	39	36	21	22	16	9
Ent-Cejiato	**32.33**	**49.52**	1	43	27	27	21	12
Ent-SP	**28.61**	**57.16**	**54.47**	1	24	24	22	15
Playon	9.92	19.49	25.04	-22.18	1	66	57	38
Nichare	9.92	18.3	22.42	-19.88	**54.53**	1	61	45
Cinco Mil	-6.91	-15.82	-20.62	-21.66	**52.88**	**50.38**	1	43
Mato	-4.44	-10.27	**-13.65**	-17.03	35.23	37.3	**42.1**	1
n	27	49	79	63	108	121	102	88
u	0	1	19	6	21	27	22	30
%u	0	2.04	24.05	9.52	19.44	22.31	21.57	34.09

La matriz de distribución total es marginalmente significantemente diferente del azar. La matriz de temperatura observada (49.93) fue 1.82 desviaciones estándar por debajo de la media de 500 simulaciones *Monte Carlo* (P=.03). Esto es enteramente debido a los taxa ampliamente distribuidos compartidos por las regiones de tierras bajas. Cuando analizamos las subregiones del Caura Superior entre ellas, obtenemos altos resultados significantes. La matriz de temperatura observada (24.09) fue más de 3.38 desviaciones estándar más frías que la media de 500 simulaciones *Monte Carlo* (P<.0004). Dados los datos de similaridad de arriba, las subregiones superiores son consistentes con un patrón de subgrupos similares. Los patrones mayormente interpretables dentro de los datos distribucionales comprenden los siguientes: 1. que existe cambios faunísticos significantes debido al Salto Pará; 2. que las subregiones del Caura Superior, están estructuradas como subgrupos relacionados; y 3. en el Bajo Caura existe más que una disyunción al final debido a la incursión de la ictiofauna del Río Orinoco. Estas conclusiones se encuentran bien resumidas por los resultados de la estructura del árbol utilizando el Análisis de Parsimonia de Camin-Sokal (Fig. 8.2). El índice de retención, el cual mide la cantidad de información debido a taxa compartidos es 0.706. Con la excepción del Río Mato siendo el más distinto de los grupos de tierras bajas, el patrón entre las subregiones Cinco Mil, Nichare y el Playón no puede ser interpretado. En el grupo subregional por arriba del Salto Pará existe un agrupamiento ribereño extendiéndose desde justamente por arriba de las cataratas al Raudal Cejiato; las muestras provenientes del Río Erebato y Río Kakada comparten secuencialmente pocas especies con el grupo del canal principal. Este patrón es confirmado por los resultados del Análisis de Coordenadas Principal (Fig. 8.3), en el cual las dos principales coordenadas (64% de la variancia) agrupa las subregiones congruentes con sus relaciones geográficas—los grupos separados por las cataratas y el Río Mato como el más distinto.

Tabla 8.4. Especies de peces de agua dulce comunmente encontrados en el Río Caura. Común es definido como haber sido capturadas en cinco o más de las ocho regiones.

Aequidens	cf. *chimantanus*
Astyanax	*integer*
Bryconops	cf. *colaroja*
Characidium	sp. A
Cyphocharax	*festivus*
Cyphocharax	sp.
Hoplias	*macrophthalmus*
Hypostomus	sp. B
Jupiaba	*atypindi*
Jupiaba	cf. *atypindi*
Jupiaba	cf. *polylepis*
Jupiaba	*zonata*
Moenkhausia	cf. *lepidura* A
Moenkhausia	cf. *lepidura* B
Moenkhausia	cf. *lepidura* C
Moenkhausia	cf. *lepidura* E
Moenkhausia	*collettii*
Moenkhausia	*grandisquamis*
Moenkhausia	*oligolepis*
Phenacogaster	sp. A
Pimelodella	sp. B
Pimelodella	sp. C
Poptella	*longipinnis*
Pseudocheirodon	sp.
Satanoperca	sp. A
Serrasalmus	*rhombeus*
Synbranchus	*marmoratus*
Tetragonopterus	sp.

Macrohábitats

La distribución de especies de peces observada no parece ser distribuidas al azar con respecto a la muestra de 20 macrohábitats. Todos los coeficientes fueron significativamente diferentes del azar (P<.005) y la matriz fue significantemente más ordenada (más fría) que el esperado al azar (P<.0001).

Las asociaciones no azarosas debido a los hábitats son evidentes por el Análisis de Parsimonia de Camin-Sokal

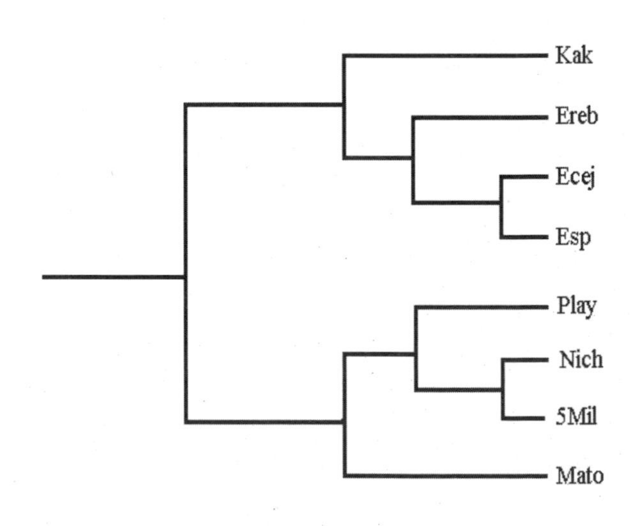

Figura 8.2. Análisis de Parsimonia de Camin-Sokal de las similaridades de Simpson entre subregions en el Río Caura, para los peces. Abreviaturas: Kak–Río Kakada; Ereb–Río Erebato; Ecej–Río Caura desde Entreríos a Cejiato; Esp–Río Caura desde Entreríos a Salto Pará; Play–El Playón; Nich–Río Nichare; 5Mil–Raudal Cinco Mil; y Mato–Río Mato.

(Fig. 8.4). El índice de retención es 0.783. En el análisis, muchos de los hábitats están cercanamente agrupados. Por ejemplo, sobre la derecha existe un agrupamiento que contiene aquellos con fondos y riberas fangosas, troncos, hojas y detritus a lo largo de márgenes boscosos. Una comunidad algo diferente es encontrada en hábitats de grandes ríos con riberas y fondos compuestos por arena y rocas. El Caura Superior está bien caracterizado por muchos hábitats de islas que comparten muchas especies con el canal principal.

El ensamblaje de especies que habitan las hierbas acuáticas macrofitas y rápidos es el más grande descubrimiento de la

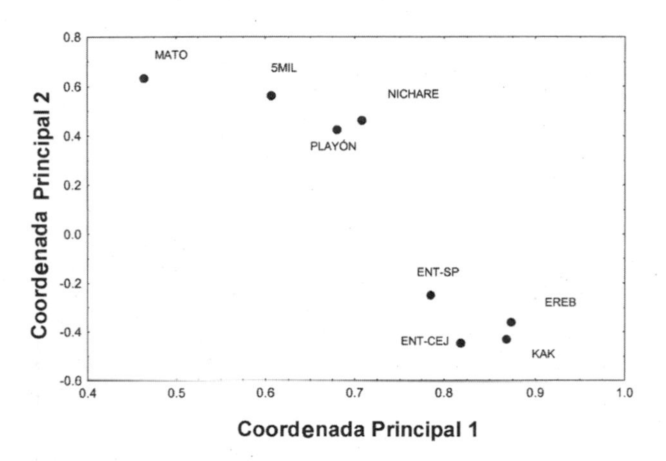

Figura 8.3. Análisis de Coordenadas Principales de las similaridades de la matriz entre las subregiones del Río Caura. Abreviaturas: Kak–Río Kakada; Ereb Río Erebato; Ent-Cej–Río Caura desde Entreríos a Cejiato; Ent-SP–Río Caura desde Entreríos a Salto Pará; Play–El Playón; Nichare–Río Nichare; 5Mil–Raudal Cinco Mil; y Mato–Río Mato.

expedición- En la región superior del Caura, existen numerosos rápidos o áreas de aguas veloces sobre grandes rocas y cantos, incluyendo las presentes sobre el Salto Pará. En estas aguas rápidas se encontraron hasta siete especies de plantas acuáticas vasculares de la Familia Podostemonaceae. Existen cerca de 130 especies de peces en total viviendo entre estas macrofitas y 120 especies en macrofitas y rápidos!

El análisis de coordinadas principales (Fig. 8.5) demuestra muy bien la transición entre los grupos de hábitats. Por ejemplo, la transición entre comunidades de fango y arena ocurre en la región superior e izquierda del gráfico. También existe una transición suave desde fondos arenosos y playas con sustratos rocosos a islas y rápidos con macrofitas. Desde la región central hacia la parte superior derecha (Fig. 8.5) representa una transición entre hábitats de remanso o baja velocidad (p.e., caños) a lagos y áreas con aguas estancadas con vegetación flotante.

Dada la diversidad de un patrón no azaroso de especies entre macrohábitats, nos preguntamos ahora cuantos macrohábitats son requeridos para proteger la mayoría de las especies. La curva de acumulación del porcentaje total de especies es mostrada en la Figura 8.6. La curva fue calculada a partir de la regresión polinomial en la cual todos los coeficientes son significativamente diferentes de cero. La ecuación de la regresión es:

$$Y = 50.9 + 24.2X - 9.7X^2 + 1.8X^3 - 0.2X^4 + .004X^5 + e$$

donde Y es el porcentaje acumulado de especies, X es el número de macrohábitats y e es el término de error. El mayor retorno del porcentaje acumulado al del número de macrohábitats es determinado a partir del punto de inflexión (colocando la segunda derivada a cero). El punto de inflexión

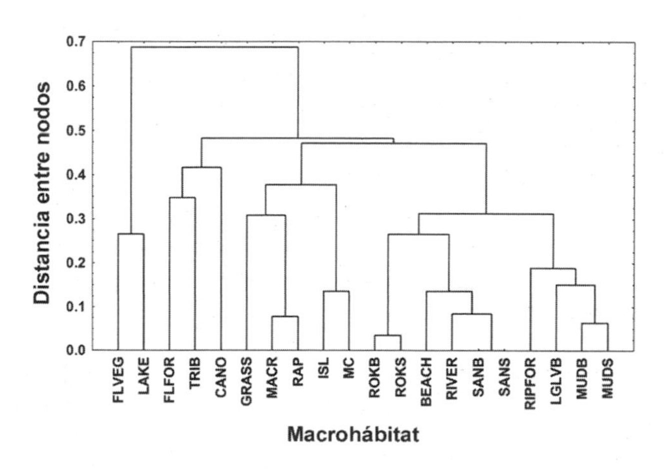

Figura 8.4. Análisis de Parsimonia de Camin-Sokal de 20 macrohábitats en el Río Caura. Abreviaturas: B–fondo; S–orilla; Flveg–vegetación flotante; Flfor–bosque inundable; Trib–tributario; Cano–caño; Macr–macrofitas; Rap–rápidos; Isl–islas; Rok–rocoso; San–arenoso; Lglv–troncos y hojas.

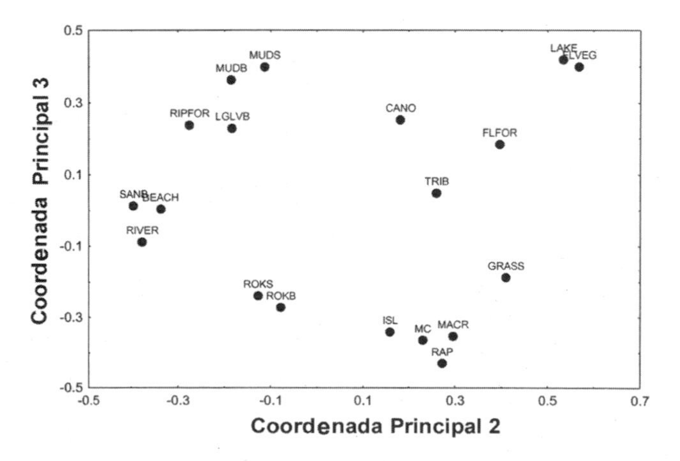

Figura 8.5. Análisis de Coordenadas Principales de los coeficientes de similaridad de entre 20 macrohábitats en el Río Caura. Abreviaturas: B–fondo; S–orilla; Flveg–vegetación flotante; Flfor–bosque inundable; Trib–tributario; Cano–caño; Macr–macrofitas; Rap–rápidos; Isl–islas; Rok–rocoso; San–arenoso; Lglv–troncos y hojas.

es 6.0 el cual corresponde al 82% de las especies (Fig. 8.6). Entonces, si una adecuada protección es dada a 6 macrohábitats, la gran mayoría de las especies de peces, puede ser protegida. Este grupo incluye 228 especies con todas las especies de importancia comercial encontradas en el Apéndice 2.1.

DISCUSIÓN

La distribución de peces en el Río Caura muestra un fuerte efecto subregional y macrohábitats y son no debidas al azar y heterogéneas. Los efectos subregionales no azarosos, son debidos a la separación por el Salto Pará. Estos resultados concuerdan muy bien con nuestros estudios en los ríos Tahuamanu y Manuripe en Bolivia, Bolivia, el sur del Pantanal en Brasil, y el Río Paraguay, Paraguay, así como también, los estudios del Río Jau, Brasil (Forsberg, 2001), la planicie de inundación del Amazonas (Cox Fernandes 1995; P. Petry, pers. comm.) y el Río Napo, Ecuador (Ibarra y Stewart 1989). Estos estudios no están de acuerdo con los comentarios generales de Lowe-McConnell (1987) y Goulding et al. (1988), quienes indican que las distribuciones de los peces de agua dulce en tierras bajas son altamente debidas al azar. Las distribuciones no azarosas y heterogéneas tienen importantes implicaciones en la conservación de la ictiofauna.

Las regiones del Caura Superior, poseen significativamente menor número de especies que las regiones por debajo del Salto Pará. Esto es debido en cierto grado a especies del Río Orinoco que penetran al Río Caura. La riqueza de especies de plantas totales es más grande arriba del río que debajo (291 *vs* 185 respectivamente). Sin embargo, hay más plantas acuáticas vasculares (macrofitas) y hierbas por arriba del Salto

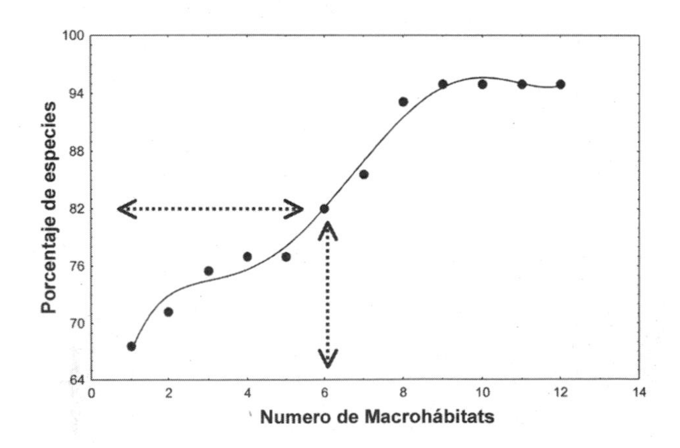

Figura 8.6. Porcentaje acumulado de especies contra el número de macrohábitats. Los valores reales son mostrados como puntos; la línea de regresión polinomial es ilustrada. Las flechas indican el punto de inflección para la ecuación de regresión, mostrando el máximo porcentaje de especies que puede ser conservado con el menor número de macrohábitats.

Pará que por debajo (Rosales et al. 2003a). Esta diferencia en riqueza de especies en plantas y animales, asociadas con cambios altitudinales, es congruente con los resultados reportados del sur del Pantanal (Chernoff y Willink 2000). Los datos provenientes de los invertebrados bénticos (Apéndice 4.1) no muestran una diferencia apreciable con 70 especies en el Bajo Caura y 75 en el Caura Sueprior.

La distribución de las especies de invertebrados bénticos difieren de la de los peces en que ellos no muestran un patrón geográfico de similaridades entre las subregiones (García y Pereira 2003). Hemos reanalizado sus datos para que correspondan a las subregiones presentadas acá y en el análisis botánico. La matriz de correlación de similaridades subregionales entre peces e invertebrados bénticos no es significante (r = 0.32, P>.05). Los resultados del Análisis de Parsimonia de Camin-Sokal y Análisis de Coordenadas Principales (Figs. 8.7, 8.8) muestran que aunque debajo de las cataratas (Salto Pará), las muestras de El Playón y el Río Mato están separadas, de las subregiones remanentes, las del Raudal Cinco Mil y Nichare comparten un número significativo de especies con las subregiones por arriba de las cataratas. Este es un contraste distinto de las distribuciones de peces y botánica. Los peces muestran un fuerte componente río arriba y río abajo (Figs. 8.2, 8.3) pero no están bien estructurados arriba de las cataratas. Las plantas por otro lado se muestran bien estructuradas de acuerdo con las subregiones (Rosales et al. 2003a,b). Las plantas muestran una marcada discontinuidad al nivel de las cataratas, pero también subregiones tales como Kakada y Erebato representan cambios florales. La distribución de los peces es intermedia entre las poco estructuradas datos de distribución de invertebrados bénticos y las bien estructuradas distribuciones de plantas.

Los efectos de los Macro-hábitat son significantes no al azar y críticos para entender propiamente las distribuciones de las plantas, invertebrados y peces dentro de la Cuenca del Río Caura (ver anteriormente, García y Pereira 2003; Rosales et al. 2003a). Es difícil comparar con presición los macrohábitats entre estos grupos. Sin embargo García y Pereira (2003), demuestran las correlaciones entre las características fisicoquímicas de los hábitats (tales como oxígeno disuelto) y las especies de Odonata y Ephemeroptera que caracterizan principalmente las aguas rápidas. Los peces fragmentan el ambiente relativo al tipo de fondo, estructura forestal, velocidad de agua y la presencia de macrofitas y otro tipo de vegetación acuática. Transiciones suaves en la composición faunística es evidente (Fig. 8.5) basadas en estas variables. Es interesante notar que muchos de los patrones de ensamblajes de peces corresponden a cambios de paisajes boscosos. Como es notado por Rosales et al. (2003a), las islas, particularmente arriba de las cataratas son hábitats únicos con un ensamblaje característicos de plantas. Estas islas son hábitats únicos para peces, también como florecimiento entre las rocas de parches de Podostemonaceae y rápidos o corrientes veloces.

El desarrollo de cualquier estrategia de conservación en la Cuenca del Río Caura debe tomar en cuenta tanto los efectos subregionales como del macrohábitat sobre la distribución de las especies. Por ejemplo, 37 de 278 especies de peces fueron encontradas en tres de cuatro subregiones y solo 28 especies son encontradas en cinco de las ocho subgregiones y consideradas comunes. Diferentes subregiones y macrohábitats son críticos como áreas de crecimiento y protección (*nursery*) para los diferentes estadios de desarrollo de los peces. Especies comercialmente importantes, tales como la palometa, *Myleus rubripinnis,* fueron encontrados en áreas de aguas negras. Disminución de la calidad del agua a través de deforestación, descargas domésticas o contaminación debido a minería eliminará las macrofitas acuáticas, invertebrados y últimamente los peces—130 especies de peces fueron encontrados cercanamente asociados con las macrofitas acuáticas. Los efectos del número de macrohábitats sobre el número de peces (Fig. 8.6) demuestra que si seis o más macrohábitats pueden ser preservados en cantidad suficiente, entonces más del 82% de la fauna de peces conocida del Río Caura puede ser salvada. Esto es particularmente crítico en incorporar el Bajo Caura, especialmente desde el Raudal Cinco Mil río abajo, debido al incremento de efectos negativos de las presiones humanas sobre el río.

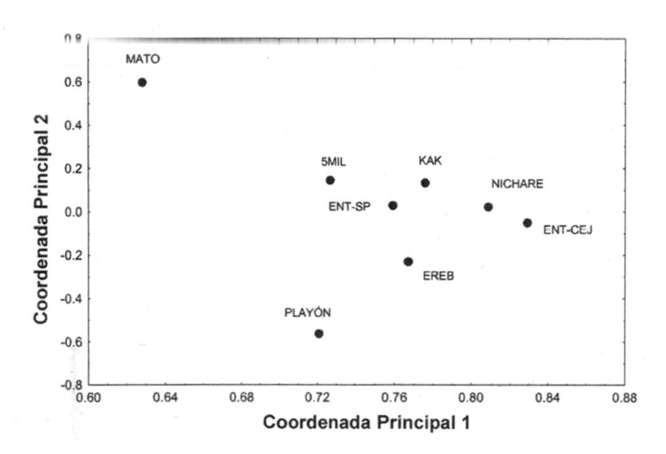

Figura 8.7. Análisis de Coordenadas Principales entre matrices de similaridades de Jaccard para subregiones del Río River para invertebrados acuáticos bénticos. Abreviaturas: Kak–Río Kakada; Ereb–Río Erebato; Ent-Cej–Río Caura desde Entreríos a Cejiato; Ent-SP–Río Caura desde Entreríos a Salto Pará; Play–El Playón; Nichare–Río Nichare; 5Mil–Raudal Cinco Mil; Mato–Río Mato.

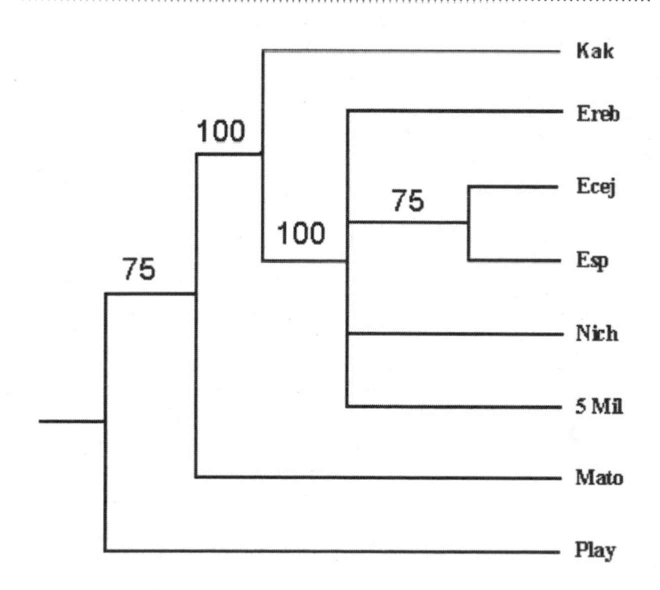

Figure 8.8. Arbol de Consenso Mayor del Análisis de Parsimonia Camin-Sokal de subregiones del Río Caura para invertebrados acuáticos bénticos. Kak–Río Kakada; Ereb–Río Erebato; Ecej–Río Caura desde Entreríos a Cejiato; Esp–Río Caura desde Entreríos a Salto Pará; Play–El Playón; Nich–Río Nichare; 5 Mil–Raudal Cinco Mil; Mato–Río Mato.

LITERATURE CITED

Alroy, J. 1992. Conjunction among taxonomic distributions and the Miocene mammalian biochronology of the Great Plains. Paleobiology 18: 326–343.

Atmar, W., y B. D. Patterson. 1993. The measure of order and disorder in the distribution of species found in fragmented habitat. Oecologia 96: 373–382.

Brualdi, R. A., y J. G. Sanderson. 1999. Nested species subsets, gaps and discrepancy. Oecologia 119: 256–264.

Chernoff, B., P. W. Willink, J. Sarmiento, A. Machado-Allison, N. Menezes y H. Ortega. 1999. Geographic and macrohabitat partitioning of fishes in the Tahuamanu-Manuripi region, Upper Rio Orthon Basin, Bolivia. *En:* Chernoff, B. y P. W. Willink (eds.). A biological assessment of aquatic ecosystems of the Upper Rio Orthon Basin, Pando, Bolivia. RAP Bulletin of Biological Assessment 15. Conservation International, Washington DC. Pp. 51–67.

Chernoff, B., y P. W. Willink. 2000. Biodiversity patterns within the Pantanal, Mato Grosso do Sul, Brasil. *En:* Willink, P., B. Chernoff, L. E. Alonso, J. R. Montambault y R. Lourival (eds.). A biological assessment of the aquatic ecosystems of the Pantanal, Mato Grosso do Sul, Brasil. RAP Bulletin of Biological Assessment 18. Conservation International, Washington, DC. Pp. 103–106.

Chernoff, B., P. W. Willink, M. Toledo-Piza, J. Sarmiento, M. Medina y D. Mandelburger. 2001a. Testing hypotheses of geographic and habitat partitioning of fishes in the Río Paraguay, Paraguay. *En:* Chernoff, B., P. W. Willink y J. R. Montambault (eds.). A Biological Assessment of the Aquatic Ecosystems of the Rio Paraguay Basin, Alto Paraguay, Paraguay. RAP Bulletin of Biological Assessment 19. Conservation International, Washington, DC. Pp. 80–98.

Chernoff, B., P. W. Willink, A. Machado-Allison, M. Fatima Mereles, C. Magalhaes, F. A. R. Barbosa, M. Callisto Faria Pereira y M. Toledo-Piza. 2001b. Congruence of diversity patterns among fishes, invertebrates, and aquatic plants within the Río Paraguay Basin, Paraguay. *En:* Chernoff, B., P. W. Willink y J.R. Montambault (eds.). A Biological Assessment of the Aquatic Ecosystems of the Rio Paraguay Basin, Alto Paraguay, Paraguay. RAP Bulletin of Biological Assessment 19. Conservation International, Washington, DC. Pp. 99–107.

Cox Fernandes, C. 1995. Diversity, distribution and community structure of electric fishes (Gymnotiformes) in the channels of the Amazon River system, Brasil. Ph.D. Dissertation. Duke University, Durham. 394 pp.

Forsberg, B., J. G. D. Castro, E. Cargnin-Ferreira y A. Rosenqvist. 2001. The structure and function of the Negro River Ecosystem: insights from the Jau Project. *En:* Chao, N. L., P. Petry, G. Prang, L. Sonneschein y M. Tlusty (eds.). Conservation and Management of Ornamental Fish Resources of the Rio Negro Basin, Amazonia, Brazil. Universidade do Amazonas, Manaus, Brazil. Pp. 125–144.

García, J. V. y G. Pereira. 2003. Diversidad de macroinvertebrados bentónicos de la Cuenca del Río Caura, Estado Bolívar, Venezuela. *En:* Chernoff, B., A. Machado-Allison, K. Riseng, y J. R. Montambault (eds.). A Biological Assessment of the Aquatic Ecosystems of the Caura River Basin, Bolívar State, Venezuela. RAP Bulletin of Biological Assessment 28, Conservation International, Washington, DC. Pp. 142–148.

Goulding, M., M. L. Carvalho y E. G. Ferreira. 1988. Rio Negro: rich life in poor water: Amazonian diversity and foodchain ecology as seen through fish communities. SPB Academic Publishing, The Hague, Netherlands. 200 pp.

Ibarra, M., y D. J. Stewart. 1989. Longitudinal zonation of sandy beach fishes in the Napo River basin, eastern Ecuador. Copeia 1989: 364–381.

Lowe-McConnell, R. H. 1987. Ecological studies in tropical fish communities. Cambridge University Press, New York. 382 pp.

Machado-Allison, A., B. Chernoff, F. Provenzano, P. Willink, A. Marcano, P. Petry y B. Sidlauskas. 2003. Inventario, Abundancia Relativa, Diversidad e Importancia de los Peces de la Cuenca del Río Caura, Estado Bolívar, Venezuela. *En:* Chernoff, B., A. Machado-Allison, K. Riseng, y J. R. Montambault (eds.). A Biological Assessment of the Aquatic Ecosystems of the Caura River Basin, Bolívar State, Venezuela. RAP Bulletin of Biological Assessment 28, Conservation International, Washington, DC. Pp. 158–169.

Rice, W. R. 1989. Analyzing tables of statistical tests. Evolution 43:223–225.

Rohlf, F. J., y R.R. Sokal. 1995. Statistical tables. W. H. Freeman and Co., NY. 199 pp.

Rosales, J., M. Bevilacqua, W. Díaz, R. Pérez, D. Rivas y S. Caura. 2003a. Comunidades de Vegetación Ribereña de la Cuenca del Río Caura, Estado Bolívar, Venezuela. *En:* Chernoff, B., A. Machado-Allison, K. Riseng, y J. R. Montambault (eds.). A Biological Assessment of the Aquatic Ecosystems of the Caura River Basin, Bolívar State, Venezuela. RAP Bulletin of Biological Assessment 28, Conservation International, Washington, DC. Pp. 127–136.

Rosales, J., N. Maxted, L. Rico-Arce y G. Petts. 2003b. Metodologías Ecohidrológicas y Ecohidrográficas aplicadas a Conservación de la Vegetación Ribereña: el Río Caura como Ejemplo. *En:* Chernoff, B., A. Machado-Allison, K. Riseng, y J. R. Montambault (eds.). A Biological Assessment of the Aquatic Ecosystems of the Caura River Basin, Bolívar State, Venezuela. RAP Bulletin of Biological Assessment 28, Conservation International, Washington, DC. Pp. 170–181.

Sneath, P. H. A., y R. R. SOCAL. 1973. Numerical Taxonomy. San Francisco, W. H. Freeman and Co. 573 pp.

Glosario

Acuacultura—Cultivo de peces u otros organismos acuáticos en áreas controladas, usualmente con fines comerciales.

Arbóreo—Perteneciente a árboles.

Béntico—Zona del fondo de un río o lago, o cualquier otro cuerpo acuático.

Biodiversidad—Descripción del número de especies, sus abundancias y el grado de diferencia entre las especies.

Carnivoros—Organismos que se alimentan de animales.

Caño—Tributario en el cual las aguas pueden correr en ambas direcciones dependiendo de la fluctuación de las mismas durante las lluvias o inundaciones.

Chinchorro—Una red, usualmente usada para capturar peces u otros organismos acúaticos de tamaño considerable.

Cretáceo—Un período de la historia de la Tierra extendiéndose desde 145 millones de años hasta 65 millones de años del presente.

Cuaternario—Un período de la historia de la Tierra extendiéndose desde 2 millones de años hasta el presente.

Cuenca—Area total del sistema de drenaje de un río.

Dead arm (madrevieja)—Un brazo del río que se extiende dentro del bosque o sabana perteneciente a un canal antiguo del río.

Endémico—Encontrado sólo en un área determinada y no en otra parte.

Endoreico—Aguas que fluyen en una cuenca cerrada.

Erosión—La acción del agua lavando el suelo y transportándolo a otro sitio.

Fisionomía—La apariencia total de constitución de un área, principalmente relacionada con vegetación.

Fluvial—Perteneciente a ríos.

Herbívoros—Organismos que se alimentan de plantas.

Heterogeneidad—Grado de diferencia entre items.

Hidrológico—Perteneciente al agua; sistema.

Insectívoros—Organismos que se alimentan de insectos.

Inundación—Aguas que pasan sobre tierra firme durante el período de lluvias.

Laguna—Cuerpo de agua cerrado parcialmente y poco profundo.

Léntico—Perteneciente a aguas estancadas; como en lagos y pozos. *ver lótico.*

Liana—Forma de crecimiento de plantas.

Litoral—La zona acuática que se extiende desde la playa hasta una profundidad que pueda soportar el crecimiento de plantas.

Lótico—Perteneciente a aguas que corren o fluyen; como en ríos y quebradas. ver *léntico.*

Macrofita—Planta acuática superior; no microscópica.

Madrevieja (Dead arm)—Un brazo del río que se extiende dentro del bosque o sabana perteneciente a un canal antiguo del río.

Mioceno—Un período de la historia de la Tierra extendiéndose desde 25 millones de años a cerca de 5 millones del presente.

Omnívoros—Organismos que se alimentan de una gran variedad de alimentos, ambos vegetales y animales.

Paisaje—Perteneciente a la estructura de una área a nivel general más que un sitio, o al menos al nivel de un área de georeferencia.

Perifiton—Algas sobre las rocas, troncos o cualquier otro sustrato debajo del aguas.

Piscívoros—Organismos que se alimentan de peces.

Pleistoceno—Un período de la historia de la Tierra extendiéndose desde 2 millones de años hasta cerca de 100.000 años del presente.

Precámbrico—Un período de la historia de la Tierra extendiéndose desde el origen hasta cerca de 580 millones de años del presente.

Riachuelo—Pequeño alfluente originándose en zonas de montaña o dentro de un bosque.

Río—Cuerpo de agua con aguas corrientes, lótico.

Riparino (Ribereño)—Encontrado en los márgenes de los ríos; usualmente en contexto de vegetación.

Sabana—Tierras planas generalmente cubiertas de gramíneas, o bosques ralos.

Sedimentación—La deposición de sedimentos finos (arcillas y limos); generalmente cubriendo estructuras existentes.

Sustrato—El suelo encontrado entre las raíces de las plantas o en el fondo de los lagos y ríos.

Appendices/Apéndices

Appendix 1

General description of georeference areas sampled during the AquaRAP expedition to the Caura River, Bolívar State, Venezuela

Antonio Machado-Allison, Barry Chernoff, Judith Rosales, Mariapia Bevilacqua, John S. Sparks, Celio Magalhaes, José Vicente García, Guido Pereira, Wilmer Díaz, y Philip W. Willink

I. UPPER CAURA. ABOVE SALTO PARÁ.

AC01. Caño-Lagoon off Caura River just downstream from Entrerios (5°55.78′ N–64°25.39′ W).

General habitat description

Temporal caño macro habitat. This caño forms a lagoon at its confluence with the main channel of the Caura River. During rainy season (August), a maximum of 5 m depth has been recorded. At this time of the year (November), water level is low with a maximum of 1 m depth. The channel was about 9.5 m wide, and the water slightly dark and turbid. In the confluence there is a clay-sand bar covered by herbaceous plants and later a dike about 1 m deep and a depression that connects with drainage. A dense forest covering the entire channel characterizes the vegetation.

The caño presents a muddy shore and bottom covered with abundant leaves. There are deep margins suggesting high current during rainy season. There is some evidence that this is a temporal water channel; the water is stagnant during this time of the year.

Aquatic habitat characterized by a communication channel between the lagoon and the river. The turbid waters have abundant suspended organic material, low dissolved oxygen, and low conductivity. This is an oligothrophic habitat.

AC02. Confluence of Caura and Erebato Rivers at Entrerios (5°56.02′N– 64°25.67′W).

General habitat description

Rapids. Vegetation characterized by shrub individuals, musgos without aquatic plants (Podostemonaceae). The water at the Caura and mouth of the Erebato has acid pH, low conductivity, turbid, no color and well oxygenated. The Caura has greater suspended material and rocky.

Backwater. Isolated form main channel by a sandy bar mixed with clay. Dense herbaceous vegetation on banks.

Caño. Flooded caño inside a dense forest, bottom covered by leaves. Turbid deep waters. No aquatic rooted plants.

AC03. Kakada River and Suajiditu Beach (5°29.86′N–64°34.76′W).

General habitat description

Sandy Beach. Located at the banks of a aluvial island covered by a dense forest. The upper area has small rapids. Left margin has a stabilized bar; dike and plain are inundated. The water is transparent with a penetration of 1.30 m (Sechi disc) and is relatively acidic with variable depths. Waters have an apparent brown (tea) coloration with abundant suspended material. After filtration a low coloration is maintained. Temperature is 24°C. Slow current. Main channel has a rocky bottom and high velocity, with rapids and vegetation over the exposed rocks.

Small caño. Located close (front) to the beach, covered by a dense forest. Maximum flooded level about 3 m. Bar covered by a low and dense forest (15 m). Clay floor. Water with low velocity channel margins muddy. Bottom muddy covered with abundant leaves.

Rapids. Fast moving waters of brown coloration in rmain river channel. Abundant organic suspended material. Margins and bottom sandy/rocky. Water with pH and conductivity low; its depth about 1.5 m.

AC04. Caño Suajiditu (5°29.59′N–64°35.16′W)

General habitat description

The caño area is characterized by the presence of wide flooded forests, and floor covered by dense levels of organic material over sandy-clay strata. The banks and bottom are muddy. We collected fish in two different caños in the area. Caño Suajiditu is characterized by clear waters. The other (no name), have turbid waters and a green-bluish coloration, pH and conductivity greater than the Kakada. Shores and bottoms covered by dense organic material. Dense forest. Low height (8 m). Waters transparent to the bottom, acidic and with temperature of 24°C.

AC05. Raudal del Perro, Caño Wididkenu (5°54.08′N–64°29.4′W; 5°54.17′N–64°29.4′W; 5°53.8′N–64°28.8′W).

General habitat description

Three areas sampled: Rapids, sandy beach in island, and a tributary (Caño Wididkenu).

Rapid. Scattered areas covered with aquatic plants (Podostemonaceae). Bottom rocky. Water velocity high.

Sandy beach. Shrubs on island with low forest growing over rocks. Fragmented rocky habitat with disperse shrubs and small trees. Roots exposed. Water transparent (1.3 m), with a large amounts of suspended brownish material, and water colorless after filtration. Acid pH (5.7–5.9). Velocity high. Conductivity low.

Caño Wididkenu. Brown waters (not black after filtration). Bottom and shores muddy. No aquatic plants; forest perturbed although still very dense.

AC06. Area close to Caño Wididikenu (5°53.9′N–64°28.7′W).

General habitat description

Depositional sandy and clay "vega," made by the confluence of the Erebato and the caño. Riverine forest with rocky island, rapids and sandy areas. Organic horizon at the surface and floor covered by a deep layer of leaves. Rapids contains Podostemonaceae. Within the caño we found a small creek with clear waters and margin, shore and bottom rocky/sandy. Area covered by forest.

AC07. Rapids in Erebato River and caño (5°52.7′N–64°29.56′W).

General habitat description

Forest area over a rocky substrate. Few shrubs and grasses toward the river margins. A rapid (raudal) covered by aquatic plants (Podostemonaceae) located in the right margin. Island on rocky habitat rocky, with a stabilized sandy bar. Waters similar to those in main channel (See previous field station).

AC08. Raudal Cejiato-Soodu (5°33.47′N–64°18.8′W).

General habitat description

Sandy beaches. Beaches characterized by fine sands mixed with fine black organic (?) residues. Shrub vegetation with rocks in surface. Water moderately acid, well oxygenated, temperature 25.5°C, turbid, with a large amount of suspended material.

Creek or small tributary. Mouth with sandy banks in the confluence with the Caura. Water transparent; bottom with great amount of detritus, leaves and logs. Water velocity moderate and transporting sand. Bottom sandy and rocky. Depth of water about 50 cm with a width average 6 m. Banks with sand and rocks. pH, 5.0. Water conductivity low; temperature 23°C ; well oxygenated.

Upper tributary. Flooded forest, dominated by *Eperua*. Forest floor with abundant leaves. Water chemistry similar to that at the mouth.

Rapids. Raudal with several different hydrographical areas and different vegetation. They are characterized by rocky elements covered by podostemonaceans plants. Water moderately acid, well oxygenated, with temperature of about 25.5°C, turbid, and tea color due to the suspended material.

AC09. Raudal Pauji (5°49.7′N–64°24.3′W).

General habitat description

The area is characterized by a series of slight and extended rapids. Various macro habitats are present such as: beaches, rapids, rocky islands, bahias and a small tributary.

Beach. Vegetation characterized by an aluvional forest with abundant leaves on the floor. Water in main channel turbid (80 cm, Sechii Disk), slightly acids, low conductivity, well oxygenated and temperature of 26°C.

Rapids. Rocky area with parches of aquatic vegetations characterized by two species of Podostemonaceae and a moss. Fast waters, and of similar conditions of the river (see above).

Rocky island. Island formed principally by rocks and patches of sand forming lateral bars. There were areas with black deposition. Water same condition of the principal channel.

Backwater. Conditions similar to the island and beaches. Water current very low. There was an inundated plain formed by a sandy lateral depositional bar.

Tributary. Small creek located at the left margin of the Raudal Pauji. Dense forest covering it. Water transparent, margins and bottom muddy, abundant logs and leaves.

AC10. Caño Jasa Kanu (5°53.02′N–64°24 a 25.6′W).

General habitat description

Tributary. Covered by a dense forest with sandy-muddy bottom and shore, abundant detritus. Water temperature 24°C (- 2 from the main channel), acidic and transparent.

Caño. Sandy bar at the confluence with the main river, flooded plain about 1.5m high. Dense forest follows by a dike characterized by an inundated forest. Actual deep 1.5 m. Sandy bottom and abundant detritus. Water temperature 24°C (- 2 from the main channel) acids, low transparency (40 cm Secchi) and conductivity.

AC11. Backwater above Salto Pará (6°16.9′N–64°29.2′W).

General habitat description

Rocky Island. Rocky front with alluvial depositions of sandy texture. Vegetation herbaceous in the shore, shrubs and scattered trees at the center backwater area, including *Montrichardia.*

Rapids. Area with abundant aquatic reophylic plants (two species of Podostemonaceae), heavily rooted between rocks. Water slightly turbid, transparency 80 cm, fast, well oxygenated, moderately acids (5.5–5.7 pH), low conductivity, and brownish coloration after filtered.

Caño. Dense flooded forest. Steep margins and water depth more than 2 m. Turbid waters, yellowish coloration.

Backwater. Characterized by a depositional bar with abundant rooted aquatic *Montrichardia arborescens.* Bottoms muddy and water slightly turbid.

AC12. Smooth-rocky beach, boulders and backwater and caño above Salto Pará (6°16.9′N– 64°29.2′W).

General habitat description

Smooth-rocky beach (boulders). Located on the right bank of the Caura River. The shore and bottom are composed o flat rocks, mixed with pools covered by sand. There are abundant aquatic plants (Podostemonaceae). Water slighty acidic.

Backwater closes the above area with abundant plants (*Montrichardia arborescens*).

Caño. Small creek covered by dense forest. Shore and bottom rocky and covered by leaves, banks very flats. Water clear and transparent, with low to moderate velocity.

AC13. Island across Yuruani River (6°8.4´N–64°25.1´W).

General habitat description

Inundated forest. Located in a flat area (confluence of the Yuruani and Caura rivers). Water deep about 4.3 m. turbid (50 cm.). River wide 60 m. Water coloration greenish, poorly oxygenated, and pH acid (5.5). In this area no fish samples were taken.

Caño. Dense forest, Step muddy banks. Bottom muddy. Turbid waters and slow current. No samples were taken

Backwater in Island in the Caura. Located in front of the mouth of the Yuruari. We collected in a backwater area covered by a small inundated forest. Water turbid and no current.

AC14. Raudal Culebra de Agua (6°04.7´N–64°26.02´W).

General habitat description

Island with backwater and beaches. Vegetation over a sandy bar of gross texture, gravel and rocky fragments in surface. Margins of the beach with disperse shrubs. Water slighty acids and transparent. Organic material deposited over and between rocks.

Rapids. Characterized by rocky shore and bottoms. Fast waters and covered by two species of the Family Podostemonaceae.

Tributary. Characterized by an inundated forest. Water slightly acids, low conductivity, and total transparency with gross suspended organic material. Depth about 1 to 12 m wide. Numerous logs over sandy bottoms.

II. LOWER CAURA, BELOW SALTO PARÁ

BC01. El Playón (Salto Pará) (6°19.5´N–64°31.6´W).

General habitat description

Sandy beach (El Playón). Sandy extended depositional bar, with scarce trees and shrubs in a large backwater just below the falls. Maximum of inundation about 100 m. Waters: acid, low conductivity, well oxygenated, without coloration after filtered.

Beach in rocky island. Island in the confluence of two main arms of the Caura. Rocky front with aluvial depositions of gross texture (gravel) forming a beach about 30 m long and 50 m wide. Water as in the main channel. A small creek in the area.

Rocky beach. Rocky front in margin of the main channel. Rocky bottom, fast waters, waves coming from rapids.

Rocks. Falls in main channel. Water very fast and strong current.

BC02. Tributary (Caño) Confluence with the Caura at El Playón (6°19.7´N–64°31.6´W).

General habitat description

Caño. A shallow stream formed by a depositional bar of gross texture. At the left margin is an erosional bank, covered with forest. Right margin very muddy and covered by shrubs. Great amount of leaves and detritus. Water quite different from the main channel of the Caura, low oxygen, no color and slightly acid. Bottom muddy with great amount of logs and detritus

BC03. Tributary o Caño Waki, several km dowstream from Salto Pará(6° 21.7´N–64°34.1´W).

General habitat description

Confluence of caño and Caura River with typical dike and pit formations. Level of inundation about 3 m. Forest moderately dense. Abundant leaves. Water acidic, turbid, low conductivity and well oxygenated. Bottom muddy, covered by logs and detritus. No rooted aquatic plants.

BC04. Tawadu (Tabaro) River Dedemai Camp (6°21.1´N–64°59.9´W).

General habitat description

Dedemai Camp. River water transparent. Deep banks scarped through a rocky-sandy bead. Dense gallery forest Fast waters except in small bahia or backwaters.

Beaches. Beaches formed by a mixture of clay and sand partially covered by shrubs and herbs continues with a gallery forest. Turbid waters. Bottom muddy with great amount of detritus.

Caño. Narrow tributary well protected by vegetation. Clear waters with abundant logs at the bottom.

BC05. Tawadu (Tabaro) River, Raudal El Pan and Raudal Dimoshi (6°19.63´N–65°02.86´W).

General habitat description

Small rapid. Raudal (El Pan). No high vegetation, few and scattered cyperaceans and abundant aquatic rooted plants of the Family Podostemonaceae. Water transparent and fast. Rapids formed by a depositional bar on the right margin of the Tawadu River. Downstream, there are several small beaches with gravel or sandy bottoms. Vegetation characterized by scattered shrubs followed by a gallery forest (less than 15 m.) We observe heavy vegetations deposits (wide and heavy logs) that suggest strong current during rainy season.

Raudal Dimoshi. Water running of a granite floor, acidic, transparent, well oxygenated and colorless. Current strong. Aquatic rooted plants present. Deposition of sand and gravel down river. Vegetation with shrubs (less than 2 m) followed by a gallery forest (15 m).

Backwater. Characterized by a depositional bar forming a "remanso," with abundant vegetation in a regenerative process, forming a dense covertures. Bottom sandy-muddy, water slightly turbid and well oxygenated.

Caño. Muddy margins and bottom with abundant leaves logs and detritus. Water transparent.

Raudal Dimoshi Downstream. Principal channel of the Tawadu River with clear and deep waters.

BC06. Tawadu (Tabaro) River. Tajañaño Rapid (6°20.25´N–65°01.50´W) .

General habitat description

Rocky Island. 75% exposed rocky area. Few shrubs and grasses on rocks. Water similar to the rapids above. Bottom rocks and sand with areas covered by leaves and mud.

Caño. Bottom and margins muddy andsandy. Water acidic but less than in the Tawadu, well oxygenated and higher temperature. Conductivity greater that in Tawadu.

Rapids. Bottom sand and mud with small, scattered rocks. Abundant logs.

BC07. Laguna 1 km from mouth Tawadu River (Tabaro) (6°21.83´N–64°58.4´W).

General habitat description

This lagoon were formed by an old arm of the Tawadu River. Banks covered by a dense shrubs, low at the beaches or borders. Shore and bottom extremely muddy. Waters acids (more than in river), warm and poor in oxygen, low conductivity, slightly transparent and currentless.

BC08. Río Icutú, Caño Wani (6°04.77´N–64°55.33´W).

General habitat description

Sandy beach. Shore and bottom sandy-mud mixed with logs and abundant leaves. Gallery forest on both shores. Acid waters, low conductivity well oxygenated and transparent.

Caño (Bori ó Wani). This is a mouth of a caño in the right shore of the Icutu River. It is well covered by a dense gallery forest. With scattered shrubs in the shoreline. The waters are well oxygenated, acidic, with low conductivity and transparent. Bottom and shore muddy and sand. Main channel with big rocks, logs and abundant detritus.

BC09. Laguna Wakawajai (6°19.35´N–64°57.22´W).

General habitat description

Beach. Area located at the left bank of the river, formed by a depositional bar of gravel without vegetation (95 %). Some shrubs and grasses scattered in the area and close to the shoreline. Bottom muddy and sandy. Water turbid and moderate velocity.

Lagoon. Acid waters, slightly turbid, no current. The area has a thermocline. Bottom muddy with abundant logs, leaves and organic matter.

BC10. Island in Nichare River, waters below mouth of Tawadu (6°21.43´–64°52.29´W).

General habitat description

The island is formed mainly by small round rocks and domes with sand. In the rocky area, there is vegetation dominated by grasses and cyperaceans and some trees of the family Rubiaceae. Dome of the island has a low forest of species adapted to periodical inundations and is completely submerged during rainy season. The floor han abundant round boulders mixed with gravel and sand and an internal backwater with sandy shores. Watera are acid and turbid, with low conductivity and are well oxygenated. Temperature about 24.5°C, cooler than the Caura.

BC11. Tawadu River Beaches (6°21.01´–64°58.44´W).

General habitat description

First beach. Characterized by lateral depositional banks forming sandy beaches with organic material at surface. During rainy season this forms a flooded area dominated by *Solanum* sp. accompanied by *Inga vera*. The bank is exposed due low water at this time of the year. The floor is sandy with abundant organic material and the bottom muddy with abundant leaves and organic detritus. Water: acidic, transparent, slightly brown and well oxygenated. Conductivity is low and temperature about 23°C. Velocity about 0.41 m/s.

Second beach. Characteristics similar to the previous one. Water acidic, transparent and slightly brown. Low conductivity and cool.

Third beach. Surrounded by a dense forest. Sandy beach with no organic material.

BC12. Ceiba Lagoon (7°05.9´–65°01.34´W).

General habitat description

Extended flooded lagoon off the Caura River surrounded by a low forest (less than 12 m). Floor sandy with mixture of clay and slime. Bottom covered by abundant leaves, logs and organic matter. Water strongly acidic, low oxygen (less than 1 mg/l), low conductivity and brown colored after filtered. No current and temperature about 27 °C.

BC13 Mato River (7°11.74´–65°09.75´W).

General habitat description

This is a typical caño characterized by a channel surrounded by a dense and flooded forest with a level of inundation about 4 m. Bottom and shores muddy, abundant leaves, sand and abundant logs. Close to the mouth are steep shorelines with some exposed beaches. Water is moderatly acid, turbid, well oxygenated, has a temperature 27°C, velocity about 0.3 m/s and no color after filtration.

BC14. "Pelona" Beach at Caura River (7°11.82´–65°08.88´W).

General habitat description

Beach formed by deposition of sand with gross texture (not gravel). No vegetation. Shores and bottom sand. Water, as in the main channel: acidic, low conductivity, well oxygenated, low transparency, no color.

BC15. Pebble Island and Tributary above Raudal Cinco Mil (06°58.29´–64°52.5´W).

General habitat description

Island. Rocky island with a backwater and a channel partially covered with shrubs. Front area has an extended sandy beach divided by backwater toward the rear.

Caño. Small shallow channel widely covered by a forest. Abundant organic material and mud in shores and bottom.

BC16. Camp at Raudal Cinco Mil (06°59.65´–64°54.82´W).

General habitat description

Rocky-Sandy beach. Zone with a mixture of flat rocks "lajas" and sandy patches. Water current fast near the shore with a depth of about 1–2 m. Low flooded forest with a clean floor. Near the shores *Cyperus* and other grasses dominate. Water acids, low conductivity well oxygenated, colorless with moderated transparency.

BC17. Tokoto River (06°53.52´–64°47.90´W).

General habitat description

Area includes rapids, falls, pools and backwaters in a very complex and heterogeneous macro habitat. The shores with scattered shrubs and grasses growing on the crevices of rocks. Both shores and bottom have a mixture of fine sand and rocks. Abundant organic material deposited in deep waters, such as in pools and backwaters. Aquatic plants (Podostemonaceae) cover the rapids. Water: slightly acidic, moderate conductivity, well oxygenated and turbid. This area has the highest pH and conductivity measured in the basin. There is abundant green algae.

Apéndice 1

Descripción general de las áreas de georeferencia estudiadas durante la Expedición AquaRAP a la Cuenca del Río Caura, Estado Bolívar, Venezuela

Antonio Machado-Allison, Barry Chernoff, Judith Rosales, Mariapia Bevilacqua, John S. Sparks, Celio Magalhaes, José Vicente García, Guido Pereira, Wilmer Díaz, y Philip W. Willink

I. CAURA SUPERIOR. AREA POR ARRIBA DEL SALTO PARÁ

AC01. Caño Laguna en el Río Caura, río abajo de Entrerios (5°55.78´N–64°25.39´W)

Descripción general de hábitat

Macro hábitat de caño temporal. Este es un caño que comunica al Río Caura con un lago de inundación. En la temporada de máximas aguas, la inundación registrada es 5 metros y esto ocurre generalmente en Agosto. En esta fecha de colecta (noviembre), el nivel del agua es bajo, con un máximo de 1 m de profundidad. Ancho 9.5 m, aguas ligeramente oscuras y turbias. En la confluencia, existe una barra lateral arcillo-arenosa cubierta con material vegetal herbáceo. Se pasa a un dique que debe tener una profundidad máxima. De 1 m y se baja hacia una depresión que conecta a un drenaje. La vegetación es boscosa, densa con bastante ramas cubriendo el canal principal del caño.

El caño presenta un fondo lodoso-arenoso con márgenes distintos cubiertos con hojarasca. Hay barrancos cortados en la ribera indicadores de gran corriente durante el período de salida de aguas. Hay evidencias que el caño es temporal. Muy poco flujo en éste período.

El habitat acuático se caracteriza por ser un canal de comunicación del río Caura con una laguna de inundación. Es un río de aguas turbias con gran cantidad de material vegetal suspendido, poca corriente, bajo contenido de oxígeno disuelto, baja conductividad, baja diversidad de organismos planctónicos. Poca profundidad, fondo lodoso, hábitat oligotrófico.

AC02. Confluencia de los ríos Caura y Erebato, en Entreríos (5°56.02´N–64°25.67´W)

Descripción general de hábitat

Rápidos. Vegetación caraterizada por individuos arbustivos, musgos sin plantas acuáticas de la Familia Podostemonaceae. El Caura y Erebato con aguas de pH ácido, baja conductividad, aguas turbias, sin color y bien oxigenadas. Caura con mayor material suspendido y más caliente.

Remanso. Formado por barra arenosa deposicional con intercalaciones de hojarasca y arcilla. Estrato herbáceo de cobertura densa.

Caño. Canal dentro de un bosque denso, inundable y aluvional. Fondos fangosos con abundante hojarasca y restos vegetales. Aguas turbias y profundas, sin plantas acuáticas.

AC03. Río Kakada y Playa Suajiditu (5°29.86´N–64°34.76´W)

Descripción general de hábitat

Playa arenosa. Se encuentra al lado de una isla aluvial con vegetación arbórea densa. Parte superior con rápidos pequeños. Margen izquierda con una barra lateral estabilizada. Dique y plano inundado. Aguas transparentes con una penetración de 1.30 m (disco de Sechi). Relativamente ácidas con una profundidad variable. Apariencia de marrón oscuro (té) con abundante materia orgánica suspendida. Después de filtrado sigue un poco de color. Temperatura de 24 grados C. Poca corriente en la cubeta. Canal principal con fondo rocoso, alta velocidad, rápidos y vegetación adherida a las rocas superficiales.

Caño. Frente de la playa, densamente sombreado. Máximo nivel de inundación de tres metros. En la barra existe un bosque bajo (hasta 15 m) denso. Suelo arcilloso. Las aguas tranquilas poca velocidad con márgenes definidos declive moderado. Fondo lodoso. Abundante hojarasca.

Rápidos. Localizados en el canal principal del río con aguas rápidas de color marrón (té). Material orgánico suspendido. En las márgenes el sustrato es arenoso y en el centro rocoso. El pH y la conductividad son bajos. Velocidad media, máxima profundidad de 1.5 metros.

AC04. Caño Suajiditu (5°29.59´N–64°35.16´W)

Descripción general de hábitat

El área del caño se caracteriza por la presencia de bosques inundables en complejo de orillar, con capas de materia orgánica sobre estratos areno-fangosos. En esta área se colectaron peces en dos caños diferentes. Caño Suajiditu con agua transparente, el otro (sin nombre), tenía agua más turbia con coloración azul verdosa. Ambos caños presentan un pH y conductividad, ligeramente mayor que el caño Kakada. Sustrato orgánico y arcilloso.

Ambos caños con alta cantidad de materia orgánica vegetal en las riberas y fondo. Ribera y fondos fangosos con abundante hojarasca. Bosque denso, regeneración natural abundante. Tamaño pequeño (8 m), angostos, aguas transparentes hasta el fondo. Aguas ligeramente ácidas pero menor que en el río. Temperatura 24 grados.

AC05. Raudal del Perro (5°54.08´N–64°29.4´W; 5°54.17´N–64°29.4´W; 5°53.8´N–64°28.8´W)

Descripción general de hábitat

Rápidos. Parches de plantas arraigadas (Podostemonaceae) presentes con cobertura intermedia y pequeña extensión. Fondos arenoso-rocoso y alta velocidad del agua, bien oxigenadas.

Playa arenosa. Isla rocosa con bosque bajo denso. Hábitat rocoso fragmentado con arbustos dispersos. Talud de socavamiento con raíces expuestas provenientes del bosque de tierra firme. Agua con transparencia elevada (1.3 m). Sin embargo, hay bastante material suspendido. Aguas poco negras y transparentes sin color después de filtrado, pH ácido cerca de. 5.7–5.9. Velocidad del agua alta. Conductividad muy baja.

Caño Wididkenu. Aguas oscuras (no negras después de filtrado). Fondos y riberas fangosas. No hay plantas acuáticas, bosque alrededor ligeramente perturbado pero todavía muy denso.

AC06. Area cercana al Caño Wididikenu (5°53.9´N–64°28.7´W)

Descripción general de hábitat

Vega deposicional arenosa arcillosa, por confluencia del río Erebato y el caño. Bosque ribereño. Isla rocosa con rápidos deposición arenosa horizonte orgánico en superficie, abundante hojarasca y bosque medio con palmas. Rápidos con Podostemaceas y margen de socavamiento. Dentro del caño se encontró una quebrada con aguas muy claras con fondos rocosos y arenosos, piedras del borde húmedas y cubiertas con material vegetal, corriente rápida, transparencia total, cubierto densamente por bosque, bastante hojarasca marginal. Aguas con baja turbidez, sin color. Las demás características similares al canal principal del río.

AC07. Raudal en Río Erebato y caño (5°52.7´N–64°29.56´W)

Descripción general de hábitat

Area boscosa de sustrato rocoso, erosional y con vegetación herbácea hacia los márgenes del río. Las márgenes del río son de socavamiento sin deposición. Un raudal con podostemonaceas en la margen derecha. Isla en hábitat rocoso, con barra arenosa estabilizada. Aguas similares al canal principal del río. Abundantes podostemonaceas.

AC08. Raudal Cejiato-Soodu (5°33.47´N–64°18.8´W)

Descripción general de hábitat

Existen varios macro hábitats con la inclusión de playas arenosas en remanso, rápidos y riachuelo.

Playa arenosa. Playa caracterizada por estar constituida por arenas muy finas y mezcladas con material negro de naturaleza aparentemente orgánica. Vegetación arbustal con afloramientos rocosos en superficie. Aguas moderadamente ácidas, bien oxigenadas, temperatura 25.5, muy turbias, mayor color debido a material suspendido.

Riachuelo boca. El riachuelo en su boca limitando con una playa arenosa por confluencia con el Caura. Aguas transparentes, con numerosos restos vegetales en el fondo. Agua con abundante hojas y transporte de arena. Fondos arenosos y presencia de grandes rocas. Corriente rápida. Profundidad de 50 cm. Ancho medio 6 m. Riberas de arena y rocas. pH, 5.0, ácido. Aguas con baja conductividad; temperatura 23°C y bien oxigenadas.

Riachuelo arriba. Bosque medio inundable sobre vega aluvial, dominado por *Eperua*. Sotobosque medio con abundante hojarasca. Agua similar a la de la boca.

Raudal. Rápidos con varias áreas diferentes en cuanto a hidrografía y formaciones vegetales. Caracterizado por elementos rocosos cubiertos por plantas acuáticas de la familia Podostemonaceae y musgos. Aguas moderadamente ácidas, bien oxigenadas, temperatura 25.5°C, muy túrbias, mayor color debido a material suspendido.

AC09. Raudal Paují (5°49.7´N–64°24.3´W)

Descripción general de hábitat

Raudal de bajo gradiente y extenso. Existen varios macro hábitats con la inclusión de playas arenosas, rápidos, islas rocosas, remansos y riachuelo.

Playa. Vegetación con bosque cercano caracterizado por un bosque bajo de isla aluvial, abundante hojarasca. Aguas turbias, (80 cm) ligeramente ácidas, baja conductividad, bien oxigenadas y cálidas (26°C).

Rápidos. Vegetación caracterizada por tres especies de plantas acuáticas (dos podostemonaceas y un musgo). Aguas de corriente rápida y de características similares a las del río.

Islas rocosas. Islas con aluviones. Barras laterales deposicionales arenosas con presencia de arenas negras.

Remansos. Similares a la de la isla, aguas idénticas al del río, tranquilas. Planicie de inundación, formada por una barra lateral deposicional.

Riachuelo. En la margen izquierda del raudal Paují. Altamente sombreado, agua clara, fondo lodoso y con poca arena, hojarasca y palos abundantes.

AC10. Caño Jasa Kanu (5°53.02´N–64°24 a 25.6´W)

Descripción general de hábitat

Tributario. Cubierto por un bosque denso con bordes y fondo arenoso–fangoso con abundante detritus. Temperatura del agua 24°C (- 2 por debajo del canal principal), ácidas y transparentes.

Caño. con barra arenosa de confluencia, vega de inundación de 1.5 m. Bosque denso. Le sigue un dique caracterizado por la presencia de un bosque de inundación del Caura. Profundidad actual 1.5 m. Fondo arenoso con bastante material vegetal sumergido. Temperatura 24°C (- 2 del canal principal). Aguas ácidas, muy turbias (40 cm disco de Secchi) y baja conductividad.

AC11. Remanso Aguas Arriba Salto Pará (6°16.9´N–64°29.2´W)

Descripción general de hábitat

Islas rocosas. Frente rocoso con deposiciones aluviales de textura arenosa. Cobertura herbácea rala. Lateral a la isla hay un remanso con *Montrichardia.*

Raudal. Plantas acuáticas reofilas con dos especies de Podostemonacea, una enraizada en fondo rocoso, la otra en superficie enraizada en sustrato rocoso. Aguas con moderada turbidez, transparencia (80 cm), rápidas, bien oxigenadas moderadamente ácidas (5.5–5.7 pH), con baja conductividad, con moderado color (marrón) después de filtradas.

Caño. Bosque inundado de confluencia con vega con cobertura media y elementos leñoso de altura. Riberas marcadamente verticales y profundas. Hojarasca abundante y regeneración abundante. Aguas turbias de color amarillento.

Remanso. Area caracterizada por una barra deposicional de remanso, con *Montrichardia arborescens* y abundante bejucos formando una densa cobertura. Fondos arenoso-fangoso, aguas ligeramente turbias.

AC12. Lajas, Remanso y Caño en Salto Pará (6°16.9´N–64°29.2´W)

Descripción general del hábitat

Lajas. Estas están localizadas en la margen derecha del Caura, Playa y fondos rocosos, con aportes aluvionales de arena. En sus áreas inundadas presenta tres especies de Podostemonaceas. Aguas ligeramente ácidas.

Barra lateral, remanso. Localizado en la margen derecha, con *Montrichardia arborescens*.

Caño. Riachuelo pequeño, sombreado, muy llano. Fondo arenoso/pedregoso, con bancos dispersos de hojarasca. Margen del talud bajo. Aguas transparentes, con corriente lenta a moderada.

AC13. Rio Yuruani. Bosque inundable, Caño, Isla (6°8.4´N–64°25.1´W)

Descripción general del hábitat

Bosque inundable. Localizado en una planicie de confluencia del Río Yuruani con el Caura. Agua con profundidad de 4.3 m, muy turbias, baja transparencia 50 cm. Ancho del río 60 m. Color del agua clara (blanquecino verdoso), bajo nivel de oxígeno y pH más ácido (5.5) y algo más frío (-1 grado). Crustáceos, bentos y peces no muestreados.

Caño. Area muy sombreada, márgenes de barrancas altas con talud. Fondos lodosos no consolidados. Abundante hojarasca y muchos palos y troncos en el caño. Aguas turbias, corriente lenta. Peces no muestreados.

Remanso en Isla en el Río Caura. Localizada frente a la Boca del Yuruari. Se analizó un área de remanso, arenosa, con un pequeño bosque inundado, con relativo margen elevado. Aguas turbias y sin corriente.

AC14. Raudal Culebra de Agua (6°04.7´N–64°26.02´W)

Descripción general del hábitat

Isla con remanso y playas. La vegetación está localizada sobre una barra arenosa de textura gruesa, grava y fragmentos de roca en superficie. El eje de playa tiene una vegetación caracterizada por un arbustal disperso. Aguas ligeramente ácidas y transparentes. Deposición de material orgánico sobre y entre las rocas.

Rápidos. Colectados en la parte superior del Raudal Culebra. Caracterizado por aguas fuertes, caudal moderado, fondos rocosos cubiertas por abundantes Podostemonacaeas (2 especies).

Riachuelo. Bosque inundable de confluencia. Presencia de numerosas especies adaptadas a áreas inundables. Aguas levemente acidez, baja conductividad, transparencia total con material suspendido grueso (hojas), profundidad 1 m. ancho 12 m. con numerosos troncos atravesados en el cauce, fondo arenosos, canal sombreado.

II. BAJO CAURA, POR DEBAJO DEL SALTO PARÁ

BC01. El Playón (Salto Pará) (6°19.5´N–64°31.0´W)

Descripción general del hábitat

Area por debajo de un gran salto, conformada por una extensa ensenada de remanso, playa arenosa, rápidos e islas rocosas.

Playa. Barra arenosa deposicional con elementos leñosos y arbustivos dispersos, formada dentro de una gran ensenada por debajo del Salto Pará. Distancia de inundación más que 100 m. Aguas ácidas, baja conductividad, saturada de oxígeno, sin color después de filtrado.

Islas Rocosa-Arenosa. Isla en la confluencia de los dos brazos del río Caura. Frente rocoso con deposiciones aluviales de textura gruesa formando una playa de 30 m de largo y 50 m de ancho. En la misma se encuentra un riachuelo con aguas claras y transparentes, llano, con algunos pozos. Fondo arenoso y pedregoso con fuerte pendiente. Hojarasca abundante.

Playa rocosa. Frente rocoso en borde del canal principal del río. Fondo con abundantes rocas. Aguas claras y rápidas con oleaje proveniente de los rápidos.

Rocas. Canal rocoso en una de las caídas principales del Salto Pará. Aguas turbulentas y muy rápidas.

BC02. Caño (Tributario) de confluencia en el Río Caura cerca del Playón (6°19.7´N–64°31.6´W)

Descripción general del hábitat

Caño. Canal poco profundo formado por una barra lateral deposicional de textura gruesa, en su margen derecha cubierto de arbustos. Margen izquierda talud de erosión. Bosque muestra evidencia de perturbación. Aporte grande de hojarasca. Aguas diferentes a las del río Caura. Sin color, baja oxígeno. Barrancos con márgenes lodoso-arenoso. Fondos lodosos con abundantes hojarasca y troncos sumergidos.

BC03. Riachuelo o Caño Waki, varios km por debajo del Salto Pará (6°21.7´N–64°34.1´W)

Descripción general del hábitat

Caño de confluencia con el Río Caura en las típicas posiciones de vega dique y cubeta de decantación. Lámina de inundación de tres metros. Bosque moderadamente denso. Abundante aporte de hojarasca al caño. Aguas ácidas, con baja conductividad, bien oxigenadas, moderadamente turbias. Caño con el fondo cubierto de elementos vegetales como hojas y ramas. Fondos lodosos, no hay plantas acuáticas.

BC04. Río Tawadu. Campamento Dedemai (6°21.1´N–64°59.9´W)

Descripción general del hábitat

Campamento Dedemai. Río de aguas claras, medianamente transparentes, profundas. Bancos profundos, cortados. Bosques de galería densos con aportes de material al río. Riberas y fondos arenosos y con abundantes lajas. Aguas rápidas excepto en las pequeñas ensenadas, moderadamente transparentes.

Playa. Playas arenoso-fangosas cerca del campamento. Vegetación arbustiva sobre las playas y bosque de galería a continuación. Aguas turbias. Fondos fangosos con abundante detritus.

Caño. Caño angosto y muy protegido por vegetación. Aguas claras con abundante material vegetal (troncos, ramas) en el fondo.

BC05. Río Tawadu (Tabaro), Raudal el Pan y Raudal Dimoshi (6°19.63´N–65°02.86´W)

Descripción general del hábitat

Pequeño raudal (El Pan). Pequeño rápido sin vegetación arbustiva, pocas cyperáceas y abundantes plantas acuáticas (Podostemonaceae), aguas claras y rápidas. Raudal Dimoshi formado por barra lateral deposicional. Margen derecho del río Tawadu, aguas abajo, presenta una barra lateral con deposiciones aluviales arenosas de textura gruesa. Ecotono con arbustos de altura menor a dos metros. En el gradiente de la transecta le sigue un bosque bajo menor de 15 m. con dosel irregular, sotobosque ralo, baja regeneración. Se observó evidencia de arrastre y deposición de ramas y troncos gruesos, arrastrados por la crecida del río Tawadu.

Raudal Dimoshi. Rápido corriendo sobre rocas sólidas de granito. Aguas ácidas, transparentes, bien oxigenadas, sin color. Plantas acuáticas presentes. Deposiciones aluviales arenosas de textura gruesa. Ecotono con arbustos de altura menor a dos metros. En el gradiente de la transecta le sigue un bosque bajo menor de 15 m. con dosel irregular, sotobosque ralo

Remanso. caracterizado por una barra deposicional de remanso, con vegetación abundante principalmente por procesos regenerativos del bosque, formando una densa cobertura. Fondos arenoso-fangoso, aguas ligeramente turbias.

Caño. Con el fondo integrado por barro, arena y abundante hojarasca. Abundantes ramas y troncos. Aguas transparentes.

Raudal Dimoshi área inferior. Canal principal de río, con aguas claras y profundas.

BC06. Río Tawadu (Tabaro) Raudal Tajañaño (6°20.25´N–65°01.50´W)

Descripción general del hábitat

Isla rocosa. Area descubierta en un 75 por ciento. Cobertura conformada por hierbas y subfructices que nacen en las fisuras de las rocas y en las depresiones donde se ha depositado arena y materia orgánica. Se observaron seis especies de subfructices y hierbas y cuatro arbustivos. Agua similar a la de los rápidos arriba. Fondos rocoso y arenoso. Algunas áreas con deposiciones fangosas y hojarasca.

Caño. Fondos arenosos y fangosos. Aguas ácidas pero más alta que en Tawadu. Bien oxigenada y más caliente y con mayor conductividad. Aguas con poca turbidez.

Rápidos. Fondos fangoso arenoso, con rocas pequeñas dispersas y con abundantes troncos y palos caídos.

BC07. Laguna 1 km de la Boca del Río Tawadu (Tabaro) (6°21.83´N–64°58.4´W)

Descripción general del hábitat

Es una laguna de inundación o una madrevieja perteneciente al río Tabaro. Areas lateral cubiertas de vegetación arbustiva densa, baja en las orillas. Fondos extremadamente fangosos. Aguas ácidas, más que en el río, cálidas, muy pobre en oxígeno y poca conductividad, transparentes y aparentemente estancadas.

BC08. Río Icutú, Caño Wani (6°04.77´N–64°55.33´W)

Descripción general del hábitat

Playa arenosa. Fondos arenosos y fangosos. Areas deposicionales con abundantes palos y hojas. Bosque de galería presente. Aguas ácidas, poca conductividad muy oxigenadas y claras.

Caño de Confluencia (Bori o Wani). Este es un caño de confluencia con el Río Icutu. Un bosque en posición de dique del lado del caño y en el otro lado en contacto con una barra lateral deposicional el Río Icutu con vegetación arbustiva. Bosque bajo, frecuencia de bejucos, presencia de palmas, aporte de hojarasca y suelo arcillo limoso en el dique. Aguas ácidas, poca conductividad muy oxigenadas, claras. Fondos arenosos/fangosos con piedras grandes, troncos y palos. Aguas deposicionales con abundante hojarasca.

BC09. Laguna Wakawajai (6°19.35´N–64°57.22´W)

Descripción general del hábitat

Playa. Localizada en la margen izquierda del canal, barra deposicional de textura gruesa desprovista de vegetación en un 95 %. Se registraron especies arbustivas de altura menor de tres metros y algunas especies herbáceas, cercanas a las orillas. Fondos arenosos y fangosos. Aguas con velocidad moderada, turbias.

Madre vieja. Aguas ácidas, cálidas no muy turbias, estancadas. Existe un termoclino. Fondo fangoso a extremadamente lodoso con abundante detritus. Palos y muchas hojas en el lado de las barrancas.

BC10. Isla en el Río Nichare, aguas abajo Boca de Tawadu (6°21.43´N–64°52.29´W)

Descripción general del hábitat

Isla con frente de cantos rodados y domo deposicional arenoso. En el área de canto rodado, vegetación herbácea dominada por gramíneas y ciperaceas y algunos arbóreos de la familia Rubiaceae. Domo de la isla se presenta un bosquete bajo con estadio sucecional temprano con elementos florísticos propios del bosque inundable ribereño. El bosque se inunda completamente en crecida. Suelos grava y canto rodados en la punta y arena gruesa formando una barra deposiconal y un remanso interno en forma de herradura. Isla con barra lateral deposicional de textura gruesa, canal con más arena. Aguas muy turbias ácidas, conductividad baja bien oxigenadas más frías (24.5°C) que en el Caura, agua sin color. Dentro del remanso las aguas son cálidas. Corriente 1 m por segundo.

BC11. Playas en el Tawadu (6°21.01´N–64°58.44´W)
Descripción general del hábitat
Primera playa. Barras laterales deposicionales formando playas arenosas con aporte orgánico en superficie que en aguas altas es un rebalse de *Solanum* sp. que forma densos matorrales menores de 2 m, acompañados de *Inga vera* y otras especies de forma de crecimiento arbóreo. La barra deposicional debido a la baja de aguas se encuentra expuesta en una longitud aproximada de 27 m. Suelo arenoso con abundante aporte orgánico en superficie. Fondos fangosos con mucha hojarasca y palos presentes. Playa deposiconal con Labiatae, fondos y riberas altamente fangosos con abundante hojarasca. Aguas muy ácidas transparente, ligeramente marrón, bien oxigenadas, muy poco turbias, conductividad muy baja, frías (23°C), 0.41 m/s.

Segunda playa. Condiciones similares a la anterior. Aguas muy ácidas transparente, ligeramente marrón, bien oxigenadas, muy poco turbias, conductividad muy baja, frías (23°C), velocidad 0.41 m/s.

Tercera playa. Bosque denso cubriendo la playa, con árboles dentro del agua. Playa arenosa, con poca hojarasca.

BC12. Laguna La Ceiba (7°05.9´N–65°01.34´W)
Descripción general del hábitat
Laguna en cubeta de decantación proveniente del desborde del río Caura. Suelos arenosos con aporte de limos y arcillas. Fondo con hojarasca, palos, fango y poco de arena. No hay talud entra directo en el agua. Agua: extremadamente ácido (ph bajo), oxígeno disuelto bajo, menor de 1mg/l, pobre conductividad, mucho detrito de hojarasca y el color del agua después de ser filtrada es de color marrón. Sin corriente y caliente (27°C).

BC13. Río Mato (7°11.74´N–65°09.75´W)
Descripción general del hábitat
El caño está caracterizado por ser un área de bosque inundable en posición de dique en meandro del río. Lámina de inundación aproximadamente de 4 metros. Fondos contienen fango, hojarasca, arena y muchos palos. Cerca de la boca, los taludes son altos y las playas están expuestas debido a la baja de agua. Agua moderadamente ácida, muy turbia, bien oxigenada, temperatura 27°C, la velocidad de corriente es 0.3 m/s y después de filtrada no hay color.

BC14. Playa Pelona, en el Río Caura (7°11.82´N–65°08.88´W)
Descripción general del hábitat
Playa formada por una barra aluvial deposicional de canal del Río Caura. Textura gruesa, desprovista de vegetación y expuesta debido a la época de bajada de aguas. Fondos arenosos. Agua en el río principal: ácido, conductividad baja, bien oxigenado, alta turbidez, sin color. El agua es la misma que arriba del salto Pará con excepción de la temperatura en los canales y áreas depresionales de la barra que fue extremadamente elevada.

BC15. Isla Rocosa y Caño arriba Raudal 5000 (06°58.29´N–64°52.5´W)
Descripción general del hábitat
Isla. Isla rocoso-arenosa con remanso y canal de aguas rápidas playas parcialmente cubierta por arbustos al igual que la zona de remansos. Frente con abundante playa arenosa y disectada por un gran remanso posterior. Bosque bajo de 8–12 m de alto, el cual presenta signos de intervención como son la presencia de tocones, esto hace que se presente una cobertura que permite la presencia de un sotobosque denso dominado por *Cyperus* sp.

Caño. Canal pequeño, poco profundo y con amplia cobertura boscosa. Aguas turbias a oscuras. Abundante material orgánico en las orillas y fondo. Orillas y fondo fangoso.

BC16. Campamento Raudal 5000 (06°59.65´N–64°54.82´W)
Descripción general del hábitat
Areas del campamento y playa arenosa/rocosa. Zona con una mezcla de frente de playa con lajas y rocas con parches de playas de arena. Corrientes rápidas a lentas en las orillas. Profundidad de 1 a 2 metros. Bosque de rebalse bajo, sotobosque ralo, conformado por especies arbóreas de porte bajo. Existe un estrato herbáceo denso dominado por *Cyperus* sp. y otras hierbas y subfructices como acompañantes. Aguas ácidas con muy baja conductividad muy oxigenadas, sin color y moderada hacia alta turbidez.

BC17. Río Tocoto (06°53.52´N–64°47.90´W)
Descripción general del hábitat
Area que comprende una zona de rápidos con caídas poco pronunciadas y amplias áreas de pozos aislados, semi-aislados, bahías y remansos, que convierten la zona en un gran complejo heterogéneo. Fondos y riberas con mezclas de arena y lajas. Abundante material orgánico en los fondos de los remansos y bahías. Plantas acuáticas (Podostemaceas) abundantes en los rápidos. Vegetación. Arbustal de 4 a 7 m de alto a orilla de lajas graníticas. Presenta cobertura rala y elementos arbóreos dispersos. Se observó una especie de Podostemonaceae en los rápidos. Aguas levemente ácidas, moderada conductividad, bien oxigenadas, muy turbias, tiene las más alta conductividad y pH de todos los ríos muestreados. Hay alta presencia de algas verdes.

Appendix/Apéndice 2

List of plants from the AquaRAP expedition to the Caura River Basin, Bolívar State, Venezuela

Lista de plantas colectadas y observadas durante la Expedición AquaRAP a la Cuenca del Río Caura, Estado Bolívar, Venezuela

Judith Rosales, Mariapia Bevilacqua, Wilmer Díaz, Rogelio Perez, Delfín Rivas y Simón Caura

Abbreviations: p/s = personal observation, wd=Wilmer Díaz, jr= Judith Rosales, and indet= undetermined.

Abreviaciones: p/s = observación personal, wd = Wilmer Díaz; jr = Judith Rosales; indet = indeterminada.

Georef	Family	Genus	Species	# Collection
Georef	**Familia**	**Género**	**Especies**	**# Colección**
BC08	Acanthaceae	Genus indet.		wdp/s
AC11	Acanthaceae	*Justicia*	*laevilinguis* (Nees) Lindau	wd4854
AC05	Acanthaceae	*Staurogyne*	*spraguei* Wassh	wd4686
AC05	Acanthaceae	*Staurogyne*	*spraguei* Wassh	wd4696
AC09	Acanthaceae	*Staurogyne*	*spraguei* Wassh	wd4824
AC11	Acanthaceae	*Staurogyne*	*spraguei* Wassh	wd4842
BC10	Acanthaceae	*Staurogyne*	*spraguei* Wassh	wdp/s
AC12	Amaranthaceae	*Cyathula*	*prostrata* (L.) Blume	wd4880
AC13	Anacardiaceae	*Anacardium*	*giganteum* Hancock ex Engler	JR P/S
BC15	Anacardiaceae	*Anacardium*	*giganteum* Hancock ex Engler	wd5015a
BC10	Anacardiaceae	*Tapirira*	*guianensis* Aubl.	wdp/s
AC03	Annonaceae	*Anaxagorea*	*dolichocarpus* Sprague & Sandwith	wd4631
BC15	Annonaceae	*Duguetia*	*lucida* Urb.	wd5028
AC06	Annonaceae	Genus indet.		wd4711
AC03	Annonaceae	*Xylopia*	*benthamii* R.E. Fries	wd4622
AC03	Annonaceae	*Xylopia*	*benthamii* R.E. Fries	wd4633
AC06	Apocynaceae	*Allamanda*	*cathartica* L.	wd4715
AC12	Apocynaceae	*Allamanda*	*cathartica* L.	wd4870
BC05	Apocynaceae	*Allamanda*	*cathartica* L.	wd4963
AC04	Apocynaceae	Genus indet.		JR P/S
AC12	Apocynaceae	Genus indet.		wd4882
BC01	Apocynaceae	Genus indet.		wd4938
BC15	Apocynaceae	*Himatanthus*	*articulatus* (Vahl) Woodson	wdp/s
BC08	Apocynaceae	*Tabernaemontana*	*macrocalyx* Muell.-Arg.	wdp/s
BC05	Apocynaceae	*Tabernaemontana*	*sananho* Ruiz & Pav.	wd4949
AC09	Araceae	*Anthurium*	*clavigereum* Poepp. & Endl.	wd4832
AC06	Araceae	*Anthurium*	*gracile* (Rudge) Lind.	wd4708

continued

Georef	Family	Genus	Species	# Collection
Georef	Familia	Género	Especies	# Colección
AC08	Araceae	*Anthurium*	sp.	wd4786
AC11	Araceae	*Anthurium*	sp.	wd4856a
BC10	Araceae	*Anthurium*	sp.	wd5000
BC03	Araceae	*Anthurium*	sp.	wd4943
BC16	Araceae	Genus indet.		wd5043
AC09	Araceae	*Heteropsis*	*flexuosa* (Kunth) G.S. Bunting	wd4830
BC05	Araceae	*Heteropsis*	*flexuosa* (Kunth) G.S. Bunting	wdp/s
AC06	Araceae	*Monstera*	*adamsonii* Schott	wd4714
BC04	Araceae	*Monstera*	*obliqua* Miq.	wd4979
BC17	Araceae	*Montrichardia*	*arborescens* (L.) Schott	wd5036
AC08	Araceae	*Montrichardia*	*arborescens* (L.) Schott	wd4751
AC09	Araceae	*Montrichardia*	*arborescens* (L.) Schott	JR P/S
AC11	Araceae	*Montrichardia*	*arborescens* (L.) Schott	JR P/S
AC12	Araceae	*Montrichardia*	*arborescens* (L.) Schott	JR P/S
AC05	Araceae	*Philodendron*	sp.	wd4687
BC01	Araceae	*Philodendron*	sp.	wdp/s
BC05	Araceae	*Spathyphylum*	*cuspidatum* Schott	wd4947
AC12	Araliaceae	*Dendropanax*	*arboreus* (L.) Decne & Planch	wd4889
BC01	Araliaceae	*Scheflera*	*morototoni* (Aubl.) Mag., Stey. & Frodim	wdp/s
AC03	Arecaceae	*Attalea*	*maripa* (Aubl.) Mart	JR P/S
AC04	Arecaceae	*Attalea*	*maripa* (Aubl.) Mart	JR P/S
AC09	Arecaceae	*Attalea*	*maripa* (Aubl.) Mart	JR P/S
AC10	Arecaceae	*Attalea*	*maripa* (Aubl.) Mart	JR P/S
AC11	Arecaceae	*Attalea*	*maripa* (Aubl.) Mart	JR P/S
AC13	Arecaceae	*Attalea*	*maripa* (Aubl.) Mart	JR P/S
AC01	Arecaceae	*Attalea*	*maripa* (Aubl.) Mart	JR P/S
AC03	Arecaceae	*Bactris*	sp.	JR P/S
AC03	Arecaceae	*Bactris*	sp.	wd4626
BC03	Arecaceae	*Bactris*	*brongniartii* Mart	wdp/s
AC14	Arecaceae	*Bactris*	*simplicifroms* Mart	wd4922
AC03	Arecaceae	*Desmoncus*	sp.	wd4624
AC11	Arecaceae	*Desmoncus*	sp.	wd4862
BC03	Arecaceae	*Euterpe*	*precatoria* Mart.	wdp/s
BC01	Arecaceae	*Euterpe*	*precatoria* Mart.	wdp/s
BC04	Arecaceae	*Euterpe*	*precatoria* Mart.	wdp/s
AC03	Arecaceae	*Euterpe*	*precatoria* Mart.	jR P/S
AC04	Arecaceae	*Euterpe*	*precatoria* Mart.	JR P/S
AC10	Arecaceae	*Euterpe*	*precatoria* Mart.	JR P/S
AC11	Arecaceae	*Euterpe*	*precatoria* Mart.	JR P/S
BC08	Arecaceae	*Euterpe*	*precatoria* Mart.	wdp/s
AC01	Arecaceae	*Euterpe*	*precatoria* Mart.	JR P/S
AC01	Arecaceae	*Geonoma*	*deversa* Poit.	wd4600

continued

Georef	Family	Genus	Species	# Collection
Georef	Familia	Género	Especies	# Colección
AC04	Arecaceae	*Geonoma*	*deversa* Poit.	JR P/S
AC06	Arecaceae	*Geonoma*	*deversa* Poit.	wd4702
BC03	Arecaceae	*Geonoma*	*deversa* Poit.	wdp/s
BC05	Arecaceae	*Geonoma*	*deversa* Poit.	wdp/s
BC08	Arecaceae	*Geonoma*	*deversa* Poit.	wdp/s
AC09	Arecaceae	*Iriartea*	sp.	JR P/S
AC03	Arecaceae	*Iriartella*	*setigera* Mart.	wd4620
AC06	Arecaceae	*Iriartella*	*setigera* Mart.	wd4706
AC08	Arecaceae	*Iriartella*	*setigera* Mart.	wd4821
BC05	Arecaceae	*Oenocarpus*	*bacaba* Mart.	wdp/s
BC10	Arecaceae	*Oenocarpus*	*bacaba* Mart.	wdp/s
BC01	Arecaceae	*Oenocarpus*	*bacaba* Mart.	wdp/s
AC09	Arecaceae	*Socratea*	*exorrhiza* (Mart.) H. Wendl.	JR P/S
AC11	Arecaceae	*Socratea*	*exorrhiza* (Mart.) H. Wendl.	JR P/S
BC05	Arecaceae	*Socratea*	*exorrhiza* (Mart.) H. Wendl	wdp/s
BC01	Arecaceae	*Socratea*	*exorrhiza* (Mart.) H. Wendl.	wdp/s
AC07	Asteraceae	*Lepidaploa*	*gracilis* (Mart.) H. Wendl	wd4734
AC08	Asteraceae	*Lepidaploa*	*gracilis* (Mart.) H. Wendl	wd4798
AC08	Asteraceae	*Piptocarpha*	*triflora* (Aubl.) Benn. Ex Baker	wd4773
AC08	Asteraceae	*Unxia*	*camphorata* L.f.	wd4800
BC01	Bignoniaceae	*Arrabidaea*	sp.	wd4932
BC04	Bignoniaceae	*Crescentia*	*amazonica* Ducke	wdp/s
BC07	Bignoniaceae	*Crescentia*	*amazonica* Ducke	wdp/s
BC09	Bignoniaceae	*Crescentia*	*amazonica* Ducke	wdp/s
BC10	Bignoniaceae	*Crescentia*	*amazonica* Ducke	wd4994
BC11	Bignoniaceae	*Crescentia*	*amazonica* Ducke	wdp/s
AC11	Bignoniaceae	*Jacaranda*	*copaia* (Aubl) D. Don	JR P/S
BC08	Bignoniaceae	*Jacaranda*	*copaia* (Aubl) D. Don	wdp/s
BC15	Bignoniaceae	*Jacaranda*	*copaia* (Aubl) D. Don	wdp/s
AC05	Bignoniaceae	*Jacaranda*	*obtusifolia* Humb. & Bonpl.	wd4678
AC07	Bignoniaceae	*Jacaranda*	*obtusifolia* Humb. & Bonpl.	wd4736
BC04	Bignoniaceae	*Jacaranda*	*obtusifolia* Humb. & Bonpl.	wdp/s
AC05	Bignoniaceae	*Jacaranda*	sp.	JR P/S
BC15	Bignoniaceae	*Jacaranda*	sp.	wd5022
BC08	Bignoniaceae	*Mussatia*	*prieurei* (A.DC.) Bureau ex K. Schum.	wdp/s
BC11	Bixaceae	*Bixa*	*urucurana* Willd.	wd5003
AC03	Bombacaceae	*Catostemma*	*commune* Sandwith	JR P/S
AC04	Bombacaceae	*Catostemma*	*commune* Sandwith	JR P/S
AC07	Bombacaceae	*Catostemma*	*commune* Sandwith	JR P/S
AC09	Bombacaceae	*Catostemma*	*commune* Sandwith	JR P/S
AC14	Bombacaceae	*Catostemma*	*commune* Sandwith	JR P/S
AC01	Bombacaceae	*Catostemma*	*commune* Sandwith	JR P/S

continued

Georef	Family	Genus	Species	# Collection
Georef	Familia	Género	Especies	# Colección
BC05	Bombacaceae	*Ceiba*	*pentandra* (L.) Gaertn	wdp/s
AC05	Bombacaceae	*Pachira*	*aquatica* Aubl.	wd4662
AC08	Bombacaceae	*Pachira*	*aquatica* Aubl.	wd4796
AC14	Bombacaceae	*Pachira*	*aquatica* Aubl.	JR P/S
BC01	Boraginaceae	*Cordia*	*nodosa* Lam.	wd4936a
AC01	Boraginaceae	*Cordia*	*sericicalyx* A. DC.	JR P/S
AC03	Bromeliaceae	*Ananas*	*parguazensis* Camargo & L.B.Sm	wd4645
AC03	Bromeliaceae	*Pitcairnia*	*caricifolia* Mart. Ex	wd4642
AC08	Bromeliaceae	*Pitcairnia*	*caricifolia* Mart. Ex	wd4761
AC12	Bromeliaceae	*Pitcairnia*	*maidifolia* (E. Morren) Decne ex Planch.	wd4890
AC05	Bromeliaceae	*Tillandsia*	*adpressiflora* Mez.	wd4674
AC05	Bromeliaceae	*Tillandsia*	*bulbosa* Hook	wd4668
AC05	Bromeliaceae	*Tillandsia*	*paraensis* Mez.	wd4700
AC05	Bromeliaceae	*Tillandsia*	sp.	wd4670
AC08	Burseraceae	*Protium*	*heptaphyllum* (Aubl.) March.	wd4791
AC05	Burseraceae	*Protium*	sp.	JR P/S
AC09	Burseraceae	*Protium*	sp.	JR P/S
AC11	Burseraceae	*Protium*	sp.	JR P/S
BC15	Burseraceae	*Protium*	sp.	wdp/s
BC15	Burseraceae	*Protium*	sp.	wdp/s
AC07	Burseraceae	*Protium*	sp.	JR P/S
AC07	Burseraceae	*Protium*	sp.	JR P/S
AC04	Burseraceae	*Protium*	*trifoliolatum* Engl.	JR P/S
BC15	Burseraceae	*Protium*	*unifoliolatum* Engl.	wdp/s
AC11	Cactaceae	*Epiphyllum*	*phyllanthus* (L.) Haw.	wd4859
BC15	Caesalpiniaceae	*Bauhinia*	sp.	wdp/s
BC16	Caesalpiniaceae	*Bauhinia*	sp.	wd5037
AC09	Caesalpiniaceae	*Brownea*	*coccinea* Jacq.	JR P/S
BC05	Caesalpiniaceae	*Brownea*	*coccinea* Jacq.	wdp/s
BC01	Caesalpiniaceae	*Brownea*	*coccinea* Jacq.	wd4935
AC12	Caesalpiniaceae	*Brownea*	sp.	wd4894
BC15	Caesalpiniaceae	*Brownea*	sp.	wdp/s
BC13	Caesalpiniaceae	*Campsiandra*	sp.	wd5017
BC15	Caesalpiniaceae	*Copaifera*	*officinalis* L.	wdp/s
BC15	Caesalpiniaceae	*Dialium*	*guianense* (Aubl.) Sandw. Ex. A.C. Sm.	wdp/s
AC11	Caesalpiniaceae	Genus indet.		wd4856
BC13	Caesalpiniaceae	Genus indet.		wd5012
AC05	Caesalpiniaceae	*Macrolobium*	*acaciaefolium* (Benth) Benth	wd4693
AC08	Caesalpiniaceae	*Macrolobium*	*acaciaefolium* (Benth) Benth	JR P/S
AC09	Caesalpiniaceae	*Macrolobium*	*acaciaefolium* (Benth) Benth	JR P/S
AC11	Caesalpiniaceae	*Macrolobium*	*acaciaefolium* (Benth) Benth	JR P/S

continued

Georef	Family	Genus	Species	# Collection
Georef	Familia	Género	Especies	# Colección
AC14	Caesalpiniaceae	*Macrolobium*	*acaciaefolium* (Benth) Benth	JR P/S
BC01	Caesalpiniaceae	*Macrolobium*	*acaciaefolium* (Benth) Benth	wdp/s
BC03	Caesalpiniaceae	*Macrolobium*	*acaciaefolium* (Benth) Benth	wdp/s
AC01	Caesalpiniaceae	*Macrolobium*	*angustifolium* ((Benth) Benth	wd4613
AC03	Caesalpiniaceae	*Macrolobium*	*angustifolium* (Benth.) R. Cowan	JR P/S
AC04	Caesalpiniaceae	*Macrolobium*	*angustifolium* (Benth.) R. Cowan	JR P/S
AC05	Caesalpiniaceae	*Macrolobium*	*angustifolium* (Benth.) R. Cowan	JR P/S
AC06	Caesalpiniaceae	*Macrolobium*	*angustifolium* (Benth.) R. Cowan	JR P/S
AC07	Caesalpiniaceae	*Macrolobium*	*angustifolium* (Benth.) R. Cowan	wd4745
AC08	Caesalpiniaceae	*Macrolobium*	*angustifolium* (Benth.) R. Cowan	JR P/S
AC09	Caesalpiniaceae	*Macrolobium*	*angustifolium* (Benth.) R. Cowan	JR P/S
AC10	Caesalpiniaceae	*Macrolobium*	*angustifolium* (Benth.) R. Cowan	JR P/S
AC10	Caesalpiniaceae	*Macrolobium*	*angustifolium* (Benth.) R. Cowan	JR P/S
AC11	Caesalpiniaceae	*Macrolobium*	*angustifolium* (Benth.) R. Cowan	JR P/S
BC12	Caesalpiniaceae	*Tachigali*	sp.	wd5008
BC02	Cecropiaceae	*Cecropia*	*peltata* L.	wdp/s
BC04	Cecropiaceae	*Cecropia*	*peltata* L.	wdp/s
BC11	Cecropiaceae	*Cecropia*	*peltata* L.	wdp/s
BC15	Cecropiaceae	*Cecropia*	*peltata* L.	wdp/s
BC04	Cecropiaceae	*Cecropia*	*sciadophylla* Mart.	wdp/s
BC09	Cecropiaceae	*Cecropia*	sp.	wd4987
BC07	Cecropiaceae	*Cecropia*	sp.	wdp/s
BC10	Celastraceae	*Goupia*	*glabra* Aubl.	wdp/s
AC07	Celastraceae	*Maytenus*	*guianensis* Klotzsch	wd4743
AC08	Celastraceae	*Maytenus*	*guianensis* Klotzsch	wd4788, 4819
AC09	Celastraceae	*Maytenus*	*guianensis* Klotzsch	JR P/S
AC14	Celastraceae	*Maytenus*	*guianensis* Klotzsch	JR P/S
BC05	Chrysobalanaceae	*Couepia*	*guianensis* Klotzsch	wdp/s
AC14	Chrysobalanaceae	Genus indet.		wd4909
AC04	Chrysobalanaceae	*Hirtella*	*racemosa* Lam.	JR P/S
AC07	Chrysobalanaceae	*Hirtella*	*racemosa* Lam.	wd4726
AC08	Chrysobalanaceae	*Hirtella*	*racemosa* Lam.	wd4806
AC09	Chrysobalanaceae	*Hirtella*	sp.	JR P/S
AC13	Chrysobalanaceae	*Hirtella*	sp.	wd4897
AC04	Chrysobalanaceae	*Parinari*	*excelsa* Sabine	JR P/S
AC12	Chrysobalanaceae	*Parinari*	*excelsa* Sabine	wd4839
AC03	Chrysobalanaceae	*Parinari*	*excelsa* Sabine	JR P/S
AC04	Clusiaceae	*Calophyllum*	*brasiliense* Camb.	JR P/S
AC04	Clusiaceae	*Calophyllum*	*brasiliense* Camb.	wd4656
BC03	Clusiaceae	*Caraipa*	*densifolia* Mart.	wdp/s
AC01	Clusiaceae	*Caraipa*	*densifolia* Mart.	JR P/S
AC05	Clusiaceae	*Caraipa*	*densifolia* Mart.	JR P/S

continued

Georef	Family	Genus	Species	# Collection
Georef	Familia	Género	Especies	# Colección
AC08	Clusiaceae	*Caraipa*	*densifolia* Mart.	wd4820
AC09	Clusiaceae	*Caraipa*	*densifolia* Mart.	wd4831
AC10	Clusiaceae	*Caraipa*	*densifolia* Mart.	JR P/S
AC11	Clusiaceae	*Caraipa*	*densifolia* Mart.	JR P/S
AC14	Clusiaceae	*Caraipa*	*densifolia* Mart.	JR P/S
BC05	Clusiaceae	*Caraipa*	*densifolia* Mart.	wdp/s
AC05	Clusiaceae	*Clusia*	sp.	JR P/S
AC05	Clusiaceae	*Clusia*	sp.	wd4667
AC05	Clusiaceae	*Clusia*	sp.	wd4677
AC07	Clusiaceae	*Clusia*	sp.	wd4725
AC12	Clusiaceae	*Clusia*	sp.	wd4891
BC05	Clusiaceae	*Clusia*	sp.	wd4958
AC06	Clusiaceae	*Garcinia*	*madruno* (Kunth) Hammel	JR P/S
BC13	Clusiaceae	Genus indet.		wd5016
AC08	Clusiaceae	*Vismia*	*cayennensis* Pers.	wd4782
AC12	Clusiaceae	*Vismia*	sp.	wd4875a
BC01	Clusiaceae	*Vismia*	sp.	wdp/s
BC01	Cochlospermaceae	*Cochlospermum*	*orinocense* Steud	wdp/s
AC12	Cochlospermaceae	*Cochlospermum*	*orinocense* Steud	wd4873
BC01	Combretaceae	*Terminalia*	*amazonia* (J.F. Gmell) Exell	wdp/s
AC14	Combretaceae	*Terminalia*	sp.	wd4915
BC16	Commelinaceae	*Commelina*	*rufipes* Seub	wd5041
AC01	Connaraceae	*Connarus*	*lambertii* (DC.) Sagott	wd4606
BC01	Connaraceae	*Rourea*	sp.	wdp/s
AC12	Convolvulaceae	*Aniseia*	*martinicensis* (Jacq.) Choisy	wd4876
AC12	Convolvulaceae	*Ipomoea*	sp.	wd4878a
BC05	Costaceae	*Costus*	*arabicus* L.	wd4954
AC12	Costaceae	*Costus*	*scaber* Ruiz & Pav.	wd4878
BC10	Costaceae	*Costus*	sp.	wdp/s
BC15	Costaceae	*Costus*	sp.	wdp/s
BC03	Costaceae	*Costus*	sp.	wdp/s
BC04	Costaceae	*Costus*	sp.	wdp/s
AC08	Cyclanthaceae	*Ludovia*	*lancifolia* Brongn.	wd4787
BC06	Cyperaceae	*Abilgaardia*	*ovata* (Burn. F.) Kral.	wd4966
AC03	Cyperaceae	*Cyperus*	*laxus* Lam.	wd4639
AC08	Cyperaceae	*Cyperus*	*laxus* Lam.	wd4804a
BC08	Cyperaceae	*Cyperus*	*laxus* Lam.	wd4982
AC11	Cyperaceae	*Cyperus*	sp.	wd4845
AC14	Cyperaceae	*Cyperus*	sp.	wd4900
BC10	Cyperaceae	*Cyperus*	sp.	wd4989
BC11	Cyperaceae	*Cyperus*	sp.	wdp/s
BC15	Cyperaceae	*Cyperus*	sp.	wd5024

continued

Georef	Family	Genus	Species	# Collection
Georef	Familia	Género	Especies	# Colección
AC11	Cyperaceae	*Cyperus*	*surinamensis* Rottb.	wd4847
AC03	Cyperaceae	*Diplazia*	*karatifolia* Rich.	wd4647
BC06	Cyperaceae	*Eleocharis*	sp.	wd4976
AC01	Cyperaceae	Genus indet.		wd4614
AC02	Cyperaceae	Genus indet.		wd4836
AC05	Cyperaceae	Genus indet.		wd4672
AC08	Cyperaceae	Genus indet.		wd4750
AC08	Cyperaceae	Genus indet.		wd4757
AC08	Cyperaceae	Genus indet.		wd4790
AC14	Cyperaceae	Genus indet.		wd4901
BC05	Cyperaceae	Genus indet.		wdp/s
AC06	Cyperaceae	*Hypolytrum*	*longifolium* (Rich.) Nees	wd4705a
AC07	Cyperaceae	*Hypolytrum*	*longifolium* (Rich.) Nees	wd4721
AC08	Cyperaceae	*Hypolytrum*	*longifolium* (Rich.) Nees	wd4805
AC08	Cyperaceae	*Hypolytrum*	*longifolium* (Rich.) Nees	wd4758
BC10	Cyperaceae	*Hypolytrum*	*pulchrum* H. Pfeifer	wd4992
AC03	Cyperaceae	*Hypolytrum*	sp.	wd4636
AC11	Cyperaceae	*Hypolytrum*	sp.	wd4855
AC01	Cyperaceae	*Hypolytrum*	sp.	wd4612
AC11	Cyperaceae	*Kyllinga*	sp.	wd4844
AC08	Cyperaceae	*Rhynchospora*	*cephalotes* (L.) Vahl	wd4775
AC03	Cyperaceae	*Rhynchospora*	*cephalotes* (L.) Vahl	wd4640
AC05	Cyperaceae	*Rhynchospora*	*pubera* (Vahl) Boeck subsp. *pubera*	wd4683
AC05	Cyperaceae	*Rhynchospora*	*pubera* (Vahl) Boeck subsp. *pubera*	wd4695
AC08	Cyperaceae	*Rhynchospora*	*pubera* (Vahl) Boeck subsp. *pubera*	wd4813
AC05	Cyperaceae	*Scleria*	sp.	wd4673
AC11	Cyperaceae	*Scleria*	sp.	wd4853
BC11	Cyperaceae	*Scleria*	sp.	wdp/s
AC08	Dilleniaceae	*Davilla*	sp.	JR P/S
AC09	Dilleniaceae	*Davilla*	sp.	JR P/S
AC01	Dilleniaceae	*Doliocarpus*	sp.	wd4607
AC07	Dryopteridaceae	*Elaphoglossum*	*styriacum* Mickel	wd4741
AC04	Dryopteridaceae	*Lomagramma*	*guianensis* Klotzsch	wd4652
BC01	Dryopteridaceae	*Tectaria*	cf. *insisa* Cav.	wdp/s
AC01	Dryopteridaceae	*Tectaria*	sp.	wd4604
AC11	Ebenaceae	*Diospyros*	*lissocarpioides* Sandwith	wd4860
AC03	Elaeocarpaceae	*Sloanea*	*guianensis* Klotzsch	JR P/S
AC14	Dryopteridaceae	*Elaphoglosum*	sp.	wd4917
AC07	Eriocaulaceae	*Paepalantus*	*lamarkii* Kunth	wd4738
AC01	Erythroxylaceae	*Erythroxylum*	*kapplerianum* Peyr.	wd 4605
AC14	Erythroxylaceae	*Erythroxylum*	*kapplerianum* Peyr.	wd4913
BC05	Erythroxylaceae	*Erythroxylum*	*kapplerianum* Peyr.	wd4955

continued

Georef	Family	Genus	Species	# Collection
Georef	Familia	Género	Especies	# Colección
AC12	Euphorbiaceae	*Amanoa*	*guianensis* Aubl	wd4838
AC04	Euphorbiaceae	*Croton*	cf. *cuneatus* Klotzsch	JR P/S
BC04	Euphorbiaceae	*Croton*	*cuneatus* Klotzsch	wdp/s
BC07	Euphorbiaceae	*Croton*	*cuneatus* Klotzsch	wdp/s
BC09	Euphorbiaceae	*Croton*	*cuneatus* Klotzsch	wdp/s
BC11	Euphorbiaceae	*Croton*	*cuneatus* Klotzsch	wdp/s
BC06	Euphorbiaceae	*Croton*	*essequiboensis* Klotzsch	wd4970
AC05	Euphorbiaceae	*Croton*	*potaroensis* Lanj.	wd4689
AC11	Euphorbiaceae	*Croton*	sp.	wd4852
AC14	Euphorbiaceae	*Croton*	sp.	JR P/S
BC10	Euphorbiaceae	Genus indet.		wd4998
AC05	Euphorbiaceae	*Mabea*	*montana* Muell.-Arg.	wd4681
AC08	Euphorbiaceae	*Mabea*	*piriri* Aubl.	wd4817
AC11	Euphorbiaceae	*Mabea*	*piriri* Aubl.	wd4850
AC14	Euphorbiaceae	*Mabea*	*tacquari* Aubl.	JR P/S
AC03	Euphorbiaceae	*Micrandra*	*minor* Benth.	JR P/S
AC04	Euphorbiaceae	*Micrandra*	*minor* Benth.	JR P/S
AC10	Euphorbiaceae	*Micrandra*	*minor* Benth.	JR P/S
AC11	Euphorbiaceae	*Micrandra*	*minor* Benth.	JR P/S
AC14	Euphorbiaceae	*Micrandra*	*minor* Benth.	JR P/S
BC01	Euphorbiaceae	*Micrandra*	*minor* Benth.	wdp/s
AC14	Euphorbiaceae	*Micrandra*	*siphonioides* Benth.	wd4907
AC05	Euphorbiaceae	*Phyllanthus*	sp.	wd4698
AC08	Euphorbiaceae	*Phyllanthus*	sp.	wd4815
AC14	Euphorbiaceae	*Phyllanthus*	sp.	wd4898
BC12	Euphorbiaceae	*Phyllanthus*	sp.	wdp/s
BC13	Euphorbiaceae	*Phyllanthus*	sp.	wd5014
AC10	Fabaceae	*Alexa*	*confusa* Pittier	JR P/S
BC01	Fabaceae	*Alexa*	*confusa* Pittier	wdp/s
AC01	Fabaceae	*Alexa*	*confusa* Pittier	JR P/S
AC03	Fabaceae	*Alexa*	*confusa* Pittier	JR P/S
AC04	Fabaceae	*Alexa*	*confusa* Pittier	JR P/S
AC13	Fabaceae	*Alexa*	*confusa* Pittier	JR P/S
AC11	Fabaceae	*Andira*	*surinamensis* (Bondt.) Splitg. Ex Pulle	JR P/S
AC14	Fabaceae	*Andira*	*surinamensis* (Bondt.) Splitg. Ex Pulle	JR P/S
AC07	Fabaceae	*Dalbergia*	*glauca* (Desv.) Amshoff	wd4744
AC05	Fabaceae	*Dalbergia*	sp.	wd4692
AC08	Fabaceae	*Dalbergia*	sp.	JR P/S
AC11	Fabaceae	*Dalbergia*	sp.	JR P/S
AC11	Fabaceae	*Dialium*	*guianensis* Aubl. Steud	JR P/S
BC04	Fabaceae	*Dialium*	*guianensis* Aubl. Steud	wdp/s
BC07	Fabaceae	*Dialium*	*guianensis* Aubl. Steud	wdp/s

continued

Georef	Family	Genus	Species	# Collection
Georef	Familia	Género	Especies	# Colección
BC03	Fabaceae	*Dialium*	*guianensis* Aubl. Steud	wdp/s
AC08	Fabaceae	*Dioclea*	*guianensis* Aubl. Steud	wd4804
AC07	Fabaceae	*Diplotropis*	sp.	JR P/S
AC11	Fabaceae	*Diplotropis*	sp.	JR P/S
AC14	Fabaceae	*Diplotropis*	sp.	JR P/S
BC05	Fabaceae	*Dipteryx*	*odorata* (Aubl.) Willd.	wdp/s
AC01	Fabaceae	*Eperua*	*jehnmanii* Oliv. ssp sandwitii Cowan	JR P/S
AC03	Fabaceae	*Eperua*	*jehnmanii* Oliv. ssp sandwitii Cowan	JR P/S
AC08	Fabaceae	*Eperua*	*jehnmanii* Oliv. ssp sandwitii Cowan	JR P/S
AC09	Fabaceae	*Eperua*	*jehnmanii* Oliv. ssp sandwitii Cowan	wd4829
AC14	Fabaceae	*Eperua*	*jehnmanii* Oliv. ssp sandwitii Cowan	JR P/S
AC05	Fabaceae	Genus indet.		wd4671
AC08	Fabaceae	Genus indet.		wd4778
AC12	Fabaceae	Genus indet.		wd4893
BC05	Fabaceae	Genus indet.		wd4962
AC14	Fabaceae	*Macrolobium*	sp.	wd4905
AC14	Fabaceae	*Mucuna*	sp.	wd4916
AC07	Fabaceae	*Sclerolobium*	*guianense* Benth.	JR P/S
AC09	Fabaceae	*Sclerolobium*	*guianense* Benth.	JR P/S
AC11	Fabaceae	*Sclerolobium*	*guianense* Benth.	JR P/S
AC14	Fabaceae	*Sclerolobium*	*guianense* Benth.	JR P/S
BC01	Fabaceae	*Swartzia*	*leptopetala* Benth.	wdp/s
AC03	Fabaceae	*Swartzia*	sp.	wd4627
AC05	Fabaceae	*Swartzia*	sp.	wd4691
AC14	Fabaceae	*Swartzia*	sp.	wd4904
BC09	Fabaceae	*Swartzia*	sp.	wd4986
BC10	Fabaceae	*Swartzia*	sp.	wdp/s
BC13	Fabaceae	*Swartzia*	sp.	wd5011
BC15	Fabaceae	*Swartzia*	sp.	wdp/s
BC07	Fabaceae	*Swartzia*	sp.	wdp/s
BC10	Flacourtiaceae	*Homalium*	*guianensis* (Aubl.) Oken	wdp/s
BC03	Flacourtiaceae	*Homalium*	*guianensis* (Aubl.) Oken	wdp/s
AC03	Flacourtiaceae	*Homalium*	*guianensis* (Aubl.) Oken	JR P/S
AC04	Flacourtiaceae	*Homalium*	*guianensis* (Aubl.) Oken	JR P/S
AC06	Flacourtiaceae	*Homalium*	*guianensis* (Aubl.) Oken	JR P/S
AC08	Flacourtiaceae	*Homalium*	*guianensis* (Aubl.) Oken	JR P/S
BC11	Flacourtiaceae	*Homalium*	*guianensis* (Aubl.) Oken	wdp/s
BC07	Flacourtiaceae	*Homalium*	*guianensis* (Aubl.) Oken	wdp/s
BC09	Flacourtiaceae	*Homalium*	*guianensis* (Aubl.) Oken	wdp/s
AC12	Gentianaceae	*Coutoubea*	sp.	wd4871
BC10	Gentianaceae	Genus indet.		wd5002
BC17	Gentianaceae	Genus indet.		wd5030

continued

Georef	Family	Genus	Species	# Collection
Georef	Familia	Género	Especies	# Colección
AC03	Gentianaceae	*Irlbachia*	sp.	wd4650
AC12	Gesneriaceae	*Chrysothemis*	*pulchella* (Donn. Ex Sims.) Docne	wd4892a
BC01	Gesneriaceae	*Chrysothemis*	*pulchella* (Donn. Ex Sims.) Docne	wdp/s
AC12	Gesneriaceae	*Codonanthe*	*crassifolia* (H. Focke) C.V. Morton	wd4892
AC08	Gesneriaceae	*Tylopsacas*	*cuneatum* (Gleason) Leeuwemb	wd4765
AC07	Grammitidaceae	*Cochlidium*	sp.	wd4747
BC05	Haemodoraceae	*Xiphidium*	sp.	wd4960
BC10	Heliconiaceae	*Heliconia*	*acuminata* L.C. Rich	wdp/s
AC12	Heliconiaceae	*Heliconia*	*hirsuta* L.f.	wd4874
AC12	Heliconiaceae	*Heliconia*	sp.	wd4895
BC05	Heliconiaceae	*Heliconia*	sp.	wdp/s
AC01	Hippocrateaceae	Genus indet.		wd4618
AC03	Hippocrateaceae	*Tontelea*	*attenuata* Miers	wd4630
AC08	Hymenophyllaceae	*Hymenophyllum*	*hirsutum* (L.) Sw.	wd4816
AC08	Hymenophyllaceae	*Hymenophyllum*	*lehmannii* Hieron	wd4764
AC04	Hymenophyllaceae	*Hymenophyllum*	sp.	wd4659
AC07	Hymenophyllaceae	*Trichomanes*	*accedens* C. Presl	wd4729
AC03	Hymenophyllaceae	*Trichomanes*	*hostmannianum*	wd4629
AC07	Hymenophyllaceae	*Trichomanes*	*pinnatum Hedw*	wd4728
AC08	Hymenophyllaceae	*Trichomanes*	*pinnatum* Hedw	wd4768
AC14	Lamiaceae	Genus indet.		wd4902
AC08	Lamiaceae	*Hyptis*	*capitata* Jacq.	wd4781
BC15	Lamiaceae	*Hyptis*	sp.	wdp/s
AC11	Lauraceae	*Endlicheria*	*dictifarinosa* C.K. Allen	JR P/S
AC07	Lauraceae	Genus indet.		JR P/S
AC07	Lauraceae	Genus indet.		JR P/S
BC09	Lauraceae	*Licaria*	sp.	wd4985
BC03	Lauraceae	*Ocotea*	*cymbarum* Kunth	wdp/s
AC14	Lauraceae	*Rhodostemonaphne*	*grandis* (Mez) Rohwer	JR P/S
AC01	Lecythidaceae	*Eschweilera*	*decolorans* Sandw.	JR P/S
AC03	Lecythidaceae	*Eschweilera*	*decolorans* Sandw.	JR P/S
AC14	Lecythidaceae	*Eschweilera*	*decolorans* Sandw.	JR P/S
AC01	Lecythidaceae	*Eschweilera*	*subglandulosa* (Steud. Ex Berg.) Miers	JR P/S
AC03	Lecythidaceae	*Eschweilera*	*subglandulosa* (Steud. Ex Berg.) Miers	JR P/S
AC04	Lecythidaceae	*Eschweilera*	*subglandulosa* (Steud. Ex Berg.) Miers	JR P/S
AC05	Lecythidaceae	*Eschweilera*	*subglandulosa* (Steud. Ex Berg.) Miers	JR P/S
AC07	Lecythidaceae	*Eschweilera*	*subglandulosa* (Steud. Ex Berg.) Miers	JR P/S
AC08	Lecythidaceae	*Eschweilera*	*subglandulosa* (Steud. Ex Berg.) Miers	JR P/S
AC09	Lecythidaceae	*Eschweilera*	*subglandulosa* (Steud. Ex Berg.) Miers	wd4834
AC10	Lecythidaceae	*Eschweilera*	*subglandulosa* (Steud. Ex Berg.) Miers	JR P/S
AC11	Lecythidaceae	*Eschweilera*	*subglandulosa* (Steud. Ex Berg.) Miers	JR P/S
AC13	Lecythidaceae	*Eschweilera*	*subglandulosa* (Steud. Ex Berg.) Miers	JR P/S

continued

Georef	Family	Genus	Species	# Collection
Georef	Familia	Género	Especies	# Colección
AC14	Lecythidaceae	*Eschweilera*	*subglandulosa* (Steud. Ex Berg.) Miers	wd4910
BC08	Lecythidaceae	*Eschweilera*	*subglandulosa* (Steud. Ex Berg.) Miers	wdp/s
BC09	Lecythidaceae	*Eschweilera*	*subglandulosa* (Steud. Ex Berg.) Miers	wdp/s
BC15	Lecythidaceae	*Eschweilera*	*subglandulosa* (Steud. Ex Berg.) Miers	wdp/s
BC08	Lecythidaceae	*Eschweilera*	*subglandulosa* (Steud. Ex Berg.) Miers	wdp/s
AC04	Lecythidaceae	*Eschweilera*	*subglandulosa* (Steud. Ex Berg.) Miers	JR P/S
BC01	Lecythidaceae	*Eschweilera*	*subglandulosa* (Steud. Ex Berg.) Miers	wdp/s
BC03	Lecythidaceae	*Eschweilera*	*subglandulosa* (Steud. Ex Berg.) Miers	wdp/s
BC04	Lecythidaceae	*Eschweilera*	*subglandulosa* (Steud. Ex Berg.) Miers	wdp/s
AC04	Lecythidaceae	*Gustavia*	*poeppigiana* Berg.	JR P/S
AC10	Lecythidaceae	*Gustavia*	*poeppigiana* Berg.	JR P/S
BC08	Lecythidaceae	*Gustavia*	*poeppigiana* Berg.	wdp/s
BC04	Lecythidaceae	*Gustavia*	*poeppigiana* Berg.	wdp/s
BC15	Lecythidaceae	*Gustavia*	sp.	wdp/s
AC01	Loganiaceae	*Strychnos*	*panurensis* Sprag. & Sandw.	wd4609
BC10	Loranthaceae	Genus indet.		wd4993
BC17	Loranthaceae	Genus indet.		wd5032
BC01	Malpighiaceae	*Bunchosia*	*mollis* Benth.	wd4933
AC07	Malpighiaceae	*Pterandra*	*sericea* W.R. And.	wd4742
AC07	Malpighiaceae	*Tetrapteris*	*mucronata* Cav.	wd4740
AC11	Malvaceae	Genus indet.		wd4865
AC04	Marantaceae	*Calathea*	sp.	wd4653
BC05	Marantaceae	*Monotagma*	sp.	wdp/s
BC05	Marcgraviaceae	*Norantea*	sp.	wd4959
BC08	Melastomataceae	*Bellucia*	sp.	wdp/s
AC08	Melastomataceae	*Clidemia*	sp.	wd4779
BC04	Melastomataceae	Genus indet.		wd4978
BC16	Melastomataceae	Genus indet.		wd5039
AC07	Melastomataceae	Genus indet.		wd4735
BC12	Melastomataceae	*Miconia*	*bubalina* (D.Don) Naudin	wd5010
AC07	Melastomataceae	*Miconia*	*myriantha* Benth.	JR P/S
AC01	Melastomataceae	*Miconia*	sp.	wd4608
AC05	Melastomataceae	*Miconia*	sp.	wd4663
AC08	Melastomataceae	*Miconia*	sp.	wd4762
AC11	Melastomataceae	*Miconia*	sp.	JR P/S
AC14	Melastomataceae	*Miconia*	sp.	wd4912
BC10	Melastomataceae	*Miconia*	sp.	wdp/s
AC08	Melastomataceae	*Miconia*	*tomentosa* Naud.	wd4783
AC07	Melastomataceae	*Mouriri*	sp.	JR P/S
BC09	Melastomataceae	*Mouriri*	sp.	wdp/s
AC03	Melastomataceae	*Tococa*	sp.	wd4649
AC06	Meliaceae	*Guarea*	*guidonia* (L.) Sleumer	wd4717

continued

Georef	Family	Genus	Species	# Collection
Georef	Familia	Género	Especies	# Colección
BC09	Meliaceae	*Trichilia*	*pleeana* (A. Juss.) C. De Candole	wdp/s
BC15	Meliaceae	*Trichilia*	*pleeana* (A. Juss.) C. De Candole	wdp/s
BC08	Meliaceae	*Trichilia*	*pleeana* (A. Juss.) C. De Candole	wdp/s
BC08	Meliaceae	*Trichilia*	*quadrijuga* Kunth	wdp/s
BC15	Menispermaceae	Genus indet.		wd5021
AC08	Metaxiaceae	*Metaxia*	*rostrata* (Kunth.) C. Presl.	wd4767
AC04	Mimosaceae	*Abarema*	*jupunba* (Willd.) Britton & Killip	JR P/S
AC08	Mimosaceae	*Calliandra*	*laxa* (Willd.) Benth.	JR P/S
BC02	Mimosaceae	*Enterolobium*	*schomburgkii* (Benth.) Benth.	wdp/s
AC01	Mimosaceae	*Inga*	*bourgonii* (Aubl.) DC.	jR P/S
AC08	Mimosaceae	*Inga*	*bourgonii* (Aubl.) DC.	wd4771
AC01	Mimosaceae	*Inga*	*coruscans* H. & B.	jR P/S
BC05	Mimosaceae	*Inga*	*laurina* (Sw.) Willd.	wd4950
AC01	Mimosaceae	*Inga*	*leiocalycina* Benth.	jR P/S
BC08	Mimosaceae	*Inga*	*oerstediana* Benth.	wdp/s
BC09	Mimosaceae	*Inga*	*oerstediana* Benth.	wdp/s
AC08	Mimosaceae	*Inga*	*pilosula* (Rich.) J.F. Macbr.	wd4770, 4802
AC04	Mimosaceae	*Inga*	sp.	JR P/S
AC05	Mimosaceae	*Inga*	sp.	JR P/S
AC14	Mimosaceae	*Inga*	sp.	wd4926
BC06	Mimosaceae	*Inga*	sp.	wd4968
BC08	Mimosaceae	*Inga*	sp.	wdp/s
BC11	Mimosaceae	*Inga*	sp.	wdp/s
BC13	Mimosaceae	*Inga*	sp.	wd5018
AC04	Mimosaceae	*Inga*	*splendens* Willd.	JR P/S
AC09	Mimosaceae	*Inga*	*vera* Willd.	JR P/S
BC02	Mimosaceae	*Inga*	*vera* Willd.	wdp/s
BC07	Mimosaceae	*Inga*	*vera* Willd.	wdp/s
BC10	Mimosaceae	*Inga*	*vera* Willd.	wdp/s
AC13	Mimosaceae	*Pentaclethra*	sp.	wd4897a
BC15	Mimosaceae	*Zygia*	*latifolia* (L.) fawc. & Rend.	wd5027
AC09	Mimosaceae	*Zygia*	*latifolia* (L.) Fawc. & Rend. Var. *Comunis* (Bar.) J.W. Grimes	JR P/S
AC14	Mimosaceae	*Zygia*	*latifolia* (L.) Fawc. & Rend. Var. *Comunis* (Bar.) J.W. Grimes	JR P/S
AC03	Mimosaceae	*Zygia*	*latifolia* (L.) Fawc. & Rend. Var. *Comunis* (Bar.) J.W. Grimes	JR P/S
BC01	Mimosaceae	*Zygia*	*latifolia* (L.) Fawc. & Rend. Var. *Comunis* (Bar.) J.W. Grimes	wdp/s
AC04	Mimosaceae	*Zygia*	*latifolia* (L.) Fawc. & Rend. Var. *Comunis* (Bar.) J.W. Grimes	JR P/S
AC12	Mimosaceae	*Zygia*	*latifolia* (L.) Fawc. & Rend. Var. *Comunis* (Bar.) J.W. Grimes	wd4885

continued

Georef	Family	Genus	Species	# Collection
Georef	Familia	Género	Especies	# Colección
AC07	Mimosaceae	*Zygia*	*latifolia* (L.) Fawc. & Rend. Var. *Comunis* (Bar.) J.W. Grimes	wd4731
AC06	Mimosaceae	*Zygia*	*latifolia* (L.) Fawc. & Rend. Var. *Comunis* (Bar.) J.W. Grimes.	JR P/S
AC09	Mimosaceae	*Zygia*	sp.	wd4833
BC01	Mimosaceae	*Zygia*	sp.	wd4930
BC08	Mimosaceae	*Zygia*	sp.	wdp/s
BC09	Mimosaceae	*Zygia*	sp.	wdp/s
BC05	Mimosaceae	*Zygia*	sp.	wdp/s
AC08	Mimosaceae	*Zygia*	sp.	JR P/S
AC05	Mimosaceae	*Zygia*	*unifoliolata* (Benth.) Pittier	wd4664
AC14	Moraceae	*Brosimum*	*rubescens* Taub.	JR P/S
AC08	Moraceae	*Ficus*	sp.	wd4785
BC13	Moraceae	*Ficus*	sp.	wd5013
BC01	Moraceae	*Ficus*	sp.	wdp/s
BC01	Moraceae	*Ficus*	sp.	wd4936
BC08	Moraceae	*Sorocea*	*muriculata Miq.*	wdp/s
BC05	Myristicaceae	*Iyianthera*	*hostmanii* (Benth.) Warb.	Wdp/s
AC14	Myristicaceae	*Iryanthera*	*hostmanii* (Benth.) Warb.	JR P/S
BC05	Myristicaceae	*Iryanthera*	*hostmanii* (Benth.) Warb.	wd4946
AC03	Myristicaceae	*Iryanthera*	*laevis* Markgr.	JR P/S
AC01	Myristicaceae	*Virola*	sp.	JR P/S
BC08	Myristicaceae	*Virola*	sp.	wdp/s
BC03	Myristicaceae	*Virola*	sp.	wdp/s
AC03	Myristicaceae	*Virola*	*surinamensis* (Rohl.) Warb.	JR P/S
BC08	Myristicaceae	*Virola*	*surinamensis* (Rohl.) Warb.	wdp/s
AC04	Myristicaceae	*Virola*	*surinamensis* (Rohl.) Warb.	JR P/S
AC10	Myristicaceae	*Virola*	*surinamensis* (Rohl.) Warb.	JR P/S
BC05	Myristicaceae	*Virola*	*surinamensis* (Rohl.) Warb.	wdp/s
AC08	Myrsinaceae	*Cybianthus*	*spicatus* (Kunth.) G. Agostini	wd4753
AC08	Myrsinaceae	*Stylogine*	*longifolia* (Mart. Ex Miq.) Mez.	wd4752
AC14	Myrsinaceae	*Stylogine*	sp.	wd4906
BC03	Myrsinaceae	*Stylogine*	sp.	wd4944
AC08	Myrtaceae	*Calycolpus*	*goetheanus* (DC.) O. Berg.	JR P/S
AC08	Myrtaceae	*Calyptranthes*	*fasciculata* O. Berg.	wd4789
AC08	Myrtaceae	*Calyptranthes*	*multiflora* O. Berg.	wd4755
BC10	Myrtaceae	*Calyptranthes*	*multiflora* O. Berg.	wd4997
BC10	Myrtaceae	*Calyptranthes*	*multiflora* O. Berg.	wd5001
BC17	Myrtaceae	*Calyptranthes*	*multiflora* O. Berg.	wd5031
AC05	Myrtaceae	*Calyptranthes*	sp.	wd4688
AC07	Myrtaceae	*Eugenia*	*biflora* (L.) DC.	wd4733
AC08	Myrtaceae	*Eugenia*	*biflora* (L.) DC.	wd4791

continued

Georef	Family	Genus	Species	# Collection
Georef	Familia	Género	Especies	# Colección
BC06	Myrtaceae	*Eugenia*	*biflora* (L.) DC.	wd4972
AC03	Myrtaceae	*Eugenia*	*ferreiraeana* O. Berg.	wd4621
AC08	Myrtaceae	*Eugenia*	*ferreiraeana* O. Berg.	wd4818
BC05	Myrtaceae	*Eugenia*	*flavescens* DC.	wd4957
AC07	Myrtaceae	*Eugenia*	*florida* DC.	JR P/S
AC06	Myrtaceae	*Eugenia*	*lambertiana* DC.	wd4703
AC03	Myrtaceae	*Eugenia*	*pseudopsidium* Jacq.	wd4632
BC01	Myrtaceae	*Eugenia*	sp.	wd4934
BC15	Myrtaceae	*Eugenia*	sp.	wd5019
BC01	Myrtaceae	*Eugenia*	*florida* DC.	wdp/s
AC04	Myrtaceae	Genus indet.		wd4655
BC05	Myrtaceae	Genus indet.		wd4953
BC06	Myrtaceae	Genus indet.		wd4971
BC12	Myrtaceae	Genus indet.		wd5006
BC12	Myrtaceae	Genus indet.		wd5007
AC05	Myrtaceae	*Myrcia*	*floribunda* (West ex Willd.) O. Berg	wd4690
BC05	Myrtaceae	*Myrcia*	*guianensis* (Aubl.) Dc	wd4964
AC03	Myrtaceae	*Myrcia*	*inaequiloba* (DC.) Legrand	wd4643
AC05	Myrtaceae	*Myrcia*	*inaequiloba* (DC.) Legrand	wd4688
AC07	Myrtaceae	*Myrcia*	*inaequiloba* (DC.) Legrand	wd4732
AC04	Myrtaceae	*Myrcia*	sp.	JR P/S
AC05	Myrtaceae	*Myrcia*	sp.	JR P/S
AC09	Myrtaceae	*Myrcia*	sp.	JR P/S
AC10	Myrtaceae	*Myrcia*	sp.	JR P/S
AC09	Myrtaceae	*Myrcia*	*splendens* Willd.	JR P/S
AC11	Myrtaceae	*Myrcia*	*splendens* Willd.	JR P/S
AC14	Myrtaceae	*Myrcia*	*splendens* Willd.	JR P/S
AC08	Myrtaceae	*Myrciaria*	*floribunda* (West ex Willd.) O. Berg	wd4801
AC08	Myrtaceae	*Myrciaria*	*floribunda* (West ex Willd.) O. Berg	wd4780
AC14	Myrtaceae	*Myrciaria*	*floribunda* (West ex Willd.) O. Berg	wd4924
AC05	Myrtaceae	*Psidium*	*personii* Mc Vaugh	wd4679
AC14	Myrtaceae	*Psidium*	*personii* Mc Vaugh	wd4903
AC02	Myrtaceaea	*Psidium*	*personii* Mc Vaugh	wd4835
AC06	Nymphaceae	Genus indet.		wd4710
AC08	Ochnaceae	*Ouratea*	*guianensis* (Aubl.) Dc	wd4754
BC15	Ochnaceae	*Ouratea*	sp.	wdp/s
AC07	Ochnaceae	*Sauvagesia*	sp.	wd4746
AC11	Ochnaceae	*Sauvagesia*	sp.	wd4850a
BC17	Ochnaceae	*Sauvagesia*	sp.	wd5034
BC06	Ochnaceae	*Sauvagesia*	sp.	wdp/s
BC10	Olacaceae	*Cathedra*	*acuminata* (Benth.) Miers.	wd4999
AC12	Onagraceae	*Ludwigia*	*affinis* (DC.) Hara	wd4875

continued

Georef	Family	Genus	Species	# Collection
Georef	Familia	Género	Especies	# Colección
AC11	Onagraceae	*Ludwigia*	*hyssopifolia* (G.Don.) Exell	wd4841
BC09	Onagraceae	*Ludwigia*	sp.	wdp/s
BC11	Onagraceae	*Ludwigia*	sp.	wdp/s
BC16	Onagraceae	*Ludwigia*	sp.	wd5038
BC17	Onagraceae	*Ludwigia*	sp.	wd5034a
AC07	Orchidaceae	*Dichaea*	*brachypoda Rchb.f.*	wd4719
AC14	Orchidaceae	*Dichaea*	sp.	wd4925
AC05	Orchidaceae	*Epidendrum*	sp.	wd4675
AC05	Orchidaceae	*Epidendrum*	sp.	wd4701
AC03	Orchidaceae	Genus indet.		wd4644
AC07	Orchidaceae	Genus indet.		wd4723
AC11	Orchidaceae	Genus indet.		wd4858
AC14	Orchidaceae	Genus indet.		wd4920
AC14	Orchidaceae	Genus indet.		wd4918
AC03	Orchidaceae	*Habenaria*	*floribunda* (West ex Willd.) O. Berg	wd4634
AC12	Orchidaceae	*Lockartia*	sp.	wd4886
AC05	Orchidaceae	*Lycaste*	sp.	wd4676
AC05	Orchidaceae	*Oncidium*	sp.	wd4685
AC11	Orchidaceae	*Oncidium*	sp.	wd4867
AC12	Orchidaceae	*Oncidium*	sp.	wd4887
AC03	Orchidaceae	*Scaphyglottis*	*cuneata* Schltr.	wd4648
AC08	Orchidaceae	*Spiranthes*	sp.	wd4809
AC09	Orchidaceae	*Spiranthes*	sp.	wd4826
AC08	Orchidaceae	*Trigonidium*	*acuminatum* Battem ex Lind.	wd4793
AC14	Orchidaceae	*Vanilla*	*odorata* C. Presl	wd4919
AC07	Oxalydaceae	*Biophytum*	*calophyllum* (Prog.) Guill.	wd4718
AC01	Piperaceae	*Piper*	sp.	wd4611
AC11	Poaceae	*Axonopus*	*longispicus* (Doell) Kuhlm.	wd4849
AC05	Poaceae	Genus indet.		wd4682
AC11	Poaceae	*Ichnanthus*	*pallens* (Sw.) Benth.	wd4857
AC09	Poaceae	*Ichnanthus*	*panicoides* Beauv.	wd4828
AC08	Poaceae	*Imperata*	*brasiliensis*	wd4774
AC01	Poaceae	*Lasiacis*	*anomala* Hitch.	wd4615
AC08	Poaceae	*Lasiacis*	sp.	wd4760
AC01	Poaceae	*Olyra*	*latifolia* L.	wd4610
AC09	Poaceae	*Olyra*	*latifolia* L.	wd4827
AC03	Poaceae	*Olyra*	*longifolia* Kunth.	wd4638
BC03	Poaceae	*Olyra*	sp.	wd4940
BC10	Poaceae	*Olyra*	sp.	wdp/s
BC15	Poaceae	*Olyra*	sp.	wdp/s
BC04	Poaceae	*Olyra*	sp.	wdp/s
AC08	Poaceae	*Panicum*	*hirtum* Lam.	wd4799

continued

Georef	Family	Genus	Species	# Collection
Georef	Familia	Género	Especies	# Colección
AC01	Poaceae	*Panicum*	*laxum* Sw.	wd46
AC11	Poaceae	*Panicum*	*laxum Sw.*	wd4843
AC03	Poaceae	*Panicum*	sp.	wd4637
BC11	Poaceae	*Panicum*	sp.	wd5003a
BC09	Poaceae	*Panicum*	sp.	wd4984
AC11	Poaceae	*Paspalum*	*conjugatum* Berg.	wd4846
AC12	Poaceae	*Paspalum*	*millegrana* Schrad.	wd4840
BC06	Poaceae	*Paspalum*	*plicatulum* Michx.	wd4973
BC06	Poaceae	*Paspalum*	sp.	wd4977a
AC08	Poaceae	*Paspalum*	sp.	wd4749, 4808
BC01	Poaceae	*Paspalum*	sp.	wd4977
BC10	Poaceae	*Paspalum*	sp.	wd4990
AC09	Podostemonaceae	*Apinagia*	*multibranchiata* (Math.) P. Royen	wd4823
AC14	Podostemonaceae	*Apinagia*	*longifolia* (Tul.)	wd4927
AC11	Podostemonaceae	*Apinagia*	*ruppioides* (Kunth.) Tulasne	wd4869
AC08	Podostemonaceae	*Apinagia*	sp.	wd4803a
AC12	Podostemonaceae	*Apinagia*	sp.	wd4881
BC06	Podostemonaceae	*Apinagia*	sp.	wd4974
BC06	Podostemonaceae	*Apinagia*	sp.	wd4975
BC06	Podostemonaceae	*Apinagia*	sp.	wd4967
BC17	Podostemonaceae	*Apinagia*	sp.	wd5029
AC08	Podostemonaceae	*Apinagia*	*staheliana* (Wenth.) P. Royen	wd4803
AC05	Podostemonaceae	Genus indet.		JR P/S
AC12	Podostemonaceae	*Mourera*	*fluviatilis* Aubl.	wd4884
AC11	Podostemonaceae	*Rhyncholacis*	sp.	wd4851
AC05	Podostemonaceae	*Rhyncholacis*	sp.	wd4669
BC12	Polygonaceae	*Coccoloba*	sp.	wd5005
BC09	Polygonaceae	*Cocoloba*	sp.	wdp/s
AC03	Polygonaceae	Genus indet.		wd4623
BC13	Polygonaceae	*Ruprechtia*	sp.	wdp/s
AC14	Polypodiaceae	*Microgramma*	*baldwinii* Brade	wd4921
AC08	Polypodiaceae	*Microgramma*	*percussa* (Cav.) de la Sota	wd4795
AC07	Polypodiaceae	*Microgramma*	*persicariifolia* (Scharad.) Presl.	wd4720
BC17	Polypodiaceae	*Microgramma*	*persicariifolia* (Scharad.) Presl.	wd5033
AC08	Polypodiaceae	*Microgramma*	sp.	wd4772
AC04	Polypodiaceae	*Niphidium*	sp.	wd4660
AC07	Polypodiaceae	*Polypodium*	*triseriale* Sw.	wd4739
AC02	Pontederiaceae	*Eichornia*	*diversifolia* (Vahl.) Urban	wd4837
AC12	Pontederiaceae	*Eichornia*	sp.	wd4883
AC12	Proteaceae	*Panopsis*	*rubescens* Pittier	wd4888
AC08	Proteaceae	*Roupala*	*complicata* Kunth.	wd4797
AC01	Pteridaceae	*Adiantum*	*latifolium* Lam.	wd4601

continued

Georef	Family	Genus	Species	# Collection
Georef	Familia	Género	Especies	# Colección
BC03	Pteridaceae	*Adiantum*	*latifolium* Lam.	wd4942
AC01	Pteridaceae	*Adiantum*	*leprieurii* Hook.	wd4602
AC01	Pteridaceae	*Adiantum*	*petiolatum* Desv.	wd4617
AC06	Pteridaceae	*Adiantum*	*petiolatum* Desv.	wd4704
AC06	Pteridaceae	*Adiantum*	*phyllitidis* J. Sm.	wd4712
AC04	Pteridaceae	*Adiantum*	sp.	wd4651
AC05	Pteridaceae	*Adiantum*	sp.	wd4697
AC14	Pteridaceae	*Adiantum*	sp.	wd4928
BC05	Pteridaceae	*Adiantum*	sp.	wdp/s
BC15	Rubiaceae	*Alibertia*	*latifolia* (Benth.) K. Schum	wd5020
AC14	Rubiaceae	*Alibertia*	sp.	wd4923
BC10	Rubiaceae	*Alibertia*	sp.	wd4991
AC11	Rubiaceae	*Amaioua*	sp.	wd4864
BC12	Rubiaceae	*Amaioua*	sp.	wd5009
AC03	Rubiaceae	*Borreria*	sp.	wd4640a
BC10	Rubiaceae	*Borreria*	sp.	wdp/s
BC01	Rubiaceae	*Cosmibuena*	*grandiflora* (Ruiz & pav.) Rusby	wd4931
AC03	Rubiaceae	*Diodia*	*apiculata* (Willd. Ex Roem. & Schult.) Shum.	wd4641
AC05	Rubiaceae	*Diodia*	sp.	wd4680
AC11	Rubiaceae	*Diodia*	sp.	wd4848
AC14	Rubiaceae	*Diodia*	sp.	wd4899
BC09	Rubiaceae	*Diodia*	sp.	wdp/s
BC10	Rubiaceae	*Diodia*	sp.	wdp/s
BC06	Rubiaceae	*Diodia*	sp.	wdp/s
AC04	Rubiaceae	*Genipa*	*spruceana* Steyerm.	JR P/S
AC04	Rubiaceae	*Genipa*	*spruceana* Steyerm.	wd4654
AC05	Rubiaceae	*Genipa*	*spruceana* Steyerm.	wd4661
AC06	Rubiaceae	*Genipa*	*spruceana* Steyerm.	JR P/S
AC08	Rubiaceae	*Genipa*	*spruceana* Steyerm.	JR P/S
BC06	Rubiaceae	*Genipa*	*spruceana* Steyerm.	wdp/s
BC07	Rubiaceae	*Genipa*	*spruceana* Steyerm.	wdp/s
BC09	Rubiaceae	*Genipa*	*spruceana* Steyerm.	wdp/s
BC10	Rubiaceae	*Genipa*	*spruceana* Steyerm.	wdp/s
BC11	Rubiaceae	*Genipa*	*spruceana* Steyerm.	wdp/s
AC03	Rubiaceae	Genus indet.		wd4619
AC04	Rubiaceae	Genus indet.		wd4658
AC05	Rubiaceae	Genus indet.		wd4666
AC05	Rubiaceae	Genus indet.		wd4684
AC05	Rubiaceae	Genus indet.		wd4699
AC07	Rubiaceae	Genus indet.		wd4727
AC08	Rubiaceae	Genus indet.		wd4748

continued

Georef	Family	Genus	Species	# Collection
Georef	Familia	Género	Especies	# Colección
AC08	Rubiaceae	Genus indet.		wd4763
AC08	Rubiaceae	Genus indet.		wd4756
AC08	Rubiaceae	Genus indet.		wd4822
AC08	Rubiaceae	Genus indet.		wd4810
AC11	Rubiaceae	Genus indet.		wd4866
AC11	Rubiaceae	Genus indet.		wd4861
AC14	Rubiaceae	Genus indet.		wd4914
AC08	Rubiaceae	*Isertia*	*hypoleuca* Benth.	wd4784
BC16	Rubiaceae	*Isertia*	*hypoleuca* Benth.	wd5042
BC15	Rubiaceae	*Isertia*	sp.	wdp/s
AC03	Rubiaceae	*Ixora*	*acuminatissima* Müll-Arg.	wd4628
AC06	Rubiaceae	*Ixora*	*acuminatissima* Müll-Arg.	wd4705
BC15	Rubiaceae	*Ixora*	*acuminatissima* Müll-Arg.	wdp/s
AC07	Rubiaceae	*Morinda*	*tenuiflora* (Benth.) Stey.	wd4730
BC15	Rubiaceae	*Morinda*	*tenuiflora* (Benth.) Stey.	wd5025
BC10	Rubiaceae	*Palicourea*	sp.	wd4996
AC01	Rubiaceae	*Psychotria*	*lupulina Benth.*	wd4601
AC03	Rubiaceae	*Psychotria*	sp.	wd4625
BC03	Rubiaceae	*Psychotria*	sp.	wdp/s
AC08	Rubiaceae	*Randia*	sp.	wd4759
BC13	Rubiaceae	*Simira*	*aristeguietae* (Stey.) Stey.	wdp/s
AC14	Rubiaceae	*Tocoyena*	sp.	wd4908
BC02	Rubiaceae	*Uncaria*	*guianensis* (Aubl.) Gmell.	wdp/s
AC03	Sapindaceae	*Cupania*	*cinerea* Poepp. & Endl.	JR P/S
AC04	Sapindaceae	*Cupania*	*cinerea* Poepp. & Endl.	JR P/S
BC08	Sapindaceae	*Cupania*	sp.	wdp/s
BC10	Sapindaceae	*Cupania*	sp.	wdp/s
BC04	Sapindaceae	*Cupania*	sp.	wdp/s
BC01	Sapindaceae	*Cupania*	*cinerea* Poepp. & Endl.	wdp/s
BC13	Sapindaceae	Genus indet.		wd5015
AC06	Sapindaceae	*Matayba*	*camptoneura* Radlk.	wd4709
AC08	Sapindaceae	*Matayba*	*camptoneura* Radlk.	wd4777
AC10	Sapindaceae	*Matayba*	sp.	JR P/S
BC15	Sapotaceae	*Ecclinusia*	*guianensis* Eyma	wdp/s
AC07	Sapotaceae	*Ecclinusia*	*guianensis* Eyma	wd4735a
AC04	Sapotaceae	Genus indet.		JR P/S
BC01	Sapotaceae	Genus indet.		wd4929
BC10	Sapotaceae	Genus indet.		wd4988
BC10	Sapotaceae	Genus indet.		wd4995
BC05	Sapotaceae	*Manilkara*	*bidentata* (A. DC.) A. Chev.	wd4951
BC08	Sapotaceae	*Micropholis*	*guyanensis* (A. DC.) Pierre	wdp/s
AC10	Sapotaceae	*Micropholis*	sp.	JR P/S

continued

Georef	Family	Genus	Species	# Collection
Georef	**Familia**	**Género**	**Especies**	**# Colección**
AC03	Sapotaceae	*Pouteria*	*cuspidata* (A. DC.) Baeni	wd4635
BC05	Sapotaceae	*Pouteria*	sp.	wd4952
AC06	Schizaeaceae	*Lygodium*	*volubile* Sw.	wd4713
AC07	Schizaeaceae	*Schizaea*	*elegans* (Vahl) Sw.	wd4724
AC08	Schizaeaceae	*Schizaea*	*elegans* (Vahl) Sw.	wd4766, 4807
AC06	Selaginellaceae	*Selaginella*	*parkeri* (Hook & Grev.) Spring.	wd4707
BC01	Selaginellaceae	*Selaginella*	sp.	wdp/s
AC08	Siparunaceae	*Siparuna*	*guianensis* Aubl.	wd4776
BC01	Smilacaceae	*Smilax*	*schomburgkiana* Kunth	wdp/s
BC05	Solanaceae	*Schwenkia*	*americana* L.	wd4961
BC06	Solanaceae	*Schwenkia*	*americana* L.	wd4969
AC12	Solanaceae	*Solanum*	*monacophyllum* Dunal	wd4877
BC09	Solanaceae	*Solanum*	*monacophyllum* Dunal	wd4983
BC05	Solanaceae	*Solanum*	sp.	wd4956
BC08	Solanaceae	*Solanum*	sp.	wdp/s
BC07	Solanaceae	*Solanum*	sp.	wdp/s
BC17	Sterculiaceae	*Helicteres*	*guazumifolia* Kunth	wd5035
AC01	Sterculiaceae	*Sterculia*	*pruriens* Schum.	JR P/S
AC10	Sterculiaceae	*Sterculia*	*pruriens* Schum.	JR P/S
BC01	Strelitziaceae	*Phenakospermum*	*guyannense* (L.C. Rich.) Endl. Ex Miq.	wdp/s
BC05	Theophrastaceae	*Clavija*	*lancifolia* Desf.	wd4948
BC01	Tiliaceae	*Apeiba*	*aspera* Aubl.	wdp/s
BC16	Tiliaceae	*Luehea*	*candida* (Moc. & Ses. Ex DC.) Mart. & Zucc.	wd5040
AC07	Turneraceae	*Turnera*	*odorata* (L.) Rich.	wd4737
BC06	Turneraceae	*Turnera*	*odorata* (L.) Rich.	wdp/s
AC12	Ulmaceae	*Trema*	*micrantha* Blume	wd4872
BC01	Urticaceae	*Urera*	*baccifera* (L.) Gaud.	wdp/s
AC08	Verbenaceae	*Amasonia*	*campestris* (Aubl.) Mold.	wd4811
BC15	Verbenaceae	*Vitex*	sp.	wdp/s
BC04	Violaceae	*Amphirrox*	sp.	wd4980
AC03	Violaceae	*Amphirrox*	*latifolia* Mart.	JR P/S
AC04	Violaceae	*Amphirrox*	*latifolia* Mart.	JR P/S
BC02	Violaceae	*Corynostylis*	*asperiifolia*	wd4939
AC11	Violaceae	Genus indet.		wd4863
AC05	Violaceae	*Hybanthus*	sp.	wd4694
AC08	Violaceae	*Hybanthus*	sp.	wd4814
AC14	Violaceae	*Hybanthus*	sp.	wd4911
BC03	Violaceae	*Payaparola*	sp.	wd4945
AC06	Violaceae	*Rinorea*	*flavescens* (Aubl.) Kuntze	wd4716
BC08	Violaceae	*Rinorea*	*flavescens* (Aubl.) Kuntze	wdp/s
BC15	Violaceae	*Rinorea*	*flavescens* (Aubl.) Kuntze	wdp/s

continued

Georef	Family	Genus	Species	# Collection
Georef	Familia	Género	Especies	# Colección
AC04	Violaceae	*Rinorea*	sp.	wd4657
BC03	Violaceae	*Rinorea*	sp.	wd4941
AC11	Vitaceae	*Cissus*	*erosa* L. C. Rich.	wd4868
BC05	Vitaceae	*Cissus*	*erosa* L. C. Rich.	wd4965
BC10	Zingiberaceae	*Renealmia*	*floribunda* K. Schum.	wdp/s
BC15	Zingiberaceae	*Renealmia*	*floribunda* K. Schum.	wdp/s

Appendix/Apéndice 3

Description of limnology sampling sites, with georeference and site numbers, for the AquaRAP expedition to the Caura River Basin, Bolívar State, Venezuela

Descripción de sitios de muestreo limnológicos con números relativos a sitios de georeferencia y estaciones durante la Expedición AquaRAP a la Cuenca del Río Caura, Estado Bolívar, Venezuela

Karen J. Riseng and John S. Sparks

Georef	Site Number	Date	Site Description
Georef	**Sitio Número**	**Fecha**	**Descripción del sitio**
AC 01	Lim 01	25-Nov-00	Tributary of Caura (caño laguna) Tributario del Río Caura (caño y laguna).
AC 02	Lim 14	29-Nov-00	Rio Caura just above confluence with Rio Erebato; rapids rocky area Río Caura arriba de la confluencia con el Río Erebato; rápidos área rocosa.
AC 02	Lim 15	29-Nov-00	Rio Erebato, mouth just above confluence with Rio Caura. Río Erebato, boca justo arriba de su confluencia con el Río Caura.
AC 03	Lim 02	26-Nov-00	Rio Kakada, beach at Raudal Suajiditu. Río Kakada, playa en Raudal Suajiditu.
AC 04	Lim 03	26-Nov-00	Caños at Suajiditu; trib. of Rio Kakada. Caños en Raudal Suajiditu; Trib. del Río Kakada.
AC 04	Lim 04	26-Nov-00	Caños at Suajidito; trib. of Rio Kakada. Caños en Raudal Sajiditu; Trib. del Río Kakada.
AC 05	Lim 05	27-Nov-00	Rio Erebato, island in main channel. Río Erebato, isla en el canal principal.
AC 06	Lim 06	27-Nov-00	Small stream through forest; tributary of Erebato River. Pequeño riachuelo en bosque. Tributario del Río Erebato.
AC 07	Lim 07	27-Nov-00	Erebato River, main channel upstream from Lim 5. Río Erebato, canal principal río arriba de Lim 5.
AC 08	Lim 08	28-Nov-00	Mouth of sm stream through forest; trib of Caura River, just below lg rapids. Boca del riachuelo en bosque; trib. del Río Caura, por abajo de rápidos.
AC 08	Lim 09	28-Nov-00	Small stream, approx. 200m upstream from Lim 8. Pequeño riachuelo, aprox. 200m río arriba de Lim 8.
AC 08	Lim 10	28-Nov-00	Caura River, main channel just below large rapids. Río Caura, canal principal bajo los rápidos.
AC 09	Lim 11	28-Nov-00	Caura River, main channel, area of slower current. Río Caura, canal principal, área de baja corriente.
AC 09	Lim 12	28-Nov-00	Caura River, main channel, oppo side of sandbar in faster water from Lim 11. Río Caura, canal principal, sit. op. de barra arenosa en rápidos de Lim 11.
AC 10	Lim 13	29-Nov-00	Medium sized tributary of Caura River. Tributario de tamaño medio del Río Caura.
AC 11	Lim 16	01-Dec-00	Caura River, just above falls, west side of river. Río Caura, justo por encima del Salto lado Oeste del río.
AC 12	Lim 17	01-Dec-00	Caura River, just above falls, East side of river. Río Caura, justo por encima del Salto lado Este del río.

continued

Georef	Site Number	Date	Site Description
Georef	Sitio Número	Fecha	Descripción del sitio
AC 12	Lim 18	01-Dec-00	Caura River, just above falls, just downstream from Lim 17. Río Caura, justo por encima del Salto, río debajo de Lim 17.
AC 13	Lim 19	01-Dec-00	Yuruani River, 5 minutes upstream from confluence with Rio Caura by boat. Río Yuruani, 5 minutos (por bote) río arriba de su confluencia con Río Caura.
AC 14	Lim 20	01-Dec-00	Caura River, approx. 1. 2 way between Entre Rios and falls. Río Caura, aprox. Mitad de camino entre Entreríos y el Salto.
AC 14	Lim 21	01-Dec-00	Stream near Lim 20, tributary of Caura River. Riachuelo cerca de Lim 20, tributario del Río Caura.
BC01	Lim 23	03-Dec-00	Caura River rapids in front of camp and at base of Salto Pará. Río Caura, rápidos en frente del campamento en base del Salto Pará.
BC01	Lim 25	04-Dec-00	Spring on island- Guido Pereira took water sample. Riachuelo en isla-Guido Pereira tomó la muestra.
BC02	Lim 22	03-Dec-00	Katuwadi Chenu - small stream (caño) very near Salto Pará camp. Pequeño riachuelo cerca del Salto Pará.
BC03	Lim 24	03-Dec-00	Caños wakikema into Caura River near camp 10 min. by boat downstream. Caño wakikema trib. Río Caura cerca del campamento 10 minutos río abajo.
BC04	Lim 26	07-Dec-00	Tawadu River in front of camp. Río Tawadu en frente campamente Dedemai.
BC05	Lim 27	05-Dec-00	Tawadu Upper (above camp and 2 rapids, below rapids where sampled fish). Río Tawadu (arriba camp. y 2 rápidos, por debajo se colectaron peces).
BC05	Lim 27a	05-Dec-00	Side stagnant water just below falls, 30m from Lim 28; no water taken. Sitio de aguas estancadas por debajo de las cataratas, 30m de Lim 28; no water taken.
BC06	Lim 28	05-Dec-00	Caño of Tawadu Middle (below the 2 falls, but above camp). Caño del Tawadu (debajo de 2 rápidos río arriba del campamento).
BC06	Lim 29	05-Dec-00	Tawadu Middle (below the 2 falls, but above camp). Río Tawadu (debajo de 2 rápidos , arriba del campamento).
BC07	Lim 30	05-Dec-00	Laguna off Nichare (~1 km downstream from confluence of Nichare-Tawadu); a floodplain or a very stagnant arm that is disconected in low water. Laguna del Nichare (~1 km río debajo de la confluencia del Nichare-Tawadu); Un área inundable con agua estancada que esta separada del río en aguas bajas.
BC08	Lim 31	06-Dec-00	Caño off of Icutú River (a tributary of the Nichare). Caño en el Río Icutú (un tributario del Nichare).
BC08	Lim 32	06-Dec-00	Icutú River (a trib of the Nichare) just upstream from where caño enters river. Río Icutú (un trib del Nichare) justamente por arriba del caño.
BC09	Lim 33	06-Dec-00	Arm of Nichare River (stagnant). Brazo del Nichare (aguas estancadas).
BC10	Lim 35	07-Dec-00	Nichare River, just upstream from Tawadu. Río Nichare, río arriba del Tawadu.
BC10	Lim 36	07-Dec-00	Nichare River, just downstream from Laguna. Río Nichare, río debajo de la Laguna.
BC11	Lim 34	07-Dec-00	Tawadu River mouth, 20m upstream from Nichare. Boca del Tawadu, 20 m río arriba del Nichare.
BC12	Lim 37	09-Dec-00	Kumaka Kujo (Lagoon); isolated lake that is connected to river in high water). Kumaka Kujo (Laguna); lago aislado que se conecta con río en aguas alta.
BC13	Lim 38	09-Dec-00	Mato River, trib. of Rio Caura. Río Mato, trib. del Río Caura.
BC14	Lim 39	09-Dec-00	Caura River—by sandy island. Río Caura—isla arenosa.

continued

Georef	Site Number	Date	Site Description
Georef	Sitio Número	Fecha	Descripción del sitio
BC17	Lim 41	10-Dec-00	Takoto River—sm river East; very dirty looking, different geology; filamentous algae present, shiny black patina on rocks, iron flock. Río Takoto—pequeño río que margen Este; parece sucio, diferente geología; Algas filamentosas presentes.
BC18	Lim 40	10-Dec-00	Caura River—in front of camp Raudal Cinco Mil on the rocks. Río Caura—en frente camp. Raudal Cinco Mil en rocas.
BC19	Lim 42	11-Dec-00	Caura River—30 minutes below camp. Sampled between island and side of river. Río Caura—30 min. Río abajo camp. Muestra entre las isla y lado del río.

Appendix/Apéndice 4

Limnological parameters of sampling sites in the Caura River Basin, State of Bolívar, Venezuela

Parámetros limnológicos de los sitios muestreados en el Río Caura, Estado Bolívar, Venezuela

Karen J. Riseng and John S. Sparks

Parameters and abbreviations: water temperature (Temp), pH, conductivity (Cond), dissolved oxygen (D.O.), alkalinity (Alk), dissolved organic carbon (DOC), sediment and water flow rate.

Abreviaturas: Temperatura del agua (Temp), pH, Conductividad (Cond), Oxígeno disuelto (D.O.), Alcalinidad (Alk), Carbono orgánico disuelto (DOC), Sedimento y flujo del agua (Flujo)

Georef	Site Number/ Sitio Número	Temp (C)	pH	Cond (uS/cm)	D.O. (mg/L)	Alk (ueq/L)	DOC (uM)	Sediment/ Sedimento (mg/L)	Flow/ Flujo (m/s)
AC 01	Lim 01	24.0	5.24	11	3.9	90		9.7	
AC 02	Lim 14	26.5	5.69	11	6.8	80	311	17.3	
AC 02	Lim 15	25.6	5.84	10	6.8			8.4	
AC 03	Lim 02	23.9	5.04	6	6.4	20	233	4.2	
AC 04	Lim 03	23.6	5.44	8	5.9	68	130	3.8	
AC 04	Lim 04	24.1	5.71	12					
AC 05	Lim 05	24.4	5.69	9	6.6	67	207	6.4	
AC 06	Lim 06	23.9	5.42	7	6.4	50	157	3.6	
AC 07	Lim 07	25.0	5.81	9	6.7	70	175		
AC 08	Lim 08	23.3	5.23	9	6.5	77	182		
AC 08	Lim 09	23.4	5.21	9	6.5	65	185	16.0	
AC 08	Lim 10	25.5	5.76	11	7.5	78	317	10.1	
AC 09	Lim 11	26.1	5.81	11	6.9				
AC 09	Lim 12	26.1	5.84	11	6.9				
AC 10	Lim 13	24.2	5.44	10	6.2	79	249	21.6	
AC 11	Lim 16	25.9	5.80	10	7.0	79	253	10.1	
AC 12	Lim 17	26.0	5.84	11	6.9				
AC 12	Lim 18	26.1	5.83	11	6.9				
AC 13	Lim 19	25.4	5.68	12	6.4	94	277	21.8	
AC 14	Lim 20	25.6	5.70	10	7.0	76	218		
AC 14	Lim 21	25.1	5.64	10	6.9				
BC01	Lim 23	26.1	5.77	11	9.1	63	234	11.8	1.8
BC01	Lim 25			35		280	200		
BC02	Lim 22	24.4	5.92	35	6.8	273	221	15.8	0.3
BC03	Lim 24	24.8	5.55	11	7.0	75	345		0.1

continued

Georef	Site Number/ Sitio Número	Temp (C)	pH	Cond (uS/cm)	D.O. (mg/L)	Alk (ueq/L)	DOC (uM)	Sediment/ Sedimento (mg/L)	Flow/ Flujo (m/s)
BC04	Lim 26	23.1	5.81	10	7.8	70	223	1.3	0.4
BC05	Lim 27	22.7	5.82	9	8.4	55	504	1.3	1.0
BC05	Lim 27a	23.1		9	8.3				
BC06	Lim 28	25.5	5.82	22	8.2	182	133		~0
BC06	Lim 29	23.3	5.64	9	8.2	57	330		
BC07	Lim 30	25.6	4.93	12	3.3	81	281	5.0	~0
BC08	Lim 31	24.4	5.91	18	7.1	165	98	3.2	0.3
BC08	Lim 32	24.8	5.87	14	6.8	126	184	13.8	0.7
BC09	Lim 33	26.4	4.87	12	4.9	92	263	5.2	~0
BC10	Lim 35	24.6	5.62	12	6.9	97	232	16.8	1.0
BC10	Lim 36	24.6	5.61	12	6.9	98	143	17.8	0.8
BC11	Lim 34	23.7		11	7.6				
BC12	Lim 37	26.6	4.34	19	0.9	86	915	10.6	~0
BC13	Lim 38	25.9	5.82	20	6.2	165	528	21.9	0.4
BC14	Lim 39	26.8	5.67	10	7.3	82	307	5.4	
BC17	Lim 41	25.4	5.99	41	6.8	375	183	16.2	0.5
BC18	Lim 40	25.9	5.69	10	7.8	74	274	13.8	
BC19	Lim 42	25.9	5.61	10	7.7	75	274	14.6	0.5

Appendix/Apéndice 5

Limnological descriptors of sampling sites in the Caura River Basin, Bolívar State, Venezuela

Descriptores limnológicos de sitios de muestreo en la Cuenca del Río Caura, Estado Bolívar, Venezuela

Karen J. Riseng and John S. Sparks

Parameters and abbreviations: depth, width, percent shade on water surface, Secchi depth (TTB= transparent to bottom), turbidity rating, filtrate color (on gradient of clear, very slightly brown (vsb), slightly brown (sb), light brown (lb) and dark brown (db)).

Abreviaturas; Profundidad (Prof.), Ancho, Porcentaje de cobertura sobre la superficie del agua (Sombra), Profundidad de Secchi (TTB= transparente al fondo), Turbidez, Color filtrado (sobre gradiente de claro, muy poco marrón (vsb), poco marrón (sb), levemente marrón (lb) y marrón oscuro (db).

Georef	Site Number	Depth (m)	Width (m)	Shade (%)	Secchi (cm)	Turbidity (low, med, high)	Filtrate Color (see above)
Georef	Sitio Número	Prof. (m)	Ancho (m)	Sombra (%)	Secchi (cm)	Turbidez (baja, med, alta)	Color filtrado (ver arriba)
AC 01	Lim 01	1	9.5		60	high/alta	clear/claro
AC 02	Lim 14				100	med to high/media a alta	sb
AC 02	Lim 15				70	med/media	vsb
AC 03	Lim 02	2			130	baja a media	db
AC 04	Lim 03	1	8		TTB	low/baja	clear/claro
AC 04	Lim 04	1.6	20		130	high/alta	
AC 05	Lim 05				130		vsb
AC 06	Lim 06	0.5	2		TTB	low/baja	clear/claro
AC 07	Lim 07				130	low/baja	sb
AC 08	Lim 08	0.5	6		TTB	low/baja	clear/claro
AC 08	Lim 09				TTB	baja a media	clear/claro
AC 08	Lim 10				TTB	high/alta	vsb
AC 09	Lim 11				80	high/alta	vsb
AC 09	Lim 12					high/alta	vsb
AC 10	Lim 13	1.5	12		40	high/alta	clear/claro
AC 11	Lim 16				80	med/media	sb
AC 12	Lim 17		50		80	med/media	sb
AC 12	Lim 18					med/media	sb
AC 13	Lim 19	4.3	70		50	high/alta	
AC 14	Lim 20					med/media	sb
AC 14	Lim 21	1	12		100	med/media	clear/claro

continued

Georef	Site Number	Depth (m)	Width (m)	Shade (%)	Secchi (cm)	Turbidity (low, med, high)	Filtrate Color (see above)
Georef	Sitio Número	Prof. (m)	Ancho (m)	Sombra (%)	Secchi (cm)	Turbidez (baja, med, alta)	Color filtrado (ver arriba)
BC01	Lim 23		200	<2	60	high/alta	sb
BC01	Lim 25						
BC02	Lim 22	1.4	10	20	65	high/alta	clear/claro
BC03	Lim 24	2.05	10	70	TTB	baja a media	clear/claro
BC04	Lim 26	3.6	25	5	360	low/baja	clear/claro
BC05	Lim 27	1.95	100	<2	TTB	low/baja	clear/claro
BC05	Lim 27a				TTB	low/baja	
BC06	Lim 28	1.5	3	80	TTB		
BC06	Lim 29	1	80	<2	200	low/baja	clear/claro
BC07	Lim 30	3.1	150	<2	160	low/baja	clear/claro
BC08	Lim 31	1	20	40	TTB	low/baja	clear/claro
BC08	Lim 32	1.7	50	<2	100	baja a media	clear/claro
BC09	Lim 33	3.6	50	<2	120	low/baja	clear/claro
BC10	Lim 35	2.5	60	5	50	med/media	clear/claro
BC10	Lim 36	3.75	70	<2	100	med/media	clear/claro
BC11	Lim 34	2.6	40	<2	215	low/baja	no water filtered/ no muestra
BC12	Lim 37	1	100	5	60	med/media	lb
BC13	Lim 38	4.4	30	<5	70	high/alta	clear/claro
BC14	Lim 39	>5m	1500	<2	90	high/alta	clear/claro
BC17	Lim 41	1.4	15	15	60	high/alta	clear/claro
BC18	Lim 40	shore		<2		high/alta	sb
BC19	Lim 42	>5m	20	10	85	high/alta	sb

Appendix/Apéndice 6

List of benthic macroinvertebrates collected during the AquaRAP expedition to the Caura River Basin, Bolívar, Venezuela

Lista de los macroinvertebrados bénticos colectados durante la Expedición AquaRAP a al Río Caura, Estado Bolívar, Venezuela

José García and Guido Pereira

See Appendix 1 for site descriptions (AC01–12, BC02–17).
Ver Apéndice 1 por descripciones de los sitios (AC01–12, BC02–17).

ARTHROPODA	**Sitios de Georeferencia**																								
	Upper Caura/Caura Superior											**Lower Caura/Bajo Caura**													
INSECTA	**AC**	**AC**	**AC**	**AC**	**AC**	**AC**	**AC**	**AC**	**AC**	**AC**	**AC**	**BC**	**BC**	**BC**	**BC**	**BC**	**BC**	**BC**	**BC**	**BC**	**BC**	**BC**	**BC**	**BC**	**BC**
	01	**03**	**04**	**05**	**06**	**08**	**09**	**10**	**11**	**12**	**14**	**02**	**03**	**04**	**05**	**06**	**07**	**08**	**09**	**10**	**11**	**12**	**13**	**15**	**17**
ODONATA																									
Corduliidae																									
Neurocordulia					X		X											X						X	
Libellulidae																									
Brechmorhoga				X	X		X				X	X				X									X
Erythrodiplax																					X				X
Macrodiplax							X			X						X									X
Macrothemis		X	X	X	X	X	X								X						X	X			
Perithemis			X		X							X					X						X		
Tauriphila			X					X			X			X	X			X			X		X		X
Tramea					X		X			X												X			
Gomphidae																									
Aphylla						X		X						X			X	X						X	
Desmogomphus							X																		
Gomphus					X										X	X		X							
Hagenius			X				X				X							X							
Ophiogomphus							X								X			X							
Progomphus										X		X						X							
Hypolestidae																									
Hypolestes					X										X			X							
Synlestidae																									
Phylolestes											X				X			X			X				
Calopterygidae																									
Hetaerina															X			X			X			X	X
Coenagrionidae																									
Argia						X									X										
Other Coenagrionidae						X		X	X		X														X

continued

ARTHROPODA	Sitios de Georeferencia																							
INSECTA	Upper Caura/Caura Superior											Lower Caura/Bajo Caura												
	AC	AC	AC	AC	AC	AC	AC	AC	AC	AC	AC	BC	BC	BC	BC	BC	BC	BC	BC	BC	BC	BC	BC	BC	BC
	01	03	04	05	06	08	09	10	11	12	14	02	03	04	05	06	07	08	09	10	11	12	13	15	17
EPHEMEROPTERA																									
Caenidae																									
Caenis																					X				
Euthyplociidae																									
Camphylocia					X	X						X			X	X		X							
Leptophlebiidae																									
Hermanella			X				X				X					X								X	
Thraulodes		X	X	X	X	X	X				X				X	X		X			X				
Ulmeritus			X	X		X	X	X			X	X	X		X			X			X			X	
Leptophlebia					X	X												X							
Hermanellopsis			X		X					X	X					X									X
Terpides																									X
Oligoneuriidae																									
Lachlania				X	X																				
Polymitarcyidae																									
Campsurus									X													X			
Siphlonuridae																									
Ameletus				X	X		X		X	X	X					X									
Tricorythidae																									
Leptohypes																X									
HEMIPTERA																									
Nepidae																									
Ranatra	X	X			X		X														X		X	X	X
Belostomatidae																									
Belostoma	X								X																X
Lethocerus										X						X						X	X	X	
Notonectidae																									
Martarega	X	X	X		X		X	X			X	X										X			
Buenoa			X							X												X	X		X
Veliidae																									
Rhagovelia		X		X	X		X					X												X	
Microvelia	X										X				X							X			
Gerridae																									
Potamobates					X			X			X														X
Limnogonus							X															X		X	
Brachymetra			X				X	X		X	X	X												X	
Naucoridae																									
Cryphocricos				X												X									X
Corixidae																									
Tenegobia										X					X							X	X		

continued

ARTHROPODA	Sitios de Georeferencia																							
INSECTA	Upper Caura/Caura Superior											Lower Caura/Bajo Caura												
	AC 01	AC 03	AC 04	AC 05	AC 06	AC 08	AC 09	AC 10	AC 11	AC 12	AC 14	BC 02	BC 03	BC 04	BC 05	BC 06	BC 07	BC 08	BC 09	BC 10	BC 11	BC 12	BC 13	BC 15	BC 17
TRICHOPTERA																									
Helicopsychidae																									
Helicopsyche				X		X					X	X													
Hydropsychidae																									
Leptonema				X	X	X	X					X				X								X	
Macronema					X																			X	
Smicridea																								X	
Other Hydropsychidae																X									
Philopotamidae																									
Chimarra											X	X				X									
Polycentropodidae																									
Polyplectropus											X														
Calamoceratidae																									
Phylloicus												X									X				
Calamoceratidae pupa																		X							
COLEOPTERA																									
Dytiscidae																									
Megadytes	X		X		X																			X	
Laccophilus																	X								X
cf. *Drovatellus*																		X							X
Gyrinidae																									
Gyretes									X	X								X			X			X	
Noteridae																									
Colpius																						X			
Elmidae																									
Planocerus											X														
Other Elmidae			X		X											X			X	X					X
Hyrdaenidae																								X	
DIPTERA																									
Chironomidae																									
Chironomini																									
Chironomus				X																		X			
Harnischia				X																					
Hyporhygma																					X				
Micropsectra		X																							
Paracladopelma		X																							
Polypedilum									X												X				
Stenochironomus						X																			
Tanytarsus				X																					

continued

ARTHROPODA	Sitios de Georeferencia																							
INSECTA	Upper Caura/Caura Superior											Lower Caura/Bajo Caura												
	AC	AC	AC	AC	AC	AC	AC	AC	AC	AC	AC	BC	BC	BC	BC	BC	BC	BC	BC	BC	BC	BC	BC	BC	BC
	01	03	04	05	06	08	09	10	11	12	14	02	03	04	05	06	07	08	09	10	11	12	13	15	17
Tanypodini																								
Ablabesmya				X					X															
Coelotanypus				X													X							
Tipulidae																								
Hexatoma																X								
Culicidae																								
Aedes				X		X	X		X	X														
Culex				X					X	X														X
PLECOPTERA																								
Perlidae																								
Anacroneura				X	X		X		X		X				X	X		X						
NEUROPTERA																								
Corydalidae																								
Corydalus						X	X				X													
Other Corydalidae																							X	
LEPIDOPTERA																								
Pyralidae				X	X	X	X	X	X	X						X								X
COLLEMBOLA																								
Entomobryidae																								
cf. *Willowsia*																					X			
CRUSTACEA																								
ISOPODA																								
Corallanidae																								
Exocorallana berbicensis																						X		
BRANCHIOPODA-CONCHOSTRACA																								
Cyclestheriidae																								
Cyclestheria hislopi																								X
MOLLUSCA																								
GASTROPODA																								
Ampullariidae											X											X	X	
Pomacea (L.) spp.																								
Melaniidae																								
Doryssa sp.						X	X	X	X	X	X					X							X	X
young *Melaniidae*									X		X													
Micro-gastropodos				X					X															
BIVALVIA																								
Sphaeriidae																								
Eupera sp.				X							X										X	X		X

continued

ARTHROPODA	Sitios de Georeferencia																								
INSECTA	Upper Caura/Caura Superior											Lower Caura/Bajo Caura													
	AC	AC	AC	AC	AC	AC	AC	AC	AC	AC	AC	BC	BC	BC	BC	BC	BC	BC	BC	BC	BC	BC	BC	BC	BC
	01	03	04	05	06	08	09	10	11	12	14	02	03	04	05	06	07	08	09	10	11	12	13	15	17
ANNELIDA																									
OLIGOCHAETA																									
Lumbriculidae							X			X		X													
Enchytraeidae				X						X											X				
Otros Oligochaeta				X	X	X	X			X	X										X		X	X	
HIRUDINEA																									
Glossiphoniidae																									
Haementeria tuberculifera				X																				X	
Other Hirudinea						X										X					X	X			
Riqueza total	5	7	13	23	24	17	26	9	14	17	23	15	1	2	14	20	4	18	1	1	16	14	9	20	23
Diversidad de Shannon (H)	0.5	1.3	1.4	2.1	1.8	2.0	2.0	1.7	1.9	2.0	2.0	1.8	0.0	0.0	1.1	2.1	1.0	1.2	0.0	0.0	1.9	1.8	1.4	2.0	2.1

Appendix 7

Ichthyological field stations and analysis of the fish fauna, sampled during the AquaRAP expedition to the Caura River Basin, Bolívar State, Venezuela

Antonio Machado-Allison, Barry Chernoff, Philip Willink, Francisco Provenzano, Paulo Petry and Alberto Marcano

I. UPPER CAURA. ABOVE SALTO PARÁ.

AC01: Caño-Lagoon off Caura River just downstream from Entrerios (Field Station ICT-01) (5°55.78´N–64°25.39´W)

General fish fauna description

We capture a total of 220 specimens placed in 18 species distributed:

Characiformes: 10 (55.5%)
Siluriformes: 4 (22.2%)
Perciformes: 2 (11%)
Others: 2

The fish diversity is low. Some components are common to this type of habitat such as: *Pimelodella* and *Corydoras*, and species of the family Cichlidae (genera *Guianacara* and *Crenicichla*). The results show that the most abundant group is the Characiformes with 55% of the species, followed by the Siluriformes (22.2%) and Perciformes (11%). Results of the relative abundance the most common species are: *Moenkhausia collettii* (71, 32.2%), *Corydoras boehlkei* (65, 29.5%), *Pimelodella* sp. B (26, 11.8%), *Pimelodella* sp. C (16, 7.2%) and *Guianacara geayi* (8, 3.6%).

AC02. Confluence of Caura and Erebato Rivers at Entrerios (Field Stations: ICT-02, ICT-06). (5°56.02´N–64°25.67´W).

General fish fauna description

Backwater (ICT-02): Fish captures by beach seines. Moderate diversity (25 species) and low abundance; 139 specimens, placed in:

Characiformes: 17 (68%)
Siluriformes: 6 (24%)
Perciformes: 2 (8%)

The richness is low compared with other field stations (see below). Some components are common to these habitats, among them specimens of *Jupiaba* (2 species) and *Corydoras* and Cichlidae (*Guianacara* and *Geophagus*). The results indicate that Characiformes (17, 68%) is the most diverse order, followed by Siluriformes (24%) and Perciformes (2%). Relative abundance showed: *Knodus* cf *victoriae* (25, 18%), Bryconops sp.A. (20, 14.4%), *Guianacara geayi* (17, 12.2%), *Corydoras boehlkei* (13, 9.3%) and *Phenacogaster* sp. B (12, 8.6%) as the best represented.

Caño (ICT-06): Fishes sampled with gillnet; low diversity (13 species) with a total of 41 specimens captured:

Characiformes: 8 (61.5%)
Siluriformes: 3 (23%)
Perciformes: 1 (7.6%)
Synbranchiformes: 1 (7.6%)

The most abundant species was a doradid catfish, preliminarily identified as *Doras* sp. with 18 (43.9%) specimens. Also collected were *Ageneiosus* sp. (5, 12.2%), *M. Rubripinnis* (4, 9.7%), *Plagioscion auratus* (4, 9.7%) and *Serrasalmus rhombeus* (2, 4.8%). In addition, two species of "pámpanos" or "pacus" *Myleus asterias* and *M. torquatus* were captured.

In general the fish fauna of the Entreríos area are characterized by 34 species inhabiting areas of low water depth and low current such as: *Jupiaba, Guianacara, Rhineloricaria, Pimelodella, Corydoras*, etc. In deep areas we found large species such as: *Ageneiosus, Doras, Myleus* and *Serrasalmus*, which are included in the local inhabitants diet. The results showed a moderate to low relative abundance.

AC03. Kakada River and Suajiditu Beach (Field Stations: ICT-03, ICT-04) (5°29.86´N–64°34.76´W).

General fish fauna description

Sandy beach (ICT-03). A total of 151 specimens were captured placed in 12 species:

Characiformes: 10 (83.3%)
Perciformes: 2 (16.6%)

Moenkhausia collettii (47, 31.1%), *Bryconops* sp. (45, 29.8%), *Moenkhausia* cf. *lepidura* B (14, 9.2%), *Jupiaba zonata* (13, 8.6%) y *Hemigrammus* sp.(11, 7.3%) are the most common species. Also we obtained specimens of: *Bryconops* cf. *colaroja, Characidium* sp. A, *Crenicichla saxatilis, Guianacara geayi, Hemigrammus* sp. B, *Jupiaba atypindi, Moenkhausia* cf. *lepidura* C, *Moenkhausia collettii* and *M. oligolepis*.

Rapids in main river channel (ICT-04). A total of 31 specimens placed in 9 species, all Characiformes.

The ichthyological results showed a relativly low richness and abundance. Groups were mainly characoids typical from sandy beaches, such as: *Moenkhausia colletti* (10, 32%), *Knodus* cf *victoriae* (5, 16%), *Hemigrammus* sp. (5, 16%), and *Moenkhausia* cf. *lepidura* (5, 16%).

AC04. Caño Suajiditu (Field Station: ICT-05) (5°29.59´N–64°35.16´W).

General fish fauna description

A total of 447 specimens were captured placed in 22 species distributed in:

Characiformes: 13 (59%)
Siluriformes: 5 (22.7%)
Perciformes: 3 (13.6%)
Synbranchiformes: 1 (4.5%)

The analysis showed a low to middle richness. Abundant specimens of characids including a possible new species of *Aphyocharax* with an exuberant color pattern. The most common species were: *Moenkhausia collettii* (256, 57.3%), *Characidium* sp. A (31, 6.9%), *Phenacogaster* sp. A (31, 6.9%), *Moenkhausia oligolepis* (25, 5.6%) and Hemigrammus sp. (15, 3.3%). Also we have specimens of: *Aequidens* cf. *chimantanus, Corydoras boehlkei, Cyphocharax* sp., *Jupiaba atypindi, J. zonata, Moenkhausia* cf. *lepidura, Rineloricaria fallax, Synbranchus marmoratus* and *Ituglanis metae*. Presence of juveniles from several species indicates that this region could be an important nursery area.

AC05. Raudal del Perro, Caño Wididkenu (Field Stations: ICT-07, ICT-08, ICT-09) (5°54.08´N–64° 29.4´W; 5°54.17´N–64° 29.4´W; 5°53.8´N–64° 28.8´W).

General fish fauna description

Rapids (ICT-07). A total 493 specimens were captured belonging to 16 species and distributed in:

Characiformes: 14 (87.5%)
Siluriformes: 1 (6.2%)
Perciformes: 1 (6.2%)

The most common species were: *Moenkhausia* cf. *Lepidura* (171, 34.7%), *Jupiaba atypindi* (161, 32.6%), *Bryconops* sp. (39, 7.9%), *Jupiaba zonata* (36, 7.3%) and *Melanocharacidium depressum* (16, 3.2%). We captured species common to fast waters and sandy bottoms such as: *Bryconops* (2 species), *Jupiaba zonata, Knodus victoriae, Melanocharacidium depressum, Moenkhausia oligolepis* y *Satanoperca* sp.

Sandy beach in island (ICT-08). A total of 220 specimens belonging to 20 species and distributed in:

Characiformes: 15 (75%)
Siluriformes: 4 (20%)
Perciformes: 1 (5%)

The most abundant species were: *Jupiaba atypindi* (73, 33.2%), *Melanocharacidium depressum* (49, 22.2%), *Moenkhausia* cf. *lepidura* (37, 16.8%), *Characidium* sp. A (11, 5%) and *Bryconops* cf. *colaroja* (8, 3.6%). Also were sampled specimens of: *Anostomus anostomus, Leporinus arcus, L. grandti, Imparfinis, Chaetostoma vasquezi* and *Rineloricaria fallax*.

Caño (ICT-09). A total of 38 specimens belonging to 15 species and distributed in:

Characiformes: 9 (60%)
Perciformes: 4 (26.6%)
Siluriformes: 1 (6.6%)
Synbranchiformes: 1 (6.6%)

The most common species were: *Moenkhausia collettii* (11, 28.9%) and *Hemigrammus* cf. *guyanensis* (7, 18.4%). Also we captured: *Aequidens chimantanus, Bryconops giacopinii, Characidium* sp., *Crenicichla alta* and *Guianacara geayi,*

The icthyological analysis showed that the sandy beach presents low abundance and richness. Only few species of caracoids represented by: *Moenkhausia*: *M. oligolepis, M. lepidura, Moenkhausia* sp1, *Moenkkausia* sp2, and abundant specimens of *Jupiaba* (cf. *zonata*) that showed a beautiful color pattern with the tip of the mouth yellow, greenish body and red fins. Finally, a *Bryconops* sp. similar to *B. colaroja.* Additionally there were some specimens of Curimatidae.

The rapids or raudales have better representation with the fishes are associated to the aquatic vegetation. Samples of three species of Anostomidae (*Anostomus anostomus, Leporinus arcus* and *Leporinus* cf. *grandti*) were obtained. We also collected specimens of Pimelodidae (*Inparfinis*) and Loricariidae such as: *Rineloricaria fallax* and *Chaetostoma vasquezi.*

AC06. Area close to Caño Wididikenu (Field Station: ICT-10) (5°53.9´N–64°28.7´W).

General fish fauna description

A total of 392 specimens belonging to 29 species and distributed in:

Characiformes: 19 (65.5%)
Siluriformes: 5 (17.2%)
Perciformes: 3 (10.3%)
Atheriniformes: 1 (3.4%)
Synbranchiformes: 1 (3.4%)

The most common species were: *Corydoras boehlkei* (103, 26.2%), *Moenkhausia oligolepis* (70, 17.9%), *M. collettii* (66, 16.8%), *Corydoras* sp. (32, 8.2%), and *Guianacara geayi* (28, 7.1%). Also samples from: *Aphyocharax* sp., *Bryconops* (2 species), *Characidium* sp., *Crenicichla alta, Jupiaba atypindi, Melanocharacidium depressum, Moenkhausia lepidura, Pimelodella* sp., *Serrasalmus rhombeus, Synbranchus marmoratus* and *Ituglanis metae.*

We made captures in two different habitats. The results showed a low abundance and richness. In the caño we captured specimens of the Family Cichlidae, among them *Guianacara geayi.* Also some species of Characidae, such as: *Bryconops* cf. *colaroja* and juveniles of other caracoids. Toward the head of the caño, where water is transparent, and the bottom with rocks, only few specimens were captured. In the rapids we collected species similar to those of the last field station (Anostomidae, Loricariidae and Characidiidae). Also, we obtained a specimen of *Serrasalmus rhombeus* captured with a gillnet in a area with slow current. In the rocks the capture was very difficult and only few *Hemigrammus minimus* were obtained.

AC07. Rapids in Erebato River and caño (Field Station, ICT-11) (5°52.7´N–64°29.56´W)

General fish fauna description

A total of 113 specimens belonging to 18 species and distributed in:

Characiformes: 16 (88.8%)
Siluriformes: 1 (5.5%)
Perciformes: 1 (5.5%)

The most abundant species were: *Jupiaba atypindi,* (24, 21.2%), *Jupiaba* sp. A (17, 15%), *Melanocharacidium depressum* (16, 14.2%), *Jupiaba* sp. B (14, 12.4%) and *Moenkhausia* cf. *lepidura* (12, 10.6%). Also were present: *Moenkhausia miangi, Guianacara geayi, Bryconops* cf. *colaroja, Leporinus arcus, L.* cf. *grandti* and *Rineloricaria fallax.*

The ichthyological analysis showed low abundance and richness. However, the fished captured showed a close association with the aquatic plants. In the rapids again we captured species of Anostomidae (*Leporinus arcus* y *L. grandti*), *Moenkhausia* and other caracoids. Also we sampled species of Characidiidae (*Characidium* and *Melanocharacidium*) and Loricariidae (*Rineloricaria fallax*). We noticed that the rooted aquatic plants were bigger and covered more extensive areas than the ones observed in previous field stations.

AC08. Raudal Cejiato-Soodu (Field Stations: ICT-12, ICT-13, ICT-14, ICT,15) (5°33.47´N–64°18.8´W).

General fish fauna description

Beaches (ICT-12): We sampled two beaches placed inside bahia Cejiato. A total of 236 specimens were captured, belonging to 28 species and distributed in:

Characiformes: 19 (67.9%)
Siluriformes: 5 (17.9%)
Perciformes: 3 (10.7%)
Synbranchiformes: 1 (3.5%)

The most abundant species were: *Corydoras osteocarus,* (82, 34.7%), *Bryconops giacopinni* (39, 16.5%), *Pimelodella* sp. B (28, 11.9%), *Guianacara geayi* (26, 11%) and *Moenkhausia* cf. *lepidura* (25, 10.6%). Also specimens from the following species were captured *Moenkhausia collettii, M oligolepis, Knodus victoriae, Jupiaba zonata, Crenicichla saxatilis Bryconops* cf. *colaroja, Hemiodus goeldii, Tetragonopterus* sp. and *Cyphocharax festivus.*

Both beaches have similar conditions with moderate abundance and richness. Fish community showed individuals generally associated to bahias or to sandy bottoms that contains large amounts of detritus, such as cichlids. Also, there are some elements associated to sandy areas such as *Corydoras.* Finally, there were some pelagic species such as three sympatric species of *Bryconops* and *Hemiodopsis.*

Mouth of Tributary (ICT-13): A total of 159 specimens were captured, belonging to 25 species and distributed in:

Characiformes: 16 (64%)
Siluriformes: 6 (24%)
Perciformes: 2 (8%)
Synbranchiformes: 1 (4%)

The most common species were: *Knodus victoriae,* (77, 48.4%), *Moenkhausia collettii* (21, 13.2%), *Bryconops* cf. *colaroja* (18, 11.3%), *Moenkhausia* cf. *lepidura* (12, 7.5%) and *Jupiaba zonata* (6, 3.8%). Also we captured: *Brycocops giacopini, Corydoras* sp., *Cregrutus* sp., *Crenicichla saxatilis, Hypostomus* sp., *Melanocharacidium melanopteron, Myleus rubripinnis* (Juvenile), *Rineloricaria fallax* and *Synbranchus marmoratus.*

Upper tributary (ICT-14): A total of 20 specimens were captured belonging to 9 species and distributed in:

Characiformes: 5 (55.5%)
Siluriformes: 2 (22.2%)
Perciformes: 2 (22.2%)

The most abundant species were: *Moenkhausia collettii,* (11, 55%), *Crenicichla saxatilis* (2, 10%). Specimens were also collected from: *Aequidens* cf. *chimantanus, Melanocharacidium melanopteron* y *Moenkhausia* cf. *lepidura.* We also sampled a specimen of *Tatia,* associated with a submerged log habitats.

Rapids and sandy beach in island (ICT-15): Species characteristic of fast waters (riacophyls) and associated to Podostemonaceae. Sampling the area was extremely difficult. A total of 28 specimens were captured belonging to 7 species, all Characiformes.

The most abundant species were: *Jupiaba atypindi,* (13, 46.4%), *Knodus victoriae* (7, 25%), and *Moenkhausia* cf. *lepidura* (4, 14.3%). Also captured were: *Jupiaba zonata, Creagrutus* sp. and *Acestrorhynchus microlepis.*

In general the ichthyology of Raudal Cejiato showed a comparativly greater richness and abundance than of previous stations. Many species are associated with particular habitats. Although, some individuals of *Bryconops* were found in beaches and at the mouth of the tributary, the great majority came from open and deep waters in the bahia. Similarly, *Corydoras* were collected mainly on sandy bottoms along the beaches with some individuals found at the mouth of the tributary. An important data and we have no clue so far is the great size of the individuals of *Moenkhausia oligolepis* (not seen in any collection so far). Although this species is very common in the Orinoco basin, it is not generally associated with the upper waters. It is important to note that several juveniles of "pacus" (*Myleus* cf *rubripinnis*), could indicate that also the area is a nursery for this species.

At the mouth of the tributary several characid and cichlid species were captured such as: *Bryconops* cf. *colaroja, B. giacopini, Crenicichla saxatilis, Jupiaba zonata, Knodus victoriae, Moenkhausia collettii, Moenkhausia* cf. *lepidura,* and the loricarids: *Hypostomus* sp., and *Rineloricaria fallax.*

Upriver, the tributary becomes narrow and deeper with numerous pools covered by logs and sandy bottoms. The common groups captured were: *Farlowella, Ancistrus,* and *Crenicichla.* Also some microcharacids.

Due to the extreme water velocity in the rapids, few individuals were captured and the great majority of them came from isolated pools in rocks.

AC09. Raudal Pauji (Field Stations: ICT-16, ICT-17, ICT-18, ICT-19, ICT-20) (5°49.7´N–64°24.3´W).

General fish fauna description

Beaches and backwater (ICT-16). A total of 178 specimens were captured belonging to 9 species and distributed in:

Characiformes: 8 (88.8%)
Perciformes: 1 (11.1%)

The most common species were: *Cyphocharax festivus,* (83, 46.6%), *Moenkhausia lepidura* (78, 43.8%) and *Bryconops* sp. A (4, 2.3%). Also we captured specimens of: *Knodus* cf. *victoriae, Aphyocharax* sp. and *Moenkhausia collettii.* Low relative abundance and richness. The majority of the species typically from sandy beach habitats.

Rapids (ICT-17). A total of 50 specimens were captured belonging to 12 species and distributed in:

Characiformes: 9 (75%)
Siluriformes: 3 (25%)

The most abundant species were: *Melanocharacidium depressum* (23, 46%), *Characidium* sp. A (6, 12%), *Hypostomus* sp. A (5, 10%) and *Jupiaba atypindi* (3, 6%). Also we collected specimens from: *Apareiodon orinocensis, Cetopsorhamdia* cf. *picklei* and *Melanocharacidium dispiloma.* Area species of the area typically from rapid habitats and with aquatic plants such as: Characidiidae, Parodontidae and Loricariidae. A interesting species of Pimelodidae (*Cetopsorhamdia* cf. *picklei*) were present. Low relative abundance and diversity.

Rocky Beach in island (ICT-18). A total of 14 specimens were captured belonging to 11 species and distributed in:

Characiformes: 8 (72%)
Perciformes: 2 (18%)
Siluriformes: 1 (9%)

The most common species were: *Crenicichla saxatilis* (2, 14%), *Jupiaba zonata* (2, 14%) and *Moenkhausia* cf. *lepidura* (2, 14%). Also present were: *Crenicichla saxatilis, Anostomus anostomus* and *Melanocharacidium depressum.* Low abundance and richness.

Sandy beach (ICT-19). A total of 52 specimens were captured, belonging to 9 species and distributed in:

Characiformes: 8 (88%)
Perciformes: 1 (12%)

The most abundant species were: *Moenkhausia* cf. *lepidura* (28, 54%), *Knodus* cf. *victoriae* (10, 19.2%) and *Bryconops* sp. A (3, 5.7%). Also we have samples of: *Aphyocharax* sp., *Creagrutus* sp., *Hemiodus unimaculatus* and *Satanoperca* sp. A. The majority of the species are typically from fast waters and sandy-gravel bottoms. Low abundance and diversity.

Backwater (ICT-20). A total of 201 specimens were captured belonging to 16 species, and distributed in:

Characiformes: 13 (81%)
Perciformes: 3 (19%)

The most common species identified were: Characidae (N.I) (99, 49%), *Guianacara geayi* (25, 12%), *Moenkhausia* cf. *lepidura* (21, 10.3%), *Geophagus* sp. (14, 7%), *Moenkhausia colletti* (14, 7%) and *Jupiaba zonata* (9, 4.4%). Also we have samples from: *Aphyocharax* sp., *Brachychalcinus opercularis, Bryconops* cf. *colaroja, B. giacopini, Jupiaba atypindi* and *Crenicichla alta.* Also there were juveniles of several species of Cichlidae and Characidae that could indicate that the area acts as a nursery habitat for those species. This area has greater abundance and diversity.

In summary we could indicate that the Pauji rapids has low abundance and diversity. Although we sampled several different habitats low species counts and specimens were present. The species captured in the rapids are associated to the aquatic plants and fast moving waters such as: *Characidium*, *Melanocharacidium*, *Apareiodon* (2 species), inparfinis, and *Hypostomus.* In the backwaters or bahias we captured several specimens of *Guianacara geayi* and a good ontogenetical series. Also we collected interesting species such as: *Poptella longipinnis*, *Bryconops* (2 species possibly news), *Moenkhausia* cf *lepidura* (several morph types) and abundant juveniles of cichlids and characids. This last data indicates recent reproductive activity in this area. In sandy beaches we captured species that could be of taxonomic importance such as the possibility of a new species of *Aphyocharax*, several curimatids and hemiodontids.

AC10. Caño Jasa Kanu (Field Stations: ICT-21, ICT-22) (5°53.02´N–64°24 a 25.6´W).

General fish fauna description

Caño (ICT-22). A total of 15 specimens were captured belonging to 5 species and distributed in:

Perciformes: 2 (40%)
Siluriformes: 2 (40%)
Characiformes: 1 (20%)

The species were: *Bryconops giacopini* (7, 46.6), *Corydoras boehlkei* (3, 20%), *Corydoras* sp. (2, 13.3%), *Guianacara geayi* (2, 13.3%) and *Aequidens chimantanus* (1, 6.6%). All species typically found in mouth of tributaries such as *Bryconops*, and protected areas such as *Guianacara geayi.* The area showed low abundance and diversity.

AC11. Backwater above Salto Pará (Field Station: ICT-23) (6°16.9´N–64°29.2´W).

General fish fauna description

Although the station had a variety of habitats we only sample fish at the rapids.

Raudal (ICT-23). A total of 429 specimens were captured belonging to 34 species distributed in:

Characiformes: 27 (79.4%)
Perciformes: 3 (8.8%)
Atheriniformes: 1 (2.9%)
Siluriformes: 1 (2.9%)

The most abundant species were: *Moenkhausia* cf. *lepidura* (108, 25.2 %), *Jupiaba zonata* (93, 21.7%), *Jupiaba atypindi* (74, 17.82%), *Cyphocharax* sp. (25, 5.8%) and *Cyphocharax festivus* (19, 4.4%). Also present were: *Acestrorhynchus falcatus, Apistogramma* sp., *Brachychalcinus opercularis, Corydoras boehlkei, Guianacara geayi, Hemiodus unimaculatus, Hoplias macrophthalmus, H. malabaricus, Moenkhausia collettii, M.*

grandisquamis, M. oligolepis, Poptella longipinnis, Rhineloricaria fallax and one species of *Rivulus.*

The icthyological analysis indicates several species typically of riacophylic environments mixed with species associated with protected areas. We note the presence of predators such as *Hoplias macrophthalmus, H. malabaricus* and *Acestrorhynchus* cf. *falcatus* as well as several ornamental species such as: *Moenkhausia* (3 spp.), *Jupiaba zonata, Bryconops, Curimata* (2 spp), *Poptella longipinnis, Hemiodus* sp, *Leporinus* sp, *Prochilodus mariae* and *Guianacara geayi.* Finally, we observe the presence of a rare species of *Rivulus* (possibly new). The relative abundance and diversity were moderated to high in this station.

AC12. Smooth-rocky beach, boulders and backwater and caño above Salto Pará (Field stations: ICT-24, ICT-25, ICT-26) (6°16.9´N–64°29.2´W).

General fish fauna description

Smooth-rocky beach (boulders) (ICT-25). A total of 359 specimens were captured belonging to 30 species and distributed in:

Characiformes: 23 (76.6%)
Siluriformes: 4 (13.3%)
Perciformes: 3 (10%)

The most common species were: *Jupiaba zonata* (205, 57.1%), *J. atypindi* (30, 8.4%), *Apareiodon* sp. (21, 5.8%), *Chaetostoma vasquezi* (16, 4.4%) and *Jupiaba* sp. B. (13, 3.6%). Specimens of the following species also were collected: *Crenicichla alta, Ctenobrycon spilurus, Guianacara geayi, Harttia* sp., *Hemiodus unimaculatus, Hoplias macrophthalmus, Leporinus arcus,* several morphotypes of the *Moenkhausia* "*lepidura*" group, *Moenkhausia cotinho, M. oligolepis, Myleus rubripinnis, Poptella longipinnis* and *Rineloricaria fallax.*

In this field station we captured species closely associated with rocky bottoms such as loricariids *Chaetostoma vasquezi, Harttia* sp., and *Rineloricaria fallax* of these, *Harttia* sp., is new and is in the process of being described (Provenzano, pers. com). Also species such as *Apareiodon* sp. (a possible new species, Machado-Allison, pers. com) were common. We also notice the presence of several species commonly found in other areas such as: *Jupiaba, Leporinus, Hemiodopsis, Astyanax, Rhineloricaria, Guianacara* y *Moenkhausia.* The area showed moderated diversity and abundance.

Backwater (ICT-26). A total of 222 specimens were captured, belonging to 13 species and distributed in:

Characiformes: 8 (61.5%)
Siluriformes: 2 (15.3%)
Perciformes: 2 (15.3%)
Gymnotiformes: 1 (7.7%)

The most common species were: *Jupiaba zonata* (120, 54%), *Moenkhausia* cf. *lepidura* (29, 13%), *Guianacara geayi* (27, 12.1%), *Cyphocharax* sp. (22, 9.9%) and *Moenkhausia* sp. (10, 4.5%). Also were collected: *Crenicichla saxatilis, Ctenobrycon spilurus, Hypostomus* sp., *Myleus rubripinnis* and *Pimelodella* sp.

Several species collected are typically associated with protected habitats such as: *Guianacara geayi* and *Crenicichla saxatilis.* We also note the first specimen of an electric fish of the genus *Gymnotus* (*G. carapo*). The area showed low diversity y moderated abundance.

AC13. Island across Yuruani River (Field Station ICT-27) (6°8.4´N–64°25.1´W).

General fish fauna description

Backwater in island (ICT-27). A total of 62 specimens were captured belonging to 14 species and distributed in:

Characiformes: 11 (78.5%)
Siluriformes: 1 (7.1%)
Perciformes: 2 (14.2%)

The most abundant species were: *Moenkhausia* cf. *lepidura* (27, 43.5%), *Pimelodella* sp. (14, 22.6%), *Guianacara geayi* (4, 6.5%) and *Jupiaba zonata* (3, 4.8%). Other species captured include: *Bryconops giacopini, Crenicichla alta, Cyphocharax festivus, Jupiaba atypindi, J. polylepis, Knodus victoriae, Poptella longipinnis* and *Rineloricaria fallax.*

The species collected are typically associated with protected habitats such as: *Guianacara* and *Crenicichla,* or inundated forests with sandy beaches: *Moenkhausia, Knodus, Poptella* and *Bryconops.* The area showed moderated diversity and low abundance.

AC14. Raudal Culebra de Agua (Field stations: ICT-28, ICT-29, ICT-30, ACT-31).

General fish fauna description

Beach (ICT-28). A total of 174 specimens were captured belonging to 16 species and distributed in:

Characiformes: 10 (62.5%)
Siluriformes: 4 (25%)
Perciformes: 2 (12.5%)

The most common species were: *Corydoras boehlkei* (43, 24.7%), *Jupiaba zonata* (31, 17.8%), *Pimelodella* sp. B. (21, 12.1%), *Moenkhausia* cf. *lepidura* (16, 9.2%) and *Guianacara geayi* (14, 8.4%). Also collected: *Bryconops* sp. A, *Cyphocharax festivus, Jupiaba atypindi, Knodus victoriae, Moenkhausia collettii, M. oligolepis, Rinbeloricaria fallax* and *Satanoperca* sp. A.

The area includes species associated to sandy beaches such as: *Corydoras boehlkei, Cyphocharax festivus, Jupiaba atypindi, Knodus victoriae.* However, there is also species found in protected areas like backwaters such as: Guianacara geayi, Satanoperca sp. and Moenkhausia oligolepis.

Rapids (ICT-29). A total of 61 specimens were captured, belonging to 16 species distributed in:

Characiformes: 11 (68.7%)
Siluriformes: 4 (25%)
Perciformes: 1 (6.3%)

The most common species were: *Jupiaba atypindi* (15, 24.6%), *Jupiaba zonata* (9, 14.7%), *Melanocharacidium depressum* (7, 11.5%), *Chaetostoma vasquezi* (6, 9.8%) and *Bryconops* sp. A. (5, 8.2%). Also collected: *Ancistrus* sp., *Bryconops giacopini, Characidium* sp. A., *Guianacara geayi,*

Hypostomus sp. A, *Leporinus grandti, Rineloricaria fallax* and *Tetragonopterus* sp.

The majority of the species are associated to aquatic plants such as: *Characidium, Leporinus Melanocharacidium depressum* and *Hypostomus.* Some species associated to submerged logs such as: *Chaetostoma* and *Rineloricaria.* Relative abundance and diversity low.

Tributary (ICT-30). A total of 20 specimens were captured belonging to 9 species and distributed in:

Characiformes: 5 (55.5%)
Perciformes: 3 (33.3%)
Siluriformes: 1 (11.1%)

The most abundant species were: *Moenkhausia collettii* (8, 40%), *Corydoras boehlkei* (2, 10%), *Aphyocharax* sp. (2, 10%) and *Characidium* sp. A. (2, 10%). Also captured: *Aequidens chimantanus, Crenicichla saxatilis, Guianacara geayi, Jupiaba zonata* and *Moenkhausia* cf. *lepidura.*

The species are typically associated to protected areas or inundated forests such as: *Guianacara geayi, Crenicichla saxatilis* and *Moenkhausia collettii.* Others are commonly found in sandy beaches such as: *Corydoras boehlkei* and *Characidium.* Low abundance and diversity.

Rapids above Raudal Culebra (ICT-31). A total of 13 specimens were captured belonging to 6 species all Characiformes. The species were: *Hoplias macrophthalmus, Jupiaba zonata, Melonocharacidium depressum, Moenkhausia* cf. *lepidura* (2 sp) and *Moenkhausia miangi.* All specimens captured associated to the aquatic plants. Low diversity and abundance.

II. LOWER CAURA, BELOW SALTO PARÁ

BC01. El Playón (Salto Pará) (Field stations: ICT-32, ICT-33, ICT-35, ICT-36, ICT-38) (6°19.5′ N–64°31.6′ W).

General fish fauna description

Sandy beach (El Playón) (ICT-32, ICT-33). We collected fish using three different kind of fishing equipent: Cast nets (ICT-32), beach seines (ICT-33) and minnow traps. The combination of the three showed the capture of 519 specimens belonging to 37 species and distributed in:

Characiformes: 27 (73%)
Siluriformes: 6 (16.2%)
Perciformes: 2 (5.4%)
Rajiformes:1 (2.7%)
Gymnotiformes: 1 (2.7%)

The most common species were: *Knodus* sp. B. (106, 20.4%), *Pimelodella* cf. *megalops* (91, 17.5%), *Anostomus ternetzi* (83, 15.9%) and *Hyphessobrycon minimus.* (23, 4.4%). Also we have samples from: *Brycon pesu, Bryconamericus cismontanus, Caenotropus laberinthicus, Creagrutus maxillaris, Curimata incompta, Cynopotamus essequibensis, Geophagus* cf. *brachybranchus, Eigenmannia macrops, Hemisorubim platyrhynchus, Homodiaetus* sp., *Microschemobrycon callops, M. casiquiare, Moenkhausia gracilima, Pimelodella cruxenti, Potamorhina altamazonica, Psectrogaster essequibensis, Schizodon* sp., *Semaprochilodus kneri, Serrasalmus* sp. A., *Synaptolaemus cingulatus, Tetragonopterus chalceus, Triportheus albus* and one parasitic catfish of the genus *Paravandellia* sp.

The area showed a mixture with species associated to backwaters such as: *Microschemobrycon callops, M. casiquiare, Moenkhausia gracilima, Geophagus* cf. *brachybranchus,* or sandy beaches: *Brycon pesu, Bryconamericus cismontanus, Caenotropus laberinthicus, Creagrutus maxillaris, Corydoras, Characidium, Curimata incompta, Cynopotamus essequibensis, Eigenmannia macrops, Hemisorubim platyrhynchus, Homodiaetus* sp., *Pimelodella cruxenti, Potamorhina altamazonica, Psectrogaster essequibensis, Schizodon* sp. This area is important for its the richness and relative abundance. The presence of parasitic catfish indicates the presence of large pimelodid catfishes. There are also food fishes, such as: *Semaprochilodus kneri* and *Piaractus brachypomus.* We observe fishermen capturing specimens of *Piaractus brachypomus.* Usind minnow traps, we collected abundant specimens of two species of anostomids: *Anostomus ternetzi* and *Synaptolaemus cingulatus. Synaptolaemus* is a very rare genus.

Island (ICT-35). A total of 344 specimens were captured belonging to 35 species, distributed in:

Characiformes: 18 (51.4%)
Siluriformes: 10 (28.6%)
Perciformes: 2 (5.7%)
Gymnotiformes: 1 (2.8%)

The most abundant species were: *Knodus* sp. B. (75, 21.8%), *Moenkhausia* cf. *lepidura* (67, 19.5%), *Moenkhausia cotinho* (60, 17.4%), *Pimelodella megalops* (27, 7.8%) and *Nanoptopoma spectabilis.* (25, 7.3%). Also captured: *Ancistrus* sp, *Anostomus ternetzi, Aphyocharax alburnus, Astyanax integer, Bryconamericus cismontanus, Bryconops* cf. *colaroja, Bunocephalus* cf. *aleuropsis, Cregrutus maxillaris, Crenicichla wallacei, Cyphocharax modestus, Farlowella vittata, Jupiaba atypindi, J. polylepis Microglanis iheringi, Moenkhausia copei, M. grandisquamis, Ochmacanthus alternus, Pimelodella cruxenti, Satanoperca* sp., *Sternopygus macrurus* and *Tatia romani.*

Many of the species are associated with rocky substrates and sandy or gravel bottoms such as: *Ancistrus, Anostomus, Bryconamerius, Knodus, Ochmacanthus, Creagrutus, Nanoptopoma, Bryconamericus.* Others are commonly found in protected habitats, such as: *Anostomus ternetzi, Aphyocharax alburnus, Astyanax integer, Bunocephalus, Moenkhausia* cf. *lepidura, M. cotinho, M. copei, M. grandisquamis, Pimelodella megalops, P. cruxenti, Satanoperca, Sternopygus macrurus* and *Tatia romani.* Moderate to high diversity and abundance.

Rocky beach (ICT-36). A total of 238 specimens were captured belonging to 13 species, all distributed in the Order Characiformes.

The most common species were: *Moenkhausia cotinho* (91, 38.2%), *Moenkhausia* cf. *lepidura* (82, 34.4%), *Jupiaba atypindi* (41, 17.2%), *Anostomus ternetzi* (7, 2.9%) and *Bryconamericus cismontanus* (3, 1.2%). Also were collected specimens from: *Apareiodon orinocensis, Bryconops giacopinii, Creagrutus maxillaris, Cyphocharax modestus* and *Jupiaba polylepis,* proper of areas with logs, roots and abundant detritus. Also there were species associated to fast waters such as *Apareidodon.* Diversity and abundance low.

Rocks (ICT-38). A total of 107 specimens were captured belonging to 10 species, distributed in:

Characiformes: 9 (90%)
Siluriformes: 1 (10%)

The most common species were: *Hyphessobrycon minimus* (97, 90.1%) and *Moenkhausia* cf. *lepidura* (2, 1.8%). Also were collected: *Ancistrus* sp. C, *Anostomus ternetzi, Bryconamericus cismontanus, Curimata incompta, Creagrutus maxillaris, Cynodon gibus, Melanocharacidium dispiloma* and *Triportheus albus.* A fisherman showed captures of two species of *Leporinus* (*L. brunneus* and *L. friderici*) and several specimens of *Piaractus brachypomus* captured with hook and line in the rapids.

BC02. Tributary (Caño) Confluence with the Caura at El Playón (Field Station ICT-34) (6°19.7′N–64°31.6′W).

General fish fauna description.

A total of 720 specimens were captured belonging to 45 species and distributed in:

Characiformes: 27 (60.0%)
Siluriformes: 11 (24.4%)
Perciformes: 4 (8.8%)
Clupeiformes: 1 (2.2%)
Gymnotiformes: 1 (2.2%)
Atheriniformes: 1(2.2%)
Synbranchiformes: 1 (2.2%)

The most abundant species were: *Ochmacanthus alternus* (81, 11.3%), *Phenacogaster* sp. A (65, 9%), *Bryconamericus cismontanus* (65, 9%), *Creagrutus maxillaris* (59, 8.2%) and *Moenkhausia cotinho* (59, 8.2%), *Pimelodella megalops* (53, 7.3%), *Hyphessobrycon minimus* (49, 6.8%), *Aphyocharax alburnus* (44, 6.1%). Also specimens from the following species were collected: *Anchoviella jamesi, Anostomus ternetzi, Astyanax integer, Brachyhypopomus occidentalis, Bryconops* sp. B, *Bujurquina mariae, Bunocephalus aleuropsis, Corydoras blochii, Crenicichla* sp., *Cyphocharax festivus, C. modestus, C. oenas, Hoplias macrophthalmus, H. malabaricus, Hypostomus plecostomus, Jupiaba atypindi, J. polylepis, J. zonata, Leporinus maculatus, Limatulichthys punctatus, Loricaria cataphracta, Loricariichthys brunneus, Microschemobrycon casiquiare, Moenkhausia* cf. *lepidura, M. copei, Odontostilbe* cf. *fugitiva, Poecilia* sp., *Ramirezella newboldi, Synbranchus marmoratus* and *Xenagoniates bondi.*

In summary the analysis showed a moderate to high relative abundance and high diversity. Some species were well represented, such as: *Pimelodella, Phenacogaster, Bryconamericus, Creagrutus* y *Ochmacanthus.* The icthyofauna in this station is comparatively one of the highest obtained, with 46 species captured. There are species typically associated with habitats found in the main Caura such as: *Bryconops, Jupiaba, Moekhausia* and *Ramirezella,* and others are common to muddy areas (caños, flooded forests and lagoons) such as: *Apistogramma, Bujurquina, Bunocephalus, Ochmacanthus, Brachyhypopomus, Hyphessobrycon* and *Microglannis.* We notice the first record of a "herring" (*Anchoviella jamesi*) in the Caura.

BC03. Tributary o caño Waki, several km dowstream from Salto Pará (Field Station ICT-37, ICT-38A) (6°21.7′N–64°34.1′W).

General fish fauna description

Caño (ICT-37). A total of 504 specimens were captured belonging to 24 species and distributed in:

Characiformes: 19 (79.2%)
Siluriformes: 2 (8.3%)
Perciformes: 2 (8.3%)
Synbranchiformes: 1 (4.2%)

The most abundant species were: *Microschemobrycon melanotus* (237, 47%), *Moenkhausia* cf. *lepidura* (75, 14.9%), Characidae sp. B (43, 8.6%), *Moenkhausia copei* (41, 8.1%) and *Jupiaba zonata* (32, 6.4%). Also were captured: *Aphyocharax alburnus, Apistogramma* sp. B, *Bujurquina mariae, Characidium* sp. A, *Cyphocharac oenas, C. spilurus, Hyphessobrycon minimus, Jupiaba polylepis, Microchemobrycon casiquiare, Moekhausia cotinho, M. grandisquamis, M. oligolepis, Ochmacanthus alternus, Poptella longipinnis, Rineloricaria* sp. and *Synbranchus marmoratus.*

The analysis of the fish fauna showed the majority of the species from this location are associated with flooded forests, lagoons, or protected habitats with muddy bottoms and abundant detritus.

Mouth of the Caño (ICT-38A). A total of 10 specimens of two species were captured with a gill net: *Hydrolicus tatauaia* (9) and *Serrasalmus rhombeus* (1).

BC04. Tawadu (Tabaro) River Dedemai Camp (Field Station ICT-39, ICT-54, ICT- 55) (6°21.1′N–64°59.9′W).

General fish fauna description

Camp. (ICT-39). A total of 164 species were captured using beach seines and traps. The 14 species identified distributed were in:

Characiformes: 10 (71.4%)
Siluriformes: 2 (14.3%)
Perciformes: 1 (7.1%)
Beloniformes: 1 (7.1%)

The most common species were: *Hyphessobrycon minimus* (126, 76.8%), *Moenkhausia copei* (11, 6.7%), *Moenkhausia* cf. *lepidura* (5, 3%), *M. oligolepis* (5, 3%) and *Knodus* sp. A (5, 3%). Also specimens of the following species were captured: *Anostomus ternetzi, Astyanax integer, Bujurquina mariae, Microschemobrycon melanotus, Moenkhausia collettii, Ochmacanthus alternus, Pimelodella* cf. *cruxenti* and *Potamorrhaphis guianensis.*

Beach (ICT-54). A total of 304 specimens were captured belonging to 22 species and distributed in:

Characiformes: 16 (72.7 %)
Perciformes: 3 (13.6%)
Siluriformes: 2 (9.1%)
Gymnotiformes: 1 (4.5 %)

The most abundant species were: *Hyphessobrycon minimus* (63, 20.7%), *Moenkhausia copei* (60, 19.7%), *Bryconamericus* cf. *cismontanus* (59, 19.1%), *Moenkhausia* cf. *lepidura* (38, 12.3%) and *Aphyocharax alburnus* (15, 4.8%). Also were collected specimes of: *Acestrorhynchus microlepis, Aequidens chimantanus, Bujurquina mariae, Characidium* sp. A, *Crenicichla wallacei, Gymnotus anguilaris, Hoplias macrophthalmus, Jupiaba polylepis, Microschemobrycon melanotus, Moenkhausia collettii, M. oligolepis, Ochmacanthus alternus, Phenacogaster* sp. A, *Poptella longipinnis, Pseudopimelodus raninus* and *Ramirezella newboldi.*

Caño (ICT-35). A total of 39 specimens were captured belonging to 9 species and distributed in:

Characiformes: 8(88.8 %)
Perciformes: 1(11.2%)

The most common species were: *Hyphessobrycon minimus* (16, 41%), *Microschemobrycon melanotus* (9, 23.1%), *Bujurquina mariae* (6, 15.4%), *Microschemobrycon casiquiare* (3, 7.7%). Also collected: *Characidium* sp., C, *Moenhausia* cf. *lepidura, Moenkhausia hemigramoides, M. oligolepis* and *Tetragonopterus chalceus.*

In summary, this area possesses moderate diversity and low abundance. The majority of the species captured are associated to areas with muddy bottoms with debris such as: *Hyphessobrycon minimus, Microschemobrycon* (3 spp.), *M. hemigramoides, M. oligolepis, Ochmacanthus alternus, Poptella longipinnis, Pseudopimelodus raninus* and *Ramirezella newboldi.* A specimen of an electric fish (*Gymnotus anguilaris*) was captured representing the first record of this species for the Caura.

BC05. Tawadu (Tabaro) River, Raudal el Pan (ICT-40) and Raudal Dimoshi ICT-41, ICT-42), (6°19.63´N–65°02.86´W).

General fish fauna description

Raudal El Pan (ICT-40). We used a small beach seine. A total of 90 specimens belonging to 4 species and distributed all in Order Characiformes

The species were: *Hyphessobrycon minimus* (48, 53.3%), *Knodus* sp. A (38, 42.2%), *Microschemobrycon callops* (3, 3.3%) and *Characidium* sp. A (1, 1.1%).

Raudal Dimoshi (ICT-41). A total of 246 specimens were captured belonging to 15 species and distributed in:

Characiformes: 9 (60%)
Perciformes: 3 (20%)
Siluriformes: 2 (13.3%)
Synbranchiformes: 1 (6.6%)

The most common species were: *Hemigrammus* cf. *guyanensis* (113, 45.9%), *Knodus* sp. A (63, 25.6%), *Bryconops* cf. *colaroja* (24, 9.7%), *Bryconops* sp. B (20, 8.1%) and *Hemibrycon metae* (17, 6.9%). Also collected: *Ancistrus* sp. C, *Bujurquina mariae, Characidium* sp., *Crenicichla geayi, Melanocharacidium depressum, Pimelodella* sp. and *Synbranchus marmoratus.* The majority of the species are typically from environments with fast moving waters, sandy beaches and associated to rooted aquatic plants.

Caño (ICT-42). A total of 112 individuals were captured belonging to 5 species and distributed in::

Characiformes: 4 (80%)
Perciformes: 1 (20%)

The species were: *Hemigrammus* cf. *guyanensis* (60, 53.6%), *Knodus* sp. A (38, 33.9%), *Bujurquina mariae* (13, 11.6%), *Astyanax integer* (5, 4.5%) and *Astyanax* sp. (4, 3.6%). This area, although, in apparently good ecological condition, showed a poorly represented fauna, possibly due to the use of "barbasco" by the local people.

Raudal Dimoshi Downstream (ICT-43). We collected this area with hook and line. Two species were sampled: *Crenicichla* cf. *lenticulata* and [illegible] sp. A.

In summary, the fish analysis showed a moderate to low diversity and abundance. The majority of the species associated to fast moving waters, sandy-gravel or rocky bottoms such as: *Ancistrus* sp. C, *Bryconops* (2 sp), *Characidium* sp. *Hemigrammus* cf. *guyanensis, Knodus* sp. A, *Hemibrycon metae Melanocharacidium depressum* and *Pimelodella* sp.

BC06. Tawadu (Tabaro) River. Tajañaño Rapid (ICT-44, ICT-45), (6°20.25´N–65°01.50´W) .

General fish fauna description

Rocky Island and rapids (ICT-44). A total of 284 specimens collected belonging to 8 species and distributed in:

Characiformes: 5 (62.5%)
Siluriformes: 2 (25%)
Perciformes: 1 (12.5%)

The most common species were: *Hyphessobrycon minimus* (223, 78.5%), *Microschemobrycon callops* (26, 9.2%), *Ochmacanthus alternus* (13, 4.6%), *Knodus* sp. B (11, 3.9%) and *Characidium* sp. A (8, 2.8%). Also, the following species were collected: *Astyanax integer, Crenicichla geayi* and *Rineloricaria* sp. The analysis showed a dominance of a single species of Characidae: *Hyphessobrycon minimus*, common to protected habits. Other species such as *Rineloricaria* are commonly found in rocky-gravel beds.

Mouth of the caño and shoreline (ICT-45). A total of 58 specimens were captured belonging to 15 species and distributed in:

Siluriformes: 8 (53.3%)
Characiformes: 5 (33.3%)
Perciformes: 1 (6.6%)
Synbranchiformes: 1 (6.6%)

The most abundant species were: *Ochmacanthus alternus* (15, 25.9%), *Ancistrus* sp. C (14, 24.1%), *Ammocryptocharax elegans* (11, 18.9%), *Farlowella vittata* (6, 10.3%) and *Characidium* sp. A (5, 8.6%). Also were collected specimens of: *Bryconops* sp., *Chasmocranus* sp., *Crenicichla wallacei, Lasiancistrus* sp., *Melanocharacidium nigrum, Microglannis poecilus, M. iheringi, Moenkhausia* cf. *lepidura, Rineloricaria* sp., and *Synbranchus marmoratus.*

In summary the area showed low abundance and diversity. Rapids and island were dominated by a single species (*Hyphessobrycon minumus*). In the caño and left banks Siluriformes associated with submerged logs, branches and muddy bottoms were common. Species captured were species of the following genera: *Ancistrus, Microglannis, Farlowella, Lasiancistrus* and *Rineloricaria.* Also we collected species associated to gravel bottoms and fast waters such as: *Ammocryptocharax elegans, Melanocharacidium nigrum* and *Characidium* sp. A. Other species common to protected habitats or backwaters, such as: *Crenicichla wallacei, Microschemobrycon, callops Microglannis poecilus, M. iheringi* and *Synbranchus marmoratus.*

BC07. Laguna 1 km from mouth Tawadu (Tabaro) River (ICT-46, ICT-51) (6°21.83´N–64°58.4´W).

General fish fauna description

We obtained two samples; one with a gillnet (ICT-46) and another using regular seines (ICT-51). In the first sample we captured a total of 84 individuals belonging to 12 species and distributed in:

Characiformes: 9 (75%)
Siluriformes: 1 (8.3%)
Perciformes: 1 (8.3%)
Synbranchiformes (8.3%)

The most abundant species were: *Hydrolicus tatauaia* (37, 44%), *Semaprochilodus laticeps* (28, 33.3%), *Serrasalmus rhombeus* (3, 3.5%) and *Triportheus albus* (3, 3.5%). Also specimens of: *Boulengerella xirekes, Hydrolicus armatus, Leporinus brunneus, Oxydoras kneri* and *Rhaphiodon vulpinus,* were obtained.

Fishes at this station are large, with low diversity and abundance. There were several specimens of "payaras" (*Hydrolicus tatauaia*) and "sapoaras" (*Semaprochilodus laticeps*). The area shows potential for fisheries.

The sample with beach seines (ICT-51) and gill net (ICT-46) combined for a total capture of 1977 specimens belonging to 50 species distributed in:

Characiformes: 35 (70%)
Siluriformes: 8 (16%)
Perciformes: 4 (8.%)
Synbranchiformes 1 (2%)
Gymnotiformes: 1 (2%)
Pleuronectiformes: 1 (2%)

The most common species were: *Pimelodella chagresi* (913, 46.2%), *Ochmacanthus alternus* (355, 18%), *Bryconamericus* cf. *cismontanus* (191, 9.7%), *Moenkhausia collettii* (69, 3.5%) and *Corydoras* cf. *bondi* (56, 2.8%). Also specimens from the following species were collected: *Acestrorhynchus microlepis, Achirus* sp., *Aphyocharax alburnus, A. erythrurus, Apistogramma* sp. A, *Astyanax integer, Boulengerella xirekes, Bujurquina mariae, Cyphocharax festivus, C. oenas, Hemiodus unimaculatus, Hydrolicus armatus, Hydrolicus tatauaia, Hyphessobrycon minimus, Jupiaba polylepis, Leporinus brunneus, Limatulichthys punctatus, Mesonauta egrerius, Microschemobrycon callops, M. casiquiare, Moenkhausia* cf. *lepidura, M. copei, M. cotinho, M. grandisquamis, M. oligolepis, Odontostilbe fugitiva, Oxydoras kneri, Phenacogaster* sp. A, *Pimelodella cruxenti, P. megalops, Pimelodella* sp., *Poptella longipinnis, Pyrrhulina brevis, Ramirezella newboldi, Rhaphiodon vulpinus, Semaprochilodus laticeps, Serrasalmus rhombeus, Steindachnerina pupula, Synbranchus marmoratus, Tetragonopterus chalceus* and *Triportheus albus.*

The analysis of the combined two stations showed a great abundance and diversity of forms related to lagoon, inundated forest and muddy bottoms. Is important to highligth the presence of commercially important and food fishes such as: *Hydrolicus armatus, Hydrolicus tatauaia, Semaprochilodus laticeps* and *Serrasalmus rhombeus.* Information obtained from local inhabitants indicates that fishermen frequently use this lagoon. Also there are a great variety of ornamental species such as: *Aphyocharax alburnus, A. erythrurus, Apistogramma* sp. A, *Astyanax integer, Boulengerella xirekes, Bujurquina mariae, Hemiodus unimaculatus, Hyphessobrycon minimus, Jupiaba polylepis, Leporinus brunneus, Limatulichthys punctatus, Mesonauta egrerius, Microschemobrycon callops, M. casiquiare, Moenkhausia* cf. *lepidura, M. copei, M. cotinho, M. grandisquamis, M. oligolepis, Poptella longipinnis, Pyrrhulina brevis, Ramirezella newboldi, Tetragonopterus chalceus* and *Triportheus albus.*

BC08. Río Icutú. Caño Wani (ICT-47, ICT-48) (6°04.77´N–64°55.33´W).

General fish fauna description

Sandy beach (ICT-47). A total of 563 specimens were captured belonging to 35 species and distributed in:

Characiformes: 24 (68.5%)
Siluriformes: 6 (17.1%)
Perciformes: 2 (5.7%)
Clupeiformes: 1 (2.8 %)
Pleuronectiformes: 1 (2.8%)
Synbranchiformes: 1 (2.8%)

The most common species were: *Moenkhausia collettii* (186, 33%), *Hyphessobrycon minimus* (73, 12.9%), *Moenkhausia cotinho* (72, 12.8%), *Pseudocheirodon* sp. (39, 6.9%) and *Characidium* sp. A (29, 5.1%). Also obtained: *Achirus* sp., *Anchoviella jamesi, Anostomus ternetzi, Aphyocharax alburnus, Astyanax integer, Bryconamericus* cf. *cismontanus, Bujurquina mariae, Cochliodon* sp., *Creagrutus maxillaris, Farlowella mariaelenae, F. vittata, Jupiaba atypindi, J. polylepis, Microschemobrycon callops, M. casiquiare, M. melanotus, Ochmacanthus alternus, Pimelodella* sp. A, *Ramirezella newboldi, Satanoperca* sp., *Steindachnerina pupula, Synbranchus marmoratus, Tetragonopterus* sp. and *Paravandellia* sp.

Caño (ICT-48). A total of 693 specimens were captured belonging to 41 species and distributed in:

Characiformes: 28 (68.3%)
Siluriformes: 7 (17.1%)
Perciformes: 3 (7.3%)
Gymnotiformes: 2 (4.9%)
Synbranchiformes: 1 (2.4%)

The most abundant species were: *Moenkhausia collettii* (204, 29.4%), *Moenkhausia cotinho* (83, 12%), *Phenacogaster* sp. A (46, 6.6%), *Bryconamericus* cf. *cismontanus* (45, 6.5%) and *Ramirezella newboldi* (42, 6.1%). Also specimens from the following species were sampled: *Aphyocharax alburnus, Apistogramma* sp. A, *Astyanax integer, Brachychalcinus occidentalis, Bujurquina mariae, Characidium* sp.A, *Corydoras* cf. *bondi, Cyphocharax oenas, Hoplias macrophthalmus, Hyphes-*

sobrycon minimus, Jupiaba atypindi, J. polylepis, J. zonata, Loricaria cataphracta, Melanocharacidium dispiloma, Microschemobrycon callops, M. casiquiare, M. melanotus, Moenkhausia cf. *lepidura, M. copei, Ochmacanthus alternus, Odontostilbe* sp., *Pimelodella* sp. A, *Poptella longipinnis, Pseudocheirodon* sp., *Rineloricaria* sp., *Satanoperca* sp., *Sternopygus macrurus, Synbranchus marmoratus* and *Tetragonopterus* sp.

The area showed a great abundance and diversity. Several species are associated to muddy-sandy bottoms such as: *Aphyocharax alburnus, Apistogramma* sp. A, *Astyanax integer, Brachychalcinus occidentalis, Bujurquina mariae, Characidium* sp. A, *Corydoras* cf. *bondi, Cyphocharax oenas, Hoplias macrophthalmus, Hyphessobrycon minimus, Microschemobrycon callops, M. casiquiare, M. melanotus* and *Sternopygus macrurus.* Others are associated with protected habitats: *Bujurquina mariae, Satanoperca* sp., *Poptella longipinnis, Pseudocheirodon* sp., *Rineloricaria* sp., and *Synbranchus marmoratus.* We note the presence of a "sole" *Achirus* sp. which represent the first record in this river basin. This area represents one of the most diverse and rich of all the Nichare River system.

BC09. Laguna Wakawajai (ICT-49, ICT-50) (6°19.35′N–64°57.22′W).

General fish fauna description

Beach (ICT-49). A total of 848 specimens were captured belonging to 33 species and distributed in:

Characiformes: 24 (72.7%)
Siluriformes: 8 (24.2%)
Perciformes: 1 (3.1%)

The most common species were: *Paravandellia* sp.(254, 30%), *Aphyocharax alburnus* (180, 21.2%), *Hyphessobrycon minimus* (110, 12.9%), *Bryconamericus* cf. *cismontanus* (56, 6.6%), *Knodus* sp. B (45, 5.3%) and *Moenkhausia* cf. *lepidura* (39, 4.6%). Also specimens from: *Astyanax integer, Bryconops* sp. B, *Bujurquina mariae, Characidium* sp., *Corydoras bondi, Creagrutus maxillaris, Cyphocharax oenas, Hoplias macrophthalmus, Jupiaba atypindi, J. polylepis, Mastiglannis asopos, Microschemobrycon casiquiare, Moenkhausia copei, M. cotinho, Myleus rubripinnis Ochmacanthus alternus, Pimelodella megalops, Pimelodella* sp. A, *Pimelodella* sp. B, *Pimelodus blochii, Ramirezella newboldi, Serrasalmus rhombeus* and *Vandellia sanguinea* were captured. The abundance of parasitic catfishes such as: *Paravandellia* sp. and *Vandellia sanguinea* is interesting

Lagoon (ICT-50). A total of 326 individuals from 22 species were captured and distributed in:

Characiformes: 17 (77.2%)
Siluriformes: 4 (18.8%)
Perciformes: 1 (4.5%)

The most common species were: *Jupiaba polylepis,* (57, 17.5%), *Moenkhausia cotinho* (51, 15.6%), *Moenkhausia copei* (50, 15.3%), *Bryconamericus* cf. *cismontanus* (50, 15.3%) and *Hyphessobrycon minimus* (45, 13.8%). Also samples of the following species: *Aphyocharax alburnus, Characidium* sp. (2 species), *Corydoras bondi, Cyphocharax meniscaporus, C. oenas, Hemigrammus tridens, Hoplias macrophthalmus, Hypostomus* sp. B, *Loricariichtys brunneus, Mesonauta egregius, Microschemobrycon casiquiare, M. melanotus, Moenkhausia collettii, Ochmacanthus alternus* and *Phenacogaster* sp. A.

The area showed a moderate diversity and low to moderate abundance. As previously mentioned, it is interesting to notice the great abundance of parasitic catfishes. Also there were several species typically associated to sandy bottoms such as: *Characidium* sp. (2 species), *Corydoras bondi, Cyphocharax meniscaporus, C. oenas, Loricariichtys brunneus, Ochmacanthus alternus* and *Phenacogaster* sp. A. Others commonly found in protected areas such as: *Mesonauta egregius, Moenkhausia copei, Bryconamericus* cf. *cismontanus* and *Hyphessobrycon minimus.*

BC10. Island in Nichare River, waters below Boca de Tawadu (ICT-52) (6°21.43′-64°52.29′W).

General fish fauna description

A total of 740 specimens belonging to 44 species were captured and distributed in:

Characiformes: 30 (68.2%)
Siluriformes: 9 (20.5%)
Perciformes: 4 (9%)
Clupeiformes: 1 (2.2%)

The most common species were: *Bryconamericus* cf. *cismontanus* (100, 13.5%), *Hyphessobrycon minimus* (93, 12.6%), *Aphyocharax alburnus* (81, 10.9%), *Moenkhausia* cf. *lepidura* (75, 10.1%) and *Knodus* sp. (73, 9.9%). Also captured: *Anchoviella guianensis, Astyanax integer, Bujurquina mariae, Characidium* sp., *Corydoras bondi, Creagrutus maxillaris, Crenicichla geayi, C. wallacei, Cyphocharax meniscaporus, C. oenas, Hemigrammus* cf. *tridens, Hemiodus unimaculatus,* Iguanodectini (unknown sp.), *Jupiaba atypindi, J. polylepis, Leporinus maculatus, Loricaria cataphracta, Mastiglannis asopos, Mesonauta egrerius, Microschemobrycon callops, M. casiquiare, Moenkhausia* cf. *lepidura* (2 sp.), *M. collettii, M. cotinho, M. grandisquamis, M. oligolepis, Ochmacanthus alternus, Phenacogaster* sp. A, *Pimelodella megalops, Pimelodella* sp. A, *Poptella longipinnis, Pseudohemiodon* sp., *Ramirezella newboldi, Rineloricaria* sp., *Satanoperca* sp., *Steindachnerina pupula, Tetragonopterus* sp. and *Vandellia sanguinea.*

In summary, the area showed moderate abundance and high diversity, with several species associated to rapids with sand or gravel bottoms: *Characidium* sp., *Corydoras bondi, Creagrutus maxillaris, Cyphocharax meniscaporus, C. oenas, Hemigrammus* cf. *tridens, Hemiodus unimaculatus,* Iguanodectini (unknown sp.), *Jupiaba atypindi, J. polylepis, Leporinus maculatus, Loricaria cataphracta, Mastiglannis asopos,* and protected habitats: *Bujurquina mariae, Crenicichla geayi, C. wallacei, Moenkshausia collettii, M. cotinho, M. grandisquamis, M. oligolepis, Pimelodella* sp.A, *Poptella longipinnis, Pseudohemiodon* sp., *Ramirezella newboldi, Rineloricaria* sp. and *Satanoperca* sp.

BC11. Tawadu River Beaches (ICT-53) (6°21.01´-64°58.44´W).

General fish fauna description

A total of 1,100 specimens were captured belonging to 29 species and distributed in:

Characiformes: 24 (82.7%)
Perciformes: 3 (10.3%)
Siluriformes: 2 (6.9%)

The most abundant species were: *Microschemobrycon melanotus* (323, 29.4%), *Corydoras bondi* (148, 13.5%), *Hyphessobrycon minimus* (129, 11.7%), *Moenkhausia collettii* (112, 10.2%) and *Ochmacanthus alternus* (73, 6.6%). Also, specimens of the following species were captured: *Aphyocharax alburnus, Apistogramma* sp. A, *Bryconamericus* cf. *cismontanus, Bujurquina mariae, Characidium* sp. A, *Crenicichla* sp. B, *Cyphocharax oenas, Hemigrammus* cf. *tridens, Hemiodus unimaculatus, Hoplias malabaricus, Jupiaba polylepis, Knodus* sp. B, *Microschemobrycon casiquiare, M. melanotus, Moenkhausia* cf. *lepidura* (2 sp.), *M. copei, M. cotinho, Odontostilbe* cf. *fugitiva, Poptella longipinnis, Ramirezella newboldi,* and *Tetragonopterus* sp.

In summary, the beaches in the Tawadu showed relative abundance, however, the diversity is moderate to low, dominated by species associated with muddy bottoms.

BC12. Ceiba Lagoon (ICT-56) (7°05.9´-65°01.34´W).

General fish fauna description

A total of 811 specimens were captured belonging to 36 species and distributed in:

Characiformes: 26 (72.2%)
Siluriformes: 7 (19.4%)
Perciformes: 2 (5.5%)
Atheriniformes: 1 (2.7%)

The most common species were: *Moenkhausia collettii* (326, 40.2%), *M.* cf. *chrysargyrea* (145, 17.9.%), *Microschemobrycon melanotus* (76, 9.4%), *Aphyocharax erythrurus* (46, 5.7%) and *Ramirezella newboldi* (35, 4.3%). Also: *Acestrorhynchus microlepis, Aequidens* cf. *chimantanus, Apistogramma iniridae, Apistogramma* sp. A, *Bryconamericus* cf. *cismontanus, Crenicichla lenticulata, Crenicichla* sp. A, *Cyphocharax oenas, Hemigrammus* cf. *tridens, Hemigrammus marginatus, Hoplias macrophthalmus, H. malabaricus, Hyphessobrycon bentosi, H. serpae, Jupiaba polylepis, Microschemobrycon melanotus, Moenkhausia* cf. *lepidura, M. copei, M. oligolepis, Nannostomus erythrurus, Orinocodoras eigenmanni, Phenacogaster* sp. A, *Platydoras armatulus, Prochilodus mariae, Pseudocheirodon* sp., *Pygocentrus cariba, Ramirezella newboldi, Rivulus* sp. A, *Satanoperca* cf. *mapiritensis, Serrasalmus elongatus,* and *Serrasalmus* sp. A.

In summary, the area showed a moderate to high abundance and diversity. We obtained the first samples of "red piranha" (*Pygocentrus cariba*). Two different species of "pinches" (*Serrasalmus elongatus* and *Serrasalmus* sp.A) and a large individuals of "guabina" (*Hoplias malabaricus*). There were also species associated to lagoons, such as the cichlids (*Apistogramma iniridae, Crenicichla* sp., *Satanoperca mapiritensis*), characoids (*Hyphessobrycon bentosi, H. serpae, Moenkhausia copei, M. oligolepis, Nannostomus erythrurus*) and siluroids (*Orinocodoras eigenmanni* and *Platydoras armatulus*).

BC13. Mato River (ICT-58, ICT-59, ICT-60) (7°11.74´-65°09.75´W).

General fish fauna description

The area was sampled using a gill net (ICT-58) and conventional seines (ICT-59 and ICT-60). The gillnet capture 15 individuals belonging to 6 species distributed in:

Characiformes: 5 (83.3%)
Clupeiformes: 1 (16.7%)

The species were: *Triportheus albus* (7, 46.6%), *Curimata cyprinoides* (3, 20%), *Pellona castelneana* (2, 13.3%), *Acestrorhynchus heterolepis* (1, 6.6%), *Pygocentrus cariba* (1, 6.6%) and *Serrasalmus rhombeus* (1, 6.6%).

Capture with beach seines at the beach (ICT-59) resulted in 590 specimens belonging to 46 species and distributed in:

Characiformes: 34 (73.9%)
Siluriformes: 7 (15.2%)
Perciformes: 4 (8.7%)
Pleuronectiformes: 1 (2.0%)

The most abundant species were: *Aphyocharax alburnus* (150, 25.4%) *Moenkhausia* sp. A (100, 16.9%), *Hyphessobrycon minimus* (68, 11.5%), *Moenkhausia* cf. *lepidura* A (30, 5.1%), *Moenkhausia* cf. *lepidura* E (30, 5.1%) and *Knodus* cf. *breviceps* (27, 4.5%). Also samples form the following species were captured: *Achirus* sp., *Ancistrus* sp. C, *Anostomus ternetzi, Apistogramma* sp. A, *Astyanax anteroides, A. integer, Brycon pesu, Bryconamericus* cf. *cismontanus, Bryconops alburnoides, Carnegiella strigatta, Corydoras* cf. *bondi, Crenicichla lenticulata, Crenicichla wallacei, Ctenobrycon spilurus, Cyphocharax* cf. *modestus, C. oenas, Farlowella vittata, Gephyrocharax valencia, Hemigrammus* cf. *tridens, Hemigrammus* sp. A, *Hoplias malabaricus, Hypoptopoma steindachneri Hypostomus* cf. *plecostomus, Jupiaba polylepis, Knodus* cf. *breviceps, Microphylipnus ternetzi, Microschemobrycon callops, M. casiquiare, M. melanotus, Moenkhausia collettii, M. grandisquamis, M. oligolepis, Ochmacanthus alternus, Phenacogaster* sp. A, *Pimelodella cruxenti, P. megalops, Poptella longipinnis, Rineloricaria* sp., *Serrasalmus rhombeus* and *Triportheus albus.*

The last station was toward the mouth (ICT-60). We capture 234 specimens belonging to 21 species distributed in:

Characiformes: 13 (62%)
Perciformes: 4 (19%)
Siluriformes: 2 (9.5%)
Atheriniformes: 2 (9.5%)

The most abundant species were: *Microphylipnus ternetzi* (74, 31.6%) *Moenkhausia* sp. A (66, 28.2%), *Bunocephalus* sp. (15, 6.4%), *Moenkhausia copei* (14, 6%), *Moenkhausia* cf. *chrysargyrea* (11, 4.7%) and *Apistogramma iniridae* (7, 3%). Also specimes from: *Acestrorhynchius microlepis, Aphyocharax erythrurus, Apistogramma* sp. A, *Bryconamericus* cf. *cismontanus, Crenicichla wallacei, Ctenobrycon spilurus, Hemigrammus* cf. *tridens, Leporinus* sp., *Moenkhausia collettii, Poptella longipinnis, Pseudopimelodus raninus, Rivulus* sp., *Steindachnerina pupula* and *Tetragonopterus argenteus.*

In summary, the analysis showed a moderate to high diversity and moderated abundance. We collected 5 species of catfishes that usually lives in habitats with logs and submerged branches such as: *Ancistrus* sp. C, *Farlowella vittata, Pimelodella cruxenti* and *P. megalops.* Also we collected species typically superficial and protected waters such as: *Carnegiella strigatta, Gephyrocharax valencia, Poptella longipinnis* and *Triportheus albus.* In the river main channel we collected "sardinatas" (*Pellona castelneana*) and other pelagic species. Additional interest is the great abundance of the eleotrld *Microphylipnus ternetzi.*

BC14. "Pelona" Beach at Caura river (ICT-61) (7°11.82´-65°08.88´W).

General fish fauna description

We used two fishing methods: Trawl net (7 hours) and beach seines (ICT-61). The combined samples showed a total of 201 specimens belonging to 20 species distributed in:

Characiformes: 10 (50%)
Siluriformes: 8 (40%)
Gymnotiformes: 1 (5%)
Other: 1 (5%)

The most abundant species were: *Aphyocharax alburnus* (128, 63.7%), *Hemigrammus tridens* (25, 12.4%), *Bryconamericus* cf. *cismontanus* (14, 7%) and *Argonectes longiceps* (8, 4%). Also collected were specimens from the following species: *Astyanax integer, Gymnorhamphichthys hypostomus, Hassar iheringi, Hemigrammus* cf. *tridens, Hemiodus unimaculatus, Hyphessobrycon minimus, Leptodoras* cf. *acipenserinus, L.* cf. *hasemani, L.* cf. *praelongus, Microschemobrycon callops, M. casiquiare, Moenkhausia collettii, Ochmacanthus orinoco, Opsodoras* cf. *trimaculatus Pimelodella megalops, Pimelodella* sp. A and *Xyliphius* cf. *melanopterus.*

In summary the analysis showed a low abundance and diversity. With the beach seines we capture 8 species of *Argonectes longiceps* that represent new records for the Caura River Basin. We capture several species of doradids with the trawl (*Hasar iheringi, Leptodoras* cf. *acipenserinus, L.* cf. *hasemani, L.* cf. *praelongus* and *Opsodoras* cf. *trimaculatus*) and one electric fish: *Gymnorhamphichthys hypostomus.* Many of these species are typical of sandy beaches. We highlight out the presence of *Xyliphius* cf. *melanopterus,* representing the first record for this river basin.

BC15. Pebble Island and Tributary above Raudal Cinco Mil (ICT-62, ICT-63) (06°58.29´-64°52.5´W).

General Fish fauna description

Island (ICT-62). A total of 1288 specimens were captured belonging to 43 species distributed in:

Characiformes: 33 (76.7%)
Siluriformes: 6 (14%)
Perciformes: 2 (4.7%)
Clupeiformes: 1 (2.3%)
Rajiformes: 1 (2.3%)

The most common species were: *Knodus* cf. *breviceps* (292, 22.7%), *Aphyocharax alburnus* (189, 14.6%), *Moenkhausia* cf. *lepidura* E (133, 10.3%) *Moenkhausia cotinho* (125, 9.7%), *Jupiaba polylepis* (90, 7%), *Curimatella inmaculata* (87, 6.7%), *Hyphessobrycon minimus* (47, 3.6%) and *Cyphocharax oenas* (29, 2.2%). Also captured were: *Anchoviella guianensis, Apareiodon orinocensis, Astyanax integer, Boulengerella lucia, Brycon pesu, Bryconamericus* cf. *cismontanus, Characidium* sp.A, *Corydoras* cf. *bondi, Creagrutus maxillaris, Curimata incompta, Curimatella immaculata, Cyphocharax festivus, C. oenas, Exodon paradoxus, Geophagus* cf. *brachybranchus, Hemigrammus* cf. *tridens, Hemigrammus* sp. A, *Hyphessobrycon minimus, Jupiaba atypindi, Knodus* sp. B, *Leporinus* cf. *maculatus, Mastiglannis asopos, Microschemobrycon callops, M. casiquiare, Moenkhausia* cf. *gracilima, Moenkhausia* cf. *lepidura* A, *M. cotinho, M. grandisquamis, M. hemigramoides, Ochmacanthus alternus, Phenacogaster* sp. A, *Pimelodus albofasciatus, Potamotrygon dorbygni, Prochilodus mariae, Pseudocheirodon* sp., *Rineloricaria* sp., *Satanoperca* sp. *Semaprochilodus kneri* and *Vandellia sanguinea* .

Caño (ICT-63). A total of 433 specimens were captured belonging to 36 species distributed in:

Characiformes: 24 (66.6%)
Siluriformes: 9 (25%)
Perciformes: 1 (2.9%)
Synbranchiformes: 1 (2.9%)
Gymnotiformes: 1 (2.9%)

The most common species were: Characidae sp. B (119, 27.5%), *Cyphocharax festivus* (72, 16.6%), *Moenkhausia* cf. *lepidura* E (55, 12.7%) *Michoschemobrycon melanotus* (38, 8.8%), and *Moenkhausia cotinho* (35, 8.1%). Also captured were: *Acestrorhynchus microlepis, Ancistrus* sp. C, *Aphyocharax alburnus, Apistogramma* sp. A, *Astyanax abrammis, A. integer, Astyanax* sp., *Bryconamericus* cf. *cismontanus, Bunocephalus* cf. *amaurus, Characidium* sp. A, *Corydoras* cf. *blochi, C. bondi, Ctenobrycon spilurus, Eigenmania virescens, Farlowella vittata, Hemigrammus* cf. *tridens, Jupiaba polylepis, Knodus* cf., *breviceps, Microschemobrycon casiquiare, Moenkhausia* cf. *chrysargyrea, M. collettii, M. grandisquamis, M. oligolepis, Ochmacanthus alternus, Phenacogaster* sp. A, *Pimelodus albofasciatus, Poptella longipinnis, Ramirezella newboldi, Synbranchus marmoratus, Tatia galaxias* and *Ituglanis* cf. *metae.*

The analysis showed an area well represented with a wide variety of species and high abundance. We would like to highlight the capture of one female of "raya de río" *Potamotrygon* and an individual of *Exodon paradoxus.* Many species are associated with fast moving waters and sandy bottoms such as: *Apareiodon orinocensis, Astyanax integer, Boulengerella lucia, Brycon pesu, Bryconamericus* cf. *cismontanus, Characidium* sp. A, *Corydoras* cf. *blochi, C. bondi, Creagrutus maxillaris, Curimata incompta, Curimatella immaculata, Cyphocharax festivus, C. oenas, Ochmacanthus alternus, Phenacogaster* sp. A and *Pimelodus albofasciatus.* Others typically live in protected habitats such as: *Aphyocharax alburnus, Apistogramma* sp. A, *Astyanax abrammis A. integer, Astyanax* sp., *Geophagus* cf. *brachybranchus, Microschemobrycon casiquiare, Moenkhausia* cf. *chrysargyrea, M. collettii, M. grandisquamis, M. oligolepis, Rineloricaria* sp. and *Satanoperca* sp.

BC16. Camp at Raudal Cinco Mil (ICT-64) (06°59.65´-64°54.82´W).

General fish fauna description

Rocky-Sandy beach (**ICT-64**). A total of 281 specimens were captured belonging to 25 species and distributed in:

Characiformes: 21 (84%)
Siluriformes: 3 (12%)
Perciformes: 1 (4%)

The most abundant species were: *Knodus* cf. *breviceps* (135, 49.1%), *Moenkhausia* cf. *lepidura* E (81, 28.8%), *Moenkhausia cotinho* (44, 15.7%), *Moenkhausia* cf. *lepidura* D (26, 9.3%) and *Jupiaba* cf. *polylepis* (23, 8.2%). Also, the following species were collected: *Anostomus ternetzi, Aphanotorulus* sp., *Aphyocharax alburnus, Astyanax integer, Chalceus macrolepidotus, Creagrutus maxillaris, Cynopotamus essequibensis, Hemiodus unimaculatus, Homodiaetus* sp., *Hydrolicus* cf. *tatauaia, Hyphessobrycon minimus, Jupiaba atypindi, Knodus* sp. B, *Leporinus* cf. *maculatus, L. friderici, Moenkhausia grandisquamis, Phenacogaster* sp. A, *Piaractus brachypomus, Pseudocetopsis* sp. and *Satanoperca* sp. A.

The area showed moderate to low diversity and abundance, with many species associated with rapids. We captured a single individual of *Chalceus macrolepidotus* and several specimens of two species of *Leporinus*. With the traps we captured a single specimen of *Pseudocetopsis* sp. With the gillnet, we captured several specimens of "sardinata" (*Pellona castelneana*), "morocoto" (*Piaractus*), "payaras" (*Hydrolicus tatauaia*) and "piranhas" (*Serrasalmus rhombeus*) that were not included in the regular surveying.

BC17. Tokoto River (ICT-65) (06°53.52´-64°47.90´W).

General fish fauna description

Rapids, pools and backwaters (ICT-65). A total of 780 specimens were captured belonging to 62 species and distributed in:

Characiformes: 46 (74.2.%)
Siluriformes: 12 (19.4%)
Perciformes: 3 (4.8%)
Gymnotiformes: 1 (1.6%)

The most abundant species were: *Knodus* cf. *breviceps* (138, 17.7%), *Aphyocharax alburnus* (79, 14.6%), *Moenkhausia cotinho* (64, 8.2%), *Moenkhausia dichroura* (53, 6.6%), *Moenkhausia* cf. *lepidura* E (52, 6.6%) and *Curimatella inmaculata* (32, 4.1%). Also specimens of the following species were collected: *Acestrorhynchus microlepis, Ancistrus* sp. C, *Apistogramma* sp. A, *Astyanax integer, Brycon bicolor, Bryconamericus* cf. *cismontanus, Bryconamericus deuterodonoides, Caenotropus labyrinthicus, Characidium* sp. A, *Cochliodon plecostomoides, Corydoras* cf. *bondi, Creagrutus maxillaris, Ctenobrycon spilurus, Curimatella immaculata, Cyphocharax festivus, C. oenas, Eigenmannia virescens, Farlowella vittata, Gephyrocharax valencia, Hoplias macrophthalmus, Hyphessobrycon minimus, Hypostomus plecostomus, Hypostomus* sp., *Imparfinis* sp. A, *Jupiaba atypindi, J. polylepis, Knodus* sp. B, *Lasiancistrus* sp., *Leporinus* cf. *maculatus, Moenkhausia* cf. *lepidura* D, *Moenkhausia collettii, M. cotinho, M. grandisquamis, M. hemigramoides, Ochmacanthus alternus, Odontostilbe* sp., *Pachypops* sp., *Phenacogaster* sp. A, *Piaractus brachypomus, Pimelodella megalops, Pimelodella* sp. B, *Poptella longipinnis, Pseudocheirodon* sp., *Pyrrhulina brevis, Raphiodon vulpinus, Rineloricaria* sp., *Roeboides affinis, Satanoperca* sp., *Serrasalmus rhombeus, Steindachnerina pupula, Tetragonopterus argenteus, T. chalceus* and *Toracocharax stellatus.*

In summary the analysis revealed an area with moderate abundance and very high diversity, includingspecies associated with rapids and aquatic plants such as: *Leporinus* cf. *maculatus, Characidium* sp. A and *Lasiancistrus* sp. Also, there were fishes typically found in sandy-rocky areas such as callichthyd, loricariid, pimelodid, and trichomycterid catfishes and specifically: *Ancistrus* sp. C, *Cochliodon plecostomoides, Corydoras* cf. *bondi, Farlowella vittata, Hypostomus plecostomus, Hypostomus* sp., *Imparfinis* sp. A, *Ochmacanthus alternus, Pimelodella megalops, Pimelodella* sp. B. The backwaters present abundant caracoids and cichlids. We notice the presence of a "Hatch Fish" (Gasteropelecidae, *Toracocharax stellatus*). In the main channel and deeper waters we captured with hook and line: *Piaractus brachypomus, Serrasalmus rhombeus* and *Brycon* sp. The Tacoto River was in the most diverse and richest area of all the field stations sampled.

Apéndice 7

Estaciones de Colecta de Peces y Análisis de la Ictiofauna durante la Expedición AquaRAP a la Cuenca del Río Caura, Estado Bolívar, Venezuela

Antonio Machado-Allison, Barry Chernoff, Philip W. Willink, Francisco Provenzano, Paulo Petry y Alberto Marcano

I. CAURA SUPERIOR. AREA POR ARRIBA DEL SALTO PARÁ

AC01. Caño Laguna en el Río Caura debajo de Entrerios (Estación ICT-01) (5°55.78'N–64°25.39'W)

Descripción general de la ictiofauna

Se capturaron un total de 220 ejemplares ubicados en 18 especies distribuidas:

Characiformes: 10 (55.5%)
Siluriformes: 4 (22.2%)
Perciformes: 2 (11%)
Otros: 2 (11%)

La diversidad de peces es baja. Algunos componentes son comunes a este tipo de ambiente, entre ellos ejemplares del género *Pimelodella* y *Corydoras* y de especies de la familia Cichlidae (géneros *Guianacara* y *Crenicichla*). Los resultados indican que el grupo más abundante y diverso fueron los Characiformes con 10 especies (55.5%), le siguen los Siluriformes (4 especies, 22.2%) y finalmente Perciformes (2, 11%). En cuanto a la abundancia reativa, las especies mejor representadas son: *Moenkhausia collettii* (71, 32.2%), *Corydoras boehlkei* (65, 29.5%), *Pimelodella* sp. B (26, 11.8%), *Pimelodella* sp. C (16, 7.2%) y *Guianacara geayi* (8, 3.6%).

AC02. Confluencia de los ríos Caura y Erebato en Entreríos (Estaciones ICT-02, ICT-06) (5°56.02'N–64°25.67'W)

Descripción general de la ictiofauna

Remanso (ICT-02): Peces capturados con chinchorro. Moderada diversidad de peces (25 especies) con baja abundancia. Se capturaron un total de 139 ejemplares:

Characiformes: 17 (68%)
Siluriformes: 6 (24%)
Perciformes: 2 (8%)

En cuanto a la abundancia reativa, las especies mejor representadas se encuentran: *Knodus* cf. *victoriae* (25, 18%), *Bryconops* sp. A (20, 14.4%), *Guianacara geayi* (17, 12.2%), *Corydoras boehlkei* (13, 9.3%) y *Phenacogaster* sp. B (12, 8.6%). La diversidad de peces es moderada a baja comparada con otras estaciones (ver adelante). Algunos componentes son comunes a este tipo de ambiente, entre ellos ejemplares del género *Jupiaba* (2 especies) y *Corydoras* y de especies de la familia Cichlidae (géneros *Guianacara* y *Geophagus*).

Caño (ICT-06): Peces capturados con red de ahorque; baja diversidad (13 especies) y baja abundancia con un total 41 ejemplares capturados incluidos en:

Characiformes: 8 (61.5%)
Siluriformes: 3 (23%)
Perciformes: 1 (7.6%)
Synbranchiformes: 1 (7.6%)

La especie más abundante fue un dorádido preliminarmente identificado como *Doras* sp. con 18 (43.9%) ejemplares, además destaca la presencia de *Ageneiosus* sp. (5, 12.2%), *Myleus rubripinnis* (4, 9.7%), *Plagioscion auratus* (4, 9.7%) y *Serrasalmus rhombeus* (2, 4.8%). Además, se capturaron dos especies de pámpanos o pacus *Myleus: M. asterias* y *M. torquatus.*

En general la ictiofauna en la zona de Entreríos se encuentra caracterizada por 34 especies que habitan zonas de aguas poco profundas y tranquilas como por ejemplo: *Jupiaba, Guianacara, Rhineloricaria, Pimelodella, Corydoras*, etc. En las zonas profundas se encuentran especies de porte mayor como por ejemplo: *Ageneiosus, Doras, Myleus* y *Serrasalmus*, los cuales generalmente forman parte de la dieta de las poblaciones humanas ribereñas. Los resultados muestran una abundancia relativa moderada a baja.

AC03. Río Kakada y Playa Suajiditu (Estaciones ICT-03, ICT-04) (5°29.86'N–64°34.76'W)

Descripción general de la ictiofauna

La playa arenosa (ICT-03). Se capturaron un total de 151 ejemplares ubicados en 12 especies distribuidas:

Characiformes: 10 (83.3%)
Perciformes: 2 (16.6%)

Destaca en esta estación la presencia de *Moenkhausia collettii* (47, 31.1%), *Bryconops* sp. (45, 29.8%), *Moenkhausia* cf. *lepidura* (14, 9.2%), *Jupiaba zonata* (13, 8.6%) y *Hemigrammus* sp. (11, 7.3%). Además, se capturaron ejemplares de dos especies de la familia Cichlidae (*Guianacara geayi* y *Crenicichla saxatilis*).

Rápidos (ICT-04). Se capturaron un total de 31 ejemplares ubicados en 9 especies ubicadas todas en el Orden Characiformes.

Los resultados ictiológicos en esta estación muestran una abundancia y diversidad relativamente bajas. Grupos principalmente caracoideos típicos de playas arenosas. Es destacar la presencia de: *Moenkhausia colletti* (10, 32%), *Knodus* cf. *victoriae* (5, 16%), *Hemigrammus* sp. (5, 16%) y *Moenkhausia* cf. *lepidura* (5, 16%).

AC04. Caño Suajiditu (Estaciones: ICT-05) (5°29.59'N–64°35.16'W)

Descripción general de la ictiofauna

Se capturaron un total de 447 ejemplares ubicados en 22 especies distribuidas:

Characiformes: 13 (59%)
Siluriformes: 5 (22.7%)
Perciformes: 3 (13.6%)
Synbranchiformes: 1 (4.5%)

Las especies más comunes fueron: *Moenkhausia collettii* (256, 57.3%), *Characidium* sp. A (31, 6.9%), *Phenacogaster* sp. (31, 6.9%), *Moenkhausia oligolepis* (25, 5.6%) y *Hemigrammus* sp. (15, 3.3%). Además, se colectaron ejemplares de *Aequidens* cf. *chimantanus, Corydoras boehlkei, Cyphocharax* sp., *Jupiaba atypindi, J. zonata, Moenkhausia* cf. *lepidura, Rineloricaria fallax, Synbranchus marmoratus* y *Ituglanis metae*. Los resultados de los análisis ictiológicos muestran una riqueza baja a media. Abundantes ejemplares de carácidos donde se destaca la presencia de una nueva especie de *Aphyocharax* con colorido exuberante. Esta zona puede ser un área importante de reproducción íctica dada la presencia de numerosos ejemplares juveniles.

AC05. Raudal del Perro (Estaciones ICT-07, ICT-08, ICT-09) (05°54.08'N–64°29.4'W; 05°54.17'N–64°29.4'W; 05°53.8'N–4°28.8'W)

Descripción general de la ictiofauna

Rápidos (ICT-07). Se capturaron un total de 493 ejemplares ubicados en 16 especies distribuidas:

Characiformes: 14 (87.5%)
Siluriformes: 1 (6.2%)
Perciformes: 1 (6.2%)

Las especies máscomunes fueron: *Moenkhausia* cf. *lepidura* (171, 34.7%), *Jupiaba atypindi* (161, 32.6%), *Bryconops* sp. (39, 7.9%), *Jupiaba zonata* (36, 7.3%) y *Melanocharacidium depressum* (16, 3.2%). Se capturaron ejemplares de especies típicas de aguas rápidas y fondos arenosos como por ejemplo: *Bryconops* (2 especies), *Jupiaba zonata, Knodus victoriae, Melanocharacidium depressum, Moenkhausia oligolepis* y *Satanoperca* sp.

Playa arenosa (ICT-08). Alrededor de la isla se colectaron un total de 220 ejemplares ubicados en 20 especies distribuidas:

Characiformes: 15 (75%)
Siluriformes: 4 (20%)
Perciformes: 1 (5%)

Las especies más comunes fueron: *Jupiaba atypindi* (73, 33.2%), *Melanocharacidium depressum* (49, 22.2%), *Moenkhausia* cf. *lepidura* (37, 16.8%), *Characidium* sp. A (11, 5%) y *Bryconops* cf. *colaroja* (8, 3.6%). También, se capturaron ejemplares de: *Anostomus anostomus, Leporinus arcus, L. grandti, Imparfinis, Chaetostoma vasquezi* y *Rineloricaria fallax*.

Caño (ICT-09). Se capturaron un total de 38 ejemplares ubicados en 15 especies distribuidas:

Characiformes: 9 (60%)
Perciformes: 4 (26.6%)
Siluriformes: 1 (6.6%)
Synbranchiformes: 1 (6.6%)

Las especies más comunes capturadas incluyen a: *Moenkhausia collettii* (11, 28.9%) y *Hemigrammus* cf. *guyanensis* (7, 18.4%). También, se capturaron ejemplares de: *Aequidens chimantanus, Bryconops giacopinii, Characidium* sp., *Crenicichla alta* y *Guianacara geayi*.

Los resultados de las observaciones icitiológicas indican que en las playas arenosas presentan una baja diversidad y abundancia de peces, sólo se capturaron pocas especies de caracoideos, donde son importantes tres especies de *Moenkhausia*: *M. oligolepis* y *M. lepidura* (2 sp). Además, se encuentran abundantes *Jupiaba* cf. *zonata* de axhuberante colaración que incluye punta del hocico amarilla, cuerpo verdoso y aletas con tonalidades rojizas. Finalmente, un *Bryconops* sp. parecido a *B. colaroja*. Adicionalmente, se capturaron ejemplares de la Familia Curimatidae.

Los raudales presentan un incremento en la diversidad, todos los peces capturados están intimamente asociados a la vegetación acuática, cabe destacar la presencia de tres especies de Anostomidae (*Anostomus anostomus, Leporinus arcus,* y *Leporinus* cf. *grandti*) de colores muy llamativos, bagres de la familia Pimelodidae (*Inparfinis*) y Loricariidae como *Rineloricaria fallax* y *Chaetostoma vasquezi*.

AC06. Area cercana al Caño Wididikenu (Estación ICT-10) (5°53.9'N–64°28.7'W)

Descripción general de la ictiofauna

Se capturaron un total de 392 ejemplares ubicados en 29 especies distribuidas:

Characiformes: 19 (65.5%)
Siluriformes: 5 (17.2%)
Perciformes: 3 (10.3%)
Atheriniformes: 1 (3.4%)
Synbranchiformes: 1 (3.4%)

Las especies más abundantes fueron: *Corydoras boehlkei* (103, 26.2%), *Moenkhausia oligolepis* (70, 17.9%), *M. collettii* (66, 16.8%), *Corydoras* sp. (32, 8.2%) y *Guianacara geayi* (28, 7.1%). También se capturaron: *Aphyocharax* sp., *Bryconops* (2 especies), *Characidium* sp., *Crenicichla alta, Jupiaba atypindi, Melanocharacidium depressum, Moenkhausia lepidura, Pimelodella* sp., *Serrasalmus rhombeus, Synbranchus marmoratus* y *Ituglanis metae*.

Se realizaron capturas de dos ambientes que resultaron ser pobres en diversidad y abundancia. En el caño se capturaron ejemplares de la familia Cichlidae, entre ellos *Guianacara geayi*. También algunas especies de Characidae, *Bryconops* cf. *colaroja* y juveniles o especies enanas de caracoideos. Avanzamos hacia las cabeceras del caño, aguas cristalinas, muy pedregoso el fondo y muy baja profundidad y anchura, pero sólo se capturaron muy pocos ejemplares de caracoideos de talla pequeña. En los rápidos se colectaron algunas especies similares a los encontrados anteriormente (Anostomidae, Loricariidae y Characidiidae). Además, se capturó un ejemplar de *Serrasalmus rhombeus* con la red de ahorque en un área de aguas relativamente tranquilas. En las piedras, la captura fue dificultosa y además no encontramos lugares con vegetación sumergida. Sólo se capturaron algunas especies de caracoideos de talla pequeña, por ejemplo: *Hemigrammus minimus.*

AC07. Raudal en Río Erebato y Caño (Estación ICT-11) (5°52.7´N–64°29.56´W)

Descripción general de la ictiofauna

Se capturaron un total de 113 ejemplares ubicados 18 especies distribuidas:

Characiformes: 16 (88.8%)
Siluriformes: 1 (5.5%)
Perciformes: 1 (5.5%)

Las especies másabundantes fueron: *Jupiaba atypindi* (24, 21.2%), *Jupiaba* sp. A (17, 15%), *Melanocharacidium depressum* (16, 14.2%), *Jupiaba* sp. B (14, 12.4%) y *Moenkhausia* cf. *lepidura* (12, 10.6%). También, se colectaron ejemplares de las especies: *Moenkhausia miangi, Guianacara geayi, Bryconops* cf. *colaroja, Leporinus arcus, L.* cf. *grandti* y *Rineloricaria fallax.*

El análisis ictiológico muestra que los peces en los rápidos poseen una diversidad y abundancia baja. Sin embargo, las especies poseen una profunda asociación con las plantas acuáticas sumergidas (Podostemonaceae). En el raudal aparecen nuevamente las especies de Anostomidae (*Leporinus arcus* y *L. grandti*) con colores llamativos, *Moenkhausia* y otros caracoideos. Además se capturaron especies de Characidiidae (*Characidium* y *Melanocharacidium*) y Loricariidae (*Rineloricaria fallax*). Cabe destacar que en este raudal la vegetación acuática sumergida esta más desarrollada, con plantas de mayor talla y áreas con mayor cobertura que en las estaciones anteriores.

AC08. Raudal Cejiato-Soodu (Estaciones ICT-12, ICT-13, ICT-14, ICT-15) (5°33.47´N–64°18.8´W)

Descripción general de la ictiofauna

Playas (ICT-12). Se muestrearon dos playas pertenecientes a una ensenada en el Cejiato. Se capturaron un total de 236 ejemplares ubicados 28 especies distribuidas:

Characiformes: 19 (67.9%)
Siluriformes: 5 (17.9%)
Perciformes: 3 (10.7%)
Synbranchiformes: 1 (3.5%)

Las especies más abundantes fueron: *Corydoras osteocarus* (82, 34.7%), *Bryconops giacopinni* (39, 16.5%), *Pimelodella* sp. B (28, 11.9%), *Guianacara geayi* (26, 11%) y *Moenkhausia* cf. *lepidura* (25, 10.6%). También se colectaron ejemplares de las especies: *Moenkhausia collettii, M. oligolepis, Knodus victoriae, Jupiaba zonata, Crenicichla saxatilis, Bryconops* cf. *colaroja, Hemiodus goeldii, Tetragonopterus* sp. y *Cyphocharax festivus.*

Ambas playas poseen similares características con altas abundancias y moderada diversidad íctica caracterizada por una mezcla de especies del río propio y de bahías con suelos con alto contenido de material vegetal sumergido como por ejemplo cíclidos. Existen elementos asociados a fondos arenosos como las *Corydoras* y la presencia de especies pelágicas como tres especies simpátricas de *Bryconops.*

Riachuelo bajo/boca (ICT-13). Se capturaron un total de 159 ejemplares ubicados 25 especies distribuidas:

Characiformes: 16 (64%)
Siluriformes: 6 (24%)
Perciformes: 2 (8%)
Synbranchiformes: 1 (4%)

Las especies más abundantes fueron: *Knodus victoriae* (77, 48.4%), *Moenkhausia collettii* (21, 13.2%), *Bryconops* cf. *colaroja* (18, 11.3%), *Moenkhausia* cf. *lepidura* (12, 7.5%) y *Jupiaba zonata* (6, 3.8%). También se colectaron ejemplares de las especies: *Brycocops giacopini, Corydoras* sp., *Cregrutus* sp., *Crenicichla saxatilis, Hypostomus* sp., *Melanocharacidium melanopteron, Myleus rubripinnis* (Juvenil), *Rineloricaria fallax* y *Synbranchus marmoratus.*

Riachuelo arriba (ICT-14). Se capturaron un total de 20 ejemplares ubicados 9 especies distribuidas:

Characiformes: 5 (55.5%)
Siluriformes: 2 (22.2%)
Perciformes: 2 (22.2%)

Las especies más abundantes fueron: *Moenkhausia collettii* (11, 55%), *Crenicichla saxatilis* (2, 10%). También se colectaron ejemplares de las especies: *Aequidens* cf. *chimantanus, Melanocharacidium melanopteron* y *Moenkhausia* cf. *lepidura.* Se capturó igualmente el primer ejemplar de *Tatia,* asociado a troncos sumergidos.

Rápidos y Playa rocosa (ICT-15). Especies características de corrientes (riacofilos) y asociados a podostemaceas. La captura en el área fue difícil. Se capturaron un total de 28 ejemplares ubicados 7 especies todas ubicadas en el Orden Characiformes. Las especies más abundantes fueron: *Jupiaba atypindi* (13, 46.4%), *Knodus victoriae* (7, 25%) y *Moenkhausia* cf. *lepidura* (4, 14.3%). También se colectaron ejemplares de las especies: *Jupiaba zonata, Creagrutus* sp. y *Acestrorhynchus microlepis.*

En general, podemos resumir que el área del Raudal Cejiato posee comparativamente gran abundancia y diversidad íctica comparada con otras regiones muestreadas. Un gran número de especies se encuentran íntimamente asociadas a hábitats particulares. Aunque algunos individuos de *Bryconops* fueron capturados cerca de la playa y boca del riachuelo, la gran mayoría fue capturada en aguas abiertas y profundas. Similarmente, las *Corydoras* fueron solamente colectadas sobre los fondos arenosos a lo largo de las playas con algunos elementos presentes en la boca del riachuelo. Un dato importante y no podemos por los momentos indicar su significancia, es el hecho de haber encontrado dos ejemplares de *Moenkhausia oligolepis* de gran tamaño (no visto en ninguna otra muestra). Aunque ésta especie es muy común en la Cuenca del Orinoco, no está íntimamente asociada a aguas superiores. Por otro lado, es importante haber encontrado juveniles de palometas o pámpanos *Myleus* cf. *rubripinnis,* lo que pudiera indicar que esta área puede estar asociada con los procesos reproductivos de esta especie.

En las cercanías de la boca del riachuelo fueron capturadas varias especies en las que cabe destacar: *Bryconops* cf. *colaroja, B. giacopini, Crenicichla saxatilis, Jupiaba zonata, Knodus victoriae, Moenkhausia collettii, Moenkhausia* cf. *lepidura,* y los loricáridos: *Hypostomus* sp. y *Rineloricaria fallax.* Después el riachuelo se hace más angosto y más profundo con numerosos pozos cubiertos por arena y detritus vegetal. Las especies asociadas a estos ambientes pertenecen a los géneros: *Farlowella, Ancistrus* y *Crenicichla* y algunos microcarácidos.

Debido a que en los rápidos la corriente fue muy fuerte, pocos ejemplares fueron capturados y la mayoría provienen de muestras en pozos aislados dentro de las riberas rocosas.

AC09. Raudal Pauji (Estaciones ICT-16, ICT-17, ICT-18, ICT-19, ICT-20) (5°49.7´N–64°24.3´W)

Descripción general de la ictiofauna

Playas (ICT-16). Se capturaron un total de 178 ejemplares ubicados 9 especies distribuidas:

Characiformes: 8 (88.8%)

Perciformes: 1 (11.1%)

Las especies más abundantes fueron: *Cyphocharax festivus* (83, 46.6%), *Moenkhausia lepidura* (78, 43.8%) y *Bryconops* sp. A (4, 2.3%). También se colectaron ejemplares de las especies: *Knodus* cf. *victoriae, Aphyocharax* sp. y *Moenkhausia collettii.* La mayoría de las especies son típicas de playas arenosas.

Rápidos (ICT-17). Se capturaron un total de 50 ejemplares ubicados 12 especies distribuidas:

Characiformes: 9 (75%)

Siluriformes: 3 (25%)

Las especies más abundantes fueron: *Melanocharacidium depressum* (23, 46%), *Characidium* sp. A (6, 12%), *Hypostomus* sp. A (5, 10%) y *Jupiaba atypindi* (3, 6%). También, se colectaron ejemplares de las especies: *Apareiodon orinocensis, Cetopsorhamdia* cf. *picklei* y *Melanocharacidium dispiloma.* Area caracterizada por especies típicas de rápidos con varios Characidiidae, Parodontidae y Loricariidae. Una especie de Pimelodidae (*Cetopsorhamdia* cf. *picklei*) interesante.

Playa rocosa en Islas (ICT-18). Se capturaron un total de 14 ejemplares ubicados 11 especies distribuidas:

Characiformes: 8 (72%)

Perciformes: 2 (18%)

Siluriformes: 1 (9%)

Las especies más abundantes fueron: *Crenicichla saxatilis* (2, 14%), *Jupiaba zonata* (2, 14%) y *Moenkhausia* cf. *lepidura* (2, 14%). También, se colectaron ejemplares de las especies: *Crenicichla saxatilis, Anostomus anostomus, Melanocharacidium depressum.*

Playa arenosa (ICT-19). Se capturaron un total de 52 ejemplares ubicados 9 especies distribuidas:

Characiformes: 8 (88%)

Perciformes: 1 (12%)

Las especies más abundantes fueron: *Moenkhausia* cf. *lepidura* (28, 54%), *Knodus* cf. *victoriae* (10, 19.2%) y *Bryconops* sp. A (3, 5.7%). También, se colectaron ejemplares de las especies: *Aphyocharax* sp., *Creagrutus* sp., *Hemiodus unimaculatus* y un Cichlidae (*Satanoperca* sp. A). La mayoría de estas especies son típicas de aguas rápidas y fondos arenoso-pedregoso.

Remansos (ICT-20). Se capturaron un total de 201 ejemplares ubicados 16 especies distribuidas:

Characiformes: 13 (81%)

Perciformes: 3 (19%)

Las especies más abundantes fueron: Un Characidae (No identificado) (99, 49%), *Guianacara geayi* (25, 12%), *Moenkhausia* cf. *lepidura* (21, 10.3%), *Geophagus* sp. (14, 7%), *Moenkhausia colletti* (14, 7%) y *Jupiaba zonata* (9, 4.4%). También, se colectaron ejemplares de las especies: *Aphyocharax* sp., *Brachychalcinus opercularis, Bryconops* cf. *colaroja, B. giacopini, Jupiaba atypindi* y *Crenicichla alta.* Muchas formas juveniles de Cichlidae y Characidae, lo que indica un proceso reproductivo en esta época. El remanso actúa como área *nursery* o de protección.

En resumen podemos indicar que el área posee poca abundancia y diversidad además de ser poco heterogéneo en cuanto a macro hábitats. Cabe destacar que los peces colectados en los rápidos son muy diferentes a los colectados en otras áreas similares. Donde la corriente es fuerte se capturaron especies de *Characidium* y *Melanocharacidium, Apareiodon* (2 especies), *inparfinis,* y *Hypostomus.* En el remanso existe abundancia del cíclido *Guianacara geayi.* Se obtuvo una buena muestra ontogenética. También se colectaron especies interesantes como: *Poptella longipinnis, Bryconops* (2 especies posiblemente nuevas), *Moenkhausia* cf. *lepidura* (varios morfotipos) y abundantes juveniles de cíclidos. La evidencia dada por la presencia de juveniles indica actividad reproductiva reciente tanto en cíclidos como peces caracoideos. En las playas arenosas del área se capturaron especies importantes desde el punto de vista sistemático como la posibilidad de una nueva especie de *Aphyocharax,* varias especies de curimátidos y de hemiodóntidos

AC10. Caño Jasa (Estaciones ICT-21, ICT-22) (5°53.02´N–64°24 a 25.6´W)

Descripción general de la ictiofauna

Caño (ICT-22). Se capturaron un total de 15 ejemplares ubicados 5 especies distribuidas:

Characiformes: 1 (20%)

Perciformes: 2 (40%)

Siluriformes: 2 (40%)

Las especies fueron: *Bryconops giacopini* (7, 46.6%), *Corydoras boehlkei* (3, 20%), *Corydoras* sp. (2, 13.3%), *Guianacara geayi* (2, 13.3%) y *Aequidens chimantanus* (1, 6.6%). El área presenta elementos típicos de peces asociados a aguas rápidas (bocas de caños) como por ejemplo *Bryconops* y de aguas protegidas como *Guianacara geayi*. El área mostró poca abundancia y diversidad.

AC11. Remanso Aguas Arriba Salto Pará (Estación ICT-23) (6°16.9´N–64°29.2´W)

Descripción general de la ictiofauna

A pesar de los diferentes macro hábitats presentes sólo se capturaron peces en el área de rápidos o raudales.

Raudal (ICT-23). Se capturaron un total de 34 especies y 429 ejemplares ubicadas en:

Characiformes: 27 (79.4%)
Perciformes: 3 (8.8%)
Atheriniformes: 1 (2.9%)
Siluriformes: 1 (2.9%)

Las especies más abundantes fueron: *Moenkhausia* cf. *lepidura* (108, 25.2 %), *Jupiaba zonata* (93, 21.7%), *Jupiaba atypindi* (74, 17.82%), *Cyphocharax* sp. (25, 5.8%) y *Cyphocharax festivus* (19, 4.4%). También, se colectaron ejemplares de las especies: *Acestrorhynchus falcatus, Apistogramma* sp., *Brachychalcinus opoercularis, Corydoras boehlkei, Guianacara geayi, Hemiodus unimaculatus, Hoplias macrophthalmus, H. malabaricus, Moenkhausia collettii, M. grandisquamis, M. oligolepis, Poptella longipinnis, Rhineloricaria fallax* y una especie de *Rivulus*.

El análisis ictiológico muestra a numerosas especies tanto de ambientes típicamente reofilos como de remanso. Cabe destacar la presencia de predadores como *Hoplias macrphthalmus, H. malabaricus* y *Acestrorhynchus* cf. *falcatus*. Además, varias especies de peces de importancia ornamental como: *Moenkhausia* (3 sp.), *Jupiaba zonata, Bryconops, Curimata* (2 sp.), *Poptella longipinnis, Hemiodus* sp., *Leporinus* sp., *Prochilodus mariae, Guianacara geayi* entre otros. Debemos destacar la captura de una especie rara como *Rivulus* sp. La abundancia relativa y diversidad es de moderada a alta.

AC12. Lajas, Remanso y Caño en Salto Pará (Estaciones ICT-24, ICT-25, ICT-26) (6°16.9´N– 64°29.2´W)

Descripción general de la ictiofauna

Las tres áreas muestreadas muestran una interesante presencia de algunas especies solamente capturadas en esta región, especialmente las correspondientes al área de las lajas.

Lajas (ICT-25). Se capturaron un total de 359 ejemplares ubicados 30 especies y distribuidos en:

Characiformes: 23 (76.6%)
Siluriformes: 4 (13.3%)
Perciformes: 3 (10%)

Las especies más abundantes fueron: *Jupiaba zonata* (205, 57.1%), *J. atypindi* (30, 8.4%), *Apareiodon* sp. (21, 5.8%), *Chaetostoma vasquezi* (16, 4.4%) y *Jupiaba* sp. B (13, 3.6%). También, fueron colectados ejemplares de las siguientes especies: *Crenicichla alta, Ctenobrycon spilurus, Guianacara geayi, Harttia* sp., *Hemiodus unimaculatus, Hoplias macrophthalmus, Leporinus arcus,* varias especies del grupo *Moenkhausia "lepidura," M. cotinho, M. oligolepis, Myleus rubripinnis, Poptella longipinnis* y *Rineloricaria fallax.*

En esta estación se capturaron especies asociadas a sustratos rocosos como por ejemplo los loricáridos (*Chaetostoma vasquezi, Harttia* sp. y *Rineloricaria fallax*). La especie de *Harttia* es posiblemente nueva y esta en proceso de descripción (Provenzano, com. per.) También se colectaron ejemplares de *Apareiodon* sp. (una nueva especie, Machado-Allison, com. pers.). Además de las especies anteriormente mencionadas, se colectaron ejemplares comunes de estas áreas como: *Jupiaba, Leporinus, Hemiodopsis,* [illegible], *Rhineloricaria, Guianacara* y *Moenkhausia.* Moderada diversidad y abundancia.

Remanso (ICT-26). Se capturaron 222 ejemplares ubicados en 13 especies y distribudos en:

Characiformes: 8 (61.5%)
Siluriformes: 2 (15.3%)
Perciformes: 2 (15.3%)
Gymnotiformes: 1 (7.7%)

Las especies más abundantes fueron: *Jupiaba zonata* (120, 54%), *Moenkhausia* cf. *lepidura* (29, 13%), *Guianacara geayi* (27, 12.1%), *Cyphocharax* sp. (22, 9.9%) y *Moenkhausia* sp. (10, 4.5%). También, fueron colectados ejemplares de las siguientes especies: *Crenicichla saxatilis, Ctenobrycon spilurus, Hypostomus* sp., *Myleus rubripinnis* y *Pimelodella* sp.

Esta área de remanso tipifica la presencia de especies de Cichlidae como: *Guianacara geayi* y *Crenicichla saxatilis.* Debemos hacer notar la captura por primera vez de una especie de pez eléctrico del género *Gymnotus* (*G. carapo*). El área presenta baja diversidad y moderada abundancia.

AC13. Rio Yuruani. Bosque inundable, Caño, Isla (Estación ICT-27) (6° 8.4´N–64° 25.1´W)

Descripción general de la ictiofauna

Remanso (ICT-27). Se colectaron un total de 62 ejemplares pertenecientes a 14 especies y ubicados en:

Characiformes: 11 (78.5%)
Siluriformes: 1 (7.1%)
Perciformes: 2 (14.2%)

Las especies más abundantes fueron: *Moenkhausia* cf. *lepidura* (27, 43.5%), *Pimelodella* sp. (14, 22.6%), *Guianacara geayi* (4, 6.5%) y *Jupiaba zonata* (3, 4.8%). También, fueron colectados ejemplares de las siguientes especies: *Bryconops giacopini, Crenicichla alta, Cyphocharax festivus, Jupiaba atypindi, J. polylepis, Knodus victoriae, Poptella longipinnis* y *Rineloricaria fallax.*

Las especies colectadas son típicas de áreas de remanso (*Guianacara* y *Crenicichla*), bosques inundados y playas arenosas con poca velocidad de corriente *(Moenkhausia, Knodus, Poptella* y *Bryconops*). Moderada diversidad y baja abundancia.

AC14. Raudal Culebra de Agua (Estaciones ICT-28, ICT-29, ICT-30, ACT-31) (6°04.7´N–64°26.02´W)

Descripción general de la ictiofauna

Playa (ICT-28). Un total de 174 ejemplares fueron capturados pertenecientes a 16 especies y ubicadas en:

Characiformes: 10 (62.5%)

Siluriformes: 4 (25%)
Perciformes: 2 (12.5%)

Las especies más abundantes fueron: *Corydoras boehlkei* (43, 24.7%), *Jupiaba zonata* (31, 17.8%), *Pimelodella* sp. B (21, 12.1%), *Moenkhausia* cf. *lepidura* (16, 9.2%) y *Guianacara geayi* (14, 8.4%). También, fueron colectados ejemplares de las siguientes especies: *Bryconops* sp. A, *Cyphocharax festivus, Jupiaba atypindi, Knodus victoriae, Moenkhausia collettii, M. oligolepis, Rinbeloricaria fallax* y *Satanoperca* sp. A.

El área incluye especies asociadas a hábitats de playa como por ejemplo *Corydoras boehlkei, Cyphocharax festivus, Jupiaba atypindi, Knodus victoriae*. Sin embargo, también hay presencia de especies asociadas a remansos como por ejemplo *Guianacara geayi, Satanoperca* sp. y *Moenkhausia oligolepis*. En el área de las playas existe una abundancia baja y diversidad ligeramente baja.

Rápidos (ICT-29). Un total de 61 ejemplares fueron capturados pertenecientes a 16 especies y ubicados en:

Characiformes: 11 (68.7%)
Siluriformes: 4 (25%)
Perciformes: 1 (6.3%)

Las especies más abundantes fueron: *Jupiaba atypindi* (15, 24.6%), *Jupiaba zonata* (9, 14.7%), *Melanocharacidium depressum* (7, 11.5%), *Chaetostoma vasquezi* (6, 9.8%) y *Bryconops* sp. A (5, 8.2%). También, fueron colectados ejemplares de las siguientes especies: *Ancistrus* sp., *Bryconops giacopini, Characidium* sp. A, *Guianacara geayi, Hypostomus* sp. A, *Leporinus grandti, Rineloricaria fallax* y *Tetragonopterus* sp.

La mayoría de las especies colectadas se encuentran asociadas a plantas acuáticas como por ejemplo: *Characidium, Melanocharacidium depressum, Leporinus* y *Hypostomus*. Algunas especies asociadas a los troncos de las plantas inundadas y depositados en el fondo tal como la presencia de *Chaetostoma* y *Rineloricaria*. El área muestra baja abundancia y moderada a baja diversidad.

Riachuelo (ICT-30). Un total de 20 ejemplares fueron capturados, pertenecientes a 9 especies y ubicadas en:

Characiformes: 5 (55.5%)
Perciformes: 3 (33.3%)
Siluriformes: 1 (11.1%)

Las especies más abundantes fueron: *Moenkhausia collettii* (8, 40%), *Corydoras boehlkei* (2, 10%), *Aphyocharax* sp. (2, 10%) y *Characidium* sp. A (2, 10%). También, fueron colectados ejemplares de las siguientes especies: *Aequidens chimantanus, Crenicichla saxatilis, Guianacara geayi, Jupiaba zonata* y *Moenkhausia* cf. *lepidura*.

Las especies presentes en el área muestreada son típicas de áreas protegidas de remansos o bosque inundable como por ejemplo *Crenicichla saxatilis, Guianacara geayi* y *Moenkhausia collettii*. Otras presentes en los fondos arenoso-pedregoso de riachuelos como por ejemplo: *Corydoras* y *Characidium*. El área muestra una pobre abundancia y riqueza.

Rápidos arriba del Raudal Culebra (ICT-31). Un total de 13 ejemplares pertenecientes a 6 especies fueron colectados. Todas las especies ubicadas en el Orden Characiformes. Las especies incluen a: *Hoplias macrophthalmus, Jupiaba zonata, Melonocharacidium depressum, Moenkhausia* cf. *lepidura* (2 spp.) y *Moenkhausia miangi*. Todos los ejemplares fueron capturados asociados a plantas sumergidas de la Familia Podostemonaceae. Los rápidos con baja abundancia y diversidad.

II. BAJO CAURA, POR DEBAJO DEL SALTO PARÁ

BC01. El Playón (Salto Pará) (Estaciones ICT-32, ICT-33, ICT-35, ICT-36, ICT-38) (6°19.5´N–64°31.6´W)

Descripción general de la ictiofauna

Playa (ICT-32, ICT-33). Se colectaron peces utilizando varios sistemas: atarraya (ICT-32), chinchorros y trampas (ICT-33). La combinación de las tres muestras muestran una alta diversidad (37 especies) y moderada a alta abundancia (519 ejemplares) ubicados en:

Characiformes: 27 (73%)
Siluriformes: 6 (16.2%)
Perciformes: 2 (5.4%)
Rajiformes:1 (2.7%)
Gymnotiformes: 1 (2.7%)

Las especies más abundantes fueron: *Knodus* sp. B (106, 20.4%), *Pimelodella* cf. *megalops* (91, 17.5%), *Anostomus ternetzi* (83, 15.9%) y *Hyphessobrycon minimus* (23, 4.4%). También, fueron colectados ejemplares de las siguientes especies: *Brycon pesu, Bryconamericus cismontanus, Caenotropus laberinthicus, Creagrutus maxillaris, Curimata incompta, Cynopotamus essequibensis, Geophagus* cf. *brachybranchus, Eigenmannia macrops, Hemisorubim platyrhynchus, Homodiaetus* sp., *Microschemobrycon callops, M. casiquiare, Moenkhausia gracilima, Pimelodella cruxenti, Potamorhina altamazonica, Psectrogaster essequibensis, Schizodon* sp., *Semaprochilodus kneri, Serrasalmus* sp. A, *Synaptolaemus cingulatus, Tetragonopterus chalceus, Triportheus albus* y una especie de bagre parásito *Paravandellia* sp.

Las especies colectadas en su mayoría están presentes en áreas de remanso tranquilas: *Microschemobrycon callops, M. casiquiare, Moenkhausia gracilima, Geophagus* cf. *brachybranchus*, y sobre playas arenosas: *Brycon pesu, Bryconamericus cismontanus, Caenotropus laberinthicus, Creagrutus maxillaris, Corydoras, Characidium, Curimata incompta, Cynopotamus essequibensis, Eigenmannia macrops, Hemisorubim platyrhynchus, Homodiaetus* sp., *Pimelodella cruxenti, Potamorhina altamazonica, Psectrogaster essequibensis, Schizodon* sp. Esta es un área importante dada la gran riqueza de especies y abundante material. La presencia de bagres parásitos indica igualmente la presencia de grandes bagres pimelódidos incluyendo especies comestibles como: *Semaprochilodus kneri* y *Piaractus brachypomus*. Observamos pescadores capturando ejemplares de gran tamaño de *Piaractus brachypomus*.

Debemos destacar la presencia de numerosos ejemplares de *Anostomus ternetzi* y *Synaptolaemus cingulatus*, capturados con trampas. Esta ultima especie muy rara en colecciones.

Isla (ICT-35). Un total de 344 ejemplares fueroin capturados y ubicados en 35 especies distribuidos en:

Characiformes: 18 (51.4%)
Siluriformes: 10 (28.6%)
Perciformes: 2 (5.7%)
Gymnotiformes: 1 (2.8%)

Las especies más abundantes fueron: *Knodus* sp. B (75, 21.8%), *Moenkhausia* cf. *lepidura* (67, 19.5%), *Moenkhausia cotinho* (60, 17.4%), *Pimelodella megalops* (27, 7.0%) y *Nanoptopoma spectabilis* (25, 7.3%). También, fueron colectados ejemplares de las siguientes especies: *Ancistrus* sp., *Anostomus ternetzi, Aphyocharax alburnus, Astyanax integer, Bryconamericus cismontanus, Bryconops* cf. *colaroja, Bunocephalus* cf. *aleuropsis, Cregrutus maxillaris, Crenicichla wallacei, Cyphocharax modestus, Farlowella vittata, Jupiaba atypindi, J. Polylepis, Microglanis iheringi, Moenkhausia copei, M. grandisquamis, Ochmacanthus alternus, Pimelodella cruxenti, Satanoperca* sp., *Sternopygus macrurus* y *Tatia romani.*

Muchas de las especies están asociadas a sustratos rocosos o fondos arenosos y de grava tales como: *Ancistrus, Anostomus, Bryconamerius, Knodus, Ochmacanthus, Creagrutus, Nanoptopoma, Bryconamericus.* Otras son caracteristicas de áreas protegidas como: *Anostomus ternetzi, Aphyocharax alburnus, Astyanax integer, Bunocephalus, Moenkhausia* cf. *lepidura, M. cotinho, M. copei, M. grandisquamis, Pimelodella megalops, P. cruxenti, Satanoperca, Sternopygus macrurus* y *Tatia romani.* Diversidad y abundancia moderada a alta.

Playa rocosa (ICT-36). Se capturaron 238 ejemplares y 13 especies ubicados todos en el Orden Characiformes.

Las especies más abundantes fueron: *Moenkhausia cotinho* (91, 38.2%), *Moenkhausia* cf. *lepidura* (82, 34.4%), *Jupiaba atypindi* (41, 17.2%), *Anostomus ternetzi* (7, 2.9%) y *Bryconamericus cismontanus* (3, 1.2%). También, fueron colectados ejemplares de las siguientes especies: *Apareiodon orinocensis, Bryconops giacopinii, Creagrutus maxillaris, Cyphocharax modestus* y *Jupiaba polylepis*, propias de elementos asociados a raíces palos, piedras y también con concentraciones de detritus en pocos remansos y aguas rápidas. Peces abundancia y diversidad baja.

Rocas (ICT-38). Se capturaron 107 ejemplares pertenecientes a 10 especies ubicados en:

Characiformes: 9 (90%)
Siluriformes: 1 (10%)

Las especies más abundantes fueron: *Hyphessobrycon minimus* (97, 90.1%) y *Moenkhausia* cf. *lepidura* (2, 1.8%). También, fueron colectados ejemplares de las siguientes especies: *Ancistrus* sp. C, *Anostomus ternetzi, Bryconamericus cismontanus, Curimata incompta, Creagrutus maxillaris, Cynodon gibus, Melanocharacidium dispiloma* y *Triportheus albus.* Además, un pescador mostró capturas de dos especies de *Leporinus* (*L. brunneus* y *L. friderici*) y ejemplares de *Piaractus brachypomus* pescados con anzuelo. Peces con abundancia y diversidad baja.

BC02. Caño (Tributario) de confluencia en el Río Caura cerca del Playón (Estación ICT-34) (6°19.7´N–64° 31.6´W)

Descripción general de la ictiofauna.

Se capturaron 720 ejemplares pertenecientes a 45 especies, ubicados en:

Characiformes: 27 (60.0%)
Siluriformes: 11 (24.4%)
Perciformes: 4 (8.8%)
Clupeiformes: 1 (2.2%)
Gymnotiformes: 1 (2.2%)
Atheriniformes: 1(2.2%)
Synbranchiformes: 1 (2.2%)

Las especies más abundantes fueron: *Ochmacanthus alternus* (81, 11.3%), *Phenacogaster* sp. A (65, 9%), *Bryconamericus cismontanus* (65, 9%), *Creagrutus maxillaris* (59, 8.2%), *Moenkhausia cotinho* (59, 8.2%), *Pimelodella megalops* (53, 7.3%), *Hyphessobrycon minimus* (49, 6.8%) y *Aphyocharax alburnus* (44, 6.1%). También, fueron colectados ejemplares de las siguientes especies: *Anchoviella jamesi, Anostomus ternetzi, Astyanax integer, Brachyhypopomus occidentalis, Bryconops* sp. B, *Bujurquina mariae, Bunocephalus aleuropsis, Corydoras blochii, Crenicichla* sp., *Cyphocharax festivus, C. modestus, C. oenas, Hoplias macrophthalmus, H. malabaricus, Hypostomus plecostomus, Jupiaba atypindi, J. polylepis, J. zonata, Leporinus maculatus, Limatulichthys punctatus, Loricaria cataphracta, Loricariichthys brunneus, Microschemobrycon casiquiare, Moenkhausia* cf. *lepidura, M. copei, Odontostilbe* cf. *fugitiva, Poecilia* sp., *Ramirezella newboldi, Synbranchus marmoratus* y *Xenagoniates bondi.*

En resumen, el análisis de la ictiofauna muestra una abundancia moderada con algunos elementos bien representados como: *Pimelodella, Phenacogaster, Bryconamericus, Creagrutus* y *Ochmacanthus.* Diversidad de esta estación es comparativamente una de las más altas obtenidas en el Río Caura, con 46 especies presentes. Hay especies típicas y provenientes del río principal tales como: *Bryconops, Jupiaba, Moekhausia* y *Ramirezella* y otros comunes de áreas fangosas, cerradas como lagunas de inundación y remansos como: *Apistogramma, Bujurquina, Bunocephalus, Ochmacanthus, Brachyhypopomus, Hyphessobrycon* y *Microglannis.* Debemos hacer notar el primer record de una "sardina" (*Anchoviella jamesi*) para el Río Caura.

BC03. Riachuelo o Caño Waki, varios km por debajo del Salto Pará (Estación ICT-37, ICT-38A) (6°21.7´N–64°34.1´W)

Descripción general de la ictiofauna

Caño (ICT-37). Un total de 504 ejemplares pertenecientes a 24 especies y ubicados en:

Characiformes: 19 (79.2%)
Siluriformes: 2 (8.3%)
Perciformes: 2 (8.3%)
Synbranchiformes: 1 (4.2%)

Las especies más abundantes fueron: *Microschemobrycon melanotus* (237, 47%), *Moenkhausia* cf. *lepidura* (75, 14.9%), Characidae sp. B (43, 8.6%), *Moenkhausia copei* (41, 8.1%) y *Jupiaba zonata* (32, 6.4%). También, fueron colectados ejemplares de las siguientes especies: *Aphyocharax alburnus, Apistogramma* sp. B, *Bujurquina mariae, Characidium* sp. A, *Cyphocharax oenas, C. spilurus, Hyphessobrycon minimus, Jupiaba polylepis, Microchemobrycon casiquiare, Moekhausia cotinho, M. grandisquamis, M. oligolepis, Ochmacanthus alternus, Poptella longipinnis, Rineloricaria* sp. y *Synbranchus marmoratus*. El análisis de la ictiofauna muestra que la mayoría de las especies son comunes a áreas de bosque inundado y/o en la boca de riachuelos, con abundantes hojarasca y fondos fangosos.

Boca del Caño (ICT-38A). Se utilizó una red de ahorque se capturaron 9 ejemplares de payara, *Hydrolicus tatauaia* y un ejemplar de *Serrasalmus rhombeus*.

BC04. Río Tawadu, Campamento Dedemai (Estación ICT-39, ICT-54, ICT- 55) (6°21.1´N–64°59.9´W)

Descripción general de la ictiofauna

Campamento (ICT-39). Se utilizaron para la colecta trampas y un chinchorro. Un total de 164 ejemplares fueron capturados pertenecientes a 14 especies y ubicados en:

Characiformes: 10 (71.4%)
Siluriformes: 2 (14.3%)
Perciformes: 1 (7.1%)
Beloniformes: 1 (7.1%)

Las especies más abundantes fueron: *Hyphessobrycon minimus* (126, 76.8%), *Moenkhausia copei* (11, 6.7%), *Moenkhausia* cf. *lepidura* (5, 3%), *M. oligolepis* (5, 3%) y *Knodus* sp. A (5, 3%). También, fueron colectados ejemplares de las siguientes especies: *Anostomus ternetzi, Astyanax integer, Bujurquina mariae, Microschemobrycon melanotus, Moenkhausia collettii, Ochmacanthus alternus, Pimelodella* cf. *cruxenti* y *Potamorrhaphis guianensis*.

Playa (ICT-54). Se utilizaron para la colecta chinchorros. Un total de 304 ejemplares pertenecientes a 22 especies y ubicados en:

Characiformes: 16 (72.7 %)
Perciformes: 3 (13.6%)
Siluriformes: 2 (9.1%)
Gymnotiformes: 1 (4.5 %)

Las especies más abundantes fueron: *Hyphessobrycon minimus* (63, 20.7%), *Moenkhausia copei* (60, 19.7%), *Bryconamericus* cf. *cismontanus* (59, 19.1%), *Moenkhausia* cf. *lepidura* (38, 12.3%) y *Aphyocharax alburnus* (15, 4.8%). También, fueron colectados ejemplares de las siguientes especies: *Acestrorhynchus microlepis, Aequidens chimantanus, Bujurquina mariae, Characidium* sp. A, *Crenicichla wallacei, Gymnotus anguilaris, Hoplias macrophthalmus, Jupiaba polylepis, Microschemobrycon melanotus, Moenkhausia collettii, M. oligolepis, Ochmacanthus alternus, Phenacogaster* sp. A, *Poptella longipinnis, Pseudopimelodus raninus* y *Ramirezella newboldi*. El análisis de la ictiofauna muestra numerosas especies típicas de playas areno-fangosas con abundante detritus.

Caño (ICT-55). Se utilizaron para la colecta chinchorros pequeños. Un total de 39 ejemplares pertenecientes a 9 especies y ubicados en:

Characiformes: 8(88.8 %)
Perciformes: 1(11.2%)

Las especies más abundantes fueron: *Hyphessobrycon minimus* (16, 41%), *Microschemobrycon melanotus* (9, 23.1%), *Bujurquina mariae* (6, 15.4%), *Microschemobrycon casiquiare* (3, 7.7%). También, fueron colectados ejemplares de las siguientes especies: *Characidium* sp. C, *Moenhausia* cf. *lepidura, Moenkhausia hemigramoides, M. oligolepis* y *Tetragonopterus chalceus*.

En resumen podemos indicar que el área cercana a la Estación Dedemai posee una moderada diversidad y poca abundancia. Las mayoria de las especies presentes son típicas de fondos fangoso-arenoso, con elementos típicos de aguas claras y ácidas tales como: *Hyphessobrycon minimus, Microschemobrycon* (3 sp.), *Bujurquina mariae, Characidium* sp. C, *Moenkhausia collettii, Moenhausia* cf. *lepidura, M. hemigramoides, M. oligolepis, Ochmacanthus alternus, Poptella longipinnis, Pseudopimelodus raninus* and *Ramirezella newboldi*. También fue capturado un ejemplar de pez eléctrico (*Gymnotus anguilaris*) el cual representa un nuevo record para el Caura.

BC05. Río Tawadu o Tabaro, Raudal el Pan (ICT-40) y Raudal Dimoshi (ICT-41, ICT-42) (6°19.63´N–65°02.86´W)

Descripción general de la ictiofauna

Raudal El Pan (ICT-40). Se utilizó para la colecta un chinchorro pequeño. Un total de 90 ejemplares pertenecientes a 4 especies todas del Orden Characiformes

Las especies fueron: *Hyphessobrycon minimus* (48, 53.3%), *Knodus* sp. A (38, 42.2%), *Microschemobrycon callops* (3, 3.3%) y *Characidium* sp. A (1, 1.1%).

Raudal Dimoshi (ICT-41). Un total de 246 ejemplares capturados pertenecientes a 25 especies y ubicados en:

Characiformes: 9 (60%)
Perciformes: 3 (20%)
Siluriformes: 2 (13.3%)
Synbranchiformes: 1 (6.6%)

Las especies más abundantes fueron: *Hemigrammus* cf. *guyanensis* (113, 45.9%), *Knodus* sp. A (63, 25.6%), *Bryconops* cf. *colaroja* (24, 9.7%), *Bryconops* sp. B (20, 8.1%) y *Hemibrycon metae* (17, 6.9%). También, fueron colectados ejemplares de las siguientes especies: *Ancistrus* sp. C, *Bujurquina mariae, Characidium* sp., *Crenicichla geayi, Melanocharacidium depressum, Pimelodella* sp. y *Synbranchus marmoratus*. La mayoría de las especies son típicas de aguas rápidas y fondos arenosos asociadas igualmente a plantas acuáticas.

Caño (ICT-42). Un total de 112 ejemplares pertenecientes a 5 especies y ubicadas en:

Characiformes: 4 (80%)
Perciformes: 1 (20%)

Las especies fueron: *Hemigrammus* cf. *guyanensis* (60, 53.6%), *Knodus* sp. A (38, 33.9%), *Bujurquina mariae* (13, 11.6%), *Astyanax integer* (5, 4.5%) y *Astyanax* sp. (4, 3.6%). El caño a pesar de su buena aparente condición ecológica, mostró muy poca ictiofauna. Posiblemente, este resultado es debido a que estos caños son frecuentemente envenenados con barbasco.

Raudal Dimoshi área inferior (ICT-43). Canal principal de río, con aguas claras y profundas. Se colectaron dos ejemplares de dos especies con anzuelo: *Crenicichla* cf. *lenticulata* y *Serrasalmus* sp. A.

En resumen el análisis de la ictiofauna presente en la región muestra una moderada a baja diversidad y abundancia. Gran número de especies asociadas a aguas rápidas, fondos arenoso-pedregoso tales como: *Ancistrus* sp. C, *Bryconops* (2 sp.), *Characidium* sp., *Hemigrammus* cf. *guyanensis*, *Knodus* sp. A, *Hemibrycon metae*, *Melanocharacidium depressum* y *Pimelodella* sp.

BC06. Río Tawadu (Tabaro), Raudal Tajañaño (ICT-44, ICT-45) (6°20.25´N–65°01.50´W)

Descripción general de la ictiofauna

Isla Rocosa y Rápidos (ICT-44). Un total de 284 ejemplares fueron capturados pertenecientes a 8 especies ubicadas en:

Characiformes: 5 (62.5%)
Siluriformes: 2 (25%)
Perciformes: 1 (12.5%)

Las especies más abundantes fueron: *Hyphessobrycon minimus* (223, 78.5%), *Microschemobrycon callops* (26, 9.2%), *Ochmacanthus alternus* (13, 4.6%), *Knodus* sp. B (11, 3.9%) y *Characidium* sp. A (8, 2.8%). También, fueron colectados ejemplares de las siguientes especies: *Astyanax integer, Crenicichla geayi* y *Rineloricaria* sp. El análisis muestra una dominancia de una sola especie de Characidae (*Hyphessobrycon minimus*), la cual se encontró en áreas protegidas. Otras formas son típicas de rápidos y sobre fondos areno-pedregoso como: *Characidium* sp. A y *Rineloricaria* sp.

Boca del caño y rápido (ICT-45). Un total de 58 ejemplares fueron capturados pertenecientes a 15 especies ubicadas en:

Siluriformes: 8 (53.3%)
Characiformes: 5 (33.3%)
Perciformes: 1 (6.6%)
Synbranchiformes: 1 (6.6%)

Las especies más abundantes fueron: *Ochmacanthus alternus* (15, 25.9%), *Ancistrus* sp. C (14, 24.1%), *Ammocryptocharax elegans* (11, 18.9%), *Farlowella vittata* (6, 10.3%) y *Characidium* sp. A (5, 8.6%). También, fueron colectados ejemplares de las siguientes especies: *Bryconops* sp., *Chasmocranus* sp., *Crenicichla wallacei, Lasiancistrus* sp., *Melanocharacidium nigrum, Microglannis poecilus, M. iheringi, Moenkhausia* cf. *lepidura, Rineloricaria* sp. y *Synbranchus marmoratus.*

En resumen el área muestra baja abundancia y diversidad. Rápidos e isla dominada por una sóla especie (*Hyphessobrycon minumus*). En el caño y margen izquierda hubo una dominancia de Siluriformes asociados a troncos y ramas sumergidos y fondos fangosos tales como: *Ancistrus, Microglannis, Farlowella, Lasiancistrus* y *Rineloricaria.* También se colectaron especies típicas de areas de rápidos con fondos arenosos y pedregosos tales como: *Ammocryptocharax elegans, Melanocharacidium nigrum* y *Characidium* sp. A. Otras especies son comunes a áreas de remanso o protegidas: *Crenicichla wallacei, Microschemobrycon callops, Microglannis poecilus, M. iheringi* y *Synbranchus marmoratus.*

BC07. Laguna 1 km de la Boca del Río Tawadu (Tabaro) (ICT-46, ICT-51) (6°21.83´N–64°58.4´W)

Descripción general de la ictiofauna

Se tomaron muestras utilizando una red de ahorque (ICT-46) y chinchorros playeros (ICT-51). La primera muestra se colectó un total de 84 ejemplares pertenecientes a 12 especies y ubicadas en:

Characiformes: 9 (75%)
Siluriformes: 1 (8.3%)
Perciformes: 1 (8.3%)
Synbranchiformes (8.3%)

Las especies más abundantes fueron: *Hydrolicus tatauaia* (37, 44%), *Semaprochilodus laticeps* (28, 33.3%), *Serrasalmus rhombeus* (3, 3.5%) y *Triportheus albus* (3, 3.5%). También, fueron colectados ejemplares de las siguientes especies: *Boulengerella xirekes, Hydrolicus armatus, Leporinus brunneus, Oxydoras kneri* y *Rhaphiodon vulpinus.*

Peces de gran porte con baja diversidad y abundancia baja excepto un número moderado de "payaras" (*Hydrolicus tatauaia*) y "sapoaras" (*Semaprochilodus laticeps*) capturadas. El área es potencialmente útil para actividades pesqueras.

La colecta con redes de playa (ICT-51) y ahorque (ICT-46) combinadas muestra una colección de 1977 ejemplares capturados pertenecientes a 50 especies ubicadas en:

Characiformes: 35 (70%)
Siluriformes: 8 (16%)
Perciformes: 4 (8.%)
Synbranchiformes 1 (2%)
Gymnotiformes: 1 (2%)
Pleuronectiformes: 1 (2%)

Las especies más abundantes fueron: *Pimelodella chagresi* (913, 46.2%), *Ochmacanthus alternus* (355, 18%), *Bryconamericus* cf. *cismontanus* (191, 9.7%), *Moenkhausia collettii* (69, 3.5%) y *Corydoras* cf. *bondi* (56, 2.8%). También, fueron colectados ejemplares de las siguientes especies: *Acestrorhynchus microlepis, Achirus* sp., *Aphyocharax alburnus, A. erythrurus, Apistogramma* sp. A, *Astyanax integer, Boulengerella xirekes, Bujurquina mariae, Cyphocharax festivus, C. oenas, Hemiodus unimaculatus, Hydrolicus armatus, Hydrolicus tatauaia, Hyphessobrycon minimus, Jupiaba polylepis, Leporinus brunneus, Limatulichthys punctatus, Mesonauta egrerius, Microschemobrycon callops, M. casiquiare, Moenkhausia* cf. *lepidura, M. copei, M. cotinho, M. grandisquamis, M. oligolepis, Odontostilbe fugitiva, Oxydoras kneri, Phenacogaster* sp. A, *Pimelodella cruxenti, P. megalops, Pimelodella* sp., *Poptella longipinnis, Pyrrhulina brevis, Ramirezella newboldi, Rhaphiodon vulpinus, Semaprochilodus laticeps, Serrasalmus rhombeus, Steindachnerina pupula, Synbranchus marmoratus, Tetragonopterus chalceus* y *Triportheus albus.*

El análisis de la ictiofauna (ambas muestras combinadas), muestra una gran diversidad de formas relacionadas a hábitats protegidos lagunares o de bosque inundado. Es importante hacer notar la presencia de especies de gran talla e importancia comercial o de consumo humano: *Hydrolicus armatus, Hydrolicus tatauaia, Semaprochilodus laticeps* y *Serrasalmus rhombeus*. Información obtenida de los pobladores muestran que esta laguna es frecuentada por pescadores comerciales. También existe un gran número de especies de importancia ornamental como: *Aphyocharax alburnus, A. erythrurus, Apistogramma* sp. A, *Astyanax integer, Boulengerella xirekes, Bujurquina mariae, Hemiodus unimaculatus, Hyphessobrycon minimus, Jupiaba polylepis, Leporinus brunneus, Limatulichthys punctatus, Mesonauta egrerius, Microschemobrycon callops, M. casiquiare, Moenkhausia* cf. *lepidura, M. copei, M. cotinho, M. grandisquamis, M. oligolepis, Poptella longipinnis, Pyrrhulina brevis, Ramirezella newboldi, Tetragonopterus chalceus* y *Triportheus albus*.

BC08. Río Icutú, Caño Wani (ICT-47, ICT-48) (6°04.77´N–64°55.33´W)

Descripción general de la ictiofauna

Playa arenosa (ICT-47). Un total de 563 ejemplares capturados pertenecientes a 35 especies ubicadas en:

Characiformes: 24 (68.5%)
Siluriformes: 6 (17.1%)
Perciformes: 2 (5.7%)
Clupeiformes: 1 (2.8 %)
Pleuronectiformes: 1 (2.8%)
Synbranchiformes: 1 (2.8%)

Las especies más abundantes fueron: *Moenkhausia collettii* (186, 33%), *Hyphessobrycon minimus* (73, 12.9%), *Moenkhausia cotinho* (72, 12.8%), *Pseudocheirodon* sp. (39, 6.9%) y *Characidium* sp. A (29, 5.1%). También, fueron colectados ejemplares de las siguientes especies: *Achirus* sp., *Anchoviella jamesi, Anostomus ternetzi, Aphyocharax alburnus, Astyanax integer, Bryconamericus* cf. *cismontanus, Bujurquina mariae, Cochliodon* sp., *Creagrutus maxillaris, Farlowella mariaelenae, F. vittata, Jupiaba atypindi, J. polylepis, Microschemobrycon callops, M. casiquiare, M. melanotus, Ochmacanthus alternus, Pimelodella* sp. A, *Ramirezella newboldi, Satanoperca* sp., *Steindachnerina pupula, Synbranchus marmoratus, Tetragonopterus* sp. y *Paravandellia* sp.

Caño (ICT-48). Un total de 693 ejemplares capturados pertenecientes a 41 especies ubicadas en:

Characiformes: 28 (68.3%)
Siluriformes: 7 (17.1%)
Perciformes: 3 (7.3%)
Gymnotiformes: 2 (4.9%)
Synbranchiformes: 1 (2.4%)

Las especies más abundantes fueron: *Moenkhausia collettii* (204, 29.4%), *Moenkhausia cotinho* (83, 12%), *Phenacogaster* sp. A (46, 6.6%), *Bryconamericus* cf. *cismontanus* (45, 6.5%) y *Ramirezella newboldi* (42, 6.1%). También, fueron colectados ejemplares de las siguientes especies: *Aphyocharax alburnus, Apistogramma* sp. A, *Astyanax integer, Brachychalcinus occidentalis, Bujurquina mariae, Characidium* sp. A, *Corydoras* cf. *bondi, Cyphocharax oenas, Hoplias macrophthalmus, Hyphessobrycon minimus, Jupiaba atypindi, J. polylepis, J. zonata, Loricaria cataphracta, Melanocharacidium dispiloma, Microschemobrycon callops, M. casiquiare, M. melanotus, Moenkhausia* cf. *lepidura, M. copei, Ochmacanthus alternus, Odontostilbe* sp., *Pimelodella* sp. A, *Poptella longipinnis, Pseudocheirodon* sp., *Rineloricaria* sp., *Satanoperca* sp., *Sternopygus macrurus, Synbranchus marmoratus* y *Tetragonopterus* sp.

El área muestra una diversidad y abundancia impresionantes. Muchas especies están asociadas a fondos fangoso-arenosos tales como: *Aphyocharax alburnus, Apistogramma* sp. A, *Astyanax integer, Brachychalcinus occidentalis, Bujurquina mariae, Characidium* sp. A, *Corydoras* cf. *bondi, Cyphocharax oenas, Hoplias macrophthalmus, Hyphessobrycon minimus, Microschemobrycon callops, M. casiquiare, M. melanotus* y *Sternopygus macrurus*. Otros son típcos de areas protegidas de remansos o bosques inundados: *Bujurquina mariae, Satanoperca* sp., *Poptella longipinnis, Pseudocheirodon* sp., *Rineloricaria* sp. y *Synbranchus marmoratus*. Debemos hacer notar la presencia de un "lenguado" *Achirus* sp. el cual representa el primer record de una especie de Pleuronectiformes para el Sistema del Río Caura. El área representa una de las ricas y abundantes de todo el Sistema del Nichare.

BC09. Laguna Wakawajai (ICT-49, ICT-50) (6°19.35´N–64°57.22´W)

Descripción general de la ictiofauna

Playa arenosa (ICT-49). Un total de 848 individuos pertenecientes a 33 especies fueron capturados y ubicados en:

Characiformes: 24 (72.7%)
Siluriformes: 8 (24.2%)
Perciformes: 1 (3.1%)

Las especies más abundantes fueron: *Paravandellia* sp. (254, 30%), *Aphyocharax alburnus* (180, 21.2%), *Hyphessobrycon minimus* (110, 12.9%), *Bryconamericus* cf. *cismontanus* (56, 6.6%), *Knodus* sp. B (45, 5.3%) y *Moenkhausia* cf. *lepidura* (39, 4.6%). También, fueron colectados ejemplares de las siguientes especies: *Astyanax integer, Bryconops* sp. B, *Bujurquina mariae, Characidium* sp., *Corydoras bondi, Creagrutus maxillaris, Cyphocharax oenas, Hoplias macrophthalmus, Jupiaba atypindi, J. polylepis, Mastiglannis asopos, Microschemobrycon casiquiare, Moenkhausia copei, M. cotinho, Myleus rubripinnis Ochmacanthus alternus, Pimelodella megalops, Pimelodella* sp. A, *Pimelodella* sp. B, *Pimelodus blochii, Ramirezella newboldi, Serrasalmus rhombeus* y *Vandellia sanguinea*. Es interesante hacer notar la presencia de abundantes ejemplares de bagres parásitos, como: *Paravandellia* sp. y *Vandellia sanguinea*.

Madrevieja o remanso (ICT-50). Un total de 326 ejemplares pertenecientes a 22 especies y ubicados en:

Characiformes: 17 (77.2%)
Siluriformes: 4 (18.8%)
Perciformes: 1 (4.5%)

Las especies más abundantes fueron: *Jupiaba polylepis* (57, 17.5%), *Moenkhausia cotinho* (51, 15.6%), *Moen-*

khausia copei (50, 15.3%), *Bryconamericus* cf. *cismontanus* (50, 15.3%) y *Hyphessobrycon minimus* (45, 13.8%). También, fueron colectados ejemplares de las siguientes especies: *Aphyocharax alburnus, Characidium* sp. (2 especies), *Corydoras bondi, Cyphocharax meniscaporus, C. oenas, Hemigrammus tridens, Hoplias macrophthalmus, Hypostomus* sp. B, *Loricariichtys brunneus, Mesonauta egregius, Microschemobrycon casiquiare, M. melanotus, Moenkhausia collettii, Ochmacanthus alternus* y *Phenacogaster* sp. A. Peces diversidad moderada y abundancia baja con excepción de caracoides pequeños. Especies típicas de aguas tranquilas como *Mesonauta egregius, Moenkhausia copei, Bryconamericus* cf. *cismontanus* y *Hyphessobrycon minimus.*

El área mostró una moderada diversidad y baja a moderada abundancia. Como indicamos anteriormente es importante indicar la presencia de abundantes bagres parásitos. También el área está caracterizada por especies asociadas a bancos arenosos como: *Characidium* sp. (2 especies), *Corydoras bondi, Cyphocharax meniscaporus, C. oenas, Loricariichtys brunneus, Ochmacanthus alternus* y *Phenacogaster* sp. A. Otros comúnmente encontrados en áreas protegidas y lagunares tales como: *Mesonauta egregius, Moenkhausia copei, Bryconamericus* cf. *cismontanus* y *Hyphessobrycon minimus.*

BC10. Isla en el Río Nichare, aguas abajo Boca de Tawadu (ICT-52) (6°21.43´N–64°52.29´W)

Descripción general de la ictiofauna

Un total de 740 individuos pertencientes a 44 especies fueron capturados y ubicados en:

Characiformes: 30 (68.2%)
Siluriformes: 9 (20.5%)
Perciformes: 4 (9%)
Clupeiformes: 1 (2.2%)

Las especies más abundantes fueron: *Bryconamericus* cf. *cismontanus* (100, 13.5%), *Hyphessobrycon minimus* (93, 12.6%), *Aphyocharax alburnus* (81, 10.9%), *Moenkhausia* cf. *lepidura* (75, 10.1%) y *Knodus* sp. (73, 9.9%). También, fueron colectados ejemplares de las siguientes especies: *Anchoviella guianensis, Astyanax integer, Bujurquina mariae, Characidium* sp., *Corydoras bondi, Creagrutus maxillaris, Crenicichla geayi, C. wallacei, Cyphocharax meniscaporus, C. oenas, Hemigrammus* cf. *tridens, Hemiodus unimaculatus,* Iguanodectini (no identificado), *Jupiaba atypindi, J. polylepis, Leporinus maculatus, Loricaria cataphracta, Mastiglannis asopos, Mesonauta egrerius, Microschemobrycon callops, M. casiquiare, Moenkhausia* cf. *lepidura* (2 sp.), *M. collettii, M. cotinho, M. grandisquamis, M. oligolepis, Ochmacanthus alternus, Phenacogaster* sp. A, *Pimelodella megalops, Pimelodella* sp. A, *Poptella longipinnis, Pkseudohemiodon* sp., *Ramirezella newboldi, Rineloricaria* sp., *Satanoperca* sp., *Steindachnerina pupula, Tetragonopterus* sp. y *Vandellia sanguinea.*

En resumen se puede indicar que el área muestra una moderada abundancia y diversidad alta con muchas especies asociadas a rápidos con fondos arenoso-pedregoso: *Characidium* sp., *Corydoras bondi, Creagrutus maxillaris, Cyphocharax meniscaporus, C. oenas, Hemigrammus* cf. *tridens, Hemiodus unimaculatus,* Iguanodectini (n.i.), *Jupiaba atypindi, J. polylepis, Leporinus maculatus, Loricaria cataphracta, Mastiglannis asopos* y áreas de remanso: *Bujurquina mariae, Crenicichla geayi, C. wallacei, Moenkshausia collettii, M. cotinho, M. grandisquamis, M. oligolepis, Pimelodella* sp. A, *Poptella longipinnis, Pkseudohemiodon* sp., *Ramirezella newboldi, Rineloricaria* sp. y *Satanoperca* sp.

BC11. Playas en el Tawadu (ICT-53) (6°21.01´N–64°58.44´W)

Descripción general de la ictiofauna

Un total de 1100 ejemplares y pertenecientes a 29 especies fueron capturados y ubicados en:

Characiformes: 24 (82.7%)
Perciformes: 3 (10.3%)
Siluriformes: 2 (6.9%)

Las especies más abundantes fueron: *Microschemobrycon melanotus* (323, 29.4%), *Corydoras bondi* (148, 13.5%), *Hyphessobrycon minimus* (129, 11.7%), *Moenkhausia collettii* (112, 10.2%) y *Ochmacanthus alternus* (73, 6.6%). También, fueron colectados ejemplares de las siguientes especies: *Aphyocharax alburnus, Apistogramma* sp. A, *Bryconamericus* cf. *cismontanus, Bujurquina mariae, Characidium* sp. A, *Crenicichla* sp. B, *Cyphocharax oenas, Hemigrammus* cf. *tridens, Hemiodus unimaculatus, Hoplias malabaricus, Jupiaba polylepis, Knodus* sp. B, *Microschemobrycon casiquiare, M. melanotus, Moenkhausia* cf. *lepidura* (2 sp.), *M. copei, M. cotinho, Odontostilbe* cf. *fugitiva, Poptella longipinnis, Ramirezella newboldi* y *Tetragonopterus* sp.

El área de playas del Tawadu muestra una ictiofauna abundante. Sin embargo, la riqueza es baja predominando especies asociadas a fondos arenoso-fangosos.

BC12. Laguna de Ceiba (ICT-56) (7° 05.9´N–65° 01.34´W)

Descripción general de la ictiofauna

Un total de 811 ejemplares pertenecientes a 36 especies fueron capturados y ubicados en:

Characiformes: 26 (72.2%)
Siluriformes: 7 (19.4%)
Perciformes: 2 (5.5%)
Atheriniformes: 1 (2.7%)

Las especies más abundantes fueron: *Moenkhausia collettii* (326, 40.2%), *M.* cf. *chrysargyrea* (145, 17.9%), *Microschemobrycon melanotus* (76, 9.4%), *Aphyocharax erythrurus* (46, 5.7%) y *Ramirezella newboldi* (35, 4.3%). También, fueron colectados ejemplares de las siguientes especies: *Acestrorhynchus microlepis, Aequidens* cf. *chimantanus, Apistogramma iniridae, Apistogramma* sp. A, *Bryconamericus* cf. *cismontanus, Crenicichla lenticulata, Crenicichla* sp. A, *Cyphocharax oenas, Hemigrammus* cf. *tridens, Hemigrammus marginatus, Hoplias macrophthalmus, H. malabaricus, Hyphessobrycon bentosi, H. serpae, Jupiaba polylepis, Microschemobrycon melanotus, Moenkhausia* cf. *lepidura, M. copei, M. oligolepis, Nannostomus erythrurus, Orinocodoras eigenmanni, Phenacogaster* sp. A, *Platydoras armatulus, Prochilodus mariae, Pseudocheirodon* sp., *Pygocentrus cariba, Ramirezella newboldi, Rivulus* sp. A,

Satanoperca cf. *mapiritensis*, *Serrasalmus elongatus* y *Serrasalmus* sp. A.

Peces con diversidad y abundancia moderada a alta, con las primeras muestras del "caribe colorado" (*Pygocentrus cariba*). Además, dos especies de "caribes pinches" (*Serrasalmus elongatus* y *Serrasalmus* sp. A) y "guabinas" (*Hoplias malabaricus*) de talla grande. Se colectaron también algunas especies de cíclidos (*Apistogramma iniridae, Crenicichla* spp., *Satanoperca mapiritensis*), caracoideos (*Hyphessobrycon bentosi, H. serpae, Moenkhausia copei, M. oligolepis, Nannostomus erythrurus*) y siluroideos (*Orinocodoras eigenmanni* y *Platydoras armatulus*) típicas de aguas tranquilas y lagunares.

BC13. Río Mato (ICT-58, ICT-59, ICT-60) (7°11.74´N–65° 09.75´W)

Descripción general de la ictiofauna

El área fue colectada con una red de ahorque (ICT-58) y redes tradicionales (ICT-59, ICT60). Con red de ahorque, se colectaron 15 ejemplares pertenecientes a 6 especies y ubicadas en:

Characiformes: 5 (83.3%)
Clupeiformes: 1 (16.7%)

Las especies fueron: *Triportheus albus* (7, 46.6%), *Curimata cyprinoides* (3, 20%), *Pellona castelneana* (2, 13.3%), *Acestrorhynchus heterolepis* (1, 6.6%), *Pygocentrus cariba* (1, 6.6%) y *Serrasalmus rhombeus* (1, 6.6%).

Con las redes de playa (ICT-59) se colectaron 590 ejemplares pertenecientes a 46 especies ubicadas en:

Characiformes: 34 (73.9%)
Siluriformes: 7 (15.2%)
Perciformes: 4 (8.7%)
Pleuronectiformes: 1 (2.0%)

Las especies más abundantes fueron: *Aphyocharax alburnus* (150, 25.4%) *Moenkhausia* sp. A (100, 16.9%), *Hyphessobrycon minimus* (68, 11.5%), *Moenkhausia* cf. *lepidura* A (30, 5.1%), *Moenkhausia* cf. *lepidura* E (30, 5.1%) y *Knodus* cf. *breviceps* (27, 4.5%). También, fueron colectados ejemplares de las siguientes especies: *Achirus* sp., *Ancistrus* sp. C, *Anostomus ternetzi, Apistogramma* sp. A, *Astyanax anteroides, A. integer, Brycon pesu, Bryconamericus* cf. *cismontanus, Bryconops alburnoides, Carnegiella strigatta, Corydoras* cf. *bondi, Crenicichla lenticulata, Crenicichla wallacei, Ctenobrycon spilurus, Cyphocharax* cf. *modestus, C. oenas, Farlowella vittata, Gephyrocharax valencia, Hemigrammus* cf. *tridens, Hemigrammus* sp. A, *Hoplias malabaricus, Hypoptopoma steindachneri, Hypostomus* cf. *plecostomus, Jupiaba polylepis, Knodus* cf. *breviceps, Microphylipnus ternetzi, Microschemobrycon callops, M. casiquiare, M. melanotus, Moenkhausia collettii, M. grandisquamis, M. oligolepis, Ochmacanthus alternus, Phenacogaster* sp. A, *Pimelodella cruxenti, P. megalops, Poptella longipinnis, Rineloricaria* sp., *Serrasalmus rhombeus* y *Triportheus albus.*

En la estación (ICT-60), se capturaron 234 individuos pertenecientes a 21 especies y ubicados en:

Characiformes: 13 (62%)
Perciformes: 5 (23.8%)
Siluriformes: 2 (9.5%)
Atheriniformes: 1 (4.7%)

Las especies más abundantes fueron: *Microphylipnus ternetzi* (74, 31.6%), *Moenkhausia* sp. A (66, 28.2%), *Bunocephalus* sp. (15, 6.4%), *Moenkhausia copei* (14, 6%), *Moenkhausia* cf. *chrysargyrea* (11, 4.7%) y *Apistogramma iniridae* (7, 3%). También, fueron colectados ejemplares de las siguientes especies: *Acestrorhynchius microlepis, Aphyocharax erythrurus, Apistogramma* sp. A, *Bryconamericus* cf. *cismontanus, Crenicichla wallacei, Ctenobrycon spilurus, Hemigrammus* cf. *tridens, Leporinus* sp., *Moenkhausia collettii, Poptella longipinnis, Pseudopimelodus raninus, Rivulus* sp., *Steindachnerina pupula* y *Tetragonopterus argenteus.* Es de destacar la abundancia de la especies de la Familia Eleotridae *Microphylipnus ternetzi.*

En resumen, el análisis de la ictiofauna mostró una diversidad moderadamente alta y abundancia regular a baja. Se colectaron 5 especies de bagres, que viven debajo de los palos o entre los palos y ramas sumergidas tales como: *Ancistrus* sp. C, *Farlowella vittata, Pimelodella cruxenti* y *P. megalops.* También se colectaron especies típicas de aguas superficiales y tranquilas como: *Carnegiella strigatta, Gephyrocharax valencia, Poptella longipinnis* y *Triportheus albus.* En el río y canal principal se colectaron ejemplares de "sardinatas" (*Pellona castelneana*) y otras especies indicadoras de ambientes pelágicos. Finalmente debemos destacar la presencia del eleótrido *Microphylipnus ternetzi.*

BC14. Playa Pelona en el Río Caura (ICT-61) (7°11.82´N–65°08.88´W)

Descripción general de la ictiofauna

La captura con redes de playa y red de arrastre (7 horas) (ICT-61). Se capturaron un total de 201 ejemplares pertenecientes a 20 especies ubicadas en:

Characiformes: 10 (50%)
Siluriformes: 9 (45%)
Gymnotiformes: 1 (5%)

Las especies más abundantes fueron: *Aphyocharax alburnus* (128, 63.7%), *Hemigrammus tridens* (25, 12.4%), *Bryconamericus* cf. *cismontanus* (14, 7%) y *Argonectes longiceps* (8, 4%). También, fueron colectados ejemplares de las siguientes especies: *Astyanax integer, Gymnorhamphichthys hypostomus, Hassar iheringi, Hemigrammus* cf. *tridens, Hemiodus unimaculatus, Hyphessobrycon minimus, Leptodoras* cf. *acipenserinus, L.* cf. *hasemani, L.* cf. *praelongus, Microschemobrycon callops, M. casiquiare, Moenkhausia collettii, Ochmacanthus orinoco, Opsodoras* cf. *trimaculatus, Pimelodella megalops, Pimelodella* sp. A y *Xyliphius* cf. *melanopterus.*

En resumen, el análisis mostró una baja abundancia y baja diversidad. Con el chinchorro se capturaron algunos ejemplares de la especie *Argonectes longiceps* que representa un nuevo record para el Caura. Con la red de arrastre se capturaron cinco especies de doráditos (*Hasar iheringi, Leptodoras* cf. *acipenserinus, L.* cf. *hasemani, L.* cf. *praelongus* y *Opsodoras* cf. *trimaculatus*) y un pez eléctrico: *Gymnorhamphichthys hypostomus.* Muchas especies típicas de playas arenosas. Debemos destacar la captura de un ejemplar de

Xyliphius cf. *melanopterus* el cual representa un nuevo record para el Río Caura.

BC15. Isla Rocosa y Caño arriba Raudal Cinco Mil (ICT-62, ICT-63)(06°58.29´N–64°52.5´W)

Descripción general de la ictiofauna

Isla (ICT-62). La captura con redes de playa, muestra una gran abundancia con 1288 ejemplares capturados y alta diversidad con 43 especies ubicadas en:

Characiformes: 33 (76.7%)
Siluriformes: 6 (14%)
Perciformes: 2 (4.7%)
Clupeiformes: 1 (2.3%)
Rajiformes: 1 (2.3%)

Las especies más abundantes fueron: *Knodus* cf. *breviceps* (292, 22.7%), *Aphyocharax alburnus* (189, 14.6%), *Moenkhausia* cf. *lepidura* E (133, 10.3%) *Moenkhausia cotinho* (125, 9.7%), *Jupiaba polylepis* (90, 7%), *Curimatella inmaculata* (87, 6.7%), *Hyphessobrycon minimus* (47, 3.6%) y *Cyphocharax oenas* (29, 2.2%). También, fueron colectados ejemplares de las siguientes especies: *Anchoviella guianensis, Apareiodon orinocensis, Astyanax integer, Boulengerella lucia, Brycon pesu, Bryconamericus* cf. *cismontanus, Characidium* sp. A, *Corydoras* cf. *bondi, Creagrutus maxillaris, Curimata incompta, Curimatella immaculata, Cyphocharax festivus, C. oenas, Exodon paradoxks, Geophagus* cf. *brachybranchus, Hemigrammus* cf. *tridens, Hemigrammus* sp. A, *Hyphessobrycon minimus, Jupiaba atypindi, Knodus* sp. B, *Leporinus* cf. *maculatus, Mastiglannis asopos, Microschemobrycon callops, M. casiquiare, Moenkhausia* cf. *gracilima, Moenkhausia* cf. *lepidura* A, *M. cotinho, M. grandisquamis, M. hemigramoides, Ochmacanthus alternus, Phenacogaster* sp. A, *Pimelodus albofasciatus, Potamotrygon dorbygni, Prochilodus mariae, Pseudocheirodon* sp., *Rineloricaria* sp., *Satanoperca* sp., *Semaprochilodus kneri* y *Vandellia sanguinea* .

Caño (ICT-63). La captura con redes de playa, muestra una moderada abundancia con 433 ejemplares capturados y moderada-alta diversidad con 36 especies ubicadas en:

Characiformes: 24 (66.6%)
Siluriformes: 9 (25%)
Perciformes: 1 (2.9%)
Synbranchiformes: 1 (2.9%)
Gymnotiformes: 1 (2.9%)

Las especies más abundantes fueron: Characidae sp. B (119, 27.5%), *Cyphocharax festivus* (72, 16.6%), *Moenkhausia* cf. *lepidura* E (55, 12.7%) *Michoschemobrycon melanotus* (38, 8.8%), *Moenkhausia cotinho* (35, 8.1%). También, fueron colectados ejemplares de las siguientes especies: *Acestrorhynchus microlepis, Ancistrus* sp. C, *Aphyocharax alburnus, Apistogramma* sp. A, *Astyanax abrammis, A. integer, Astyanax* sp., *Bryconamericus* cf. *cismontanus, Bunocephalus* cf. *amaurus, Characidium* sp. A, *Corydoras* cf. *blochi, C. bondi, Ctenobrycon spilurus, Eigenmania virescens, Farlowella vittata, Hemigrammus* cf. *tridens, Jupiaba polylepis, Knodus* cf. *breviceps, Microschemobrycon casiquiare, Moenkhausia* cf. *chrysargyrea, M. collettii, M. grandisquamis, M. oligolepis, Ochmacanthus alternus, Phenacogaster* sp. A, *Pimelodus albofasciatus, Poptella longipinnis, Ramirezella newboldi, Synbranchus marmoratus, Tatia galaxias* y *Ituglanis* cf. *metae.*

El análisis mostró un área ictiológicamente bien representada con una variedad de especies y abundancia apreciables. Debemos hacer notar la captura de una hembra "raya de río *Potamotrygon* y un individuo de *Exodon paradoxus.* Muchas especies están asociadas a aguas rápidas y fondos arenosos tales como: *Apareiodon orinocensis, Astyanax integer, Boulengerella lucia, Brycon pesu, Bryconamericus* cf. *cismontanus, Characidium* sp. A, *Corydoras* cf. *blochi, C. bondi, Creagrutus maxillaris, Curimata incompta, Curimatella immaculata, Cyphocharax festivus, C. oenas, Ochmacanthus alternus, Phenacogaster* sp. A y *Pimelodus albofasciatus.* Otros típicos de áreas protegidas tales como: *Aphyocharax alburnus, Apistogramma* sp. A, *Astyanax abrammis, A. integer, Astyanax* sp., *Geophagus* cf. *brachybranchus, Microschemobrycon casiquiare, Moenkhausia* cf. *chrysargyrea, M. collettii, M. grandisquamis, M. oligolepis, Rineloricaria* sp. y *Satanoperca* sp.

BC16. Campamento Raudal Cinco Mil (Kuekue) (ICT-64) (06°59.65´–64°54.82´W)

Descripción general de la ictiofauna

Playa rocosa-arenosa (ICT-64). Un total de 281 ejemplares fueron capturados pertenecientes a 25 especies ubicadas en:

Characiformes: 21 (84%)
Siluriformes: 3 (12%)
Perciformes: 1 (4%)

Las especies más abundantes fueron: *Knodus* cf. *breviceps* (135, 49.1%), *Moenkhausia* cf. *lepidura* E (81, 28.8%), *Moenkhausia cotinho* (44, 15.7%), *Moenkhausia* cf. *lepidura* D (26, 9.3%) y *Jupiaba* cf. *polylepis* (23, 8.2%). También, fueron colectados ejemplares de las siguientes especies: *Anostomus ternetzi, Aphanotorulus* sp., *Aphyocharax alburnus, Astyanax integer, Chalceus macrolepidotus, Creagrutus maxillaris, Cynopotamus essequibensis, Hemiodus unimaculatus, Homodiaetus* sp., *Hydrolicus* cf. *tatauaia, Hyphessobrycon minimus, Jupiaba atypindi, Knodus* sp. B, *Leporinus* cf. *maculatus, L. friderici, Moenkhausia grandisquamis, Phenacogaster* sp. A, *Piaractus brachypomus, Pseudocetopsis* sp. y *Satanoperca* sp. A.

Peces con diversidad moderada-baja, con especies asociados a rápidos. Se capturó un sólo ejemplar de *Chalceus macrolepidotus* y numerosos ejemplares de dos especieks de *Leporinus* (*Leporinus* cf. *maculatus, L. friderici*). En las trampas se capturó el único ejemplar de *Pseudocetopsis* sp. En un caño lateral se colocó una red de ahorque con el cual se capturaron algunas especies de gran porte como "Sardinata" (*Pellona castelneana*), "morocoto" (*Piaractus brachypomus*), "payaras" (*Hydrolicus tatauaia*) y "caribes" (*Serrasalmus rhombeus*), los cuales no fueron incluidos en la muestra.

BC17. Río Tokoto (ICT-65) (06°53.52´N–64° 47.90´W)

Descripción general de la ictiofauna

Rápidos y remansos (ICT-65). Un total de 780 ejemplares fueron capturados pertenecientes a 62 especies ubicadas en:

Characiformes: 46 (74.2.%)

Siluriformes: 12 (19.4%)
Perciformes: 3 (4.8%)
Gymnotiformes: 1 (1.6%)

Las especies más abundantes fueron: *Knodus* cf. *breviceps* (138, 17.7%), *Aphyocharax alburnus* (79, 14.6%), *Moenkhausia cotinho* (64, 8.2%), *Moenkhausia dichroura* (53, 6.6%), *Moenkhausia* cf. *lepidura* E (52, 6.6%) y *Curimatella inmaculata* (32, 4.1%). También, fueron colectados ejemplares de las siguientes especies: *Acestrorhynchus microlepis, Ancistrus* sp. C, *Apistogramma* sp. A, *Astyanax integer, Brycon bicolor, Bryconamericus* cf. *cismontanus, Bryconamericus deuterodonoides, Caenotropus labyrinthicus, Characidium* sp. A, *Cochliodon plecostomoides, Corydoras* cf. *bondi, Creagrutus maxillaris, Ctenobrycon spilurus, Curimatella immaculata, Cyphocharax festivus, C. oenas, Eigenmannia virescens, Farlowella vittata, Gephyrocharax valencia, Hoplias macrophthalmus, Hyphessobrycon minimus, Hypostomus plecostomus, Hypostomus* sp., *Imparfinis* sp. A, *Jupiaba atypindi, J. polylepis, Knodus* sp. B, *Lasiancistrus* sp., *Leporinus* cf. *maculatus, Moenkhausia* cf. *lepidura* D, *Moenkhausia collettii, M. cotinho, M. grandisquamis, M. hemigramoides, Ochmacanthus alternus, Odontostilbe* sp., *Pachypops* sp., *Phenacogaster* sp. A, *Piaractus brachypomus, Pimelodella megalops, Pimelodella* sp. B, *Poptella longipinnis, Pseudocheirodon* sp., *Pyrrhulina brevis, Raphiodon vulpinus, Rineloricaria* sp., *Roeboides affinis, Satanoperca* sp., *Serrasalmus rhombeus, Steindachnerina pupula, Tetragonopterus argenteus, T. Chalceus* y *Toracocharax stellatus.*

En resumen el análisis muestra un área con abundancia y diversidad alta, con especies asociados a rápidos con plantas acuáticas como *Leporinus* cf. *maculatus, Characidium* sp. A y *Lasiancistrus* sp. Remansos y regiones arenosas con abundantes bagres pimelódidos y calíctidos y tricomictéridos: *Ancistrus* sp. C, *Cochliodon plecostomoides, Corydoras* cf. *bondi, Farlowella vittata, Hypostomus plecostomus, Hypostomus* sp., *Imparfinis* sp. A, *Ochmacanthus alternus, Pimelodella megalops* y *Pimelodella* sp. B. Zonas de remanso y bahía hacia abajo con abundantes carácidos y cíclidos y la presencia de peces hacha (Gasteropelecidae, *Toracocharax stellatus*). Canal principal con especies de mayor talla los cuales se capturaron con anzuelo *Piaractus brachypomus, Serrasalmus rhombeus* y *Brycon.* El Río Tacoto resultó ser el área más rica en éste estudio del sistema del Río Caura.

Appendix/Apéndice 8

Fishes collected during the AquaRAP expedition to the Caura River Basin, Bolívar State, Venezuela

Peces colectados durante la Expedición Aqua RAP a la Cuenca del Río Caura, Estado Bolívar, Venezuela

Antonio Machado-Allison, Barry Chernoff, Francisco Provenzano, Philip W. Willink, Alberto Marcano, Paulo Petry, Tracy Jones and Brian Sidlauskas

"X" indicate species collected in each subregion.
Taxa marked by ** are new records for the Caura River basin.

"X" indican especies colectadas en cada subregión.
Taxa marcados con ** representan nuevos registros para el Río Caura.

Kakada–Río Kakada
Erebato–Río Erebato
Ent-Cej–Río Caura between Cejiato and Entrerios
Ent-Sp–Río Caura between Entrerios and Salto Pará
Playon–Río Caura at El Playon
Nichare–Río Nichare
Cinco Mil–Río Caura at Raudal Cinco Mil
Mato–Río Mato

Taxa	Upper Caura River Río Caura Superior				Lower Caura River Bajo Río Caura			
	Kakada	Erebato	Ent-Cej	Ent-Sp	Playon	Nichare	Cinco Mil	Mato
Rajiformes								
Potamotrygonidae								
Potamotrygon cf. *schoederi*					X			
Potamotrygon dorbygni							X	
Clupeiformes								
Engraulidae								
Anchoviella guianensis						X	X	
Anchoviella jamesi **					X	X		
Clupeidae								
Pellona catelnaeana								X
Characiformes								
Anostomidae								
Anostomus anostomus			X					
Anostomus ternetzi					X	X	X	X
Caenotropus labyrinthicus					X		X	
Leporinus arcus **		X	X	X				
Leporinus brunneus						X		
Leporinus cf. *granti* **		X	X	X				
Leporinus cf. *maculatus*					X	X	X	
Leporinus friderici							X	
Schizodon sp.					X			
Synaptolaemus cingulatus **					X			

continued

Taxa	Upper Caura River				Lower Caura River			
	Río Caura Superior				Bajo Río Caura			
	Kakada	Erebato	Ent-Cej	Ent-Sp	Playon	Nichare	Cinco mil	Mato
Characidae								
Acestrorhynchus cf. *apurensis* **			X					
Acestrorhynchus falcatus			X	X			X	
Acestrorhynchus heterolepis **								X
Acestrorhynchus microlepis **			X			X	X	X
Ammocryptocharax elegans						X		
Aphyocharax alburnus					X	X	X	X
Aphyocharax erythrurus						X		X
Aphyocharax sp. **	X	X	X	X				
Astyanax abramis **							X	
Astyanax cf. *anteroides* **								X
Astyanax integer **			X		X	X	X	X
Astyanax sp.						X	X	X
Brachychalcinus orbicularis			X	X				
Brycon cf. *bicolor*							X	
Brycon pesu					X		X	X
Bryconamericus cf. *breviceps* **						X		
Bryconamericus cf. *cismontanus* **					X	X	X	X
Bryconamericus deuterodonoides							X	
Bryconops alburnoides								X
Bryconops n. sp. **	X	X	X		X	X		
Bryconops giacopinii		X	X	X	X			
Bryconops sp. A **	X	X	X	X				
Bryconops sp. B **					X	X		
Chalceus microlepidodus **							X	
Characidae sp. A		X		X				
Characidae sp. B					X	X	X	X
Characidae sp. C		X	X			X		
Characidium sp. A	X	X	X	X	X	X	X	
Characidium sp. B	X	X	X					
Characidium sp. C						X		
Characinae sp. A			X					
Characinae sp. B						X		
Creagrutus cf. *maxillaris*					X	X	X	
Creagrutus sp.			X	X				
Ctenobrycon spilurus			X	X			X	X
Ctenobrycon spilurus?				X				
Cynopotamus essequibensis					X		X	
Exodon paradoxus **							X	
Gephyrocharax valencia							X	X
Hemibrycon metae **						X		

continued

Taxa	Upper Caura River Río Caura Superior				Lower Caura River Bajo Río Caura			
	Kakada	Erebato	Ent-Cej	Ent-Sp	Playon	Nichare	5mil	Mato
Hemigrammus cf. *guyanensis* **	X	X	X			X		
Hemigrammus cf. *mimus* **					X			
Hemigrammus cf. *tridens* **					X	X	X	X
Hemigrammus marginatus								X
Hemigrammus sp. A		X		X	X			X
Hemigrammus sp. B	X	X						
Hyphessobrycon bentosi **								X
Hyphessobrycon minimus					X	X	X	X
Hyphessobrycon serpae **								X
Hyphessobrycon sp.	X		X		X			
Iguanodectini sp.						X		
Jupiaba atypindi **	X	X	X	X	X	X	X	
Jupiaba cf. *atypindi*		X	X	X	X	X	X	
Jupiaba cf. *polylepis* **				X	X	X	X	X
Jupiaba cf. *zonata* **		X		X				
Jupiaba polylepis					X	X	X	X
Jupiaba sp. A		X	X	X				
Jupiaba sp. B		X		X				
Jupiaba zonata	X	X	X	X	X	X		
Knodus cf. *breviceps*							X	X
Knodus cf. *victoriae*	X	X	X	X				
Knodus sp. A						X		
Knodus sp. B					X	X	X	
Knodus sp. C		X	X					
Megalamphodus sp. **						X		
Melanocharacidium cf. *dispilomma*					X			
Melanocharacidium depressum		X	X	X		X		
Melanocharacidium dispilomma			X			X		
Melanocharacidium melanopterum			X					
Melanocharacidium nigrum **						X		
Microschemobrycon callops					X	X	X	X
Microschemobrycon casiquiare					X	X	X	X
Microschemobrycon melanotus **					X	X	X	X
Moenkhausia cf. *chrysargyrea* **							X	X
Moenkhausia cf. *gracilima* **				X	X		X	
Moenkhausia cf. *grandisquamis*			X					
Moenkhausia cf. *lepidura* A		X	X	X	X		X	X
Moenkhausia cf. *lepidura* B	X	X	X	X	X			
Moenkhausia cf. *lepidura* C	X	X	X	X	X			
Moenkhausia cf. *lepidura* D					X	X	X	
Moenkhausia cf. *lepidura* E			X	X	X	X	X	X

continued

Taxa	Upper Caura River Río Caura Superior				Lower Caura River Bajo Río Caura			
	Kakada	Erebato	Ent-Cej	Ent-Sp	Playon	Nichare	Cinco mil	Mato
Moenkhausia cf. *miangi* **		X		X				
Moenkhausia collettii	X	X	X	X		X	X	X
Moenkhausia copei					X	X		X
Moenkhausia cotinho				X	X	X	X	
Moenkhausia dichroura							X	
Moenkhausia grandisquamis				X	X	X	X	X
Moenkhausia hemigrammoides **						X	X	
Moenkhausia oligolepis	X	X	X	X	X	X	X	X
Moenkhausia sp. A								X
Moenkhausia sp. B				X				
Myleus asterias			X					
Myleus rubripinnis **			X	X		X		
Myleus torquatus **			X					
Odontostilbe cf. *fugitiva* **					X	X		
Odontostilbe sp.						X	X	
Phenacogaster sp. A		X			X	X	X	X
Phenacogaster sp. B	X		X					
Piaractus brachypomus							X	
Poptella longipinnis **			X	X	X	X	X	X
Potamorhina altamazonica **					X			
Pseudocheirodon sp.		X		X	X	X	X	X
Pygocentrus cariba								X
Ramirezella newboldi **					X	X	X	X
Roeboides affinis							X	
Serrasalmus elongatus **								X
Serrasalmus rhombeus		X	X		X	X	X	X
Serrasalmus sp. A					X	X		X
Tetragonopterus argenteus							X	X
Tetragonopterus chalceus					X	X	X	
Tetragonopterus sp.			X	X	X	X	X	
Triportheus albus **					X	X		X
Xenagoniates bondi **					X			
Ctenoluciidae								
Boulengerella lucia							X	
Boulengerella xyrekes						X		
Curimatidae								
Curimata cyprinoides **								X
Curimata incompta **					X	X	X	
Curimatella immaculata							X	
Cyphocharax cf. *festivus* **				X				
Cyphocharax cf. *modestus* **			X		X			X

continued

Taxa	Upper Caura River				Lower Caura River			
	Río Caura Superior				Bajo Río Caura			
	Kakada	Erebato	Ent-Cej	Ent-Sp	Playon	Nichare	Cinco mil	Mato
Cyphocharax cf. *oenas* **							X	
Cyphocharax festivus		X	X	X	X	X	X	
Cyphocharax meniscaprorus						X		
Cyphocharax oenas					X	X	X	X
Cyphocharax sp.	X	X	X	X	X	X		
Cyphocharax spilurus					X			
Psectrogaster essequibensis **					X			
Steindachnerina pupula						X	X	X
Cynodontidae								
Cynodon gibbus					X			
Hydrolycus armatus						X		
Hydrolycus tatauaia					X	X	X	
Rhaphiodon vulpinus						X	X	
Erythrinidae								
Hoplias macrophthalmus	X	X	X	X	X	X	X	X
Hoplias malabaricus				X		X		X
Gasteropelecidae								
Carnegiella strigata **								X
Thoracocharax stellatus **							X	
Hemiodontidae								
Argonectes longiceps								X
Hemiodus cf. *unimaculatus* **			X	X				
Hemiodus goeldii **			X					
Hemiodus unimaculatus						X	X	X
Lebiasinidae								
Lebiasina uruyensis **					X			
Nannostomus erythrurus **								X
Pyrrhulina brevis				X		X	X	
Parodontidae								
Apareiodon orinocensis			X		X		X	
Apareiodon sp.				X				
Prochilodontidae								
Prochilodus mariae				X			X	X
Semaprochilodus kneri					X		X	
Semaprochilodus laticeps						X		
SILURIFORMES								
Ageneiosidae								
Ageneiosus sp.			X					
Aspredinidae								
Bunocephalus aleuropsis					X			
Bunocephalus cf. *aleuropsis* **					X			

continued

Taxa	Upper Caura River Río Caura Superior				Lower Caura River Bajo Río Caura			
	Kakada	Erebato	Ent-Cej	Ent-Sp	Playon	Nichare	Cinco mil	Mato
Bunocephalus cf. *amaurus* **						X	X	
Bunocephalus sp.								X
Xyliphius cf. *melanopterus* **								X
Auchenipteridae								
Tatia galaxias **							X	
Tatia romani			X		X			
Callichthyidae								
Corydoras boehlkei	X	X	X	X				
Corydoras cf. *blochi*					X		X	
Corydoras cf. *bondi*					X	X	X	X
Corydoras cf. *osteocarus*			X					
Corydoras sp.	X	X	X					
Cetopsidae								
Pseudocetopsis sp.							X	
Doradidae								
Doras? **			X					
Hassar iheringi **								X
Leptodoras cf. *acipenserinus* **								X
Leptodoras cf. *hasemani* **								X
Leptodoras cf. *praelongus* **								X
Opsodoras cf. *trimaculatus* **								X
Orinocodoras eigenmanni **								X
Oxydoras kneri **						X		
Platydoras armatulus								X
Loricariidae								
Ancistrus sp. A			X					
Ancistrus sp. B			X	X				
Ancistrus sp. C					X	X	X	X
Aphanotorulus sp.							X	
Chaetostoma vasquezi			X	X				
Cochliodon plecostomoides **							X	
Cochliodon sp. **						X		
Farlowella mariaelenae **						X		
Farlowella oxyryncha **			X					
Farlowella vittata					X	X	X	X
Harttia sp. **				X				
Hypoptopoma steindachneri **								X
Hypostomus cf. *plecostomus*					X		X	X
Hypostomus cf. *ventromaculatus* **			X					
Hypostomus sp. A			X	X	X			
Hypostomus sp. B		X	X		X	X	X	

continued

Taxa	Upper Caura River Río Caura Superior				Lower Caura River Bajo Río Caura			
	Kakada	Erebato	Ent-Cej	Ent-Sp	Playon	Nichare	Cinco mil	Mato
Lasiancistrus sp.						X	X	
Limatulichthys punctatus **					X	X		
Loricaria cf. *cataphracta* **					X	X		
Loricariichthys cf. *brunneus*					X	X		
Nanoptopoma spectabilis **					X			
Pseudohemiodon sp.						X		
Rineloricaria fallax **	X	X	X	X				
Rineloricaria sp.					X	X	X	X
Pimelodidae								
Cetopsorhamdia cf. *picklei* **			X					
Chasmocranus sp.						X		
Hemisorubim platyrhynchos **					X			
Imparfinis sp. A **							X	
Imparfinis sp. B **			X					
Mastiglanis asopos **						X	X	
Microglanis cf. *poecilus* **						X		
Microglanis iheringi **					X	X		
Pimelodella cf. *chagresi* **						X		
Pimelodella cf. *cruxenti* **					X	X		X
Pimelodella cf. *megalops* **					X	X	X	X
Pimelodella sp. A						X		X
Pimelodella sp. B		X	X	X		X	X	
Pimelodella sp. C	X	X	X	X		X		
Pimelodus albofasciatus							X	
Pimelodus blochii					X	X		
Pimelodus cf. *ornatus* **			X					
Pseudopimelodus raninus **						X		X
Sorubim lima					X			
Sorubim sp.					X			
Trichomycteridae								
Haemomaster venezuelae					X			
Homodiaetus sp. **					X		X	
Ituglanis sp.	X	X					X	
Ochmacanthus alternus **					X	X	X	X
Ochmacanthus orinoco **								X
Paravandellia sp. **					X	X	X	
Vandellia sanguinea **					X	X	X	
Gymnotiformes								
Electrophoridae								
Electrophorus electricus						X		

continued

Taxa	Upper Caura River Río Caura Superior				Lower Caura River Bajo Río Caura			
	Kakada	Erebato	Ent-Cej	Ent-Sp	Playon	Nichare	Cinco mil	Mato
Gymnotidae								
Gymnotus anguillaris **						X		
Gymnotus carapo **				X				
Hypopomidae								
Brachyhypopomus cf. *occidentalis*					X	X		
Rhamphichthyidae								
Gymnorhamphichthys hypostomus								X
Sternopygidae								
Eigenmannia macrops					X			
Eigenmannia virescens							X	
Sternopygus macrurus					X			
Sternopygus macrurus?						X		
Beloniformes								
Belonidae								
Potamorrhaphis guianensis					X	X		
Cyprinodontiformes								
Poeciliidae								
Poecilia sp.		X	X	X	X			
Rivulidae								
Rivulus sp. A				X				X
Synbranchiformes								
Synbranchidae								
Synbranchus marmoratus	X	X	X		X	X	X	
Perciformes								
Cichlidae								
Aequidens cf. *chimantanus* **	X	X	X	X		X		X
Aequidens sp.		X						
Apistogramma cf. *inidirae* **								X
Apistogramma sp. A						X	X	X
Apistogramma sp. B					X			
Bujurquina mariae					X	X		
Crenicichla alta **		X	X	X				
Crenicichla cf. *geayi*						X		
Crenicichla cf. *lenticulata* **						X		
Crenicichla cf. *wallacei* **					X			
Crenicichla johanna **						X		
Crenicichla lenticulata					X			X
Crenicichla saxatilis **	X	X	X	X				
Crenicichla sp. A								X
Crenicichla sp. B					X	X		
Crenicichla sp. C					X	X		

continued

Taxa	Upper Caura River Río Caura Superior				Lower Caura River Bajo Río Caura			
	Kakada	Erebato	Ent-Cej	Ent-Sp	Playon	Nichare	Cinco mil	Mato
Crenicichla wallacei						X		X
Geophagus cf. *brachybranchus* **					X		X	
Geophagus sp.			X					
Guianacara cf. *geayi*			X	X				
Guianacara geayi	X	X	X	X				
Mesonauta egregius						X		
Satanoperca cf. *mapiritensis* **								X
Satanoperca sp. A **		X	X	X	X	X	X	
Eleotridae								
Microphilypnus cf. *ternetzi* **								X
Sciaenidae								
Pachypops sp. **							X	
Plagioscion cf. *auratus* **			X					
Pleuronectiformes								
Achiridae								
Achirus sp.						X		X
Total = 278 taxa	**27**	**49**	**79**	**63**	**109**	**122**	**102**	**88**

Appendix/Apéndice 9

Leguminosae collections from the riparian corridor of the Caura River Basin, Bolívar State, Venezuela

Colecciones de Leguminosae del corredor ribereño de la Cuenca del Río Caura, Estado Bolívar, Venezuela

Judith Rosales, Nigel Maxted, Lourdes Rico-Arce and Geoffrey Petts

Abbreviations: L1, L2 and L3 represent the functional riparian landscapes, the numbers below the functional sectors. RCM= Raudal Cinco Mil.

Abreviaciones: L-1, L-2 and L-3 representan los ambientes funcionales ribereños, los números abajo representan los sectores funcionales

Species/Especies	Boca - La Mura			La Mura - Salto Pará			Salto Pará-Kanaracuni				
	L-1			L-2			L-3				
	1	2	3	4	5	6	7	10	11	12	13
Abarema jupunba (Willd.) Britton&Rose var					1				2	1	
Acacia multipinnata Ducke										1	
Acacia paniculata Willd.					1						
Acosmium nitens (Vog.) Yakovlev	4			3							
Alexa canaracunensis Pittier								1			2
Alexa confusa Pittier							2				
Andira surinamensis (Bondt.) Splitg.ex Pu					2		1	1			
Apuleia leiocarpa (Vogel) JF. Macbr.				1							
Balizia pedicellaris (DC) Barneby&Grimes					1						
Bauhinia guianensis Aublet		1			2		1				
Bauhinia rutilans Spruce ex Benth.					1						
Bauhinia ungulata L.				1							
Brownea coccinella ssp. *capitella* (Jacq.)				2	1	4	4	1		2	
Brownea longipedicellata Huber				1	1						
Calliandra laxa Benth.									1		
Calliandra magdalenae Benth.											1
Campsiandra comosa var.*laurifolia* (Bent.)		2		2			1				
Campsiandra taphornii Sterg.		1				1					
Cassia moschata H.B.K.		1		1							
Clathrotropis brachypetala (Tul.) Kleinh.		1			2						
Crudia glaberrima (Steud.) Macbr.					2						
Cynometra bauhiniifolia Benth.		2				1	1				

continued

Species/Especies	Boca - La Mura			La Mura - Salto Pará			Salto Pará-Kanaracuni				
	L-1			L-2			L-3				
	1	2	3	4	5	6	7	10	11	12	13
Cynometra marginata Benth.				1	2						
Dalbergia amazonica (Radlk.) Ducke		2			1	1					
Dalbergia ecastophylla (L.) Taub.								1			1
Dalbergia glauca (Desv.) Amshoff	3	2	1		1	1			1		
Dalbergia hygrophylla (Mart.) Hoehm					1		3				
Dalbergia monetaria L.fil.									1		
Derris negrensis Benth.		1	1		1						
Dialium guianense (Aubl.) Sandw.					1	1	1		4	2	1
Dioclea guianensis Benth.			1				1	1			
Dioclea malacocarpa Ducke						1	2				2
Dioclea reflexa Hook f.				1			1				
Dioclea rudii R.H.Maxwell					2						
Diplotropis purpurea (Rich.) Amsh						2	1				
Dipteryx odorata (Aubl.) Willd.					1						
Dipteryx punctata (S.F. Blake) Amshoff		1									
Enterolobium barinense Cardenas			1								
Enterolobium schomburgkii Benth.?					1						
Eperua jehmanii Oliver ssp. *sandwithii* Cowan							2	1	2		
Etaballia dubia (Benth.) Rudd.	1	2	3	1							
Hydrochorea corymbosa (Rich.)Barn&Grim.		2				1					
Hymenaea courbaril L.		1			1	1					
Inga alba (Sw.) Willd.				1	1						
Inga bourgoni (Aubl.) DC.					1			1	1		
Inga capitata Desv.	1				1						
Inga edulis Mart.							1				
Inga fastuosa (Jac.) Willd.				1			1				
Inga ingoides (Rich.) Willd				1							
Inga laurina (SW) Willd.					1						
Inga leiocalycina Benth.					1				1		
Inga macrophylla Humb.&Bonpl. ex Willd.					1						
Inga nobilis Willd.					2	1	2	2	1		
Inga pilosula (Rich.) Macbr.					1		1		1		1
Inga sertulifera DC.					1		1	1			
Inga splendens Willd.					3		3		1		
Inga stenoptera Benth.	1		4	1							
Inga thibaudiana D.C.					1					1	
Inga ulei Harms								1			
Inga umbellifera (Vahl) Steud. ex DC					1					1	
Inga vera Willd.		1		2	1		1		1	1	1
Lonchocarpus JR-130	1										

continued

Species/Especies	Boca - La Mura			La Mura - Salto Pará			Salto Pará-Kanaracuni				
	L-1			L-2			L-3				
	1	2	3	4	5	6	7	10	11	12	13
Machaerium altiscandens Ducke										1	
Machaerium macrophyllum Benth.											2
Machaerium madeirense Pittier											2
Machaerium quinata (Aubl.) Sandw. var. *quinata*											1
Machaerium quinata (Aubl.) Sandwich					1						
Macrolobium acaciaefolium (Benth.) Benth.	1	1	1	1			3				
Macrolobium angustifolium (Benth.) Cowan			1		1	1			2		1
Macrolobium bifolium (Aubl.) Pers.							1			1	1
Macrolobium multijugum (DC.) Benth.	1	5									
Mucuna altissima (Jacq.) DC							1				
Mucuna urens (L.) DC							1				
Myrocarpus venezuelensis Rudd.		1									
Myroxylon balsamum (L.) Harms				1							
Newtonia suaveolens (Miq.) Brenan					2						
Parkia pendula (Willd.) Benth. ex Walp.				1	1	1	2				
Peltogyne paniculata (Benth.) Silva subsp				2							
Pentaclethra macroloba (Willd.) O.Kuntze					1						
Piptadenia peregrina (L.) Benth.		1									
Piptadenia viridifolia (Kunth.) Benth.			1								
Pithecellobium cauliflorum (Willd) Benth.							2				
Pithecellobium divaricatum (Bons) Benth.							1				
Platymiscium trinitatis Benth.				1							
Pterocarpus acapulcensis Rose				2							
Pterocarpus rohri Vahl					1		1				
Rhynchosia phaseoides (Sw.) DC.					1						
Senna bacillaris (Linn.F) Irwin y Barneby				2			2				
Senna macrophylla Kunth. var. *gigantifolia*										1	1
Senna pendula (Humb. & Bonpl. ex Willd.)		1									
Senna pilifera (Vogel) Pittier				2							
Senna quinquangulata (Rich.) Irw. & Barn							1				
Senna reticulata (Willd.) I.&B.					1						
Senna silvestris (Vell.) I.&B. var. *silvestris*					2						
Senna silvestris (Vell.) Irwin & Barneby							3	1	1	1	
Swartzia arborescens (Aubl.) Pittier								2			
Swartzia dipetala Willd. ex Vogel	1				1						
Swartzia leptopetala Benth.	2	1			3	3	2				
Swartzia panacoco (Aubl.) Cowan								2		1	1
Swartzia panacoco var. *cardonae* Cowan								1			
Swartzia picta Spruce ex Benth. var. *picta*					1	1					
Swartzia schomburgkii Benth.											2

continued

Species/Especies	Boca - La Mura			La Mura - Salto Pará			Salto Pará-Kanaracuni				
	L-1			L-2			L-3				
	1	2	3	4	5	6	7	10	11	12	13
Tachigali (=*Sclerolobium*) *guianensis* (Benth.)	2	1								1	
Tachigali plumbea Ducke				2							
Zollernia paraensis Hub.					1	1	1				
Zygia cataractae (Kunth.) L.Rice	1			1	1	1				1	
Zygia cauliflora (Willd) Killip		1									
Zygia claviflora (Spruce ex Benth.) Barn							1				
Zygia latifolia var. *communis* Barn.&Grim.	2	1		3	1	2		2			
Zygia unifoliolata (Benth.) Pittier	2	2				1					

Collectors: Aymard, Cardona, Elcoro, Fernandez, Horner, Knab-Vispo, Liesner, Morillo, Rodriguez, Rosales, Sanoja, Stergios, Steyermark, Tillet, Velazco, Diaz, Williams and Winfried

Colectores: Aymard, Cardona, Elcoro, Fernandez, Horner, Knab-Vispo, Liesner, Morillo, Rodriguez, Rosales, Sanoja, Stergios, Steyermark, Tillet, Velazco, Diaz, Williams and Winfried.

Appendix/Apéndice 10

Species of Ingeae ordered by TWINSPAN classification

Especies de Ingeae ordenadas por una clasificación TWINSPAN

Judith Rosales, Nigel Maxted, Lourdes Rico-Arce and Geoffrey Petts

The widely used cluster analysis algorithm TWINSPAN (Two Way INdicator SPecies ANalysis; Hill 1979) was designed specifically for ecological data sets in which the abundance values for each species in each sample are not regularly distributed and where there are many zero values (species are absent from many samples). TWINSPAN is a hierarchical divisive technique that divides samples on the basis of their similarity or dissimilarity in terms of species composition. It identifies "indicator species" that are strongly positively associated with samples in one group (occurs in >80% of the samples) and is strongly negatively associated with samples in the other group (occurs in <20% of the samples).

Abbreviations of rivers: Negro (N), Japura (J), Madeira (M), Amazon (A), Upper Orinoco Casiquiare (U), Caroni (C), Caura (c).

Abreviaciones de ríos: Negro (N), Japura (J), Madeira (M), Amazon (A), Upper Orinoco Casiquiare (U), Caroni (C), Caura (c).

Ingeae species	Rivers NJMAUCc	Species classes
Especies de Ingeae	**Ríos NJMAUCc**	**Especies clases**
Abarema curvicarpa var. *rodriguezii* Barn. & Grimes	1------	00000
Abarema villifera (Ducke) Barn. &Grimes	1------	00000
Abarema adenophora (Ducke) Barn. & Grimes	1------	00000
Zygia juruana (Harms) L. Rico	1------	00000
Abarema leucophylla var. *vaupesiana* Barn. & Grimes	11-----	00001
Zygia ampla (Spruce ex Benth.) Pittier	11-----	00001
Abarema floribunda (Spr. ex Benth.) Barn. & Grimes	1--1---	0001
Abarema microcalyx var. *enterolobioides* Barn. & Grimes	1--1---	0001
Abarema barbouriana (Sta.) Bar.&Grim. var. *barbouriana*	-1-----	00100
Zygia longifolia (H.& B. ex Willd.) Britton & Rose	-1-----	00100
Balizia elegans (Ducke) Barn. & Grimes	--1----	001010
Abarema macradenia (Pittier) Barneby & J.W. Grimes	--1----	001010
Abarema mataybifolia (Sandwith) Barneby & J.W. Grimes	---1---	001010
Abarema campestris (Spruce ex Benth.) Barn. & Grimes	--11---	001010
Abarema cochleata (Willd.) Barn. & Grimes	111----	001011
Macrosamanea spruceana (Benth.) Killip ex Record	11-1---	001100
Hydrochorea marginata (S.ex B.) Barn.& Grim. var. *marginata*	1111---	001101
Abarema auriculata (Benth.) Barn. & Grimes	1111---	001101
Zygia racemosa (Ducke) Barn. & Grimes	1111---	001101
Abarema piresii Barn. & Grimes	1-11---	00111
Macrosamanea duckei (Huber) Barn. & Grimes	1-11---	00111
Albizia multiflora (Kunth) Barn. & Grimes	1-11-1-	01000
Macrosamanea pubiramea (St.) Barn.& Grim. var. *pubiramea*	1-111--	01001

continued

Ingeae species Especies de Ingeae	Rivers NJMAUCc Ríos NJMAUCc	Species classes Especies clases
Abarema jupunba (Willd.) Britton & Killip var. *jupunba*	11111--	01010
Abarema laeta (Benth.) Barn. & Grimes	11111--	01010
Zygia ramiflora (Benth.) Barn. & Grimes	11111--	01010
Macrosamanea pubiramea var.*lindsaeifolia* (S.exBenth.) B&G	111-1--	01011
Macrosamanea discolor (H & B. ex Willd.) Br. & Rose ex Britton & Killip var. *discolor*	1--11--	0110
Macrosamanea consanguinea (R.S. Cowan) Barn. & Grimes	11--1--	0111
Macrosamanea simabifolia (Spruce ex Benth.) Pittier	11--1--	0111
Macrosamanea amplissima (Ducke) Barn. & Grimes	11--1--	0111
Zygia inaequalis (Humb. & Bonpl. ex Willd.) Pittier	111111-	1000
Hydrochorea corymbosa (Rich.) Barn. & Grimes	11111-1	1001
Zygia latifolia var. *lasiopus* (Benth.) Barn. & Grimes	1-1111-	10100
Cedrelinga cateniformis (Ducke) Ducke	1-111-1	10101
Zygia cataractae (Kunth) L. Rico	1111111	1011
Zygia latifolia var. *communis* Barn. & Grimes	1111111	1011
Zygia unifoliolata (Benth.) Pittier	1111111	1011
Abarema levelii (R.S. Cowan) Barn. & Grimes	---11--	11000

Hill, M.O. 1979. TWINSPAN - A FORTRAN program for arranging multivariate data in an ordered two-way table by classification of the individuals and attributes. Cornell University, Ithaca, NY.

Appendix/Apéndice 11

Inventory of decapod crustaceans, by regions, georeference sites and sampling stations, collected during the AquaRAP expedition to the Río Caura Basin, Bolívar State, Venezuela

Inventario de crustáceos decápodos por región, georeferencia y localidades colectados durante la Expedición AquaRAP a la Cuenca del Río Caura, Estado Bolívar, Venezuela

Celio Magalhaes and Guido Pereira

Region	Georeference station	Sampling station	Species collected	Number of specimens
Region	**Estación georeferencia**	**Muestra**	**Especies colectadas**	**Numero de especimenes**
Upper Caura/ Caura Superior				
	AC-01	V2000-CR-01	*Macrobrachium brasiliense*	2
	AC-01	V2000-CR-01	*Poppiana dentata*	4
	AC-01	V2000-CR-01	*Valdivia serrata*	3
	AC-02	V2000-CR-14	*Fredius stenolobus*	2
	AC-03	V2000-CR-02	*Macrobrachium brasiliense*	3
	AC-03	V2000-CR-03	*Macrobrachium brasiliense*	16
	AC-03	V2000-CR-04	*Poppiana dentata*	1
	AC-04	V2000-CR-05	*Macrobrachium brasiliense*	60
	AC-04	V2000-CR-05	*Macrobrachium* n. sp. 1	5
	AC-04	V2000-CR-05	*Poppiana dentata*	26
	AC-04	V2000-CR-05	*Valdivia serrata*	2
	AC-05	V2000-CR-06	-	0
	AC-06	V2000-CR-07	*Macrobrachium* n. sp. 1	7
	AC-06	V2000-CR-07	*Fredius stenolobus*	2
	AC-06	V2000-CR-07	*Valdivia serrata*	1
	AC-06	V2000-CR-08	*Fredius stenolobus*	2
	AC-06	V2000-ICT-10	*Poppiana dentata*	2
	AC-07	V2000-CR-09	*Fredius stenolobus*	1
	AC-08	V2000-CR-10	*Macrobrachium brasiliense*	12
	AC-08	V2000-CR-10	*Valdivia serrata*	1
	AC-08	V2000-CR-10	*Fredius stenolobus*	3
	AC-08	V2000-CR-11	*Macrobrachium brasiliense*	40
	AC-08	V2000-CR-11	*Valdivia serrata*	6
	AC-08	V2000-CR-12	-	
	AC-09	V2000-CR-13	*Fredius stenolobus*	2
	AC-09	V2000-CR-15	-	
	AC-09	V2000-CR-16	*Macrobrachium brasiliense*	19

continued

Region	Georeference station	Sampling station	Species collected	Number of specimens
Region	Estación georeferencia	Muestra	Especies colectadas	Numero de especimenes
	AC-09	V2000-CR-18	*Macrobrachium brasiliense*	5
	AC-09	V2000-CR-18	*Valdivia serrata*	1
	AC-10	V2000-CR-17	*Macrobrachium brasiliense*	12
	AC-10	V2000-CR-17	*Poppiana dentata*	2
	AC-11	V2000-CR-19	-	
	AC-11	V2000-CR-20	*Poppiana dentata*	1
	AC-12	V2000-CR-21	-	
	AC-12	V2000-CR-22	*Fredius stenolobus*	1
	AC-13	V2000-CR-23	-	
	AC-14	V2000-CR-24	*Macrobrachium brasiliense*	27
	AC-14	V2000-CR-24	*Valdivia serrata*	2
	AC-14	V2000-CR-25	-	
Lower Caura/ Bajo Caura				
	BC-01	V2000-CR-26	-	
	BC-01	V2000-CR-28	*Macrobrachium brasiliense*	110
	BC-01	V2000-CR-28	*Fredius stenolobus*	8
	BC-02	V2000-CR-27	*Macrobrachium brasiliense*	23
	BC-02	V2000-CR-27	*Valdivia serrata*	3
	BC-02	V2000-CR-27	*Poppiana dentata*	4
	BC-03	V2000-CR-29	-	
	BC-04	V2000-CR-30	*Macrobrachium brasiliense*	18
	BC-05	V2000-CR-31	*Fredius stenolobus*	2
	BC-05	V2000-CR-32	*Macrobrachium brasiliense*	10
	BC-05	V2000-CR-33	*Valdivia serrata*	2
	BC-05	V2000-CR-33	*Macrobrachium brasiliense*	7
	BC-06	V2000-CR-34	-	
	BC-06	V2000-CR-35	*Valdivia serrata*	1
	BC-06	V2000-CR-35	*Macrobrachium brasiliense*	18
	BC-06	V2000-CR-35	*Forsteria venezuelensis*	6
	BC-06	V2000-CR-35	*Fredius stenolobus*	1
	BC-06	V2000-CR-35	*Pseudopalaemon* n. sp.	13
	BC-06	V2000-CR-36	-	
	BC-07	V2000-CR-41	-	
	BC-07	V2000-CR-42	-	
	BC-08	V2000-CR-37	*Macrobrachium brasiliense*	82
	BC-08	V2000-CR-37	*Forsteria venezuelensis*	15
	BC-08	V2000-CR-37	*Valdivia serrata*	5
	BC-08	V2000-CR-37	*Pseudopalaemon* n. sp.	10
	BC-08	V2000-CR-38	*Forsteria venezuelensis*	4
	BC-08	V2000-CR-38	*Macrobrachium brasiliense*	?
	BC-09	V2000-CR-39	*Macrobrachium amazonicum*	44

continued

Region	Georeference station	Sampling station	Species collected	Number of specimens
Region	Estación georeferencia	Muestra	Especies colectadas	Numero de especimenes
	BC-10	V2000-CR-43	*Macrobrachium brasiliense (larvas)*	1
	BC-10	V2000-CR-43	*Valdivia serrata*	2
	BC-10	V2000-CR-43a	-	
	BC-11	V2000-CR-44	-	
	BC-11	V2000-CR-45	*Valdivia serrata*	2
	BC-11	V2000-CR-45	*Poppiana dentata*	2
	BC-12	V2000-CR-46	*Macrobrachium brasiliense*	4
	BC-12	V2000-CR-46	*Macrobrachium amazonicum*	7
	BC-12	V2000-CR-46	*Poppiana dentata*	1
	BC-13	V2000-CR-47	*Macrobrachium amazonicum*	10
	BC-13	V2000-CR-48	*Macrobrachium amazonicum*	10
	BC-13	V2000-CR-48	*Palaemonetes carteri*	5
	BC-13	V2000-CR-48	*Palaemonetes mercedae*	3
	BC-14	V2000-CR-49	-	
	BC-15	V2000-CR-50	-	
	BC-15	V2000-CR-51	*Macrobrachium brasiliense*	49
	BC-15	V2000-CR-51	*Poppiana dentata*	15
	BC-15	V2000-CR-51	*Valdivia serrata*	9
	BC-15	V2000-CR-52	*Poppiana dentata*	2
	BC-16	V2000-CR-53	-	
	BC-17	V2000-CR-54	*Macrobrachium amazonicum*	32
	BC-17	V2000-CR-54	*Palaemonetes mercedae*	18
	BC-17	V2000-CR-54	*Macrobrachium brasiliense*	7
	BC-17	V2000-CR-54	*Poppiana dentata*	2
	BC-17	V2000-CR-54	*Valdivia serrata*	1
	BC-17	V2000-CR-55	*Fredius stenolobus*	2